Biogeochemistry and the Environment

Michael O'Neal Campbell

Biogeochemistry and the Environment

Springer

Michael O'Neal Campbell
Simon Fraser University
Burnaby, BC, Canada

ISBN 978-3-031-47016-5 ISBN 978-3-031-47017-2 (eBook)
https://doi.org/10.1007/978-3-031-47017-2

Cover image: Contribution by Maria Elena Torres Campbell

This Springer imprint is published by the registered company Springer Nature Switzerland AG
The registered company address is: Gewerbestrasse 11, 6330 Cham, Switzerland

Paper in this product is recyclable.

Preface

This book examines the complex science of biogeochemistry from a novel perspective, examining its comparatively recent development, while also emphasizing its interlinked relationship with the earth sciences (including the complementary science of geochemistry), the geographical sciences (biogeography, oceanography, and geomatics), and the chemical sciences (including environmental chemistry, and overlapping with geochemistry). Biogeochemistry may be defined as the science that combines biological and chemical perspectives for the examination of the Earth's surface, including the relations between the biosphere, lithosphere, atmosphere, and hydrosphere. Biogeochemistry is a comparatively recently developed science that incorporates scientific knowledge and findings, research methodologies, and models linking the biological, chemical, and earth sciences. Therefore, while it is a definitive science with a strong theoretical core, it is also dynamically and broadly interlinked with other sciences. Biogeochemistry is highly scaled, from the smallest environmental units (chemical elements) through larger units (chemical compounds) to the largest units (hydrological, biological, geological, and atmospheric systems such as oceanic differentiation, climate change, and global ecosystems such as biomes). It is also highly temporalized, as past and current environments may necessarily be studied as a continuum. It is also highly spatial, as concepts such as regions, global energy flows, and latitudes assume relevance. This creates close links with the sciences of biogeography and biological oceanography, which are fundamentally spatial sciences. Biogeochemistry is also increasingly technological, linking to scientific developments such as geomatics and computing, and chemical analyses, which have increased the analytical and explanatory power of derived models and applied concepts. Links to the earth sciences, especially geochemistry and chemical oceanography, are also essential to the origins and current developments of biogeochemistry. These characteristics make biogeochemistry a vast subject, the full extent of which is beyond the coverage of a single volume.

Several detailed studies have been published on biogeochemistry and related sciences, to enhance the knowledge of inquirers and stakeholders of the essence of this fascinating discipline, which is emerging from an integrated cluster of disciplines. However, few studies have examined the current and developing links between biogeochemistry and related sciences, and how these complex and frequently even interdisciplinary sciences may assist in the development of biogeochemistry. This book covers cutting-edge topics on the science of biogeochemistry, examining its development, structure, interdisciplinary, multidisciplinary, and transdisciplinary relations, and the future of the current complex knowledge systems, especially in the context of technological developments. It also explores the links between this subject and the related sciences of geochemistry, biogeography, oceanography, earth systems science and systems ecology, environmental pollution, conservation biology, urban development, and technological developments such as geomatics. The origins of these highly complicated subjects are explored, as well as their roots in the basic sciences of chemistry, biology, and physics, using literature sources and research findings, as well as academic and scientific developments.

Writing such a book is an interesting challenge and reflects the author's background in the ecological sciences, conservation biology and biogeography, geomatics, environmental chemistry, and the issues of the Anthropocene. This book emerges as an acknowledgment of the importance of environmental change in current and future scenarios, and a desire to communicate some of the exciting new developments between biogeochemistry and its founding, associated, and codeveloping disciplines. Although acknowledging the strong focus of biogeochemistry as a defined discipline, it also argues that the borders and wider relationships of this discipline are fluid, conditioned by technological and scientific developments, social and political requirements, and environmental change. Concepts emerge and decline in importance, and technological advancements may increase the analytical possibilities in scientific research. In writing this volume, the author acknowledges the current context of computerization, Big Data, Big Tech, modeling, global interconnectedness, and the rapidity of scientific development of current scientific developments.

This book is divided into eight chapters. Chapter 1 introduces biogeochemistry as a clearly defined but broad subject, within a larger multidisciplinary, cross-disciplinary, and interdisciplinary complementary network. It is argued that biogeochemistry sits between the biological and geological sciences, hence the prefixes "geo" and "bio" which imply a link with both the earth sciences (e.g., geography, geology, geophysics, and geochemistry) and the life sciences (e.g., biology, biochemistry, and biophysics). This gives biogeochemistry its fundamentally biogeographical character, the latter science comprising the spatial and networked issues of the biosphere, lithosphere, and hydrosphere. The main stem word of biogeochemistry, "chemistry" also implies an allegiance to the chemical sciences. The history of the discipline is explored, which emerged with studies of chemical elements and compounds and their spatial and temporal movements through various materials, networks, and spatial scales. It developed from the basic, earth, and environmental sciences during the nineteenth and twentieth centuries, into a discipline that studies the cycling of chemical elements, namely carbon (C), nitrogen (N), phosphorous (P), potassium (K), oxygen (O), iron (Fe), calcium (Ca), selenium (Se), sulfur (S), and mercury (Hg). These components of compounds move through the lithosphere, atmosphere, hydrosphere, and biosphere. The chapter examines case studies of such developments and explains how the breadth of the subject contributes to knowledge of integrated environmental research. This provides a background for the remaining chapters.

Chapter 2 examines the environmental and earth science of geochemistry as the background discipline for biogeochemistry and for studies of the lithosphere and biosphere. The placing of a chapter covering geochemistry just after the introduction to biogeochemistry is justified on the grounds that geochemistry is among the principal sciences from which biogeochemistry emerged, and a basic cover of this founding science is therefore essential for the study of biogeochemistry. Geochemistry is defined as the study of the interfaces of geology and chemistry, using the methodology of the chemical sciences to investigate the Earth's surface and interior composition. Geochemistry is presented as an extremely broad subject, as it uses a chemistry-based methodology to examine any part of the lithosphere and the links with the hydrosphere, atmosphere, and even the biosphere. Biogeochemistry is partly an offshoot of geochemistry, combined with biological inputs, as much of biogeochemistry lies between the biological sciences and geochemistry. The chapter also examines selected topics of chemistry, and branches of chemistry are explored for their relevance to geochemistry. The development of chemistry as a subject has had significant developments for geochemistry and consequently advanced the science of biogeochemistry.

Chapter 3 examines the related, emerging science of Earth systems science (ESS) and refers to the application of systems science to the earth's surface, and its links with biogeochemistry. The modern subject of Earth systems science is important for any consideration of biogeochemistry, as it considers the systems within which the biological and chemical interactions of biogeochemistry exist (the biosphere, lithosphere, hydrosphere, and atmosphere). The bases of Earth systems science, including systems science, provide spatial and integrated methodologies that support the study of biogeographical and ecological relations, and the wider disciplines of the environmental sciences. Earth systems science is a comparatively new discipline, generally within or linked to physical geography. The breadth of Earth systems science is strongly relevant to biogeochemistry, especially due to the shared multidisciplinary, interdisciplinary, and transdisciplinary methodologies and subject matter, that overrides the artificial disciplinary boundaries between biological, chemical, and physical sciences. Earth systems science develops broader knowledge systems and integrates this knowledge while offering specialization to solve environmental issues. Earth systems science and biogeochemistry also relate to changes in the approaches of the basic sciences (chemistry, physics, mathematics, biology) and the applied environmental sciences (geochemistry, environmental chemistry, biochemical and chemical engineering, biochemistry, geophysics, environmental and atmospheric physics, civil, geological, and environmental engineering, biogeography, oceanography, statistics, etc.). These analyses contribute to the development of biogeochemistry as a strong, emerging discipline, and its role in wider environmental research.

Chapter 4 examines the contribution of the modern discipline of geography to biogeochemistry, especially through the geographical subdisciplines of biogeography and geomatics, these being expanding, interconnected, hybrid branches of life, earth, conservation, spatial, and technical sciences. The chapter notes that the geographical subdisciplines contribute to biogeochemistry as a regionally and globally scaled subject, by acknowledging spatial scales from the local to the global, the interconnected linking of multiple ecosystems through which the biogeochemical relations interact, and the use of increasingly sophisticated, computer-generated research methodologies. The chapter also acknowledges the lack of studies that examine the increasingly complex relationship between biogeochemistry and geographical subdisciplines. Biogeography is defined and compared with biogeochemistry, using studies that compare the two disciplines. Geomatics-based studies mostly center on the use of GIS, radar, satellite imaging, and LiDAR-based research to uncover new assessments of biogeochemical parameters such as vegetation, water, and soil. Case studies are taken from the global scale of biomes and earth systems and local contexts, including the role of human action on these ecosystems.

Chapter 5 examines the biogeochemistry of the oceans and marine life, within the larger context of global biogeochemistry. The science of oceanography is posited as a major scientific contribution to biogeochemistry. As the links between oceanography are largely through biological and chemical (including geological) systems, the four branches of oceanography (biological, chemical, physical, and geological oceanography) are examined in relation to marine biogeography and biology, marine chemistry, marine physics and marine geology, and marine applications of geomatics techniques, especially LiDAR. Biological oceanography is contrasted with the overlapping science of marine biology, the former being more focused on broader marine ecosystems, and the latter more focused on narrower, species-focused studies. Marine biogeography is also examined, as the links between the sciences of biogeochemistry and biogeography apply strongly in the oceanic context. Chemical oceanography or marine chemistry is argued to be situated within the larger global biogeochemical cycles, with the ocean being an important or even dominant component and storage of these systems. The relationship between biogeochemistry, biogeography, and oceanic sciences is still evolving, but more studies are required that seek explanations of this complex relationship at global levels. The biogeochemistry of oceanic systems enables the existence and variables in animal and plant biogeography, and the biogeography of the ocean also plays a role in the biogeochemistry of these oceans. Living and dead organisms interact strongly with the chemical elements and compounds of the oceans, with compounds enabling and constraining the distributions of each other. Case studies are cited from all the Earth's oceans. It is concluded that the biogeochemistry and the biogeography of the Earth's oceans are linked as explanatory factors, within the wider scenario of the Earth, life, and environmental sciences.

Chapter 6 examines the interface of biogeochemistry and environmental chemistry, from conservation biology and conservation biogeography perspectives. Natural processes and human-constructed systems have changed the environmental chemistry of many ecosystems, and hence the biogeochemistry and the biogeography of several species at all levels of the food webs. This has compromised the conservation biogeography of these species. Topics covered include the deliberate and accidental poisoning of animals with artificial toxicants, heavy metals, and other compounds such as anti-inflammatory drugs (NSAIDs), polychlorinated biphenyls (PCBs), and pesticides such as herbicides, fungicides, zoocides, insecticides, and rodenticides. The chapter argues that the information on these relations is largely scattered, focusing on the impacts of particular chemicals on particular species, rather than on combined chemical groups on ecosystems, and the possibilities for conservation science and policy, within the larger biogeochemical scenario. Additionally, current research findings indicate an increasing shift from case studies of global-scale chemical flows to smaller scales at the regional, local, and micro-level scenarios. Case studies are presented on the impacts of lead, pesticides, and veterinary compounds on large raptorial and scavenging birds, toxic pollution of aquatic and marine life, and the associated impacts of land cover change on ecosystems. Conservation policies are also assessed, with case studies from several continents. As chemical systems are increasingly studied as linked to conservation biology, some positive results have emerged.

Chapter 7 examines the field of urban biogeochemistry, with Asian and South American examples. The perspective is the examination of the relations between urban biogeochemistry and the related biological, hydrological, chemical, and geological sciences, especially urban biogeography and urban geochemistry. It is argued that human population increases in a coevolutionary relationship with economic development, and developmental activities (especially that of linear infrastructure) impact the human environment, and some studies examine the ecological aspects of linear infrastructure developments while neglecting the biogeochemical issues of these developments. Examples of biogeochemical issues linked to linear infrastructural developments include changes in soil hydrology, invasions of exotic species with road and highway construction, changes in biological composition and ecological balances, complicated synergistic interactions, and feedback loops for surface runoff from roads, hydrocarbon pollution from groundwater and oil spills from pipelines, and the Urban Heat Island effect. Local environmental chemistry, hydrology, and plant nutrient levels and productivity may be negatively impacted by construction work. Urban infrastructure development may contribute to drainage disruption and rapid slope runoff, varied water discharge, slope instability, soil compaction and disturbance, the creation of impermeable ground layers, drainage disruption, deforestation, and pollution, which affect chemical cycling, especially elements such as carbon and nitrogen, and pollutants. Therefore, the biogeochemistry of such developments must be taken into consideration, in addition to profits and sustainable planning. This contributes to the investigation of the biogeochemical bases of linear development projects.

Chapter 8 covers the possible future of biogeochemistry, including its supporting research methods and relations with supporting and complementary sciences. Biogeochemistry has a bright and even decisive future as a discipline and as a scientific background for applied research, partly due to its association with other complementary disciplines, which may feed the biogeochemical studies with research methods and scientific advancements. Examples of relevant developments include the increasing sophistication of the geographical and geomatics-based research methodologies, such as LiDAR and computer techniques within the biogeographical and other environmental sciences, and the growing international acknowledgment of the relevance of multidisciplinary and interdisciplinary methods in scientific research on the environment. Predictive modeling is mentioned as an important development for future research methodologies. Additionally, developments based on human impacts on the environment are increasingly relevant. The increasingly complex nature of biogeochemical changes indicates that future assessments may be dominated by models, some analytical, others predictive, and even retrospective.

Burnaby, BC, Canada — Michael O'Neal Campbell
Project Associates, Wildlife Institute of India — Adil Khan
Dehradun, India — Indushree Maharana

The original version of the book has been revised. A correction to this book can be found at https://doi.org/10.1007/978-3-031-47017-2_10

Contents

1 **Biogeochemistry and Its Complexity** ... 1
Introduction ... 2
Links Between Biogeochemistry and Other Sciences ... 9
Biogeochemistry and Physical Geography ... 10
Biogeochemistry and Ocean Sciences ... 11
Urban and Pollution Biogeochemistry ... 14
Conclusions ... 35
References ... 35

2 **Geochemistry as the Core of Biogeochemistry** ... 41
Introduction ... 42
Links of Geochemistry with Biogeochemistry ... 45
Subfields of Geochemistry ... 47
Organic Geochemistry ... 48
Inorganic Geochemistry ... 63
Isotope Geochemistry ... 68
Aqueous Geochemistry ... 72
Cosmochemistry ... 74
Trace-Element Geochemistry ... 77
Igneous and Metamorphic Rock Geochemistry ... 82
Photogeochemistry ... 88
Environmental or Low-Temperature Geochemistry ... 89
Geochemical Techniques ... 91
Conclusions ... 96
References ... 97

3 **Earth Systems Science (ESS) and Systems Ecology** ... 113
Introduction ... 114
Earth Systems Science (ESS) and Systems Ecology ... 119
The Lithosphere or Geosphere ... 123
The Hydrosphere ... 125
Origin of Water on Earth ... 125

The Cryosphere 133
The Atmosphere 136
The Biosphere 140
Conclusions 149
References 150

4 **Biogeochemistry, Biogeography, and Geomatics** 167
Introduction 168
The Development of Biogeography 169
Links Between Biogeography and Other Sciences 170
Links Between Biogeography and Ecology 172
Links Between Biogeochemistry and Biogeography 176
Global Biomes and Carbon Sinks 179
Forest Vegetation and Biogeochemistry 181
Savanna/Grassland Vegetation and Biogeochemistry 185
The Tools of Geomatics 194
Geodesy 195
Radio Detection and Ranging (RADAR) 200
Light Detection and Ranging (LiDAR) Technologies 204
Conclusions 214
References 215

5 **Biogeochemistry and Oceanography** 227
Introduction 227
Links Between Biogeochemistry, Oceanography, and Biogeography 229
Biological Oceanography 229
Chemical Oceanography or Marine Chemistry 240
Physical Oceanography 249
Geological Oceanography or Marine Geology 255
Geomatics and Ocean Biogeochemistry 259
Conclusions 263
References 264

6 **Biogeochemistry and Conservation Biology** 279
Introduction 279
Deliberate Poisoning of Animals 283
Unintentional Poisoning with Chemical Elements 287
Poisoning with a Range of Heavy Metals 293
Poisoning from Lead 298
Poisoning from Mercury 311
Poisoning from Chromium 313
Poisoning from Nickel 314
Poisoning from Cadmium 317
Poisoning from Copper 319
Poisoning from Selenium 321

Poisoning from Zinc 326
Poisoning from Arsenic. 328
Unintentional Poisoning from Chemical Compounds 331
Poisoning from Pesticides. 331
Poisoning from Diclofenac Nonsteroidal Anti-Inflammatory Drug (NSAID) 335
Poisoning from Polychlorinated Biphenyls (PCBs) and Perfluoroalkyl Substances (PFASs) 337
Conclusions. 339
References. 340

7 Urban Biogeochemistry and Development: The Biogeochemical Impacts of Linear Infrastructure 365
Introduction. 366
Urban Biogeochemistry and Other Sciences 371
Urban Biogeography. 371
Urban Geochemistry. 374
Urban Biogeochemistry and Linear Infrastructure Development. 376
Focus: Pipelines and Their Effects 382
Effects of Linear Infrastructure Development 383
Pollution 383
Effect on Hydrology 385
Effects on Soil Biogeochemistry and Physical Properties 387
Conclusion 393
References. 394

8 The Future Developments in Biogeochemistry 405
Introduction. 405
Disciplinary Developments. 407
Technological Developments 408
Anthropogenic Developments. 410
Conclusions. 414
References. 414

Correction to: Urban Biogeochemistry and Development: The Biogeochemical Impacts of Linear Infrastructure. C1

Correction to: Biogeochemistry and the Environment C3

Index. 421

Chapter 1
Biogeochemistry and Its Complexity

Michael O'Neal Campbell

Abstract This chapter provides a background to the science of biogeochemistry, examining its history, content, and relationship with other disciplines. This supports the novel approach of this book, which is to examine biogeochemistry from the perspective of its relationship with other disciplines, and thus illustrates its importance in current scientific research in the biological, environmental, and earth sciences. Biogeochemistry is an extremely broad, yet detailed subject, with a multidisciplinary, cross-disciplinary, and interdisciplinary focus. Although it is fundamentally a biological science, it is deeply rooted in the study of chemical elements and compounds and their spatial and temporal movements through global, regional, and local physical spaces. Its name, including the prefixes "geo" and "bio," implies a concern with the earth sciences (e.g., geography, geology, geophysics, and geochemistry), the life sciences (e.g., biology, biogeography, biochemistry, and biophysics), and pure and environmental chemistry. With the word "chemistry" as the main stem word, biogeochemistry is conceived as a branch or an allied discipline of the chemical sciences. Currently, biogeochemistry has developed into a vast field of study, including studies of biological, chemical, and geological aspects of the environment, their interactions, and the chemical cycling between these spheres. This chapter examines the historical development of biogeochemistry and the chemical cycling of the main chemical elements. Biogeochemistry developed from the basic, earth, and environmental sciences during the nineteenth and twentieth centuries, and later included topics such as atmospheric chemistry, carbon cycling, climate change and weathering, biological sciences, and links between climate and the solid earth. Chemical elements studied included carbon (C), nitrogen (N), phosphorous (P), potassium (K), oxygen (O), iron (Fe), calcium (Ca), selenium (Se), sulfur (S), and mercury (Hg). Each of these elements is examined as a component of compounds that comprise and move through the lithosphere, atmosphere, hydrosphere, and biosphere. For example, common compounds that contain carbon include diamond, graphite, Carbon-14 (pure), calcium carbonate ($CaCO_3$), and dolomite

M. O. Campbell (✉)
Simon Fraser University, Burnaby, BC, Canada
e-mail: evaz@torontomu.ca

M. O. Campbell, *Biogeochemistry and the Environment*,
https://doi.org/10.1007/978-3-031-47017-2_1

($CaMg(CO_3)_2$). Compounds containing oxygen include carbon monoxide (CO), carbon dioxide (CO_2), calcium hydroxide (Ca $(OH)_2$, and calcium oxide (CaO), while those with nitrogen include ammonium ($NH+_4$), nitrite (NO−2), nitrate (NO−3), nitrous oxide (N_2O), and nitric oxide (NO). In recent decades, biogeochemistry has evolved into a complex discipline beyond the chemical bases of its development, including the broader issues of global chemical cycling, and the interfacial relations between the biosphere, lithosphere, hydrosphere, and atmosphere. This breadth is shared with the related sciences. This chapter, therefore, explores the rudiments of these relationships, for the greater detail of the later chapters.

Introduction

Biogeochemistry is a broad, yet deep, discipline, with a regional and global focus, serving as the backbone of some other large-scale spatial sciences like biogeography and oceanography, and is strongly linked to its background subjects including geochemistry, environmental chemistry, and ecology (Butcher et al., 1992; Libes, 1992; Dobrovolsky, 1994; Malyuga, 1995; Schlesinger, 1997; Kabata-Pendias & Pendias, 1999; Bashkin & Howart, 2002; Campbell, 2017, 2018; Schlesinger & Bernhardt, 2020). With this breadth, the definitions of the science of biogeochemistry may vary (Campbell, 2020). The only common terms in the definition are usually based on the biological, geological and chemical bases of the subject (bio, geo, chemistry), and the global focus (e.g., global chemical cycling), with varying focus on these components. The study of chemical element movements through the Earth's solid, water, and air systems (the lithosphere, hydrosphere, and atmosphere, respectively) has come to dominate some research developments (Butcher et al., 1992; Libes, 1992; Dobrovolsky, 1994; Malyuga, 1995; Campbell, 2017, 2018; Schlesinger & Bernhardt, 2020). Britannica (2020) refers to the term biogeochemical "as a contraction that refers to the consideration of the biological, geological, and chemical aspects of each cycle" and biogeochemical cycling as "any of the natural pathways by which essential elements of living matter are circulated." Zavarzin (2008) writes that the biogeochemical cycles "represent the major chemical machinery on the Earth." These chemical movements are also fundamental to the understanding of modern physical geography, environmental science, and ecology (Campbell, 2018).

Citing the location of biogeochemistry between the physical/chemical sciences and the life sciences, up to the global level, Cutter (2005) notes that biogeochemistry can be defined "as the mutual interactions (two-way) between the biology and chemistry of the Earth system, and as such is clearly an important component of the broader discipline of geobiology." Karl and Schlesinger (2014) give an extremely wide definition, as biogeochemistry refers to "the holistic study of the Earth system," and biogeochemical research includes "theoretical, observational, experimental, and modeling components, and now spans from molecular studies to the largest scales in space and time that affect our planet. Therefore, as a scientific discipline, biogeochemistry is both integrative and transdisciplinary, and it is attracting a new and diverse spectrum of investigators each year." According to this definition, an important distinguishing point is the focus on the interactions of living organisms within the lithosphere, atmosphere, and hydrosphere. There is also an acknowledgment of the increasing relevance of biogeochemistry to the current global climate changes and anthropogenic impacts (Karl & Schlesinger, 2014). It is further argued that "while there are no rigid intellectual boundaries between biogeochemistry and other fields that also seek a fundamental understanding of our universe, a distinguishing characteristic is the interaction of living organisms within the crust of the Earth and its atmosphere and hydrosphere" (Karl and Schlesinger, 2014).

Schlesinger (2004) also gives a very broad definition, mostly from chemical perspectives, but inclusive of other sciences, as basically, "the chemistry of the arena of life – that is Earth's biogeochemistry – will be at the center of how well we do, and all biogeochemists should strive to articulate that message clearly and forcefully to the public and to leaders of society, who must know our message to do their job well." Biogeochemistry is defined as a recently developed discipline, while also being an integrative subject that attempts an understanding of the functioning of the planet Earth, which is a closed chemical system supportive of life for possibly over 3.5 billion years. The interdisciplinary nature of biogeochemistry is important, as the subject is studied by different specialists; "for example, it is pursued by physicists who want to understand what determines Earth's climate, also by molecular biologists who want to understand what controls the gene expression for certain biochemical pathways, and by geologists who want to understand what controls the breakdown of rocks and the composition of the oceans through geologic time" (Schlesinger, 2004). It is further noted that "biogeochemistry is often described as the chemistry of the surface of the Earth, where the imprint of life is pervasive. Biogeochemistry is at the heart of global-change biology – trying to predict the human impact on the surface chemistry of the planet" (Schlesinger, 2021a).

Recent developments in biogeochemistry have engaged with dynamic changes in global, regional, and local ecological, biogeographical, and environmental issues, and according to Karl and Schlesinger (2014), there have been "collaborations among individuals who do not normally interact, where the whole of the coordinated effort is much greater than the sum of its parts," the reduction of the "rigid intellectual boundaries between biogeochemistry and other fields," new sections in biogeosciences in the Ecological Society of America and the American Geophysical Union, new professional journals, and increased links between biogeochemistry and geochemistry (Karl & Schlesinger, 2014). The authors also defend the "bio" in biogeochemistry, as "biota – ranging from bacteria to higher plants" relating to "the level of CO_2 in the atmosphere, the amount of precipitation that falls on land, the content of nitrogen and phosphorus in rivers, and the silicon that is deposited in ocean sediments," which therefore indicates that "life's diversity – the bio in biogeochemistry – performs a great service to us, which we must understand better if we are to preserve its function" (Karl & Schlesinger, 2014).

Academic institutions have also cited definitions that support their research, and relate to current realities, which are important as such definitions may reflect the rapidly changing world of scientific work and student education in the twenty-first century. The Scripps Institution of Oceanography (2022) states that biogeochemistry studies the "chemical interactions between living things and the natural environment, ranging from how organisms incorporate and respond to elements in their environment to the alterations biological systems make to the chemical environment of the Earth." The Department of Marine Sciences (2022) of the University of Georgia also prioritizes the biological aspect of the discipline, as "biogeochemistry examines the role of biological processes in mediating the chemical and geological dynamics of the Earth's atmosphere, hydrosphere, and lithosphere. It is an inherently interdisciplinary science… The integrative nature of this work requires the application of tools from a wide range of disciplines such as microbiology, geochemistry, ecology, hydrology, mathematics, and physics." This integration between the life and physical sciences was acknowledged by the Woods Hole Oceanographic Institution (2022), which offers a broader definition: biogeochemistry "explores the physical, chemical, biological, and geological processes and reactions that govern the composition of and changes to the natural environment… the cycles of crucial elements, such as carbon and nitrogen, and their interactions with other substances and organisms as they move through Earth's atmosphere, hydrosphere (water and ice), biosphere (life), and lithosphere (rock)." Additionally, biogeochemistry "focuses especially on the diverse and interlinked chemical cycles that are either driven by or have an impact on biological activity, in particular carbon, nitrogen, sulfur, and phosphorus" (Woods Hole Oceanographic Institution, 2022)."

Historically, biogeochemistry traces its roots to the developments of the basic, earth, and environmental sciences, mostly during the nineteenth and twentieth centuries (Bianchi, 2021) (Table 1.1). For example, Laane and Middleburg (2011) describe the concept of biogeochemistry as about one hundred years old, being originated by the Russian scientist Vernadsky in 1926. Bianchi (2021, see also Gorham, 1991; Butcher et al., 1992; Libes, 1992; Dobrovolsky, 1994; Malyuga, 1995; Schlesinger, 1997; Schlesinger & Bernhardt, 2020) notes that biogeochemistry evolved with new methodologies and understandings of earth science topics such as atmospheric chemistry, carbon cycling, and climate change and weathering, and biological (including microbial) links between climate and the solid earth, with links to ecology, evolution and biogeography. Unlike Ecosystem Science and Earth System Science, "the roots of biogeochemistry are… not… as centrally linked with ecology and geology, respectively" (Bianchi, 2021). Twentieth-century developments included increased knowledge of carbon cycling, agricultural technology, human demographic change, and environmental events such as eutrophication. By the twenty-first century, molecular technologies, genetics, geomatics, and Earth system models had also made important contributions (Bianchi, 2021) (Table 1.1).

At the beginning of the twentieth century, there was a pronounced tendency for scientific disciplines to divide into variably related, specialized subdisciplines. The biosphere concept, originated by the Austrian geologist Eduard Suess in his work Die Entstehung der Alpen (1875), was further developed by Vernadsky, who by contrast, sought to create a trend toward increased integration of the biological and physical sciences. Bianchi (2021) notes that biogeochemistry was "born from multidisciplinary interactions between biological, geological, and chemical sciences

Table 1.1 History of biogeochemistry

Time cluster	Dominant themes	References
2433–2374 BP	Physical world divided into air, water, fire, and earth	Empedocles (483–424 BC)
Seventeenth to nineteenth centuries	Geology and earth science links, spatial biochemistry, including elementary chemical modeling, Earth as a superorganism	Hutton (1795), Gorham (1991)
Late nineteenth century	Increased links between geochemistry and the biosphere, chemistry as a sophisticated science.	Suess (1875), Arrhenius (1896), Vernadsky (1926)
Twentieth century	Broader studies of links between earth, atmospheric, oceanic and life sciences	Vernadsky (1926, 1944, 1945), Bianchi et al. (2018)
Twentieth to twenty-first century	Increased recognition of human role in changing biogeochemical relations	Campbell (2020)

Sources: Bianchi (2021). The scientists listed in the reference column participated in the foundation of some sciences that later were relevant to the development of biogeochemistry. For example, Arrhenius was among the founders of the science of physical chemistry, including the theory of electrolytic dissociation, and the hypothesis of the greenhouse effect as a result of burning fossil fuels. Hutton, a geologist, and polymath, sometimes referred to as the "Father of Modern Geology," helped in the development of geology as a modern science. Vernadsky was a mineralogist and geochemist, and a founding influence on geochemistry, biogeochemistry, and radiogeology. Empedocles developed the theory of four elements composing the Earth: fire, air, water, and earth

early in the 19–20th centuries," and "has continued to expand its scope in the twenty-first century on scales that range from microbiological/'omics approaches (genomics, transcriptomics, proteomics, and metabolomics) to global elemental flux transfers in Earth System models" (see also Steffen et al., 2020), and that biogeochemistry enabled clustering of geobiology and geochemistry. Other researchers have published related pieces on the development of biogeochemistry within the earth, environmental, and even socio-economic sciences (Campbell, 2017, 2018).

Laane and Middelburg (2011) note that eventually, biogeochemistry gained much greater attention, "especially because the footprint of man on his own environment is increasing and integrated information and knowledge is necessary to understand man's impact," and "an enormous explosion has been observed in the number of scientific papers containing the keyword biogeochemistry," making the examination of the content and changing trends of biogeochemistry a historically timely endeavor (Laane & Middleburg, 2011). These studies may involve chemical as well as biological foundations. Numerous chemical elements may be studied, and these may form complex compounds (Tables 1.2, 1.3, 1.4, 1.5, 1.6, 1.7, 1.8, 1.9, 1.10, 1.11, and 1.12). These tables show some of the chemical elements and the compounds that may be formed between these compounds. Further, it is noted that "elements within biogeochemical cycles flow in various forms from the nonliving (abiotic) components of the biosphere to the living (biotic) components and back" (Britannica (2020). See Figs. 1.1, 1.2, 1.3, 1.4, and 1.5 for examples of the components of the Earth's system.

Table 1.2 Examples of chemical elements of earth systems, involved in biogeochemical cycling

Element	Symbol	A few examples of related compounds
Carbon	C	Diamond, graphite, carbon-14 (pure); calcium carbonate ($CaCO_3$), dolomite ($CaMg(CO_3)_2$)
Nitrogen	N	Ammonium ($NH+_4$), nitrite (NO−2), nitrate (NO−3), nitrous oxide (N_2O), nitric oxide (NO)
Phosphorous		Phosphophyllite Zn_2 (Fe, Mn)$(PO_4)_2$ $4H_2O$, apatite group $Ca_5(PO_4)_3$(F,Cl,OH), Triphylite Li(Fe,Mn)PO_4
Potassium	K	Potassium chloride (KCl), sylvite (KCl), carnallite ($MgCl_2$.KCl), polyhalite ($2CaSO_4.K_2SO_4.2H_2O$)
Oxygen	O	Carbon monoxide (CO), carbon dioxide (CO_2), calcium hydroxide (Ca $(OH)_2$, calcium oxide (CaO)
Iron	Fe	Hematite (Fe_2O_3), magnetite (Fe_3O_4), siderite ($FeCO_3$), pyrite (FeS_2), ferrous oxide (FeO), ferric oxide (Fe_2O_3)
Calcium	Ca	Calcium chloride ($CaCl_2$), calcium carbonate ($CaCO_3$), calcium oxide (CaO), calcium sulfate ($CaSO_4$)
Selenium	Se	Selenium dioxide ($SeO_{2),}$ selenous acid (H_2SeO_3), selenium oxychloride (SeO_2Cl_2), selenic acid (H_2SeO_4)
Sulfur	S	Lead sulfide (PbS), zinc sulfide (ZnS), iron disulfide, (FeS_2), barium sulfate ($BaSO_4$)
Mercury	Hg	Methylmercury (CH_3Hg), mercury chloride (Hg_2Cl_2), mercuric oxide (HgO), mercury hydride (HgH)

Sources: Klein and Hurlbut Jr. (1985), Thomas and Weller (1992), Greenwood and Earnshaw (1997), Appl (1999), Schultz et al. (2000), Warren (2000), Wildermuth et al. (2000), Rytuba (2003), Hartikainen (2005), Lide (2005), Smart and Moore (2005), Nehb and Vydra (2006), Devillanova and Mont (2007), House (2008), Hoffmann and King (2010), Jäger et al. (2010), Bodnar et al. (2012), Böcker et al. (2013), Calvo and Lamberg-Allardt (2015), Rieuwerts (2015), Gandhi and Sarkar (2016), Kirschner et al. (2017), Grenne and Slack (2019)
This table lists the main chemical elements that occur in global biogeochemical cycling, and compounds that are formed from combinations of the elements. Carbon, nitrogen, and oxygen are among the most important elements, and may form the most important compounds for life. These elements may move through the lithosphere, atmosphere, biosphere, and hydrosphere, the last of which is illustrated below

Biogeochemical cycles transform energy and matter into forms usable for ecosystem functioning and relate to the movement of matter between the Earth's main reservoirs (the atmosphere, the terrestrial biosphere, the oceans, and the solid soil and rock geosphere) (Brusseau et al., 2019). The study of biogeochemical cycles assumes the existence of two main components. These are "a reservoir (nutrient) pool – a larger, slow-moving, usually abiotic portion- and an exchange (cycling) pool – a smaller but more-active portion concerned with the rapid exchange between the biotic and abiotic aspects of an ecosystem" (Britannica, 2020). The biogeochemical cycles can be defined as gaseous, in which the reservoir is either the air or the oceans (through evaporation), or sedimentary, where the reservoir is the Earth's crust. The gaseous cycles include those of nitrogen, oxygen, carbon, and water, while the sedimentary cycles comprise those of iron, calcium, phosphorus, sulfur, and other more-earthbound elements (Britannica, 2020). These biogeochemical cycles are crucial for biological systems and influence global, regional, and local

Table 1.3 Carbon-included compounds that are involved in biogeochemical cycling

Associated element	Examples of compounds
Nitrogen	Cyanogen (C_2N_2), dicyanoacetylene (C_4N_2), hydrogen cyanide (HCN), tetracyanoethylene ($C_2(CN)_4$)
Phosphorous	Carbon phosphate ($C_3O_{16}P_4$), adenosine triphosphate ($C_{10}H_{16}N_5O_{13}P_3$), organophosphorus compounds
Potassium	Potassium carbonate (K_2CO_3), potassium bicarbonate ($KHCO_3$)
Oxygen	Carbon dioxide (CO_2), carbon monoxide (CO), ethanol (C_2H_6O), glucose ($C_6H_{12}O_6$)
Iron	Siderite ($FeCO_3$), ferric carbonate ($Fe_2(CO_3)_3$, iron carbide (Fe_3C), Iron pentacarbonyl ($Fe(CO)_5$)
Calcium	Calcium carbonate ($CaCO_3$), calcium bicarbonate ($Ca(HCO_3)_2$)
Selenium	Carbon diselenide (CSe_2), selenomethionine ($C_5H_{11}NO_2Se$), dimethyl selenide ($(CH_3)_2Se$)
Sulfur	Carbon sulfur (CH_6S), organosulfur compounds (e.g., sulfides – carbon disulfide CS_2, carbon monosulfide CS
Mercury	Organomercury compounds: methylmercury (II) CH_3Hg+; ethylmercury(II) C_2H_5Hg+; dimethylmercury, $(CH_3)_2Hg$)

Sources: Klein and Hurlbut Jr. (1985), Thomas and Weller (1992), Greenwood and Earnshaw (1997), Appl (1999), Schultz et al. (2000), Warren (2000), Wildermuth et al. (2000), Rytuba (2003), Hartikainen (2005), Lide (2005), Smart and Moore (2005), Nehb and Vydra (2006), Devillanova and Mont (2007), House (2008), Hoffmann and King (2010), Jäger et al. (2010), Bodnar et al. (2012), Böcker et al. (2013), Calvo and Lamberg-Allardt (2015), Rieuwerts (2015), Gandhi and Sarkar (2016), Kirschner et al. (2017), Grenne and Slack (2019)
This table lists the main chemical compounds that contain carbon and the elements listed in the first column. The National Ocean Service (2023) notes that "Carbon is the foundation for all life on Earth," and "carbon atoms continually travel from the atmosphere to the Earth and then back into the atmosphere" and "carbon is continually exchanged between the ocean's surface waters and the atmosphere or is stored for long periods of time in the ocean depths"

biogeography (Campbell, 2018). The commonest elements for organic molecules, which are the basis of life forms are carbon, nitrogen, hydrogen, oxygen, phosphorus, and sulfur (Table 1.2). Boero et al. (2019) add the importance of biogeochemical studies for the "understanding of primary producers, these often termed 'nutrients'… and hence biogeochemical cycling may be an indication of ecosystem functioning and ecological processes" (Boero et al., 2019) and a measure of ecological processes (Campbell, 2022).

Biogeochemical studies, as noted in several definitions of the research focus (e.g., Karl & Schlesinger, 2014), may also examine the impact of biological organisms on the Earth's chemistry, a complex subject due to the vast number of biological species which may impact variously on the chemical environment, increasing the size of the biogeochemical subject area (Hellweger, 2008; Campbell, 2017, 2018; Schlesinger & Bernhardt, 2020). Tartowski and Howarth (2013) support this definition, stating that biogeochemistry is the "discipline which studies biotic controls on the chemistry of the environment and geochemical controls on the structure and function of ecosystems." Woods Hole Oceanographic Institution (2022) points

Table 1.4 Nitrogen-included compounds that are involved in biogeochemical cycling

Associated element	Compounds
Carbon	Cyanogen (C_2N_2), dicyanoacetylene (C_4N_2), hydrogen cyanide (HCN), tetracyanoethylene ($C_2(CN)_4$)
Phosphorous	Nitrogen phosphorus carbonate (C_3NO_9P), ammonium dihydrogen phosphate ($(NH_4)(H_2PO_4)$), diammonium phosphate ($(NH_4)_2HPO_4$)
Potassium	Potassium nitrate (KNO_3), potassium nitride (K_3N), potassium nitrite (KNO_2)
Oxygen	Nitric oxide (NO), nitrogen dioxide (NO_2), nitrogen trioxide (NO_3), nitrous oxide (N_2O).
Iron	Iron nitrides (Fe_2N, Fe_3N_4, Fe_4N, Fe_7N_3 and $Fe_{16}N_2$), Fe-NH(2) to Fe-NH(3), ferric nitrate ($Fe(NO_3)_3 \cdot (H_2O)_n$)
Calcium	Calcium nitride (Ca_3N_2), calcium nitrite ($Ca(NO_2)_2$), calcium nitrate ($(Ca(NO_3)_2(H_2O)_x$
Selenium	Nitrogen selenide (N_4Se_4), selenium nitride (Se_2N_2), selenium-nitrogen halides $[N(SeCl_n)_2]^+$
Sulfur	Sulfur mononitride (NS), tetrasulfur tetranitride, S_4N_4, disulfur dinitride, S_2N_2
Mercury	Mercury-nitrogen (Hg-N), mercury(I) nitride $(Hg_2)3N_2$, mercury II nitrate ($Hg(NO_3)_2$)

Sources: Klein and Hurlbut Jr. (1985), Thomas and Weller (1992), Greenwood and Earnshaw (1997), Appl (1999), Schultz et al. (2000); Warren (2000), Wildermuth et al. (2000), Rytuba (2003), Hartikainen (2005), Lide (2005), Smart and Moore (2005), Nehb and Vydra (2006), Devillanova and Mont (2007), House (2008), Hoffmann and King (2010), Jäger et al. (2010), Bodnar et al. (2012), Böcker et al. (2013), Calvo and Lamberg-Allardt (2015), Rieuwerts (2015), Gandhi and Sarkar (2016), Kirschner et al. (2017), Grenne and Slack (2019)

This table presents the nitrogen-included compounds that occur through biogeochemical cycling. As noted by Kuypers et al. (2018: 263), "Nitrogen is an essential component of all living organisms and the main nutrient limiting life on our planet" and "by far, the largest inventory of freely accessible nitrogen is atmospheric dinitrogen"

to the importance of biological phenomena, as "biogeochemistry is the study of the nearly limitless "transactions" that drive the entire planetary system, including life on Earth. Understanding these fundamental processes provides crucial insights into how life formed, has evolved, is sustained, and is threatened on our planet, and how the various chemical cycles govern and regulate Earth's climate and environment."

Table 1.2 shows the common chemical elements and some of the compounds created when these elements combine with other elements. Tables 1.2, 1.3, 1.4, 1.5, 1.6, 1.7, 1.8, 1.9, 1.10, 1.11, and 1.12 show more details on the combinations of elements into compounds. These tables are provided to illustrate the chemical elements and compounds that compose the materials of biogeochemical cycling. The models in Figs. 1.6, 1.7, 1.8, 1.9, 1.10, 1.11, 1.12, 1.13, 1.14, 1.15, and 1.16 show simplified examples of the biogeochemical cycles. All the cycles include stages in the lithosphere, biosphere, hydrosphere, and atmosphere. All the compounds described may be formed naturally or through human scientific action, but frequently occur in the environment and may have variable impacts on plant and animal ecologies. Some compounds are relatively insoluble solids, while others may be

Table 1.5 Phosphorus compounds that are involved in biogeochemical cycling

Associated element	Compounds, including phosphorus
Carbon	Carbon phosphate ($C_3O_{16}P_4$), adenosine triphosphate ($C_{10}H_{16}N_5O_{13}P_3$), organophosphorus compounds
Nitrogen	Nitrogen phosphorus carbonate (C_3NO_9P), ammonium dihydrogen phosphate ($(NH_4)(H_2PO_4)$), diammonium phosphate ($(NH_4)_2HPO_4$)
Potassium	Potassium phosphorus (H_3KP), tripotassium phosphate (K_3PO_4), potassium phosphate (K_3PO_4)
Oxygen	Phosphorus pentoxide (P_4O_{10}), phosphoryl chloride ($POCl_3$), triphenylphosphine oxide ($OP(C_6H_5)_3$)
Iron	Ferric phosphate ($FePO_4$), iron phosphide (FeP), ferrous ammonium phosphate ($FeNH_4PO_4$)
Calcium	Hydroxyapatite ($Ca_5(PO_4)_3(OH)$, calcium phosphate ($Ca_3(PO_4)_2$)
Selenium	Phosphorus selenide ($P_4\ Se_3$), triphenylphosphine selenide $(C_6H_5)_3PSe$
Sulfur	Phosphorus pentasulfide (P_2S_5), phosphorus sesquisulfide (P_4S_3)
Mercury	Mercury phosphate ($Hg_3O_8P_2$), mercury(ii) phosphite (Hg_3O_6P), mercury(2+);phosphate (HgO_4P^-)

Sources: Klein and Hurlbut Jr. (1985), Thomas and Weller (1992), Greenwood and Earnshaw (1997), Appl (1999), Schultz et al. (2000), Warren (2000), Wildermuth et al. (2000), Rytuba (2003), Hartikainen (2005), Lide (2005), Smart and Moore (2005), Nehb and Vydra (2006), Devillanova and Mont (2007), House (2008), Hoffmann and King (2010), Jäger et al. (2010), Bodnar et al. (2012), Böcker et al. (2013), Calvo and Lamberg-Allardt (2015), Rieuwerts (2015), Gandhi and Sarkar (2016), Kirschner et al. (2017), Grenne and Slack (2019)
This table describes the phosphorus-included compounds that may circulate in global biogeochemical cycles. Phosphorus cycling is nonatmospheric as phosphorus-based compounds are mostly solids (Turner et al., 2005). "Organic phosphorus is involved in almost every biological process." (CABI Digital Library, 2023)

dissolved into water or gaseous form, especially those containing nitrogen, oxygen, and carbon (Campbell, 2022).

Links Between Biogeochemistry and Other Sciences

Biogeochemistry is strongly linked to numerous environmental and earth sciences, as it is scaled from the study of microbial ecologies and chemical elements to the characteristics of vast Earth systems (Figs. 1.17, 1.18, 1.19, and 1.20). At the most basic level, the primary science is chemistry (Campbell, 2017, 2018). The basic components of biogeochemistry include the atmospheric, solid earth, biological and water sciences. Branches of biogeochemistry may link with associated disciplines, or even disciplines less related, but relevant to chemistry or even biology, such as urban planning and pollution. The spatial, global perspective of biogeochemistry

Table 1.6 Potassium compounds that are involved in biogeochemical cycling

Associated element	Compounds
Carbon	Potassium carbonate (K_2CO_3), potassium bicarbonate ($KHCO_3$)
Nitrogen	Potassium nitrate (KNO_3), potassium nitride (K_3N), potassium nitrite (KNO_2)
Phosphorus	Potassium phosphorus (H_3KP), tripotassium phosphate (K_3PO_4), potassium phosphate (K_3PO_4).
Oxygen	Potassium Oxide (K_2O), potassium peroxide (K_2O_2), potassium hydroxide (KOH).
Iron	Potassium ferrate (K_2FeO_4), potassium ferrioxalate ($K_3[Fe(C_2O_4)_3]$.) or $K_3[Fe(C_2O_4)_3]\cdot 3H_2O$ (trihydrate)
Calcium	Potassium calcium citrate ($C_6H_5CaKO_7$), calcium potassium phosphate ($CaKO_4P$).
Selenium	Potassium selenocyanate (CNKSe), potassium; selenium2-(KSe_2^{-3}), potassium selenate (K_2SeO_4)
Sulfur	Potassium sulfate (K_2SO_4), potassium sulfite (K_2SO_3), potassium metabisulfite ($K_2S_2O_5$).
Mercury	Potassium mercuric iodide (K_2HgI_4), mercuric potassium cyanide (C_4HgN_4.2K)

Sources: Klein and Hurlbut Jr. (1985), Thomas and Weller (1992), Greenwood and Earnshaw (1997), Appl (1999), Schultz et al. (2000), Warren (2000), Wildermuth et al. (2000), Rytuba (2003), Hartikainen (2005), Lide (2005), Smart and Moore (2005), Nehb and Vydra (2006), Devillanova and Mont (2007), House (2008), Hoffmann and King (2010), Jäger et al. (2010), Bodnar et al. (2012), Böcker et al. (2013), Calvo and Lamberg-Allardt (2015), Rieuwerts (2015), Gandhi and Sarkar (2016), Kirschner et al. (2017), Grenne and Slack (2019)
This table lists some potassium-included compounds that may circulate in global biogeochemical cycles. Schlesinger (2021b: 427) notes that "Potassium presents a conundrum for biogeochemists. Potassium is cycled wastefully at the plant level, but it appears to be conserved in the nutrient budgets of entire ecosystems, where it sometimes limits net primary productivity"

also links it with the geographical sciences, including oceanography (Figs. 1.26, 1.27, 1.28, 1.29, 1.30, and 1.31).

Biogeochemistry and Physical Geography

Biogeochemistry is also closely linked to physical geography, and in some cases may even be seen as almost synonymous to physical geography (Campbell, 2017, 2018). As pointed out by Warf (2010), "biogeochemistry is the chemistry of Earth's surface. Or, in other words, it is physical geography through the lens of chemistry" and a "biogeochemical cycle examines how a particular chemical element moves through the physical geographic spheres (i.e., the atmosphere, biosphere, lithosphere, and hydrosphere)." While some classifications of physical geography divide the discipline into the sections of geomorphology, climatology, pedology, biogeography and sometimes oceanography and geomatics, the perspective of

Table 1.7 Oxygen compounds that are involved in biogeochemical cycling

Associated element	Compounds
Carbon	Carbon dioxide (CO_2), carbon monoxide (CO), ethanol (C_2H_6O), glucose ($C_6H_{12}O_6$)
Nitrogen	Nitric oxide (NO), nitrogen dioxide (NO_2), nitrogen trioxide (NO_3), nitrous oxide (N_2O)
Phosphorus	Phosphorus pentoxide (P_4O_{10}), phosphoryl chloride ($POCl_3$), triphenylphosphine oxide ($OP(C_6H_5)_3$)
Potassium	Potassium oxide (K_2O), potassium peroxide (K_2O_2), potassium hydroxide (KOH)
Iron	Iron oxide (Fe_2O_3), iron (II, III) oxide (Fe_3O_4), iron (II) sulfate ($FeSO_4{\cdot}7H_2O$)
Calcium	Calcium hydroxide (Ca(OH)2, calcium oxide (CaO), Calcium;hydron;oxygen(2-) (CaH_2O_2)
Selenium	Selenium dioxide (SeO_2), selenium trioxide (SeO_3), selenic acid (H_2SeO_4), sodium selenate (Na_2SeO_4)
Sulfur	Thionyl dichloride (SCl_2O), sulfuric acid (SH_2O_4 (s), sulfur dioxide (SO_2), sulfur trioxide (SO_3)
Mercury	Mercury (II) oxide (HgO), mercury(I) acetate ($C_2H_3HgO_2$), hydroxymethylmercury (CH_3HgOH or CH_4HgO)

Sources: Klein and Hurlbut Jr. (1985), Thomas and Weller (1992), Greenwood and Earnshaw (1997), Appl (1999), Schultz et al. (2000), Warren (2000), Wildermuth et al. (2000), Rytuba (2003), Hartikainen (2005), Lide (2005), Smart and Moore (2005), Nehb and Vydra (2006), Devillanova and Mont (2007), House (2008), Hoffmann and King (2010), Jäger et al. (2010), Bodnar et al. (2012), Böcker et al. (2013), Calvo and Lamberg-Allardt (2015), Rieuwerts (2015), Gandhi and Sarkar (2016), Kirschner et al. (2017), Grenne and Slack (2019)
This table lists some oxygen-included compounds that may circulate in global biogeochemical cycles. Oxygen is one of the most important elements and is included in many biogeochemical cycles. Testa and Kemp (2011) note that "the biogeochemical cycling of many important inorganic elements is highly oxygen dependent. Consequently, spatial, and temporal variations in oxygen, and the forces that control such variation, are of central importance in the study of coastal ecology and biogeochemistry"

biogeochemistry would be to integrate these disciplines or subdisciplines to examine their interrelationships, the better to study and explain the dynamic and highly holistic nature of the Earth's environment, without artificial, disciplinary boundaries (Campbell, 2020). Figures 1.21, 1.22, 1.23, 1.24, 1.25, 1.26, 1.27, 1.28, 1.29, 1.30, and 1.31 show the links between the components of physical geography and biogeochemistry including some urban issues.

Biogeochemistry and Ocean Sciences

The oceans are an important study area for both biogeochemistry and physical geography/oceanography. They represent a large portion of global biogeochemistry, and they are an important focus of such studies, both within and outside the

Table 1.8 Iron compounds that are involved in biogeochemical cycling

Associated element	Compounds
Carbon	Siderite ($FeCO_3$), ferric carbonate ($Fe_2(CO_3)_3$, iron carbide (Fe_3C), iron pentacarbonyl ($Fe(CO)_5$)
Nitrogen	Iron nitrides (Fe_2N, Fe_3N_4, Fe_4N, Fe_7N_3, and $Fe_{16}N_2$), Fe-NH(2) to Fe-NH(3), ferric nitrate ($Fe(NO_3)_3{\cdot}(H_2O)_n$)
Phosphorus	Ferric phosphate ($FePO_4$), iron phosphide (FeP), ferrous ammonium phosphate ($FeNH_4PO_4$)
Potassium	Potassium ferrate (K_2FeO_4), potassium ferrioxalate ($K_3[Fe(C_2O_4)_3]$.) or $K_3[Fe(C_2O_4)_3]{\cdot}3H_2O$ (trihydrate)
Oxygen	Iron oxide (Fe_2O_3), iron (II, III) oxide (Fe_3O_4), iron(II) sulfate ($FeSO_4{\cdot}7H_2O$)
Calcium	Calcium iron oxide ($CaFe_2H_8O_4$), calcium aluminoferrite ($Ca_2(Al,Fe)_2O_5$)
Selenium	Iron (II) selenide (FeSe), bis(selanylidene)iron ($FeSe_2$), iron(III) selenide (Fe_2Se_3)
Sulfur	Iron sulfide (FeS), ferrous sulfate ($FeSO_4$), ferrous sulfate heptahydrate ($H_{14}FeO_{11}S$), ferric sulfate $Fe_2(SO_4)_3$
Mercury	Rare alloy

Sources: Klein and Hurlbut Jr. (1985), Linderoth and Mørup (1992), Thomas and Weller (1992), Greenwood and Earnshaw (1997), Appl (1999), Schultz et al. (2000), Warren (2000), Wildermuth et al. (2000), Rytuba (2003), Hartikainen (2005), Lide (2005), Smart and Moore (2005), Nehb and Vydra (2006), Devillanova and Mont (2007), House (2008), Hoffmann and King (2010), Jäger et al. (2010), Bodnar et al. (2012), Böcker et al. (2013), Calvo and Lamberg-Allardt (2015), Rieuwerts (2015), Gandhi and Sarkar (2016), Kirschner et al. (2017), Grenne and Slack (2019)
This table lists some iron-included compounds that may circulate in global biogeochemical cycles. Kappler et al. (2021: 360) note that "biogeochemical cycling of iron is crucial to many environmental processes, such as ocean productivity, carbon storage, greenhouse gas emissions and the fate of nutrients, toxic metals and metalloids"

discipline of physical geography (e.g., biological, chemical, physical, and geological oceanography, and marine biology/ecology, marine geology, marine chemistry, and marine physics) (Scripps Institution of Oceanography, 2022). Such oceanographic studies may focus on both physical and biological parameters. For example, Brasseur et al. (2009, 206) hold that "monitoring and predicting the biogeochemical state of the ocean and marine ecosystems is an important application of operational oceanography that needs to be expanded... most of these applications require accurate estimates of both physical and biogeochemical ocean conditions over a wide range of spatial and temporal scales."

In another example, ocean ecology and biogeochemistry may be integrated, as indicated in recent university led research (College of Earth, Ocean, and Atmospheric Sciences, 2022). Ocean ecology and biogeochemistry is defined as "the study of the ocean's biological, chemical, and geological processes, and their interplay" and the study of the "structure and function of ecosystems across space and time, including feedbacks between land, atmosphere and ocean" (College of Earth, Ocean, and Atmospheric Sciences, 2022). Falkowski (2003: 163) notes that a major characteristic of the Earth is that it is the only planet in the solar system with known water

Table 1.9 Calcium compounds that are involved in biogeochemical cycling

Associated element	Compounds
Carbon	Calcium carbonate ($CaCO_3$), calcium bicarbonate ($Ca(HCO_3)_2$)
Nitrogen	Calcium nitride (Ca_3N_2), calcium nitrite ($Ca(NO_2)_2$), calcium nitrate ($(Ca(NO_3)_2(H_2O)_x$
Phosphorus	Hydroxyapatite ($Ca_5(PO_4)_3(OH)$, calcium phosphate ($Ca_3(PO_4)_2$)
Potassium	Potassium calcium citrate ($C_6H_5CaKO_7$), calcium potassium phosphate ($CaKO_4P$)
Oxygen	Calcium hydroxide (Ca (OH)2, calcium oxide (CaO), Calcium;hydron;oxygen(2-) (CaH_2O_2)
Iron	Calcium iron oxide ($CaFe_2H_8O_4$), calcium aluminoferrite ($Ca_2(Al,Fe)_2O_5$)
Selenium	Calcium selenide (CaSe)
Sulfur	Calcium sulfate ($CaSO_4$)
Mercury	Calcium–mercury (3/1) (Ca_3Hg)

Sources: Klein and Hurlbut Jr. (1985), Thomas and Weller (1992), Greenwood and Earnshaw (1997), Appl (1999), Schultz et al. (2000), Warren (2000), Wildermuth et al. (2000), Rytuba (2003), Hartikainen (2005), Lide (2005), Smart and Moore (2005), Nehb and Vydra (2006), Devillanova and Mont (2007), House (2008), Hoffmann and King (2010), Jäger et al. (2010), Bodnar et al. (2012), Böcker et al. (2013), Calvo and Lamberg-Allardt (2015), Rieuwerts (2015), Gandhi and Sarkar (2016), Kirschner et al. (2017), Grenne and Slack (2019)

This table lists some calcium-included compounds that may circulate in global biogeochemical cycles

Fantle and tipper (2014: 148) argue that "at the Earth's surface, calcium (Ca) is a critical element at a variety of scales: it is soluble in natural waters, a biological nutrient, and a major constituent of the dominant mineral sink for carbon in the ocean ($CaCO_3$)"

supplies, and with an atmosphere with concentrations of free molecular oxygen. It is argued that these two features, oceans, and an atmosphere, were derived from the photobiologically catalyzed splitting of water by unicellular photosynthetic organisms that existed in the oceans for millions of years, during which the organisms used the hydrogen atoms from water and Earth materials to form organic matter from carbon dioxide (CO_2) and other hydrated equivalents, this being the process by which organic matter was formed from inorganic carbon. This primary production "is the basis for all life on Earth" (Falkowski, 2003: 163).

Achterberg (2011) justifies the development of an ocean-focused biogeochemistry: The ocean is vital for the Earth's climate system, ecosystem services, chemical element cycling between the sea floor, ocean, land and atmosphere, and human actions increasing emissions of nitrogen, phosphorus, carbon, and trace elements. Hence, a "detailed understanding of biogeochemical processes, including their rates, is essential to the identification and assessment of climatic and chemical feedbacks associated with changes in the chemical and physical environment that are mediated through ocean biology, chemistry and physics" (Achterberg, 2011). Research areas include "the cycling of organic and inorganic forms of carbon, nitrogen and phosphorus, the cycling and biological roles of essential trace

Table 1.10 Selenium compounds that are involved in biogeochemical cycling

Associated element	Compounds
Carbon	Carbon diselenide (CSe_2), selenomethionine ($C_5H_{11}NO_2Se$), dimethyl selenide ($(CH_3)_2Se$)
Nitrogen	Nitrogen selenide (N_4Se_4), selenium nitride (Se_2N_2), selenium-nitrogen halides $[N(SeCl_n)_2]^+$
Phosphorus	Phosphorus selenide (P_4 Se_3), triphenylphosphine selenide ($(C_6H_5)_3PSe$.)
Potassium	Potassium selenocyanate (CNKSe), potassium;selenium2-(KSe_2^{-3}), potassium selenate (K_2SeO_4)
Oxygen	Selenium dioxide (SeO_2), selenium trioxide (SeO_3), selenic acid (H_2SeO_4), sodium selenate (Na_2SeO_4)
Calcium	Calcium selenide (CaSe)
Iron	Iron (II) selenide (FeSe), bis(selanylidene)iron ($FeSe_2$), Iron(III) selenide (Fe_2Se_3)
Sulfur	Selenium disulfide (S_2Se)
Mercury	Bis(methylmercuric) selenide ($(CH_3Hg)2Se$)

Sources: Klein and Hurlbut Jr. (1985), Linderoth and Mørup (1992), Thomas and Weller (1992), Greenwood and Earnshaw (1997), Appl (1999), Schultz et al. (2000), Warren (2000), Wildermuth et al. (2000), Rytuba (2003), Hartikainen (2005), Lide (2005), Smart and Moore (2005), Nehb and Vydra (2006), Devillanova and Mont (2007), House (2008), Hoffmann and King (2010), Jäger et al. (2010), Bodnar et al. (2012), Böcker et al. (2013), Calvo and Lamberg-Allardt (2015), Rieuwerts (2015), Gandhi and Sarkar (2016), Kirschner et al. (2017), Grenne and Slack (2019), Nogara et al. (2021)

This table lists some selenium-included compounds that may circulate in global biogeochemical cycles

Sharma et al. (2015: 49) note that "Selenium levels and speciation in environmental compartments and the dynamics of global Se cycling continue to be a subject of intense interest largely because Se is both an essential element and a toxicant at elevated levels"

elements, and the fate and climatic impact of marine produced trace gases" (Achterberg, 2011).

Urban and Pollution Biogeochemistry

Urban biogeochemistry is another recently developed and continually developing branch of biogeochemistry that links this science to urban planning, ecology, and management (Bianchi, 2021). Bianchi (2021) observes that the subdisciplines of biogeochemistry are increasing in number, and the emergence of urban biogeochemistry partially represents an integration of urban ecology and environmental sciences (see also Grimm et al., 2008; Pickett et al., 2011; Schulze, 2015). Bianchi (2021) further notes that urban biogeochemistry is related to the early developments of ecology, such as that of Odum, who noted that cities are more wasteful of energy consumption than are natural systems, and urban areas are more dependent on external matter and energy, with higher human impacts than other ecosystems.

Table 1.11 Sulfur compounds that are involved in biogeochemical cycling

Associated element	Compounds
Carbon	Carbon sulfur (CH_6S), organosulfur compounds (e.g., sulfides – carbon disulfide CS_2, carbon monosulfide CS
Nitrogen	Sulfur mononitride (NS), tetrasulfur tetranitride, S_4N_4, disulfur dinitride, S_2N_2
Phosphorus	Phosphorus pentasulfide (P_2S_5), phosphorus sesquisulfide (P_4S_3)
Potassium	Potassium sulfate (K_2SO_4), potassium sulfite (K_2SO_3), potassium metabisulfite ($K_2S_2O_5$)
Oxygen	Thionyl dichloride (SCl_2O), sulfuric acid (SH_2O_4 (s), sulfur dioxide (SO_2), sulfur trioxide (SO_3)
Iron	Iron sulfide (FeS), ferrous sulfate ($FeSO_4$), ferrous sulfate heptahydrate ($H_{14}FeO_{11}S$), ferric sulfate $Fe_2(SO_4)_3$
Calcium	Calcium sulfate ($CaSO_4$)
Selenium	Selenium disulfide (S_2Se)
Mercury	Mercury sulfide (HgS)

Sources: Klein and Hurlbut Jr. (1985), Thomas and Weller (1992), Greenwood and Earnshaw (1997), Appl (1999), Schultz et al. (2000), Warren (2000), Wildermuth et al. (2000), Rytuba (2003), Hartikainen (2005), Lide (2005), Smart and Moore (2005), Nehb and Vydra (2006), Devillanova and Mont (2007), House (2008), Hoffmann and King (2010), Jäger et al. (2010), Bodnar et al. (2012), Böcker et al. (2013), Calvo and Lamberg-Allardt (2015), Rieuwerts (2015), Gandhi and Sarkar (2016), Kirschner et al. (2017), Grenne and Slack (2019)
This table lists some sulfur-included compounds that may circulate in global biogeochemical cycles. Fox et al. (2022) point out that "Sulfur (S) is an essential macronutrient and important component of the earth's crust, and its cycling has critical impacts on trace metal mobility, water quality, and human health. Pyrite weathering is the primary pathway by which sulfur enters surface waters. However, biogeochemical cycling of sulfur in soils and the river corridor mediates sulfate exports"

Urban areas are also linked with human constructed drivers such as engineering, urban demographic trends and household-scale actions which may be important for future biogeochemical research (Bianchi (2021, see also Odum, 1971; Kaye et al. 2006).

One factor for the increasing importance of urban biogeochemistry is the increased human populations living in urban areas (Gardner et al., 2014). From 1950 to 2014, the proportion of the world's population living in urban centers increased from 30% to 54%, and the United Nations' World Urbanization Prospects estimates this number will reach 66%, or 6.3 billion, by 2050. Most of this growth is projected to occur in Africa and Asia, which today are mostly rural in cultural land cover. Cities in more developed countries will also have to combat aging infrastructure and the spread of toxic metals, organic contaminants, and "emerging" contaminants, just to name a few issues" (Gardner et al., 2014). These authors describe an academic meeting on urban biogeochemistry under the International Association of Geochemistry's (IAGC) Urban Geochemistry Working Group. This meeting emphasized "quantifying anthropogenic and natural sources of CO_2 in urban areas, the transformation and fate of everyday chemicals in urban watersheds, and geochemistry and urban health" (Gardner et al., 2014).

Table 1.12 Mercury compounds that are involved in biogeochemical cycling

Associated element	Compounds
Carbon	Organomercury compounds: methylmercury (II) CH_3Hg+; ethylmercury(II) C_2H_5Hg+; dimethylmercury, $(CH_3)_2Hg)$
Nitrogen	Mercury-nitrogen (Hg-N), mercury(I) nitride $(Hg_2)3N_2$, mercury II nitrate $(Hg(NO_3)_2)$
Phosphorus	Mercury phosphate $(Hg_3O_8P_2)$, mercury(ii) phosphite (Hg_3O_6P), Mercury(2+);phosphate (HgO_4P^-)
Potassium	Potassium mercuric iodide (K_2HgI_4), mercuric potassium cyanide $(C_4HgN_4.2K)$
Oxygen	Mercury (II) oxide (HgO), mercury(I) acetate $(C_2H_3HgO_2)$, hydroxymethylmercury (CH_3HgOH or CH_4HgO)
Iron	Rare alloy
Calcium	Calcium–mercury (3/1) (Ca_3Hg)
Selenium	Bis(methylmercuric) selenide $((CH_3Hg)2Se)$
Sulfur	Mercury sulfide (HgS)

Sources: Klein and Hurlbut Jr. (1985), Linderoth and Mørup (1992), Thomas and Weller (1992), Greenwood and Earnshaw (1997), Appl (1999), Schultz et al. (2000), Warren (2000), Wildermuth et al. (2000), Rytuba (2003), Hartikainen (2005), Lide (2005), Smart and Moore (2005), Nehb and Vydra (2006), Devillanova and Mont (2007), House (2008), Hoffmann and King (2010), Jäger et al. (2010), Bodnar et al. (2012), Böcker et al. (2013), Calvo and Lamberg-Allardt (2015), Rieuwerts (2015), Gandhi and Sarkar (2016), Kirschner et al. (2017), Grenne and Slack (2019)
This table lists some sulfur-included compounds that may circulate in global biogeochemical cycles. Singh et al. (2023) write that mercury "is among the naturally occurring heavy metal with elemental, organic, and inorganic distributions in the environment… These toxicants have high persistence, toxicity, and widespread contamination in the soil, water, and air resources"

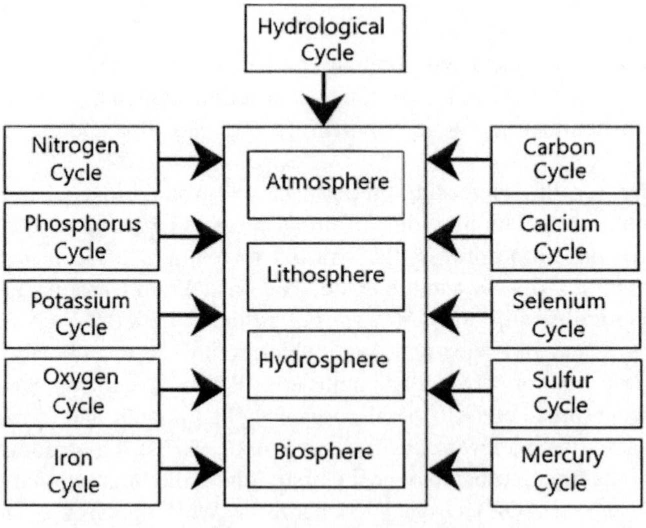

Fig. 1.1 Biogeochemical cycles. Some more elements have been added to the six usually included. The 11 circles illustrated flow through the four main global systems, interactively, as the basis of all life and ecologies

Fig. 1.2 Images of the biosphere. Clockwise from top left: deciduous shrub; coniferous stand; deer; cormorants. (Sources: Author)

Additionally, there are "eight key issues to be addressed in urban geochemical research" these comprising the alteration of global biogeochemical cycles in the urban setting, the evolution of urban infrastructure and impacts on geochemical processes, the measurement of urban systems to assess chemical issues, the role of urban geochemists in policy decisions, the impacts of geochemistry of urban environments on human and ecosystem health, the impact of "episodic perturbations" including natural disasters on the urban environment, impacts of the past disturbances on present geochemical systems and how the "geochemical 'footprint' of urban areas may be defined" (Gardner et al., 2014).

There is also the biogeochemistry of pesticides, mostly linked to the agricultural supports of urban populations (Campbell, 2015, 2022). Barbash (2014: 535) argues that the use of synthetic pesticides from the early twentieth century has created "the most extensive release of synthetic compounds that Earth's ecosystems have ever experienced," and the assessment and analysis of this

Fig. 1.3 Images of the lithosphere. Clockwise from top left: rocky ground, Canada; flat coastal land, Canada; volcano, El Salvador; mountain range, El Salvador. (Sources: Author)

development requires knowledge of the physical, chemical, biological, and hydrologic factors that may contribute to the concentrations of pesticide compounds, which may include pesticide parent compounds, pesticide adjuvants or inerts, and compounds created by the transformation of pesticides or adjuvants. These may occur in air, water, soil, or biota (Barbash, 2014: 535). Effective management requires an approach that investigates the factors for the pesticide compounds in the environments, and the transformations, as well as the use, management, and media attention (Barbash, 2014: 535). The Figures 1.27, 1.28, 1.29, 1.30 and 1.31 above illustrate some of the features of urban and related environmental changes that may create chemical and, hence, biogeochemical changes.

Fig. 1.4 Images of the hydrosphere. Clockwise from top left: rain clouds, Canada; waterfall, El Salvador; Pacific Ocean, El Salvador; river, El Salvador. (Sources: Author)

Fig. 1.5 Images of the atmosphere. Clockwise from top left: low cumulus clouds over Costa Rica; high cirrus clouds over Canada; nimbostratus clouds over Canada; cumulus clouds over Central America. (Sources: Author)

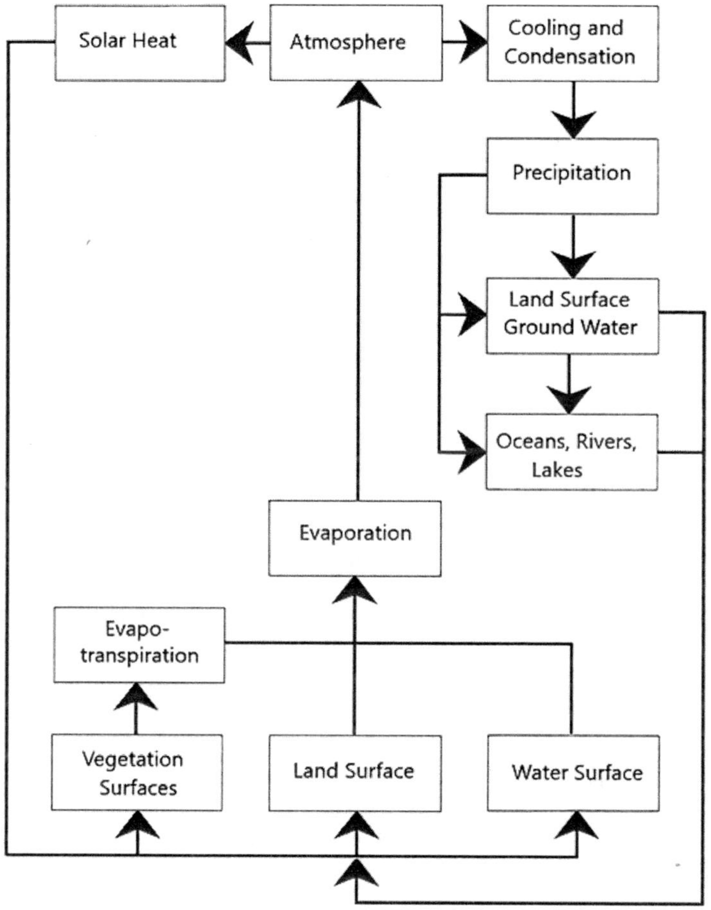

Fig. 1.6 The hydrological cycle

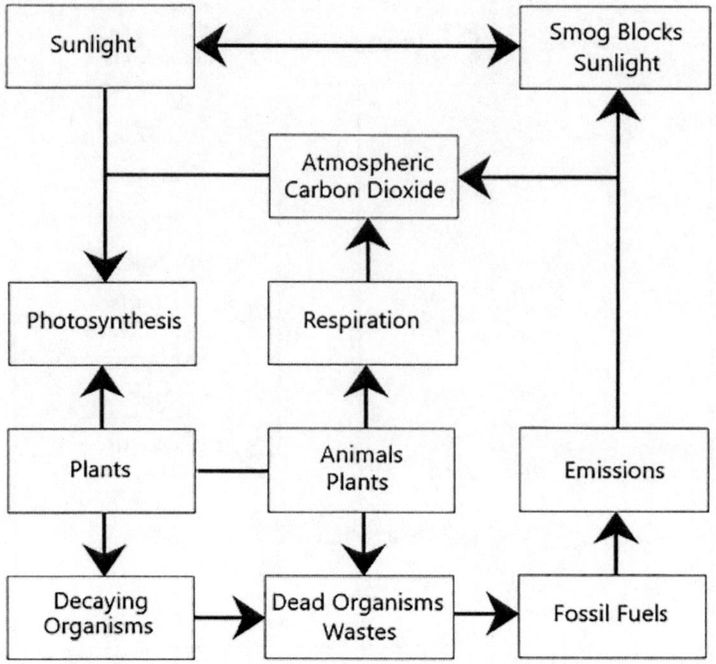

Fig. 1.7 The carbon cycle

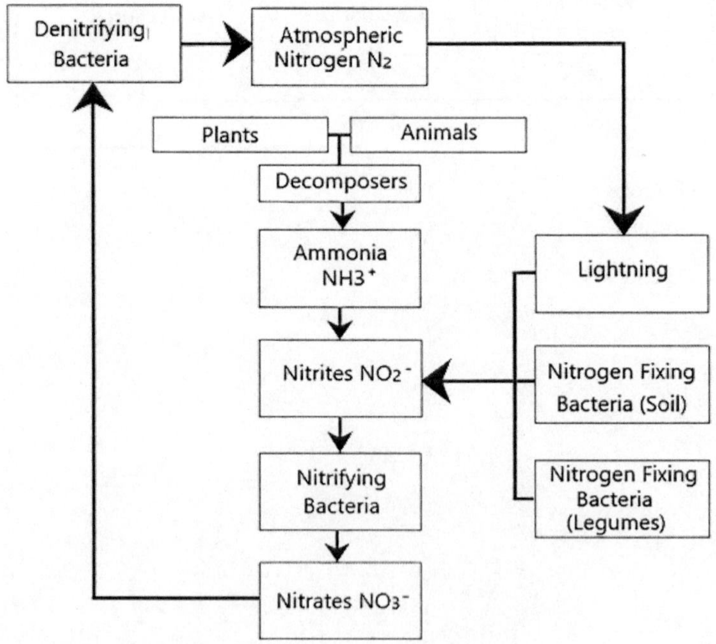

Fig. 1.8 The nitrogen cycle

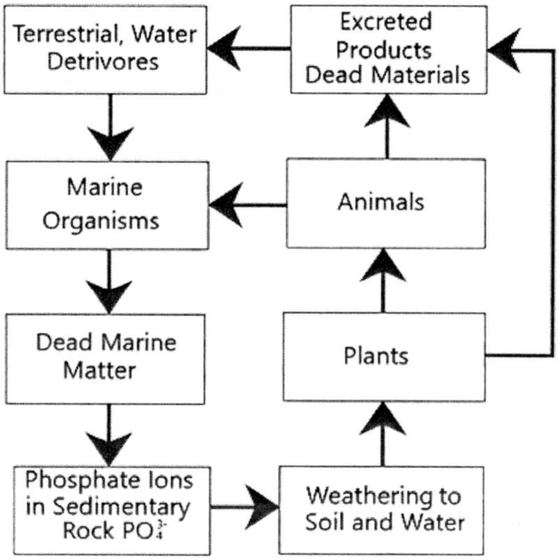

Fig. 1.9 The phosphorous cycle. This is one of the most important cycles. Kvakić et al. (2020) note that "phosphorus (P) is the second most important nutrient after nitrogen (N) and can greatly diminish plant productivity if P supply is not adequate. Plants respond to soil P availability by adjusting root biomass to maintain uptake and productivity due to P use"

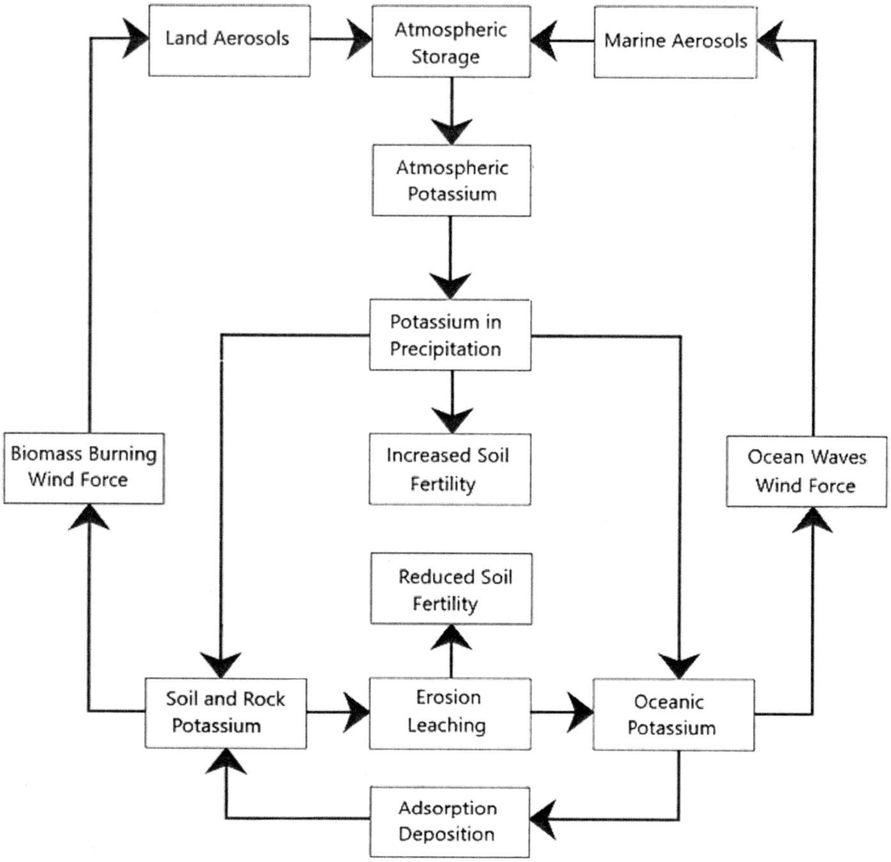

Fig. 1.10 The potassium cycle

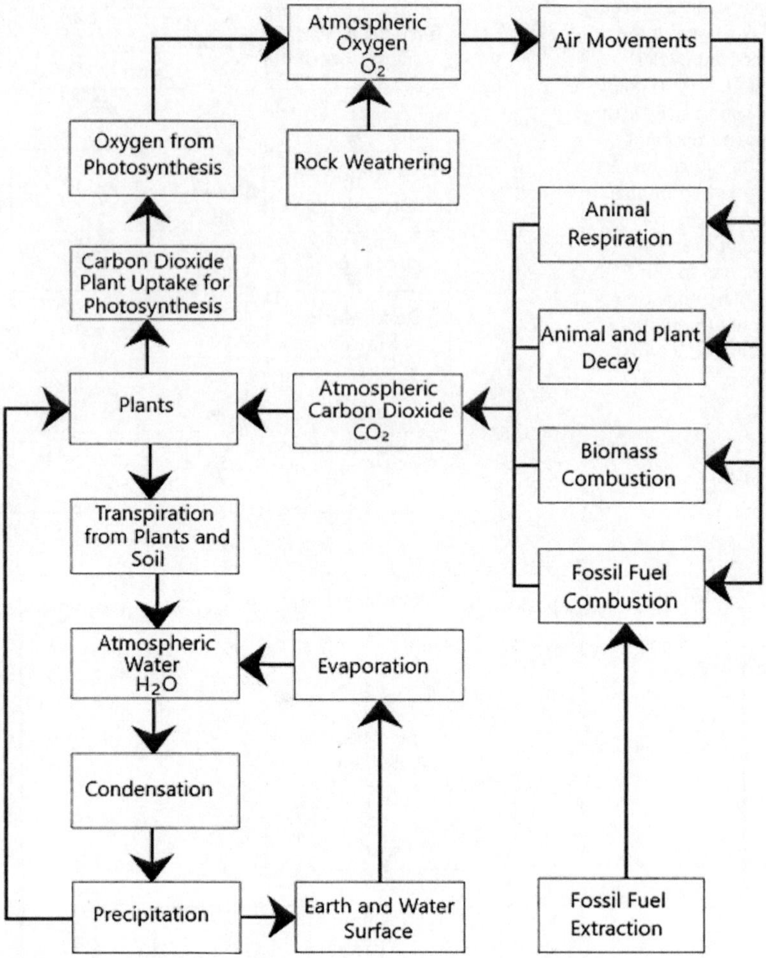

Fig. 1.11 The oxygen cycle

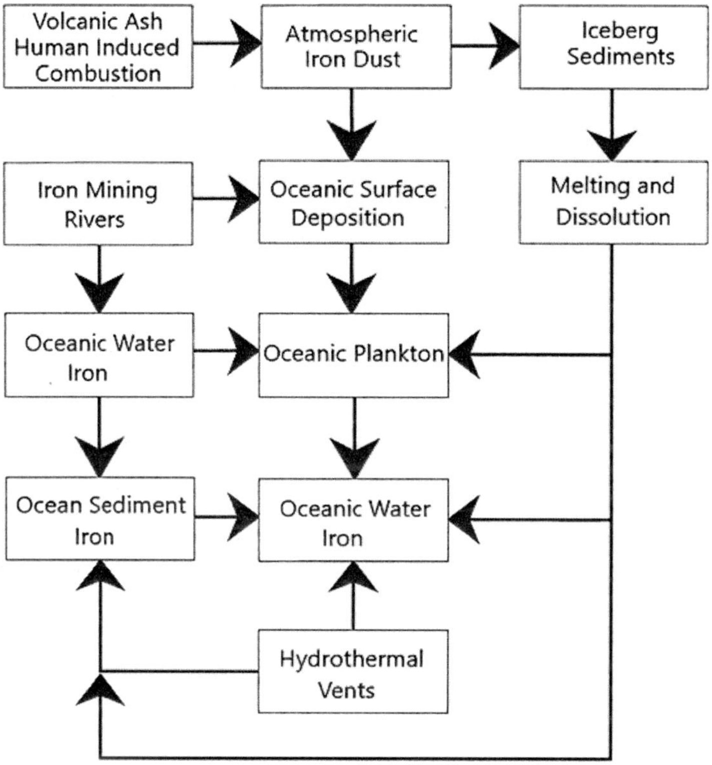

Fig. 1.12 The iron cycle

Fig. 1.13 The calcium cycle

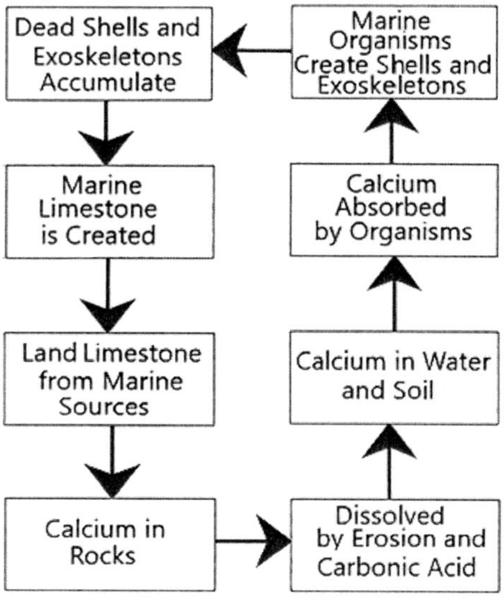

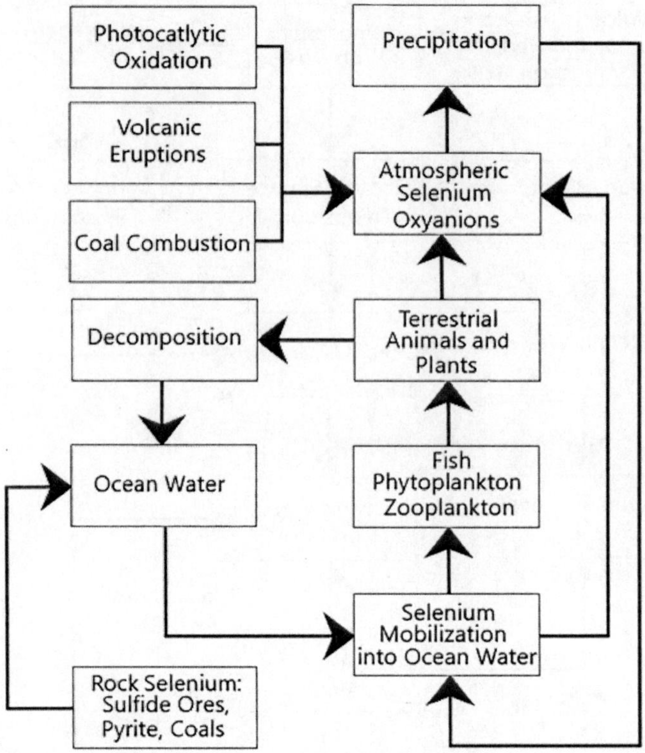

Fig. 1.14 The selenium cycle

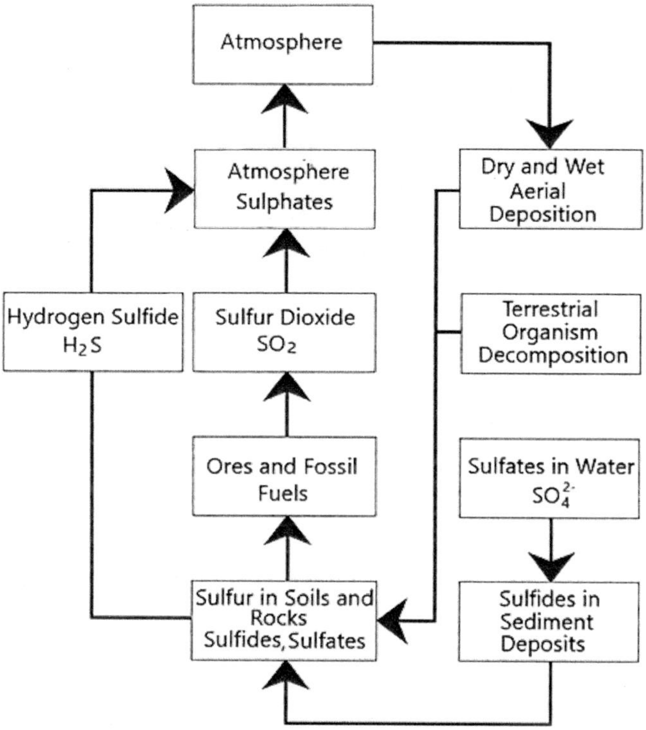

Fig. 1.15 The sulfur cycle

Fig. 1.16 The mercury cycle

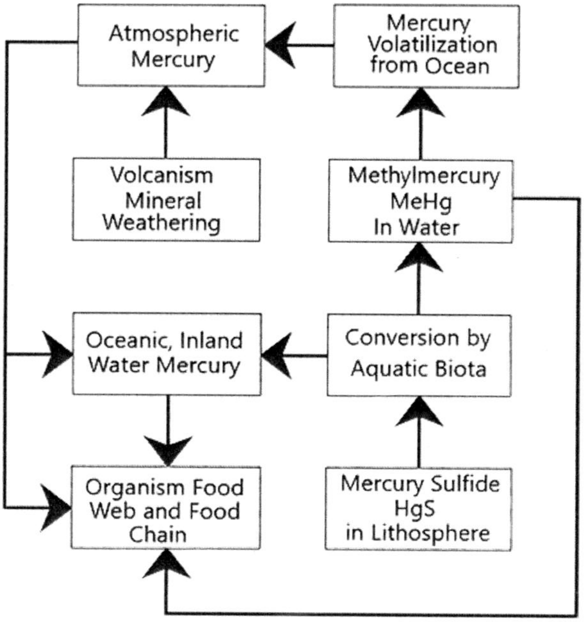

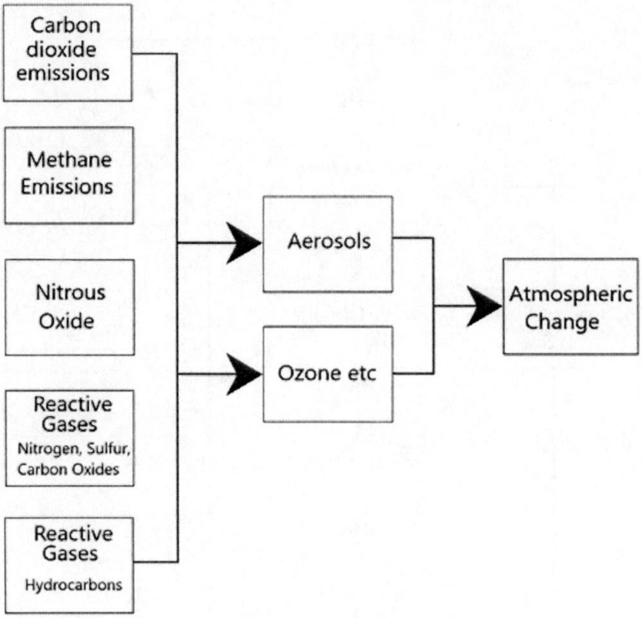

Fig. 1.17 Relationship between biogeochemistry and atmospheric sciences, based on changing emissions. The chemical compounds change the chemical composition of the air and increase the aerosols, contributing to atmospheric change. (Source: Denman et al., 2007)

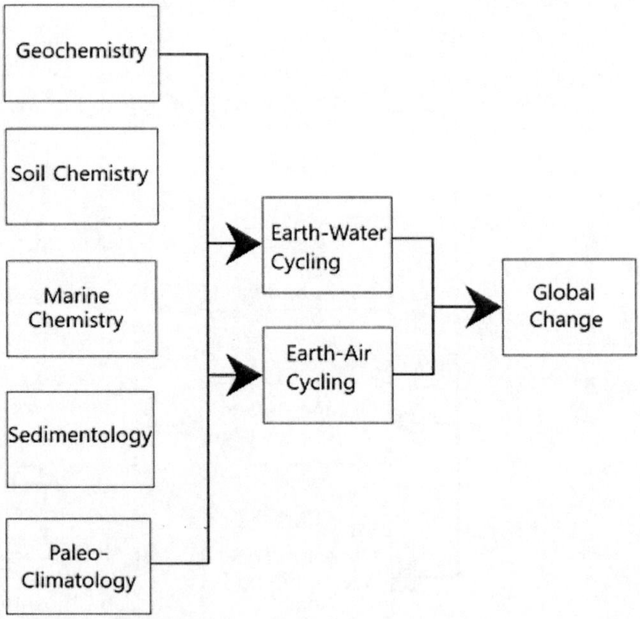

Fig. 1.18 Global change. Relationship between biogeochemistry and solid earth sciences. Earth, water, and air cycling are influenced by chemical changes, which are studied by the disciplines in the first column. (Adapted from Bianchi et al., 2020)

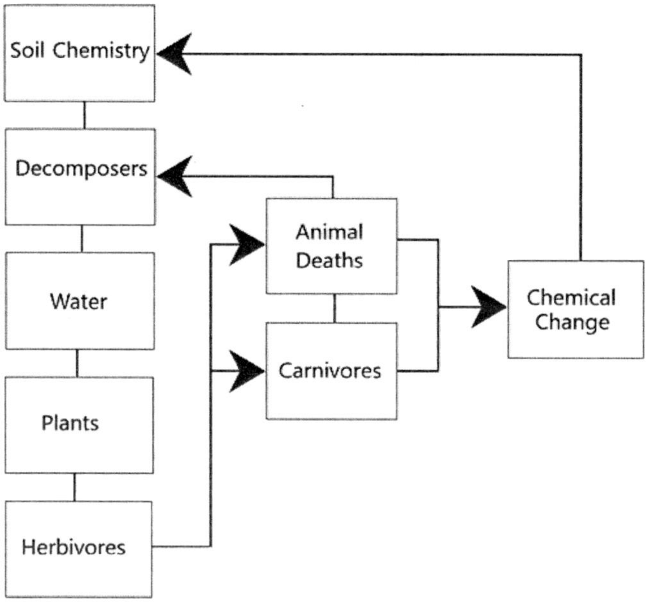

Fig. 1.19 Relationship between biogeochemistry and biological sciences. Here, chemical cycling occurs as carnivores at the top of the food chain kill and eat herbivores, which also eat plants and associated water. Decomposers in association with soil chemistry start the process again. The remains of animals and plants. (Source: Example derived from Schmitz (2020))

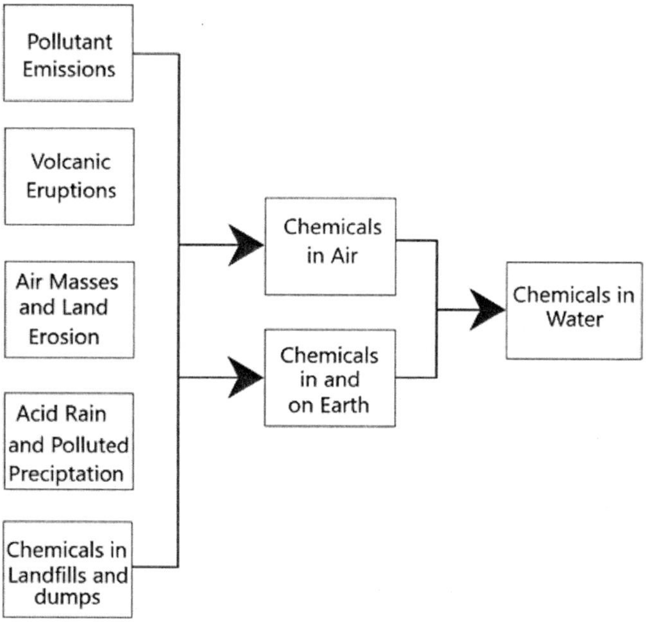

Fig. 1.20 Relationship between biogeochemistry and water sciences. Chemical elements and compounds in the air are strongly influenced by both nature and human influenced hazards. (Example derived from Campbell (2020))

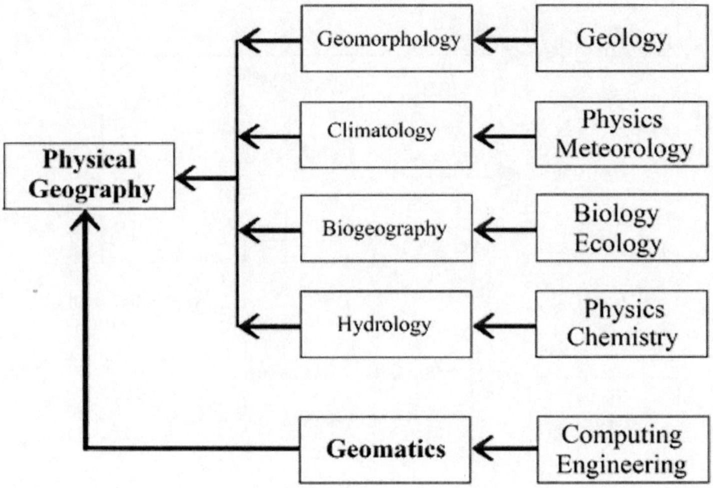

Fig. 1.21 The background subjects of physical geography. Note that these subjects cover the focus of biogeochemistry

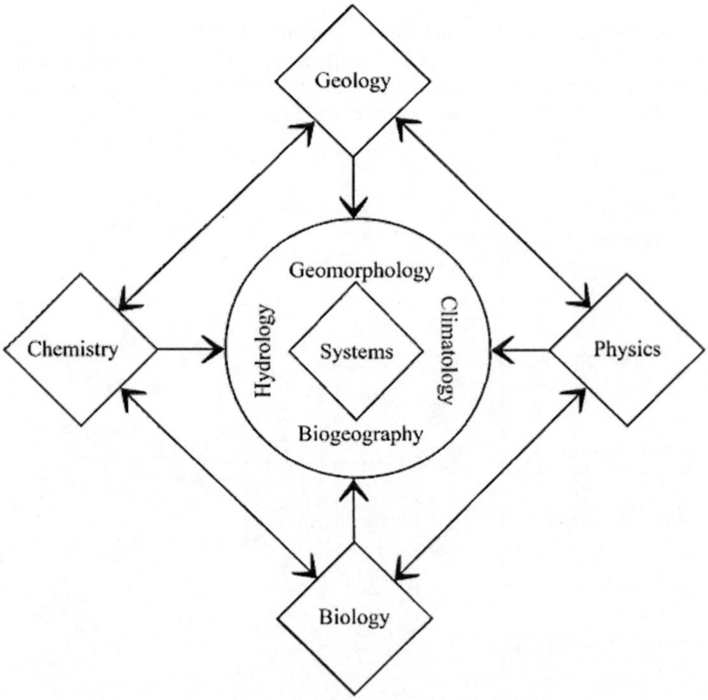

Fig. 1.22 The spatial systems approach of physical geography, with related subjects

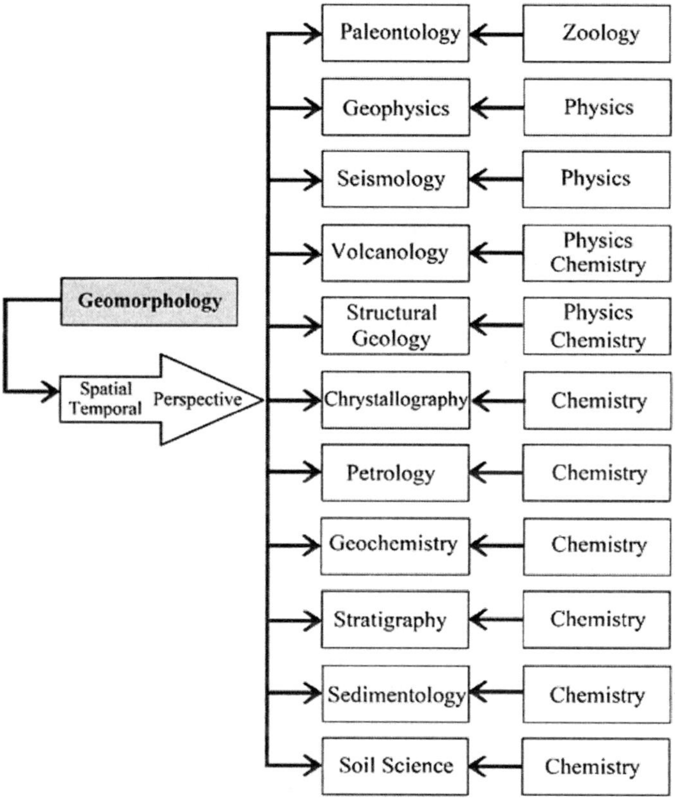

Fig. 1.23 Background of geomorphology, which includes geochemical bases of biogeochemistry. The pure sciences are listed in the column on the right, these forming the backbone of the applied earth sciences listed to the left, which may be part of the "geo" in biogeochemistry, and link strongly to the geomorphological part of physical geography

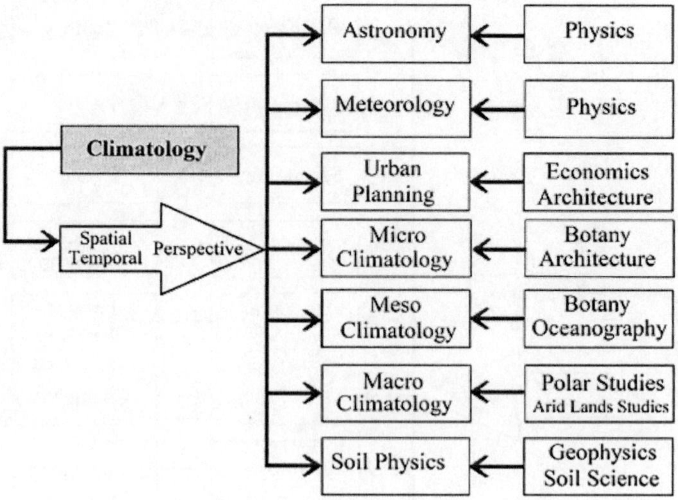

Fig. 1.24 Background of climatology, the bases for biogeochemical atmospheric cycling. The sciences on the right provide inputs to the more applied sciences on the left

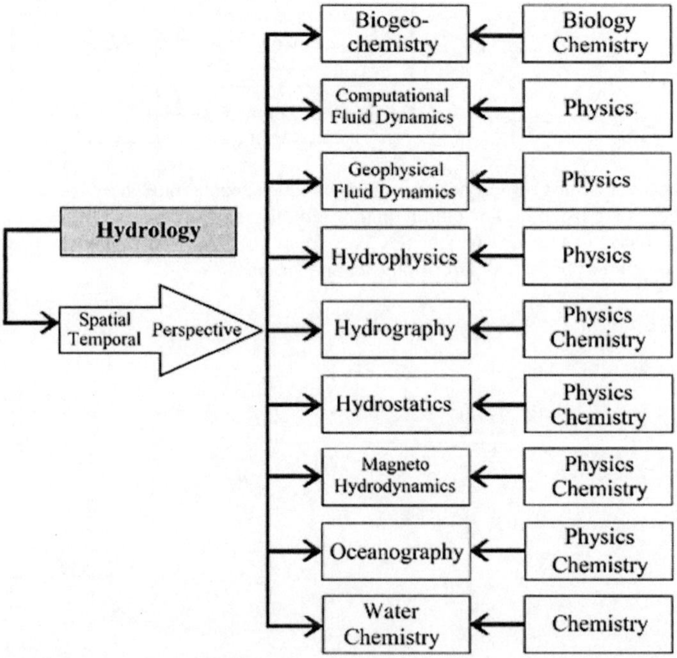

Fig. 1.25 Links between hydrology and other sciences, including biogeochemistry. All the hydrological sciences are vital to biogeochemistry, as water is a principal means of movement for chemical elements and compounds within biogeochemical systems

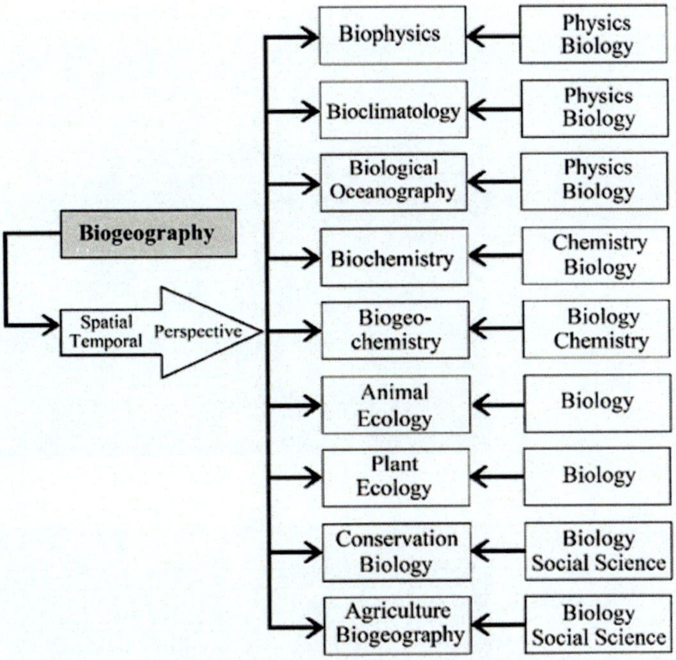

Fig. 1.26 Links between biogeography and other sciences, including biogeochemistry. The pure science of biology is basic for the study of the applied biological sciences listed in the left column. These study the dynamics of the "bio" area of biogeochemistry

Fig. 1.27 Factories in Poznań, Poland. (Source: Marcin Jozwiak-Unsplash)

Fig. 1.28 Factories in North Bend, OH, United States. (Source: Kelly – Pexels)

Fig. 1.29 Decaying Port, Jakarta, Jakarta, Indonesia. (Source: Tom Fisk – Pexels)

Fig. 1.30 Pollution in Myanmar. (Source: Stijn Dijkstra-Pexels)

Preface

We are delighted to introduce the proceedings of the second edition of the 2019 European Alliance for Innovation (EAI) International Conference on Big Data Innovation for Sustainable Cognitive Computing (BDCC 2019). This conference brought researchers, developers, and practitioners around the world who are leveraging and developing Big Data technology for a smarter and more resilient data. The theme of BDCC 2019 was "Big Data Analytics for Sustainable Computing."

The technical program of BDCC 2019 consisted of 35 full papers in oral presentation sessions at the main conference tracks. The conference tracks were Main Track—Big Data Analytics for Sustainable Computing and three workshop tracks were Track 1—Workshop on Big Data and Society, Track 2—Workshop on Big Data Visualization and Analytics, and Track 3—Workshop on Big Data and Machine Learning. Aside from the high-quality technical paper presentations, the technical program also featured two keynote speakers designed for professionals working in the early stages of building an advancement program, as well as those with more mature operations. The two keynote speakers were Dr. Mu-Yen Chen, Professor, Department of Information Management, National Taichung University of Science and Technology, Taiwan, and Dr. M.G. Sumithra, Professor and Head, Electronics and Communication Engineering, KPR Institute of Engineering and Technology, Coimbatore, Tamil Nadu, India.

Coordination with the steering chair Dr. Imrich Chlamtac and Dr. Anandakumar Haldorai was essential for the success of the conference. We sincerely appreciate his constant support and guidance. It was also a great pleasure to work with such an excellent organizing committee for their hard work in organizing and supporting the conference. In particular, the Technical Program Committee, led by our TPC Chair Dr. Arulmurugan Ramu, Publication Chair Prof. M. Suriya, and Local Committee Chairs Prof. K. Karthikeyan and Prof. K. Aravindhan who have completed the peer-review process of technical papers and made a high-quality technical program. We are also grateful to Conference Managers Ms. Karolina Marcinova and Mr. Lukas, Venue Manager Ms. Katarina Srnanova, and Publication and Managing Editor Ms. Eliska Vlckova for their support and guidance. We thank all the authors who submitted and presented their papers to the BDCC 2019 conference and workshops.

We strongly believe that BDCC 2019 conference provides a good forum for all researchers, developers, and practitioners to discuss all science and technology aspects that are relevant to Big Data technology. We also expect that the future BDCC 2020 conference will be as successful and stimulating, as indicated by the contributions presented in this series.

Coimbatore, Tamil Nadu, India — Anandakumar Haldorai
Bengaluru, Karnataka, India — Arulmurugan Ramu
Coimbatore, Tamil Nadu, India — Sudha Mohanram
Taichung, Taiwan — Mu-Yen Chen

Conference Organization

Steering Committee	
Imrich Chlamtac	Bruno Kessler Professor, University of Trento, Italy
Dr. Sudha Mohanram	Sri Eshwar College of Engineering, Coimbatore, India
Organizing Committee	
General Chair	
Dr. Anandakumar Haldorai	Sri Eshwar College of Engineering, Coimbatore, India
TPC Chair	
Dr. Arulmurugan Ramu	Presidency University, Bangalore, India
Sponsorship and Exhibit Chair	
Dr. Akshaya V. S.	Sri Eshwar College of Engineering, Coimbatore, India
Local Chair	
Prof. Sivakumar K.	National Institute of Technology, Karnataka, India
Workshops Chair	
Prof. Aravindhan K.	SNS College of Engineering, Coimbatore, India
Publicity and Social Media Chair	
Dr. Michaelraj Kingston Roberts	Sri Eshwar College of Engineering, Coimbatore, India
Publications Chair	
Prof. Suriya Murugan	KPR Institute of Technology, Coimbatore, India
Web Chair	
Prof. Karthikeyan K.	SNS College of Engineering, Coimbatore, India
Conference Manager	
Ms. Karolina Marcinova	EAI

Technical Program Committee

Dr. Chan Yun Yang	National Taipei University, Taiwan
Dr . Shahram Rahimi	Southern Illinois University, Illinois, USA
Dr. Marie Nathalie Jauffret	Director of BECOM Program, Principality of Monaco
Dr. Mohan Sellappa Gounder	Al Yamamah University, Saudi Arabia
Dr. Vani Vasudevan Iyer	Al Yamamah University, Saudi Arabia
Prof. Rojesh Dahal	Kathmandu University, Kathmandu, Nepal
Dr. Ram Kaji Budhathoki	Nepal Engineering College, Nepal
Dr. Hooman Samani	National Taipei University, Taiwan
Prof. M.D. Arif Anwary	United International University, Bangladesh
Prof. Fachrul Kurniawan	Universitas Islam Negeri Maulana Malik Ibrahim Malang, Indonesia
Dr. Deepak B. Dhami	Nepal Engineering College, Nepal
Dr. K.R. Baskaran	Kumaraguru College of Technology, Coimbatore, India
Dr. C. Malathy	SRM Institute of Science and Technology, Tamil Nadu, India
Dr. G.K.D. Prasanna Venkatesan	Karpagam University, Coimbatore, India
Dr. M.G. Sumithra, Professor	Bannari Amman Institute of Technology, Tamil Nadu, India
Dr. U. Dinesh Acharya	Manipal Institute of Technology, Manipal, India.
Dr. Latha Parameswaran	Amrita Vishwa Vidyapeetham, Coimbatore, India
Dr. Bhabesh Nath	Tezpur University Assam, India
Dr. K.V. Prema, Professor	Manipal Institute of Technology, Manipal, India
Dr. E. Poovammal, Professor	SRM Institute of Science and Technology, Tamil Nadu, India
Dr. Ashalatha Nayak	Manipal Institute of Technology, Manipal, India

Contents

Convergence Potential of IT and RTI Act 2005: A Case Study 1
S. Ravi Kumar, K. Chandra Sekharaiah, and Y. K. Sundara Krishna

Sentiment Analysis Using Long Short-Term Memory Model in Deep Learning 15
Anushree Goud and Bindu Garg

Minimization of the Wind Turbine Cost of Energy Through Probability Distribution Functions 25
P. S. Divya, M. Lydia, G. Manoj, and S. Devaraj Arumainayagam

Investigating the Performance of Planar Inverted-F Antenna (PIFA) in Dual Resonant Mode 33
T. Prabhu, S. Chenthur Pandian, and E. Suganya

Fog Computing for Smart Grid Transition: Requirements, Prospects, Status Quos, and Challenges 47
Md. Muzakkir Hussain, Mohammad Saad Alam, and M. M. Sufyan Beg

Multispectral Data Processing for Agricultural Applications Using Deep Learning Classification Methods 63
Anuj Rapaka and Arulmurugan Ramu

Lecture Video Summarization Using Subtitles 83
Ramamohan Kashyap Abhilash, Choudhary Anurag, Vaka Avinash, and D. Uma

Retinal Image Enhancement by Intensity Index Based Histogram Equalization for Diabetic Retinopathy Screening 93
Arun Pradeep, X. Felix Joseph, and K. A. Sreeja

A PGF-Mediated Social Media Approach for Cyberpolicing India Abusers in Telangana Academics 107
S. Ravi Kumar, K. Chandra Sekharaiah, and Y. K. Sundara Krishna

Quantitative Performance Analysis of Hybrid Mesh Segmentation 115
Vaibhav J. Hase, Yogesh J. Bhalerao, Mahesh P. Nagarkar, and Sandip N. Jadhav

Enhancing the Performance of Efficient Energy Using Cognitive Radio for Wireless Networks 143
D. Seema Dev Aksatha, R. Pugazendi, and D. Arul Pon Daniel

An Internet of Things Inspired Approach for Enhancing Reliability in Healthcare Monitoring 155
G. Yamini and Gopinath Ganapathy

Design of Deep Convolutional Neural Network for Efficient Classification of Malaria Parasite 169
M. Suriya, V. Chandran, and M. G. Sumithra

Data Security and Sensitive Data Protection using Privacy by Design Technique 177
M. Suresh Babu, K. Bhavana Raj, and D. Asha Devi

Effects of Customer Brand Engagement and Online Brand Experience: An Empirical Research 191
R. Mary Metilda and A. Grace Antony Rose

Bayesian Personalized Ranking-Based Rank Prediction Scheme (BPR-RPS) 203
J. Sengathir, M. Deva Priya, A. Christy Jeba Malar, G. Aishwaryalakshmi, and S. Priyadharshini

Optimal Placement Techniques of Mesh Router Nodes in Wireless Mesh Networks 217
S. Seetha, Sharmila Anand John Francis, and E. Grace Mary Kanaga

Predicting Replication Time in Cassandra 227
Abhin S. Lingamaneni, Adarsh Mishra, and Dinkar Sitaram

A 600 mV + 12 dBm IIP3 CMOS LNA with G_m Smoothening Auxiliary Path for 2.4 GHz Wireless Applications 241
D. Sharath Babu Rao and V. Sumalatha

A Survey on Non-small Cell Lung Cancer Prediction Using Machine Learning Methods 255
S. Shanthi

A Study of Internet-Based Cognitive Platforms for Psychological and Behavior Problems 267
Aashish A. Gadgil, Pratijnya S. Ajawan, and Veena V. Desai

Virtual Machine Migration and Rack Consolidation for Energy Management in Cloud Data Centers 279
I. G. Hemanandhini, R. Pavithra, and P. Suggantha Priyadharshini

Adaptive Uplink Scheduler for WiMAX Networks 289
M. Deva Priya, A. Christy Jeba Malar, N. Kiruthiga, R. Anitha, and G. Sandhya

Mask R-CNN for Instance Segmentation of Water Bodies from Satellite Image 301
S. Dhyakesh, A. Ashwini, S. Supraja, C. M. R. Aasikaa, M. Nithesh, J. Akshaya, and S. Vivitha

Dynamic Spectrum Access for Spectrum Handoff in Cognitive Radio Networks Using Optimization Techniques 309
M. Kalpana Devi and K. Umamaheswari

VLSI Implementation of Color Demosaicing Algorithm for Real-Time Image Applications 317
S. Allwin Devaraj, Stalin Jacob, Jenifer Darling Rosita, and M. M. Vijay

DNN-Based Decision Support System for ECG Abnormalities 331
S. Durga, Esther Daniel, and S. Deepakanmani

A Hybrid Artificial Bee Colony Algorithmic Approach for Classification Using Neural Networks 339
C. Mala, Vishnu Deepak, Sidharth Prakash, and Surya Lashmi Srinivasan

Performance Analysis of Polar Codes for 5G Wireless Communication Network 361
Jyothirmayi Pechetti, Bengt Hallingar, P. V. N. D. Prasad, and Navin Kumar

Performance Analysis of Spectrum Sharing in mmWave Cellular Networks 379
Bandreddy Suresh Babu and Navin Kumar

Performance Analysis of VLC Indoor Positioning and Demonstration 395
S. Rishi Nandan and Navin Kumar

Ranking of Educational Institutions Based on User Priorities Using Multi-criteria Decision-Making Methods 411
A. U. Angitha and M. Supriya

Evaluation of Graph Algorithms for Mapping Tasks to Processors 423
Sesha Kalyur and G. S. Nagaraja

Smart Trash Segregator Using Deep Learning on Embedded Platform ... 449
Namrata Anilrao Mahakalkar and Radha D.

Efficient Graph Algorithms for Mapping Tasks to Processors 467
Sesha Kalyur and G. S. Nagaraja

Index 493

Convergence Potential of IT and RTI Act 2005: A Case Study

S. Ravi Kumar, K. Chandra Sekharaiah, and Y. K. Sundara Krishna

1 Introduction

As our great leaders said, the society has to move in the right way for the development. If a society moves in the wrong way or ill society, it leads to an ill-development society/cybercriminal society. Nowadays, the entire world is moving toward digitization. Smart cities are developing through the digitization technology. Smart cities must be free from cybercrimes. As the population increases, the crimes also increase in the same way. We must defuse these types of cybercrimes in the same way.

Fig. 1 shows the cybercrime website through a Wayback Machine, a web archive tool captured as on November 12, 2011. On the right side of the home page, there is a cybercriminal content included; hence it committed nearly four cybercrimes: firstly, The Improper Use Act Violation of State Emblem of India, secondly, the law violation of sedition, thirdly, the nation crime of cheating and fourthly, identity theft crimes. These multiple cybercrimes are identified and under our research work can be done by a case study. In this cybercrime more than 2014 members registered in the cybercriminal website jntuhjac.com [1–20].

S. Ravi Kumar (✉)
Krishna University, Machilipatnam, AP, India

K. Chandra Sekharaiah
Department of Computer Science and Engineering, JNTUH University, Hyderabad, Telangana, India

Y. K. Sundara Krishna
Department of Computer Science and Engineering, Krishna University, Machilipatnam, AP, India

A. Haldorai et al. (eds.), *2nd EAI International Conference on Big Data Innovation for Sustainable Cognitive Computing*, EAI/Springer Innovations in Communication and Computing, https://doi.org/10.1007/978-3-030-47560-4_1

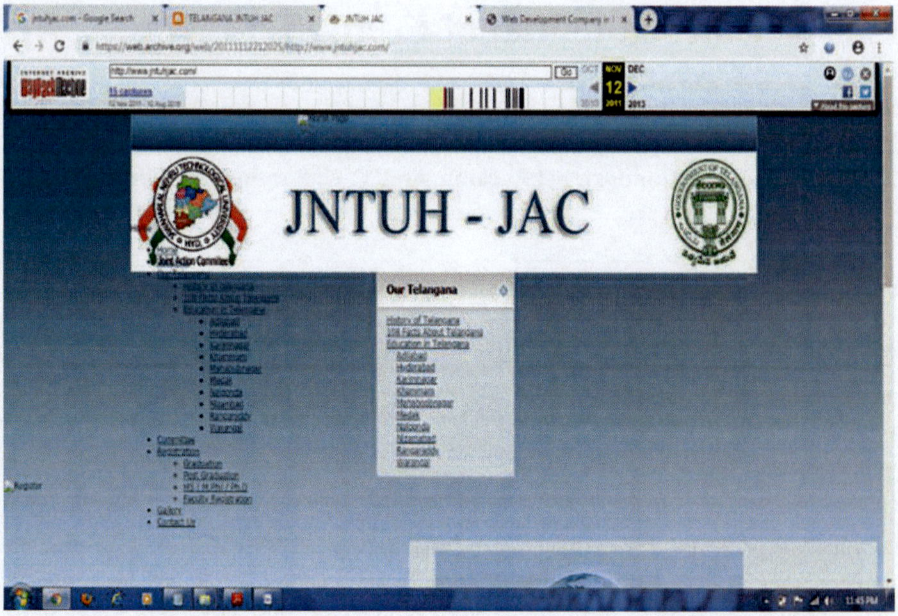

Fig. 1 Screenshot of the JNTUHJAC website as on November 12, 2011 [18, 19]

1.1 *Motivation*

The study of the cybercrimes is unchecked in the academic environment under the popular universities is against the nation building. In a developing country like India these types of cybercrimes are not allowed for nation building. For defusing the same, we are approaching different methodologies, creating awareness through a People Governance Forum (PGF) [18], social media approach, etc. As a part of development, India is divided into smart cities and these cities must be free from cybercrimes. So we plan to bring the information from Right to Information Act (RTI Act 2005) from the society in a reputed organization in India. To defuse it, we approach RTI Act 2005 together with ICT solution that could have a deleterious effect on cybercrimes.

This chapter is organized as follows: Section 2 describes the methodology of cybercriminal GoT. Section 3 describes data analysis and collection of cybercrime remedial steps. Section 4 describes the results and Sect. 5 concludes the chapter and future scope.

2 Methodology

Right to information is an act to provide information to the general public in the society. A society is well developed when there are no cybercrimes. An ill-informed society always has problems like cybercrimes. For defusing, it we approach a People Governance Forum (PGF). It creates awareness to the society on a cybercrime and creates a motivation for defusing the same for remedial solution.

2.1 *People Governance Forum (PGF) as a Cyber Remedial Forum*

The PGF was created in the year 2017 without any personal benefit and profit; it is a nonprofit remedial forum for giving awareness on a cybercrime and nation building. We utilize the PGF as a platform for giving awareness to the general public and motivating to defuse the cybercrimes in the same.

2.2 *Public Information Officers (PIOs) Selected for our RTI Act 2005 Implementation Analysis*

We selected seven Public Information Officers (PIOs) for RTI Act Implementation analysis, among which four are national-level organizations: (1) Lok Sabha Secretariat, (2) Principal Secretary to Hon'ble Minister of HRD, (3) All India Council Technical Education, (4) Council of Scientific and Industrial Research, (5) Andhra Pradesh Public Service Commission, (6) the Public Information Officer and Chief Secretary, and (7) the Principal Secretary to President.

Lok Sabha Secretariat

The Lok Sabha Secretariat is selected by us for our RTI Act 2005 implementation analysis w.r.t. the TBDCOs with the following objectives:

1. To check the Governor of Telangana State for his awareness about TBDCOs that committed the abovementioned cybercrimes.
2. To verify the Governor for his initiatives to take remedial measures against the deleterious effects of these cybercrimes on Telangana society.
3. To verify the Governor for his coordination with concerned ministers of Telangana State, Telangana State universities, colleges, and researchers and his initiatives to root out the TBDCOs in the JNTUH academic environment.

4. To verify the Governor for his recommendations of appropriate policies to the Telangana State cabinet in order to monitor Telangana State universities and colleges such that these types of cybercrime incidents do not recur in JNTUH or any other university or college in Telangana State in future.

Principal Secretary to Hon'ble Minister of HRD

The Principal Secretary to Hon'ble Minister of HRD is selected by us for our RTI Act 2005 implementation analysis w.r.t. the TBDCOs with the following objectives:

1. To check the Governor of Telangana State for his awareness about TBDCOs that committed the abovementioned cybercrimes.
2. To verify the Governor for his initiatives to take remedial measures against the deleterious effects of these cybercrimes on Telangana society.
3. To verify the Governor for his coordination with concerned ministers of Telangana State, Telangana State universities, colleges, and researchers and his initiatives to root out the TBDCOs in the JNTUH academic environment.
4. To verify the Governor for his recommendations of appropriate policies to the Telangana State cabinet in order to monitor Telangana State universities and colleges such that these types of cybercrime incidents do not recur in JNTUH or any other university or college in Telangana State in the future.

All India Council Technical Education

The All India Council Technical Education is selected by us for our RTI Act 2005 implementation analysis w.r.t. the TBDCOs with the following objectives:

1. To check the Governor of Telangana State for his awareness about TBDCOs that committed the abovementioned cybercrimes.
2. To verify the Governor for his initiatives to take remedial measures against the deleterious effects of these cybercrimes on Telangana society.
3. To verify the Governor for his coordination with concerned ministers of Telangana State, Telangana State universities, colleges, and researchers and his initiatives to root out the TBDCOs in the JNTUH academic environment.
4. To verify the Governor for his recommendations of appropriate policies to the Telangana State cabinet in order to monitor Telangana State universities and colleges such that these types of cybercrime incidents do not recur in JNTUH or any other university or college in Telangana State in the future.

Council of Scientific and Industrial Research

The Council of Scientific and Industrial Research is selected by us for our RTI Act 2005 implementation analysis w.r.t. the TBDCOs with the following objectives:

1. To check the Governor of Telangana State for his awareness about TBDCOs that committed the abovementioned cybercrimes.
2. To verify the Governor for his initiatives to take remedial measures against the deleterious effects of these cybercrimes on Telangana society.
3. To verify the Governor for his coordination with concerned ministers of Telangana State, Telangana State universities, colleges, and researchers and his initiatives to root out the TBDCOs in the JNTUH academic environment.
4. To verify the Governor for his recommendations of appropriate policies to the Telangana State cabinet in order to monitor Telangana State universities and colleges such that these types of cybercrime incidents do not recur in JNTUH or any other university or college in Telangana State in the future.

Andhra Pradesh Public Service Commission

The Andhra Pradesh Public Service Commission is selected by us for our RTI Act 2005 implementation analysis w.r.t. the TBDCOs with the following objectives:

1. To check the Governor of Telangana State for his awareness about TBDCOs that committed the abovementioned cybercrimes.
2. To verify the Governor for his initiatives to take remedial measures against the deleterious effects of these cybercrimes on Telangana society.
3. To verify the Governor for his coordination with concerned ministers of Telangana State, Telangana State universities, colleges, and researchers and his initiatives to root out the TBDCOs in the JNTUH academic environment.
4. To verify the Governor for his recommendations of appropriate policies to the Telangana State cabinet in order to monitor Telangana State universities and colleges such that these types of cybercrime incidents do not recur in JNTUH or any other university or college in Telangana State in the future.

The Public Information Officer and Chief Secretary

The Public Information Officer and Chief Secretary is selected by us for our RTI Act 2005 implementation analysis w.r.t. the TBDCOs with the following objectives:

1. To check the Governor of Telangana State for his awareness about TBDCOs that committed the abovementioned cybercrimes.
2. To verify the Governor for his initiatives to take remedial measures against the deleterious effects of these cybercrimes on Telangana society.

3. To verify the Governor for his coordination with concerned ministers of Telangana State, Telangana State universities, colleges, and researchers and his initiatives to root out the TBDCOs in the JNTUH academic environment.
4. To verify the Governor for his recommendations of appropriate policies to the Telangana State cabinet in order to monitor Telangana State universities and colleges such that these types of cybercrime incidents do not recur in JNTUH or any other university or college in Telangana State in the future.

The Principal Secretary to President

The Principal Secretary to President is selected by us for our RTI Act 2005 implementation analysis w.r.t. the TBDCOs with the following objectives:

1. To check the Governor of Telangana State for his awareness about TBDCOs that committed the abovementioned cybercrimes.
2. To verify the Governor for his initiatives to take remedial measures against the deleterious effects of these cybercrimes on Telangana society.
3. To verify the Governor for his coordination with concerned ministers of Telangana State, Telangana State universities, colleges, and researchers and his initiatives to root out the TBDCOs in the JNTUH academic environment.
4. To verify the Governor for his recommendations of appropriate policies to the Telangana State cabinet in order to monitor Telangana State universities and colleges such that these types of cybercrime incidents do not recur in JNTUH or any other university or college in Telangana State in the future.

3 Data Analysis and Collection for Cybercrime Remedying

We gathered information through RTI Act 2005 from Public Information Offices shown in Table 1 and analyzed in Table 2 toward analysis of the remedy solution to the cybercrime.

3.1 Detailed Analysis and Interpretation

We gathered data through RTI Act 2005 from different organizations in India and got the response according to it. We analyzed the data according to the questions posed. Table 2 describes detailed analysis and interpretation.

Table 1 RTI request-response communication details w.r.t. various PI offices in India

S No	PIO details	Response receive for RTI request?	Response satisfactory?	First appeal sent?	Response received from F.A.A.?
					Satisfactory?
PIO 1	The CPIO of the Lok Sabha Secretariat, Parliament House Annexe, New Delhi, India	Yes	RNS[a]	No	–
PIO 2	Principal Secretary to Hon'ble Minister of HRD, Shastri Bhawan, New Delhi, India	Yes	RS[a]	No	–
PIO 3	The PIO of All India Council for Technical Education, New Delhi, India	Yes	RS[a]	No	–
PIO 4	The PIO of Council of Scientific and Industrial Research, AnusandhanBhawan, New Delhi, India	Yes	RS[a]	No	–
PIO 5	The PIO of Andhra Pradesh Public Service Commission, Vijayawada, Andhra Pradesh, India	Yes	RS[a]	No	–
PIO 6	The PIO of Chief Secretary, Telangana Secretariat, Hyderabad, Telangana	Yes	RNS[a]	Yes	Yes
					RS
PIO 7	The PIO of Principal Secretary to President, Rashtrapati Bhavan, New Delhi, India	Yes	RNS[a]	Yes	Yes
					RS

[a]The abbreviations used are: *RS* response satisfactory, *RNS* response not satisfactory

Table 2 Analysis of RTI Act 2005 documents

S. No	RTI questionnaire	RTI Response (from PIO/D.A.A.)	Interpretation and conclusion
PIO 1	1. List of members of Lok Sabha whoever held deliberations with any governor of FGoT 2. List of members of Lok Sabha whoever held deliberations with any chief minister of FGoT 3. List of members of any parliamentary committee such that any member in the committee is a Lok Sabha MP and held deliberations with the governor of FGoT 4. List of members of any parliamentary committee such that any member in the committee is a Lok Sabha MP and held deliberations with the chief minister of FGoT	I am directed to state that such type of information is not maintained in this Secretariat. Under the RTI Act 2005, only such information can be supplied which already exists and is held by public authority or held under the control of the public authority. As per the RTI Act, CPIO is not supposed to do research work on behalf of the applicant. All the details regarding Lok Sabha Secretariat are freely available in the public domain in the website, i.e., https//loksabha.nic.in	As of my knowledge, no record or information maintained at Lok Sabha. Because it is an underground and GoT under FGoT
PIO 2	1. Did FGoT seek any funds from MHRD? 2. Did MHRD release any funds to FGoT? 3. Did PFoT seek any funds from MHRD? 4. Did MHRD release any funds to PGoT?	In this regard, it is to inform that both FGoT and PGoT have neither sought any funds from M/o HRD nor M/o HRD has released any funds to them	As of my knowledge, it was clear that MHRD did not give any type of funds to FGoT
PIO 3	1. Did any university covered under FGoT seek any funds from AICTE? 2. Did AICTE release any funds to any university covered under FGoT? 3. Did any university covered under PGoT seek any funds from AICTE? 4. Did AICTE release any funds to any university covered under PGoT? 5. List of universities covered by FGoT according to AICTE	The DPIO has provided the information as AICTE does not provide any fund to any university covered under PGoT and FGoT	As of my knowledge, it was clear that AICTE did not give any type of funds to FGoT
PIO 4	1. Did FGoT seek any funds from CSIR? 2. Did CSIR release any funds to FGoT? 3. Did PGoT seek any funds from CSIR? 4. Did CSIR release any funds to PGoT?	1. No funds were sought by aforesaid FGoT from CSIR 2. Not applicable in light of "1" above 3. No funds were sought by aforesaid PGoT from CSIR 4. Not applicable in light of "3" above	As of my knowledge, it was clear that CSIR did not provide any type of funds to FGoT

PIO 5	1. Is "cybercrime" concept included in the syllabus for competitive exams conducted by APPSC? 2. List the competitive exams conducted by APPSC in which "cybercrime" concept is included in its syllabus	For point Nos. 1 and 2: The cybercrime concept is not found in the syllabus for competitive exams conducted by APPSC. Further, it is informed that the examiner can ask all current topics under general studies	Their response is clear that with respect to the questions asked to APPSC
PIO 6	1. For FGoT, who was the chief minister? 2. Are FGoT chief minister and PGoT chief minister one and the same? 3. Did chief minister of PGoT ever hold deliberations with the chief minister of FGoT? 4. Was FGoT a state government such as Karnataka/Kerala/Maharashtra/Gujarat . . . .? 5. Was FGoT a national government such as Bangladesh/Sri Lanka/Maldives/U.K/U.S.A . . . ? 6. For FGoT who was the prime minister? 7. Did the chief minister or finance minister of PGoT take oath of office from the governor/president of FGoT? 8. List the actions taken by the chief minister of PGoT to prevent recurrence of the criminal actions by JNTUHJAC during 2011–2018	It is informed that, as per the RTI Act 2005, the public Information Officer (PIO) has to supply the material in the form held by the public authority. The PIO is not supposed to create information that is not a part of the record of the public authority. The PIO is also not required to furnish information which requires drawing of inference and/or making assumptions; or to interpret information; or to solve the problems raised by the applicants; or to furnish replies to hypothetical questions	As of my knowledge, it was clear that did not have any type of funds to FGoT
PIO 7	1. For FGoT, who was the president? 2. Did the president of India ever hold deliberations with the president of FGoT? 3. Did the president of India ever hold deliberations with the governor of FGoT? 4. Did the president of India ever hold deliberations with the chief minister of FGoT? 5. Did the president of India ever hold deliberations with the prime minister of FGoT? 6. Did the president of India sign any order/proceedings w.r.t. FGoT? 7. Did the president of India give oath of office to any minister under FGoT?	The information sought by the applicant in the referred application does not come under the definition of "information" as per Sect. 2 (f) of RTI Act, 2005	As of my knowledge, it was clear that the president of India did not provide any type of funds to FGoT

4 Results and Discussion

We approached the right to information (RTI Act 2005) to get the related information from the different categories of organizations in India. We chose seven PIO organizations and requested information like whether any of these government organizations either release funds to PFoT or seek funds from FGoT (JNTUHJAC).

We got a reply from MHRD; both FGoT and PGoT have neither sought funds from MHRD nor did MHRD release any funds to them. Figure 2 depicts the RTI application and the response from MHRD. Table 1 depicts four PIO satisfactory responses and three are not a satisfactory response. Hence our research results are good in this chapter and our efforts are up to the mark. Here Fig. 3 depicts the information (the data and results of a PGF) of a particular website.

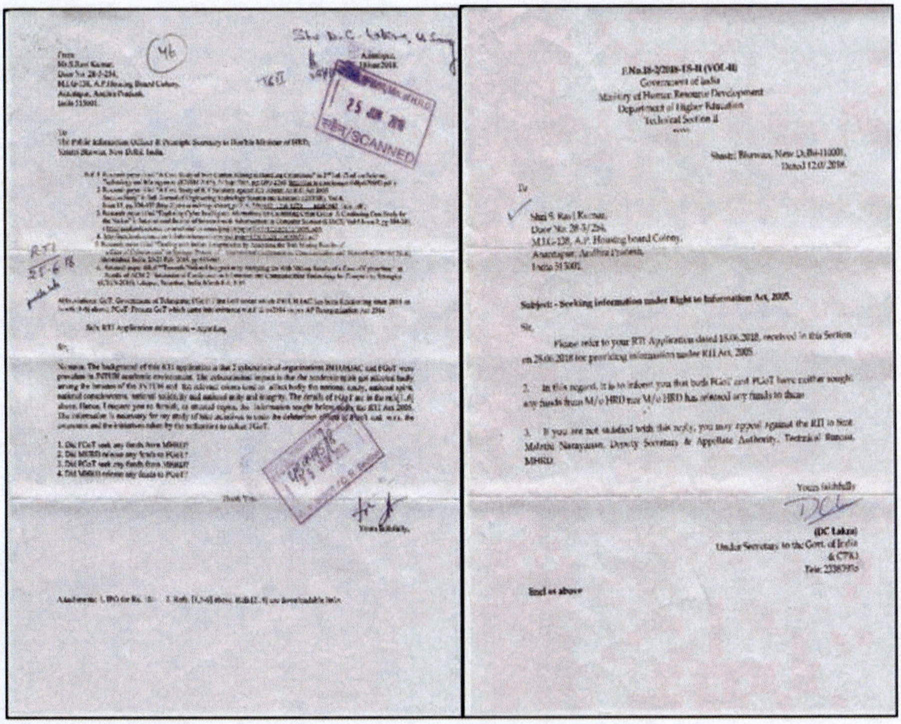

(a) (b)

Fig. 2 (a) Image of RTI application to MHRD, Shastri Bhavan, New Delhi. (b) Image of RTI document response received from MHRD, New Delhi

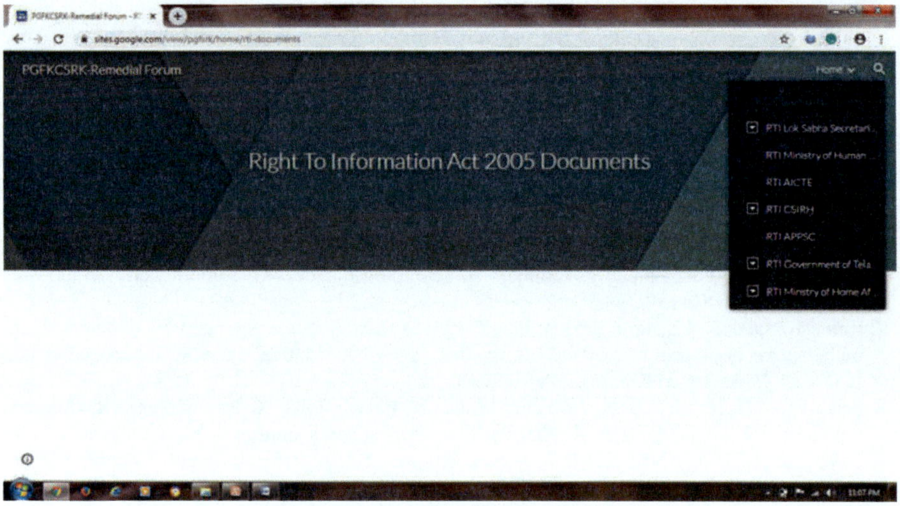

Fig. 3 Screenshot of the People Governance Forum (PGF) website [18]

5 Conclusion and Future Work

Our research methodology aims to change the ill-informed society to a well-informed society. Thus, our work gives awareness of cybercrimes for changing to the well-informed society. Only continuing effort and motivation can ensure a good national development, national spirit, national consciousness, etc.

Here the PGF provides some awareness to the remedial solution to the cyber-crime. Thus, we also approached the RTI Act for giving an exact solution to the problem. Our future research work will be a better remedial solution to these types of cybercrimes.

References

1. Ravi Kumar S, Chandra Sekharaih K, Sundara Krishna YK (2018) Impact of the RTI act within a public authority organization towards employee-employer engagement: a case study. In Proceedings of the international conference on science and technology at MRIT, Telangana, India, 19–20 Jan 2018
2. Ravi Kumar SR, Chandra Sekharaih K, Sundara Krishna YK (2018) Cybercrimes- trends and challenges in achieving swachch and digital India using a public cloud: a case study. In Proceedings of the national conference on role of law enforcement authorities government in upholding justice, SOL, Pondicherry, India, 2-3 Mar. 2018, p. 149
3. Srihari Rao N, Chandra Sekharaiah K, Ananda Rao A (2018) An approach to distinguish the conditions of flash crowd versus DDoS attacks and to remedy a cyber crime. In: Proceedings of national conference on role of law enforcement authorities and government in upholding justice, SOL, Pondicherry, India, 2-3 Mar. 2018, p. 146

4. Usha Gayatri P, Chandra Sekharaiah K, Premchand P (2018) Analytics of a judicial case study of multiple cybercrimes against the Union of India. In: Proceedings of national conference on role of law enforcement authorities and government in upholding justice, SOL, Pondicherry, India, 2–3 Mar. 2018, pp. 151
5. Ravi Kumar S, Chandra Sekharaih K, Sundara Krishna YK (2018) Cybercrimes- trends and challenges in achieving Swachch digital India using a public cloud: a case study. In Proceedings of the international conference on science and technology at MRIT, Telangana, India, 19–20 Jan 2018
6. Santhoshi N, Chandrasekharaiah K, Madan Mohan K, Nagalaxmi N (2018) ICT based social policy for Swatchh digital India. In: Proceedings of national conference on role of law enforcement authorities and government in upholding justice, SOL, Pondicherry, India, 2–3 Mar. 2018, pp. 151–152
7. Ramesh Babu J, Chandra Sekharaiah K (2018) Adaptive management of cybercriminal, maladaptive organizations, in the offing, that imperil the nation. In: CSI JC special issue, ICDMAI, Pune, 19–21 Jan 2018, April 2018
8. Srihari Rao N, Ramesh Babu J, Madan Mohan K, Pavana Johar K, Ravi Kumar SR, Chandra Sekharaiah K, Ananda Rao A (2018) A wolf in sheep's clothing- Fake Government of Telangana (FGoT), JNTUHJAC:Why 'not Prohibited'? In: 3rd International conference on research trends in engineering, applied science and management (ICRTESM-2018) 4 Nov, 2018
9. Srihari Rao N, Ramesh Babu J, Madan Mohan K, Pavana Johar K, Ravi Kumar SR, Chandra Sekharaiah K, Gouri Sankar M, Punitha P, Santhoshi N, Malathi B (2018) Cyberpolicing the multifaceted cybercriminal, Fake Government of Telangana: what is sauce for the goose is sauce for the gander. In: 3rd International conference on research trends in engineering, applied science and management (ICRTESM-2018), 4 Nov 2018
10. Madan Mohan K, Chandra Sekharaiah K (2017) A Case Study of ICT Solutions against ICT Abuse: An RTI Act 2005 success story. In national seminar on science and technology for national development, Manipur University, Imphal, March 2017
11. Gouri Shankar M, Usha Gayatri P, Niraja S, Chandra Sekharaiah K(2017) Dealing with Indian Jurisprudence by analyzing the web mining results of a case of cybercrimes. In: Proceedings of international conference on communication and networks. Advances in intelligent systems and computing, vol 508. Springer, Singapore, pp. 655–665
12. Usha Gayatri P, Chandra Sekharaiah K (2017) A case study of multiple cybercrimes against the Union of India. In: National conference on innovations in science and technology NCIST'17, Manipur University, Imphal, 20–21 March 2017 & International Journal of Computer & Mathematical Sciences (IJCMS) ISSN 2347–8527 Vol. 6, Issue 3
13. Tirupathi Kumar B, Chandra Sekharaiah K, Suresh Babu D (2016) Towards national integration by analyzing a case study of cybercrimes. ICTCS '16: Proceedings of the ACM second international conference on information and communication technology for competitive strategies, Udaipur, Rajasthan, India, 4–5 Mar 2016, Article No 79
14. Tirupathi Kumar B, Chandra Sekharaiah K, Mounitha P (2015) A case study of web content mining in handling cybercrime. In: 2nd international conference on science, technology and management, University of Delhi, Conference Center, New Delhi, India, 27 Sep 2015 & International Journal of Advance Research in Science and Engineering, Vol. 4, Spl. Issue 1
15. Usha Gayatri P, Chandra Sekharaiah K (2014) Exploring cyber intelligence alternatives for countering cyber crime: a continuing case study for the Nation. In: Proceedings of the International conferenc.at BharatiVidyapeeth's Institute of Computer Applications and Management (BVICAM), New Delhi, India
16. Usha Gayatri P, Neeraja S, Leela Poornima CH, Chandra Sekharaiah K, Yuvaraj M (2014) Exploring cyber intelligence alternatives for countering cyber crime. International conference on computing for sustainable global development (INDIACom'14) 2014, BVICAM, New Delhi, pp. 900–902

17. Usha Gayatri P, Chandra Sekharaiah K (2013) Encasing the baneful side of internet. In: National conference on computer science & security" (COCSS-2013), 5–6 April 2013, SardarVallabhbhai Patel Institute of Technology, Vasad, Gujarat, India
18. https://sites.google.com/view/pgfsrk
19. https://sites.google.com/site/sekharaiahk/apeoples-governanceforumwebpage
20. https://archive.org/web/

Sentiment Analysis Using Long Short-Term Memory Model in Deep Learning

Anushree Goud and Bindu Garg

1 Introduction

Multiple ways exist to recognize the emotion of a person. We can detect the emotion of an individual under which author writes the personal web blog which is a website that consist of series often updated on frequent basis [1]. Many people use the internet to share their views. Blog is the simplest way to share the view of a person. This work focuses on a tracking system for sentiment analysis at different touch points based on the textual content expression extracted from dataset. A sentiment allows conveying a person's emotions through expression.

Emotion detection could be of different types:

1. Facial emotion detection [2]
2. Speech emotion detection
3. Text emotion detection [3]

Here our concentration is to detect emotion from text expression [4] extracted from a file, more precisely a dataset such as open-source biometric recognition data, Google AudioSet, Yelp Open Dataset, and Kaggle dataset page. To detect the emotion of a user from the feature of face file is called face-based emotion detection. From face-based emotion detection, we can identify the emotion state of a person. Other authors use the method when a person uses webcam, video conversation, etc. It could help us to understand the emotion of a person in an effective way. Machine Learning techniques determines individual mood on the basis of his/her emotions. Let's consider an example of video conferencing, skype, whatsapp calling one can

A. Goud (✉) · B. Garg
Computer Engineering Department, Bharati Vidyapeeth College of Engineering, Pune, India
e-mail: brgarg@bvucoep.edu.in

A. Haldorai et al. (eds.), *2nd EAI International Conference on Big Data Innovation for Sustainable Cognitive Computing*, EAI/Springer Innovations in Communication and Computing, https://doi.org/10.1007/978-3-030-47560-4_2

determine easily the mood of a person. Moreover intrusion detection system plays a vital role in deep learning [5] because when we fetch data from the existing dataset there can be possibility of intruders as it is very common in computer networks or there can be some form of noise in a channel which can affect our overall system performance. Many researchers focus on detecting emotion from text. They use different mechanisms such as keyword spotting techniques and statistical method to detect emotion but they all have some limitations to detect emotion effectively from text files. Most of them are unable to identify the semantic in the text file. They all detect emotion syntactically but not semantically.

Artificial intelligence is a broad area for research. If we are moving to its specific field, i.e., machine learning [6], it again has one subset known as deep learning technique [7] where extraction of pattern from data is performed in an automated way. Deep learning technique [8] has a very little human effort. It is one of the classes of machine learning algorithm for transformation and feature extraction that uses different layers of non-processing units where each successive layer depends on its previous layer for input. Deep learning [9] and generalization have seen a boom of research for the last 50 years. Nowadays, deep learning techniques and neural networks gain major importance compared to other fields of research. Deep learning architectures constitute the state of the art in all those areas and multiple efforts have managed to increase the values of existing performance measures (Fig. 1).

Sentiment recognition [10] is a new era of research where while interacting with customers employees have high level of emotion intelligence. Previous work has shown that over 90% of our communication can be because of emotions [11]. But affective computing also called sentiment recognition [12] is becoming a boom for

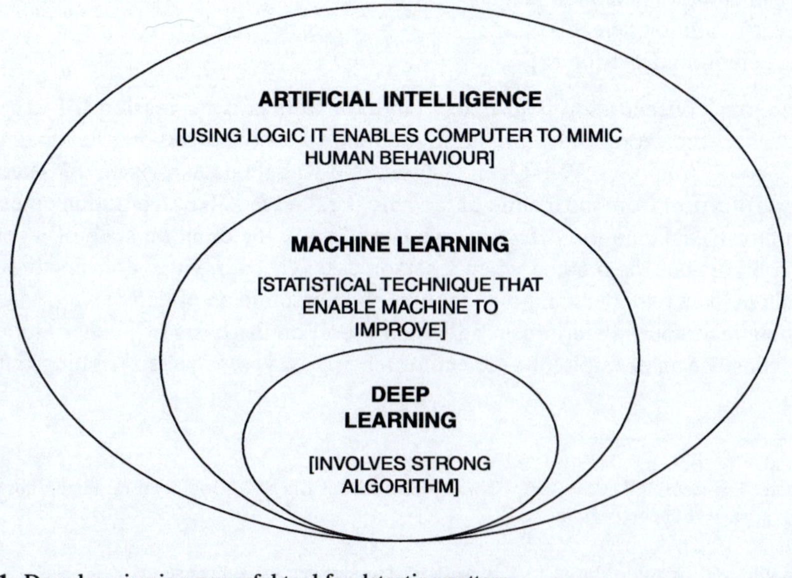

Fig. 1 Deep learning is a powerful tool for detecting pattern

research areas in computer science and information technology. Deep learning, a subset of machine learning, is a type of technique that gains popularity over the last few years. Additionally these algorithms are helpful in emotion recognition and feature engineering automatically.

The rest of the chapter is organized as follows: we explain about the literature review in Sect. 2 and give a brief description of long short-term memory, system model, and AWS dataset in Sect. 3, and we give a brief of result in further Sect. 4. However, Sect. 5 is about conclusion and future work and last section specifies the references.

2 Literature Review

In a neural network, one can input the data (text, image, speech, etc.) and the data get forwarded to several different "layers" known as convolution layer in CNN or hidden layer in RNN. Every layer modifies the input values depending on the output of the previous one and calculates predictive output in the model. For better results, we can input our data and tweak the model to output which we actually needs. Deep belief networks (DBNs), deep convolution neural networks (DCNNs), and recurrent neural networks (RNN) [13] are specific fields of architecture that are absolutely trained to get rich internal representations and achieve high performances in all major domains. Convolution neural networks (CNNs) are the more specific special type of neural network that are very effective for the use of images as processing unit. These networks are useful for engineers which are responsible for input images and have greater accuracy in recognition of emotion. CNN has a number of filters to detect patterns like shape, image, and texture.

The methods that are carried out in the field of sentiment detection so far are keyword spotting technique [14], statistical method, etc. Facial expression plays a vital role in human networking. Human faces can have a variety of expressions that are helpful for specifying emotions. Emotions that prove to be universally recognizable across different societies and cultures are happy, anger, fear, sad, disgust, surprise, and contempt. Additionally, even for complex expressions where a combination of emotions could be used as descriptors, cross-cultural agreement is identified.

Statistical method is based on word counting and word frequencies. A new statistical method named latent semantic analysis (LSA) is used for classification of texts. Latent semantic analysis [15] (LSA) is a technique in natural language processing, in particular with vector semantics, for analyzing relationships between a set of documents and the terms they contain by producing a set of concepts related to the documents and terms. LSA can use a term–document matrix which describes the occurrences of terms in documents; it is a sparse matrix whose rows correspond to terms and whose columns correspond to documents. A typical example of the weighting of the elements of the matrix is termed as inverse document frequency; the element of the matrix is proportional to the number of times the terms appear in each

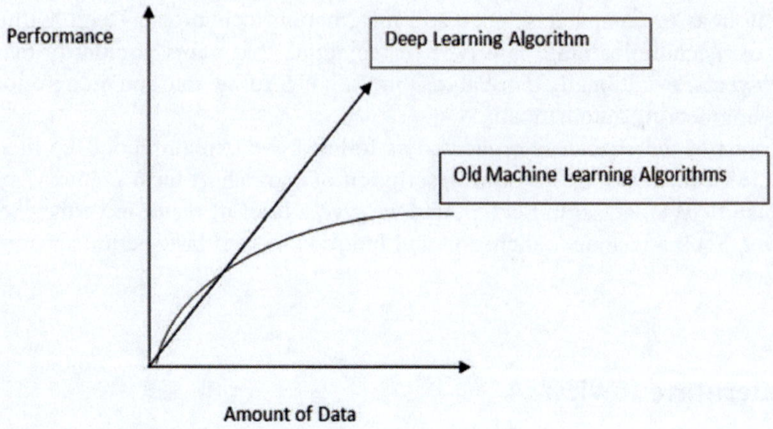

Fig. 2 Deep learning algorithm performance based on dataset

document, where rare terms are up-weighted to reflect their relative importance. This matrix is also common to standard semantic models, though it is not necessarily explicitly expressed as a matrix, since the mathematical properties of matrices are not always used (Fig. 2).

LSA cannot capture multiple meanings of a word. Feature extraction is treated as having the identical meaning due to the feature being represented as a single instance in space. The occurrence of the word "House" in a document containing "The House of the Card" and in a separate document containing "The house keeper" is treated as identical. This results in the average of all the words, which can make it difficult for differentiation. In terms of emotion detection however, statistical methods are still generally semantically weak because it relies on obvious keywords and other lexical analyses in a statistical model. As a result, statistical text classifiers only work with an acceptable accuracy when given a sufficiently large text input. Estimation of arousal of sentiment feeling [16] while reading comics is based on the analysis of physiological signal.

3 LSTM and System Model

The system model shows the implementation of sentiment analysis using RNN with LSTM model.

We have used sequential model in Keras and layers include dense, embedding, and LSTM [17] and we have included SSL for security purpose. SSL is an inbuilt package in Jupyter notebooks, only we have to create default https context using secure socket layer.

Sentiment detection framework can be diagrammatized as (Fig. 3):

In this section we propose a new framework to detect sentiment from text using recurrent neural network with LSTM model. Framework takes input from

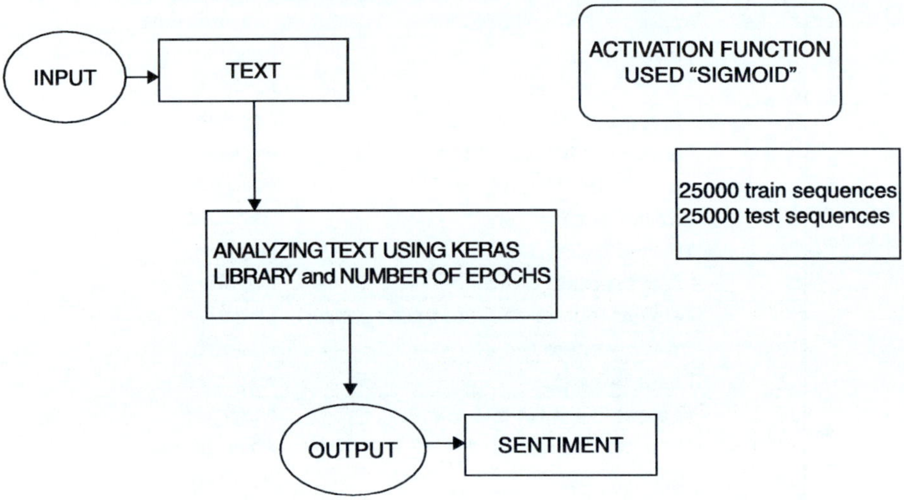

Fig. 3 Sentiment detector framework

internet users. Input could be any text. Framework contains sentiment csv file which analyzes sentiment by using text.

3.1 Proposed Framework Contains Ten Main Components

1. Import packages: During the initial phase of implementation we have used numpy and pandas library. Then we have imported Countvectorizer and train_test_split from sklearn library. From Keras library preprocessing model is implemented and preprocessing_text tokenizer and pad_sequences are imported. Model specification is sequential and layers we have imported are dense, embedding, and LSTM.
2. Preparing dataset: Model includes dataset downloaded from Amazon aws using imdb package [18]. Then we trained and tested the following sequence of dataset.
3. Visualize the data: We have to visualize the dataset from getword index function defined in imdb package.
4. Building a model: The next phase is to build a model, so for that purpose we have used sequential, embedding, LSTM, and dense function defined in Keras library.
5. Model training using training set: A next important phase is to model the training set using compile function which includes parameters like loss, optimizer, and metric. Then we have to fit the model using batch size and the number of epochs is set as 2 to validate the dataset.

id	text	sentiment
1	I love this car	Positive
2	This view is amazing	Positive
3	I feel great this morning	Positive
4	I am so excited about the concert	Positive
5	He is my best friend	Positive
6	I do not like this car	Negative
7	This view is horrible	Negative
8	I feel tired this morning	Negative
9	I am not looking forward to the concert	Negative
10	He is my enemy	Negative
11	I hate this girl.	Negative
12	I do not like this boy	Negative
13	I love this flower	Positive
14	He is my friend	Positive
15	I feel great this night	Positive
16	I do not like this girl	Negative
17	She is ugly.	Negative
18	She likes kurtis	Positive
19	This view is fantastic	Positive
20	She is beautiful	Positive

Fig. 4 Sentiment detector framework

6. Testing: Now evaluate the model using x_test, y_test and batch size. This is done using evaluate function in model library.
7. Prediction: Prediction is done using predict function defined in model library.
8. Load data: Read only necessary column using read_csv function.
9. Format data using test samples: Using tokenizers and pad sequences we have to format the dataset and find out positive and negative samples.
10. Text analysis for sentiment analysis (Fig. 4).

The main purpose of the algorithm is to identify emotions based on different text. This calculation requires some parameters like dropout and Long Short Term Memory [17] which will be used for sentiment analysis purpose. Hence the first step is calculating the parameters. Frequency is also an important parameter as the more is the frequency, the more will be the importance of that term. This value is calculated by parsing the text document and searching for occurrences of the features. The model compiles using "binary cross entropy" loss function.

This is a step where with the help of different parameters such as number of epochs, recurrent dropout, activation function, and train sequence from text each individual feature is calculated.

The main purpose of the algorithm is to calculate sentiment [19] to be assigned to different emotion features extracted from the document so that they can be sorted according to it.

Frequency is also an important parameter as the more is the frequency, the more will be the importance of that term. This value is calculated by parsing the text document and searching for occurrences of the features.

Algorithm: Illustration of Proposed Framework

1.	We have used LSTM and embedding from Keras layer and imdb from Keras dataset.
2.	For dataset preprocessing we have used tokenizer.
3.	For intrusion detection system this work includes SSL, i.e., secure socket layer. For best-fit security purpose SSL/TLS can be used on dataset.
4.	Extract Amazon Web Services dataset using certain control policies and fetch the corresponding state transition profiles.
5.	We have worked on 25,000 train sequences and 25,000 test sequences as shown in Fig. 5.
6.	After training phase, validate all the required samples with two epochs.
7.	We have built a model using activation function sigmoid and recurrent dropout parameter and implemented in Microsoft Azure notebooks Jupyter as shown in Fig. 6.
8.	Using binary cross entropy method and Adam optimization algorithm in LSTM which have tremendous benefits as compared to classical stochastic gradient descent procedure, we achieved an accuracy of up to 80 for this model as depicted in Fig. 7.

4 Results

Using long short-term memory analysis results are improved as compared to basic neural network model. As the number of epochs is increased we achieve better results. Figures 5, 6, and 7 show the results and total accuracy of the model:

```
Loading data...
Downloading data from https://s3.amazonaws.com/text-datasets/imdb.npz
17301504/17464789 [============================>.] - ETA: 0s25000 train sequences
25000 test sequences
Pad sequences (samples x time)
x_train shape: (25000, 80)
x_test shape: (25000, 80)
```

Fig. 5 Train sequences and test sequences

```
Build model...
_________________________________________________________________
Layer (type)                 Output Shape              Param #
=================================================================
embedding_1 (Embedding)      (None, None, 8)           8000
_________________________________________________________________
lstm_1 (LSTM)                (None, 16)                1600
_________________________________________________________________
dense_1 (Dense)              (None, 1)                 17
=================================================================
Total params: 9,617
Trainable params: 9,617
Non-trainable params: 0
_________________________________________________________________
```

Fig. 6 Embedding, LSTM, and dense layer

```
25000/25000 [==============================] - 37s
Test score: 0.43470786940574646
Test accuracy: 0.80236
```

Fig. 7 Accuracy of the model

5 Conclusion

In this chapter we have done a survey of framework for text-based detection of sentiment. In our further review we will use convolution deep learning technique with special parameters for better accuracy. This approach is based on long short-term memory analysis where features are extracted using CountVectorizer. In future we can consider other axioms of sentiment to understand emotion more accurately for specific domain.

References

1. Ceccacci S, Generosi A, Giraldi L, Mengoni M (2018) Tool to make shopping experience responsive to customer emotions. Int J Autom Technol 12(3):319–326
2. Lv Y, Feng Z, Xu C (2015) Facial expression recognition via deep learning. IETE Tech Rev 32(5):347–355
3. Arora P (2013) Sentiment analysis for Hindi language. Diss., International Institute of Information Technology, Hyderabad
4. Elliott C (1992) The affective reasoned: a process model of emotions in a multi-agent system. Doctoral thesis, Northwestern University, Evanston, IL, May 1992
5. Madhoushi Z, Hamdan AR, Zainudin S (2015) Sentiment analysis techniques in recent works. In: 2015 Science and information conference, pp 288–291
6. Affective deep learning. https://www.affectiva.com/how/deep-learning-at-affectiva/. Retrieved 21 May 2018
7. Matsubara M, Augereau O, Sanches CL, Kise K (2016) Emotional arousal estimation while reading comics based on physiological signal analysis. In: Proceedings of the 1st international workshop on comics analysis, processing and understanding, ser. MANPU'16. ACM, New York, pp 7:1–7:4

8. Hochreiter S, Schmidhuber J (1997) Long short-term memory. Neural Comput 9(8):1735–1780
9. Balahur A, Hermida JM, Montoyo A (2012) Detecting implicit expressions of emotion in text: a comparative analysis. Decision Support Syst 53:742–753
10. Elmadany NED, He Y, Guan L (2016) Multiview emotion recognition via multi-set locality preserving canonical correlation analysis. In: 2016 IEEE international symposium on circuits and systems (ISCAS), Montreal, QC, Canada, pp 590–593
11. Dos Santos CN, Gatti M (2014) Deep convolutional neural networks for sentiment analysis of short texts. In: COLING, pp 69–78.
12. Sarawgi K, Pathak V (2015) Opinion mining: aspect level sentiment analysis using SentiWordNet and Amazon Web Services. Int J Comput Appl 158:31–36
13. Liu W, Zheng W-L, Lu B-L (2016) Emotion recognition using multimodal deep learning. In: Hirose A, Ozawa S, Doya K, Ikeda K, Lee M, Liu D (eds) Neural information processing. Springer International Publishing, Cham, pp 521–529
14. Liu H, Lieberman H, Selker T (2003) A model of textual affect sensing using real-world knowledge. In: Proceedings of the 7th international conference on intelligent user interfaces, pp 125–132
15. Ebrahimi Kahou S, Michalski V, Konda K, Memisevic R, Pal C (2015) Recurrent neural networks for emotion recognition in video. In: Proceedings of the 2015 ACM international conference on multimodal interaction, pp 467–474
16. Schouten K, Frasincar F (2015) Survey on aspect-level sentiment analysis. IEEE Trans Knowl Data Eng 28:813–830
17. LeCun Y, Bengio Y, Hinton G (2015) Deep learning. Nature 521(7553):436–444
18. Vateekul P, Koomsubha T (2016) A study of sentiment analysis using deep learning techniques on Thai Twitter data
19. Sadegh M, Ibrahim R, Othman ZA (2012) Opinion mining and sentiment analysis. Int J Comput Technol 2:171–178

Minimization of the Wind Turbine Cost of Energy Through Probability Distribution Functions

P. S. Divya, M. Lydia, G. Manoj, and S. Devaraj Arumainayagam

1 Introduction

The total cost of a wind energy mainly hinges on the selection of wind turbine based on wind features. Wang and Stelson [1] have stated that meager selection of turbine will increase the cost of investment; disparately, a more efficient wind turbine means that more functional power can be produced from the wind. This extra power will adjust the expense of venture. The total efficiency (η) is composed of three efficiencies called a mechanical efficiency which will transfer the energy to the axis of the electric generator, an aerodynamic efficiency to transmit the dynamic energy of the wind into motorized energy in the axis of the rotor, and finally an electric efficiency to produce electrical usable power. Many research works have been done to explore the financial characteristics of wind turbines.

In [2], Hdidouan and Staffell have proposed a framework to measure the effect of climate change on the cost of wind energy and they have introduced a new Weibull transfer function to characterize the climate signal. Selection of a cost-effective wind

P. S. Divya (✉)
Department of Mathematics, Karunya Institute of Technology and Sciences, Coimbatore, Tamil Nadu, India
e-mail: divya_deepam@karunya.edu

M. Lydia
Department of EEE, SRM University, Sonipat, Haryana, India

G. Manoj
Department of ECE, Karunya Institute of Technology and Sciences, Coimbatore, Tamil Nadu, India

S. Devaraj Arumainayagam
Department of Statistics, Government Arts College, Coimbatore, Tamil Nadu, India

A. Haldorai et al. (eds.), *2nd EAI International Conference on Big Data Innovation for Sustainable Cognitive Computing*, EAI/Springer Innovations in Communication and Computing, https://doi.org/10.1007/978-3-030-47560-4_3

turbine for a wind project is one of the critical tasks. To overcome this, Andres and Gilberto [3] have introduced a method for the comparison of wind turbines based on the cost of energy. As the feasible hub height and the total efficiency are the two key variables which were not easily obtainable at the initial phase, they have established the regression models for the estimation of those variables. In [4], the turbine CoE is exemplified with a use of eight variables: diameter of the turbine rotor, number of blades, height of the hub H, speed of the rotor, V_r, P_r, type of regulation, and type of generator. The optimization process has become a difficult one as it has numerous variables. In [5], the turbine CoE model is reduced to be a function of four variables: R, P_r, tip-speed ratio (TSR), and H. Among these four variables, the TSR is an operational parameter and the others are physical parameters of the turbine. In [6], Eminoglu and Ayasun have modified a commonly used method to describe turbine output power which is known as S-type curve. In this model, the turbine CoE is further reduced to be a function of three variables: turbine capacity factor, H, and rotor diameter. In [7], Chen et al. have proposed a mathematical model to minimize the CoE which is of only two variables, V_r and P_r. Thus the minimization of CoE model is being gradually reduced to only two variables from eight variables.

2 Wind Turbine Configurations

There are various configurations of wind turbines used these days. The wind turbines are classified based on their rotor shaft rotation, mode of operation, and its power level [8–10].

The four important operating sections of a present-day wind turbine are displayed in Fig. 1. As wind speed in region 1 is lesser than the cut-in speed (V_c), there will be a low power in that region and hence the turbine will be in backup mode. Wind speed in region 2 will be above V_c and lesser than the rated speed (V_r) and thus the turbine can produce extreme power in this region. However in the third region, the wind speed is above V_r and below the cutout speed (V_f) and so output power of the turbine is restricted to the rated power (P_r). If the wind speed were more than V_f, the turbine would be shut down to avoid destruction in the fourth region [11]. Maximum turbine yield power can be obtained by improving the performance of turbine in region 2.

3 Methodology for Cost Minimization

The target of this examination is to give a scientific way to deal with the limits of CoE of a turbine with the use of various probability distributions and to find the appropriate distribution to minimize the CoE for the particular region. The turbine CoE is the ratio of the total turbine cost and the turbine annual energy production (AEP).

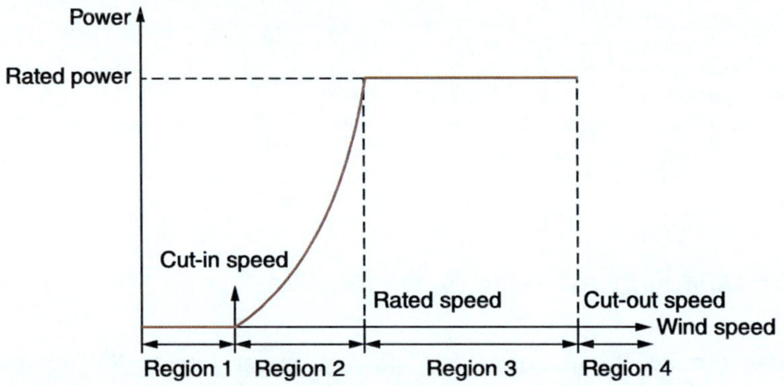

Fig. 1 Operating regions of a wind turbine

Thus

$$\begin{aligned} \text{CoE} &= f\,(P_r, V_r) \\ &= \frac{\text{Cost}\,(P_r, V_r)}{\text{AEP}\,(P_r, V_r)} \end{aligned} \quad (1)$$

The overall cost of wind turbine is obtained from NREL cost model [12] as

$$\text{Cost} = \text{ICC} \times \text{FCR} + \text{AOE} \quad (2)$$

where ICC and FCR are the initial capital cost and fixed charge rate of the turbine, and AOE is the turbine annual operating expense obtained from NREL [7].

The turbine AEP is given as

$$\text{AEP} = P_{ave} \times 8760 \times (1 - \mu) \quad (3)$$

The average turbine yield power P_{ave} can be calculated by convoluting the turbine yield power and the probability density function at each wind speed:

$$P_{ave} = \int_0^{\infty} P f(V) dV \quad (4)$$

3.1 *Rotor Radius of a Turbine*

The rotor radius R of the turbine is an expression of two variables P_r and V_r:

$$R = \sqrt{\frac{2P_r}{\rho \pi C_{pr} \eta_{mf} \eta_{gf} V_r^3}} \quad (5)$$

where

ρ—air density (1.225 kg/m^3)
C_{pr}—aerodynamic efficiency of the blade (0.45)
η_{mf}—gearbox efficiency (0.96)
η_{gf}—generator efficiency (0.97)

3.2 *Methodology for Wind Speed Models*

The mean turbine output power P_{ave} plays a major role in the minimization process of CoE. In this chapter, the wind speed data observed from three different stations have been modelled using three different distributions: Dagum, Gamma, and Weibull distribution.

Dagum Distribution

Dagum distribution is an incessant probability distribution defined over the positive real numbers. It is termed after Camilo Dagum. The Dagum distribution has three parameters: $k > 0$ and $\alpha > 0$ are continuous shape parameters and $\beta > 0$ is a scale parameter. The Dagum pdf for the wind speed variable is expressed as

$$f(V) = \frac{\alpha k \left(\frac{V}{\beta}\right)^{\alpha k - 1}}{\beta \left(1 + \left(\frac{V}{\beta}\right)^{\alpha}\right)^{k+1}}, \quad \text{the wind speed } V > 0 \tag{6}$$

Gamma Distribution

The Gamma is a continuous probability distribution family of two parameters $\alpha > 0$, a continuous shape parameter, and $\beta > 0$, a continuous scale parameter. Its pdf is given as

$$f(V) = \frac{1}{\Gamma(\alpha)\,\beta^{\alpha}} V^{\alpha-1} \exp\left(-\frac{V}{\beta}\right), \quad \text{the wind speed } V > 0 \tag{7}$$

Weibull Distribution

In the field of modelling of the wind speed data, Weibull is the most generally used continuous probability distribution. The Weibull pdf is a simplification of the Rayleigh pdf. When comparing the models used for wind speed, i.e., the Rayleigh pdf and Weibull pdf, the better result is provided by the Weibull pdf. The Weibull pdf for the wind speed variable is expressed as

$$f(V) = \frac{\alpha}{\beta}\left(\frac{V}{\beta}\right)^{\alpha-1} \exp\left[-\left(\frac{V}{\beta}\right)^{\alpha}\right], \quad \text{the wind speed } V > 0 \tag{8}$$

Thus by substituting the probability density functions (6)–(8) in Eq. (4), we get the average turbine yield power P_{ave}, for the Dagum, Gamma, and Weibull distributions, respectively.

3.3 *Methodology for Wind Power Model*

In the analysis of cost minimization, the turbine yield power among the region V_c and V_r is denoted through a mathematical expression of a polynomial function, four-parameter logistic or five-parameter logistic function. In this chapter the polynomial function of linear model is employed to describe the output power of the turbine.

Linear Model

The linear model is a very simple one and it needs only the V_c, V_r, and P_r. There will be an increase of turbine yield power with the wind speed linearly in region 2. The linear power model expression is shown in Eq. (9):

$$P(V) = \frac{V - V_c}{V_r - V_c} P_r \tag{9}$$

By substituting the wind power of linear model (9) in (4) we get the linearly modelled mean turbine output power as

$$\begin{aligned} P_{\text{ave}} &= \int_{V_c}^{V_r} \frac{V - V_c}{V_r - V_c} P_r f(V) dV + \int_{V_r}^{V_f} P_r f(V) dV \\ &= \frac{P_r}{V_r - V_c} \int_{V_c}^{V_r} (V - V_c) f(V) dV + \int_{V_r}^{V_f} P_r f(V) dV \\ &= \frac{P_r}{V_r{}^3 - V_c{}^3} \int_{V_c}^{V_r} \left(V^3 - V_c{}^3\right) f(V) dV + \int_{V_r}^{V_f} P_r f(V) dV \end{aligned} \tag{10}$$

4 Results and Discussion

The results of the work done are discussed below.

4.1 Data Description

The real-time datasets were obtained from the National Renewable Energy Laboratory, which is operated by the Alliance for Sustainable Energy for the U.S. Bureau of Energy. The hourly data of October–December 2006 of three different wind farms have been utilized.

For the CoE minimization of a turbine, the inputs provided are the shape and scale parameters of each distribution at the particular station. Hence the parameters are obtained using maximum likelihood estimation and are given in Table 1.

4.2 Estimation of Cost of Energy

By giving the scale and shape parameters of the distributions of each data as inputs and by varying the V_r and P_r, the minimum CoE has been found out and the optimized P_r and V_r are also obtained. The minimum CoE for the three stations modelled using Dagum, Gamma, and Weibull distributions with linear power modelling is presented in Table 2.

From Table 2, it is observed that among the three Dagum, Gamma, and Weibull distributions, the minimum CoE is obtained when the wind speed is modelled by Dagum distribution. The optimum rotor radius R of the turbine is obtained by the optimized P_r and V_r of the individual data. While minimizing the CoE, the optimal P_r and V_r are obtained for each data. With the use of these P_r and V_r, the turbine rotor radius of each data is obtained and listed in Table 3.

A three-dimensional (3D) map for minimum turbine CoE of all the discussed data is shown in Fig. 2.

Table 1 Parameters of three data

	Dagum distribution			Gamma distribution		Weibull distribution	
Data	k	α	β	α	β	α	β
Data 1	0.1892	10.27	9.9515	5.9494	1.1536	2.3435	7.8114
Data 2	0.1537	12.775	13.321	6.9478	1.2924	2.5738	10.183
Data 3	0.2041	9.2986	13.509	5.9584	1.5401	2.4472	10.4

Table 2 Minimum cost of energy

	Minimum cost of energy ($)		
Data	Dagum distribution	Gamma distribution	Weibull distribution
Data 1	**31.70**	33.13	31.88
Data 2	**20.74**	21.85	21.61
Data 3	**20.58**	21.58	21.28

Table 3 Rotor radius of the turbine

Data	Optimized P_r and V_r		Rotor radius$R$ (m)
	P_r (MW)	V_r (m/s)	
1	1	11	30.52
2	1.3	12.5	28.72
3	1.4	13	28.11

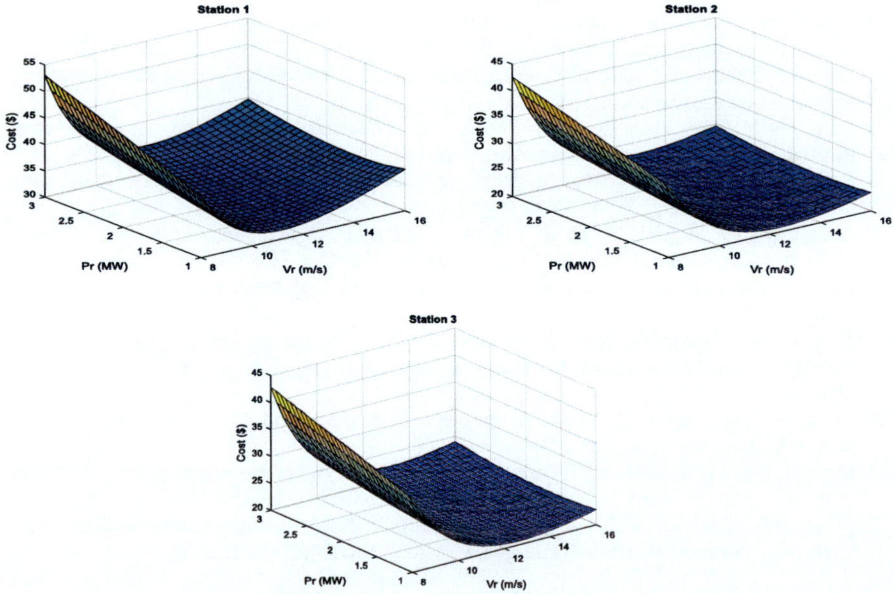

Fig. 2 Turbine CoE minimization results of six stations by modelling the wind speed using Dagum distribution and modelling the wind power using linear function

5 Conclusion

In this chapter, a scientific methodology is presented to minimize the turbine CoE. For the analysis of turbine cost, a cost model which was developed by the US National Renewable Energy Laboratory (NREL) has been used. In the minimization of CoE methodology, the observed wind speed data is modelled using Dagum, Gamma, and Weibull distributions and the wind power is modelled by the linear function. The analysis is performed for the observations obtained from three different stations. Comparative study has been done using the three statistical distributions and the estimation of minimum CoE is carried out. The outcomes from the statistical distributions while modelling the wind speed concluded that Dagum distribution produces the minimum turbine CoE. The proposed method also gives the optimum size of the rotor radius of the turbine for each station. This will help to choose the correct size of the turbine to produce maximum power at the lowest cost.

References

1. Wang F, Stelson KA (2011) Model predictive control for power optimization in a hydrostatic wind turbine. In: The 13th Scandinavian international conference on fluid power, Linköping, Sweden
2. Hdidouan D, Staffell I (2017) The impact of climate change on the levelised cost of wind energy. Renew Energy 101:575–592
3. Andres AS, Gilberto OG (2018) Wind turbine selection method based on the statistical analysis of nominal specifications for estimating the cost of energy. Appl Energy 228:980–998
4. Diveux T, Sebastian P, Bernard D, Puiggali JR (2001) Horizontal axis wind turbine systems: optimization using genetic algorithms. Wind Energy 4:151–171
5. Mirghaed MR, Roshandel R (2013) Site specific optimization of wind turbines energy cost-iterative approach. Energy Convers Manage 73:167–175
6. Eminoglu U, Ayasun S (2014) Modeling and design optimization of variable-speed wind turbine systems. Energies 7:402–419
7. Chen J, Wang F, Stelson AK (2018) A mathematical approach to minimizing the cost of energy for large utility wind turbines. Appl Energy 228:1413–1422
8. Burton T, Sharpe D, Jenkins N, Bossanyi E (2011) Wind energy handbook, 1st edn. John Wiley & Sons, New York
9. Savino MM, Manzini R, Selva VD, Accorsi R (2017) A new model of environmental and economic evaluation of renewable energy systems: the case of wind turbines. Appl Energy 189:739–752
10. Pérez JMP, Márquez FPG, Tobias A, Papaelias M (2013) Wind turbine reliability analysis. Renew Sustain Energy Rev 23:463–472
11. Dutta R, Wang F, Bohlmann BF, Stelson KA (2014) Analysis of short-term energy storage for midsize hydrostatic wind turbine. J Dyn Syst Meas Control 136:1–9
12. Fingersh L, Hand M, Laxson A (2006) Wind turbine design cost and scaling model, vol 50. Tech. Rep., National Renewable Energy Laboratory, NREL/TP, Golden, CO

Investigating the Performance of Planar Inverted-F Antenna (PIFA) in Dual Resonant Mode

T. Prabhu, S. Chenthur Pandian, and E. Suganya

1 Introduction

The transmission and reception of information is done by converting electrical energy to electromagnetic (EM) energy and vice versa. Transceiver antennas are applicable to all communicating devices that start from TV (television) communication to space communication. Antenna plays a vital role in every wireless communication device. However, designing a proper antenna and providing a good bandwidth within the available radio-frequency (RF) spectrum give much importance for 5G technology to speed the data rates. Recently, the low-profile antennas (i.e., low power, low cost, and small volume) are much suitable for connecting the wireless terminal devices that support both circular and linear polarization [1–4]. The traditional patch antenna designs are suitable for narrow bandwidth because of its high intrinsic Q-factor that ascends from their thin substrate. Various research groups have designed their planar antennas to show the improvement of bandwidth. In [5, 6], the researcher experimentally demonstrated the improvement in bandwidth to 7% by electromagnetically coupling the lower and upper parasitic element. However, the dimensions of the antenna noticeably increased by adding the extra parasitic patch. In [7, 8], the impedance bandwidth was enriched to 87% and 42.3% by forming the L- and M-shaped probes. As reported in [9, 10], impedance bandwidth was enhanced to 20% by creating series resonant element for U-shaped slot in the upper patch and combining with the probe feeder. However, some of the researcher had demonstrated and reported the antenna with the total height of $0.1\lambda_0$

T. Prabhu (✉) · S. Chenthur Pandian
SNS College of Technology, Coimbatore, Tamil Nadu, India

E. Suganya
SNS College of Engineering, Coimbatore, Tamil Nadu, India

A. Haldorai et al. (eds.), *2nd EAI International Conference on Big Data Innovation for Sustainable Cognitive Computing*, EAI/Springer Innovations in Communication and Computing, https://doi.org/10.1007/978-3-030-47560-4_4

(where λ_0 is the free space wavelength), but the earlier studies revealed that the antenna height must be at least $0.06\lambda_g$ (λ_g is the guided wavelength) to obtain other excitation mode in order to maintain its wider bandwidth [9–19].

This paper proposes a new PIFA which can be in two different resonant frequencies of $f_{0,1/2}$ and $f_{2,1/2}$ to enhance the bandwidth and increase the gain. It entails rectangular patch with narrow slot, ground plane, two shorting pins, and a shorting wall. Bandwidth and gain are improved in the following three methods. Initially, PIFA is made by inserting the shorting wall at the corner of the patch to consider only the odd-order modes in patch antenna. Secondly, a couple of shorting (conducting) pins are inserted at the nodal lines of the radiating element to reallocate the resonant frequencies between $f_{0,1/2}$ and $f_{2,1/2}$ mode. Finally, a linear (narrow) slot with the size of $L_s = 13$ mm and $W_s = 1$ mm is cut out from the upper radiating patch to minimize the corresponding inductance initiated by the coaxial feeding probe and a couple of connecting pins. Introducing a narrow slot in the patch and a couple of shorting pins below the patch provides an small improvement in gain and bandwidth.

2 Construction and Working Mechanism

2.1 *Configuration of Antenna*

The physical construction of the proposed shorted (electric wall) PIFA is shown in Fig. 1. The details of the structure depict with the substrate (Rogers RT5880) and their material properties of relative permittivity (ε_r) = 2.2, dielectric loss tangent ($\tan \delta$) = 0.0009, and a height (H) = 1.575 mm. The rectangular radiating element with the dimension of $L_p \times W_p = 19.65$ mm × 49.41 mm and the ground plane with the dimension of $L_g \times W_g = 48.7$ mm × 108.3 mm, are taken into consideration. The radiating patch is placed above the dielectric substrate, and ground plane is placed below the bottom of the dielectric substrate, and they are interconnected at one end of the radiating edge through a rectangular shorting post (electric wall) and a couple of shorting pins, respectively. The power is fed by the coaxial probe from one conductor (ground plane) to another conductor (radiating patch). The schematic of all the key geometrical parameters of the proposed patch antenna is shown in Fig. 2. A couple of shorting pins are established between two conducting planes and are placed at a distance of D_2 from one radiating edge of the patch (i.e., shorting wall side); both the shorting pins are placed at the centric point with the radius value of $R = 0.54$ mm and fixing the appropriate spacing between the pins as $S_1 = 28$ mm. Next, a narrow slot with length $L_s = 13$ mm and $W_s = 1$ mm is cut from the radiating element to minimize the corresponding inductance caused by a couple of connecting pins and feeding probes. The optimized parametric values of the proposed antenna are shown in Table 1. In our analysis, Ansoft High Frequency Structure Simulator (HFSS 13.0) is used to design and validate the results.

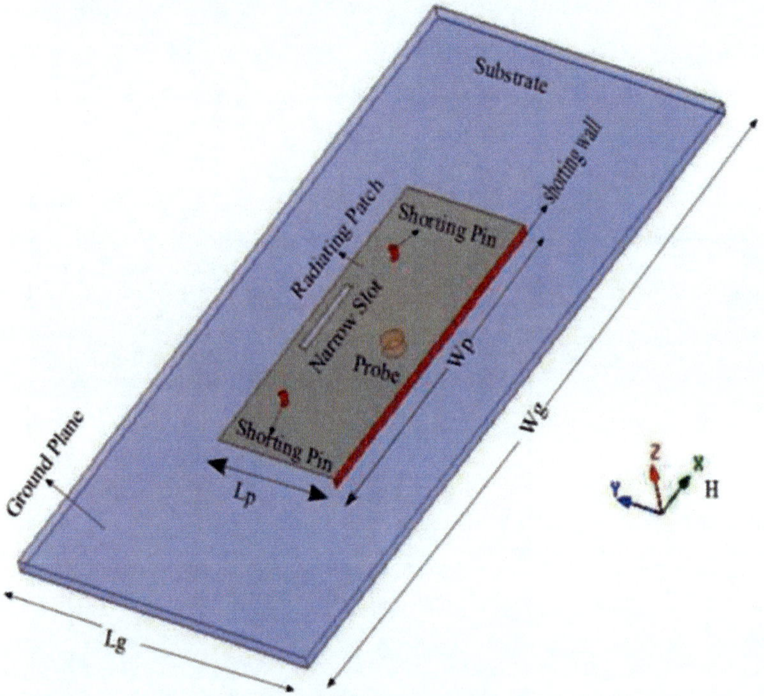

Fig. 1 Schematic representation of the shortened PIFA

2.2 *Working Mechanism*

The main objective of this design is to focus on suppressing the even-order modes and to reallocate the resonant frequencies. In this framework, the resonant frequencies of $f_{0,1/2}$ and $f_{2,1/2}$ are reallocated to each other. The dual resonant frequency works together within a frequency band to show their improvement in the gain and bandwidth without changing the polarization and radiation pattern characteristics.

1. Initially the design starts with the conventional patch antenna. Based on the cavity model theorem, the resonant modes and wave number can be determined; it can be expressed as

$$f_{mn} = \frac{k_{mn}c}{2\pi\sqrt{\varepsilon_r}} \tag{1}$$

$$K_{mn}^2 = \left(\frac{m\pi}{a}\right)^2 + \left(\frac{n\pi}{a}\right)^2 \tag{2}$$

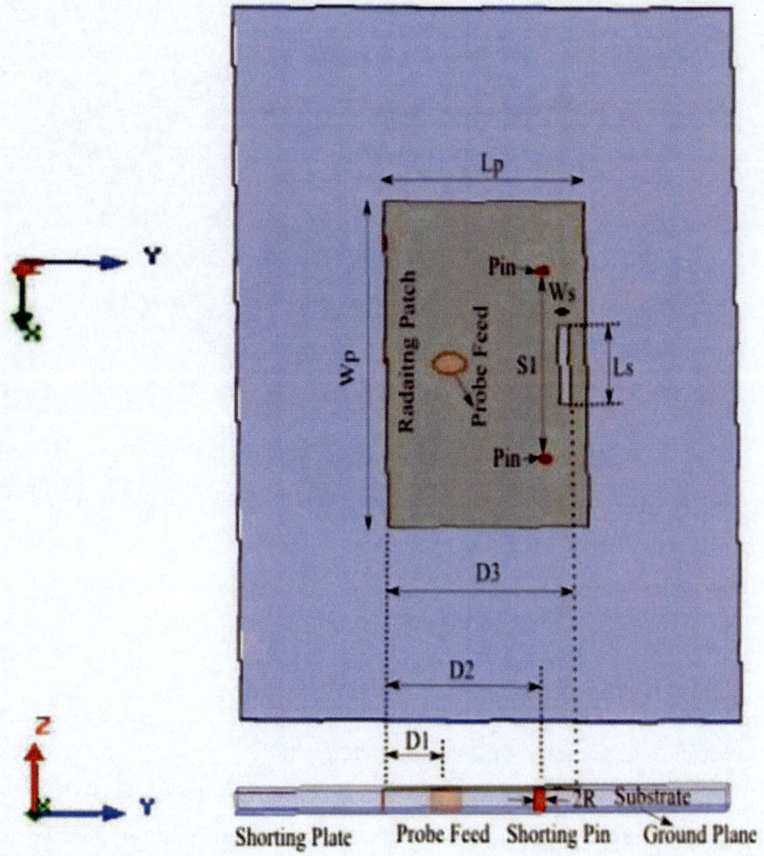

Fig. 2 Geometrical parameters of the shortened PIFA

Table 1 Optimized parametric value of the proposed patch antenna

Parameters	L_p	W_p	L_g	W_g	L_s	W_s
Optimized values (mm)	19.65	49.41	48.7	108.3	13	1
Parameters	D	D_1	D_2	S_1	H	R
Optimized values (mm)	4.55	14.58	16.6	28	1.57	0.54

where a and b are defined as the width and length of the patch, c is the velocity of the light, and m and n are the integer with 0, 1/2, 1, and 3/2.

The modified wave number with respect to patch length (L_p) and width (W_p) can be stated as

$$K_{mn}^2 = \left(\frac{m\pi}{L_p}\right)^2 + \left(\frac{n\pi}{W_p}\right)^2 \tag{3}$$

2. In order to remove the even-order modes like TM_{02} and TM_{22}, a shorting wall has been introduced exactly at the mid of the square patch antenna. Now, the conventional patch antenna has become a conventional shortened patch antenna (i.e., conventional PIFA). The proposed antenna has dual resonant frequencies of $f_{0,1/2}$ and $f_{2,1/2}$ at 2.38 and 4.81 GHz.
3. Later on, a couple of shorting pins have been initiated in order to move the resonant frequencies of $f_{0,1/2}$ and $f_{2,1/2}$. A couple of pins are placed beneath the rectangular radiating element as shown in Fig. 3b. As a result, the PIFA loaded with connecting pins moved the lower resonant frequency (from 2.48 to 3.49 GHz) and higher resonant frequency (from 4.81 to 4.49 GHz), and also it reduces the frequency ratio from 1.93% to 1.28%.
4. Finally, a small rectangular slot is created on the radiating patch antenna at one edge (i.e., opposite to shorting post) in order to minimize the effective inductance produced by the probe and shorting pins, it decreases the return loss, but still it maintains the frequency ratio. Further introducing the narrow slot on the radiating element will produce a hike in return loss and reduces the frequency ratio to 1.3%.

3 Analysis and Final Design

The parametric study of the shortened patch antenna is analyzed and validated in HFSS 13.0. This section highlights the need of shorting wall, shorting pins, and narrow slot. The basic study helps in reallocating the dual resonant frequencies and their improvement in bandwidth and gain. Initially the design starts with conventional microstrip patch antenna (MPA), based on the cavity model theorem, and resonance occurs only when the microstrip patch antenna is fed properly in a suitable position. Electromagnetic energy is radiated at both the edges of the patch antenna; however, placing a shorting wall at one edge of the patch will reduce all the even-order modes. This makes the conventional square patch antenna to rectangular PIFA. We know that electric field distribution remains zero at the mid of the radiating element for both the modes. This gives a clear idea that placing the shorting pins at the center will not affect the resonant modes. From Fig. 4b, c, it is clear that higher resonant modes are constant and there is a change in lower resonant mode. In order to obtain the higher resonant modes more closely to lower resonant modes, then the width of the patch can be further increased. Finally, a small rectangular slot is created on the radiating patch in order to diminish the corresponding inductance originated by the conducting pins and feeding probes. Further it shrinks the size of the patch and increases the current distribution.

From Table 2 it was concluded that placing the narrow slot near the shorting wall and dual slot in the radiating patch will reduce the gain drastically and etching the single slot at the far distance will improve both the gain and bandwidth.

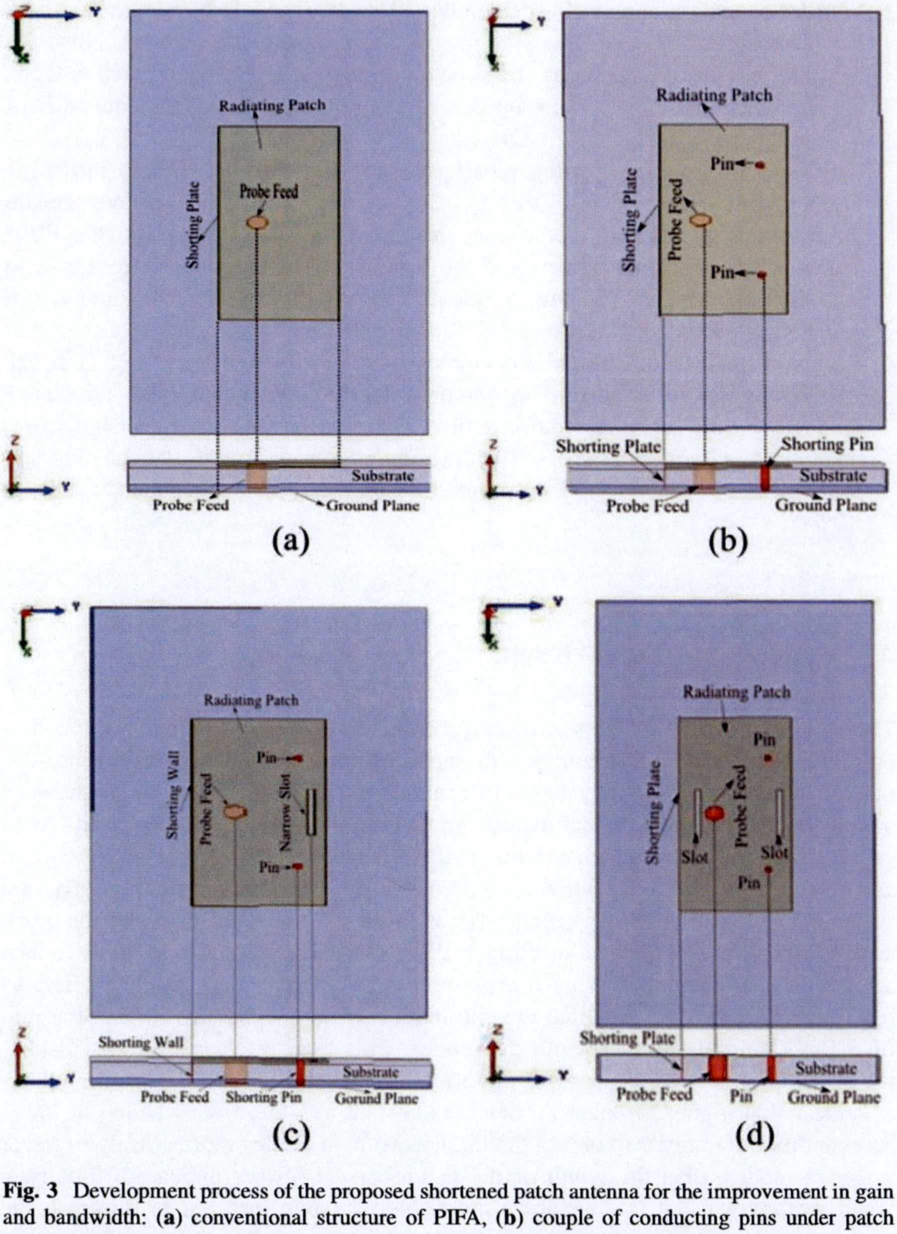

Fig. 3 Development process of the proposed shortened patch antenna for the improvement in gain and bandwidth: (**a**) conventional structure of PIFA, (**b**) couple of conducting pins under patch antenna, (**c**) couple of conducting pins under patch and dual small rectangular slot on the patch, and (**d**) couple of conducting pins under patch and single small rectangular slot on the patch

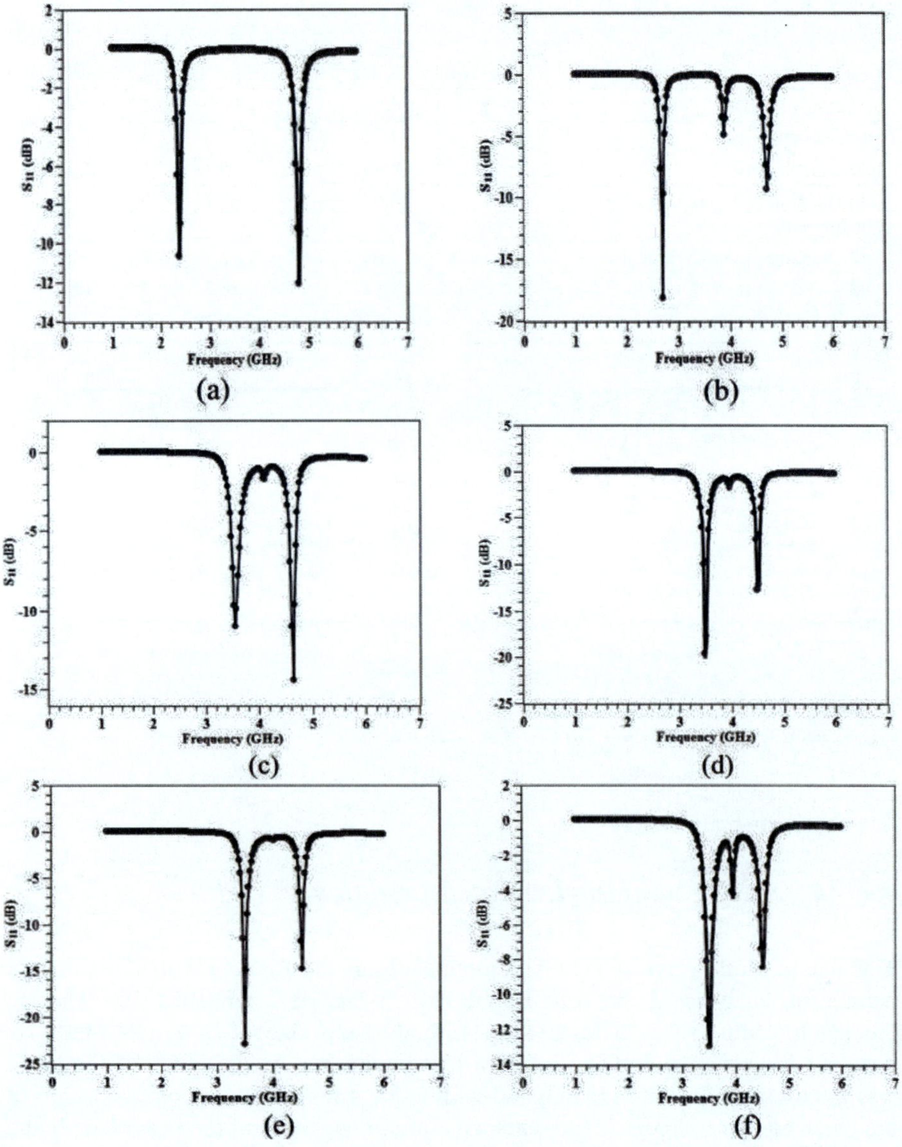

Fig. 4 (**a**) Conventional patch, (**b**) patch under single shorting pin, (**c**) patch under dual shorting pin, (**d**) patch under dual pin with slot near shorting wall, (**e**) patch under dual pin with slot far shorting wall, and (**f**) patch under dual pin with dual slot

Table 2 Performance values of antenna to improve the gain and bandwidth

Design	f_{R1} (GHz)	RL (dB)	f_{R2} (GHz)	RL (dB)	BW (MHz)	Gain (dB)	VSWR
Conventional patch	2.38	−10	4.81	−12	14.9	5.72	1.8961
Single shorting pin	2.68	−18	4.71	−9.4	44	5.42	1.4326
Dual shorting pin	3.49	−19	4.49	−12.7	75.1	2.1	1.2361
Dual shorting pin with slot near shorting wall	3.51	−12	4.52	−8.4	65.1	0.55	1.5921
Dual shorting pin with dual slot	3.56	−11	4.64	−14.4	42	1.2	1.8052
Dual shorting pin with single slot	3.51	−22	4.54	−14.8	76.8	6.45	1.9

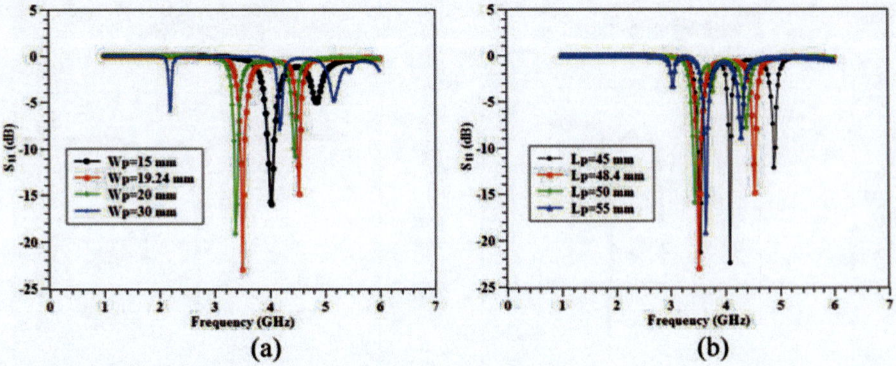

Fig. 5 (**a**) Effect of patch width. (**b**) Effect of patch length

4 Parametric Study

4.1 Effect of Changing the Patch Length and Width

The patch width plays a vital part in the design of PIFA. Figure 5a shows a parametric analysis of the patch width W_p. It can be established that, varying the patch width from 15 to 30 mm with the step size of 5 mm, it shifts its resonant frequency and changes return loss parameter. As the width of the patch was maximized, the patch length and width give a higher impact on the design of the patch element. Figure 5a presents a parametric analysis of the patch length and width. It can conclude that, varying the patch length and width from 45 to 45 mm and 15 to 30 mm in steps of 5 mm, it shows the variations in both frequency and return loss. As the patch length maximizes, the reflection coefficient increases from −22 to −19 dB, and as the patch width maximizes, the reflection coefficient increases from −16 to −6 dB and reduces the frequency ratio of $f_{2,1/2}/f_{0,1/2}$ to 1.3%.

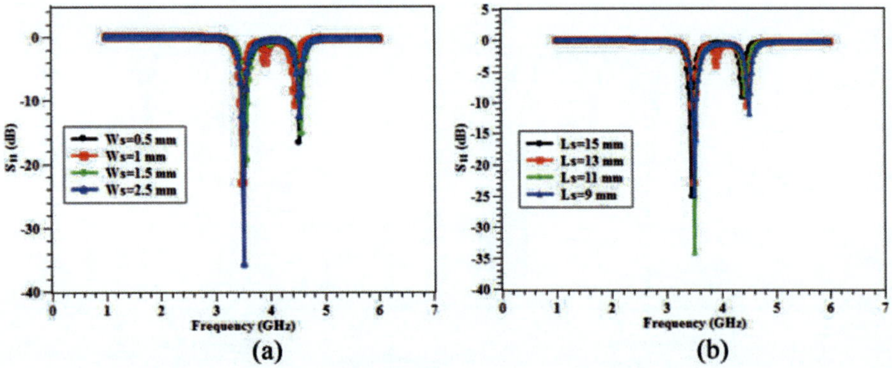

Fig. 6 (**a**) Impact on slot width. (**b**) Impact on slot length

4.2 *Effect of Narrow Slot Width and Length*

The slot length and width give a minor impact on the design of the patch element. Figure 6 presents a parametric analysis of narrow slot length and width. It can conclude that, by changing the narrow slot length and width from 9 to 15 mm and 0.5 to 2.5 mm with the step size of 2 and 0.5 mm, it modifies both the resonant frequency and return loss parameter. As the length of slot maximizes, the return loss minimizes from −25 to −26 dB, and as the slot width maximizes, the return loss reduces from −22 to −33 dB, and there was a minimum frequency shift.

4.3 *Effect of Pin Position (S_1) and D_2 from the Shorting Wall*

The spacing S_1 is the gap between two shorting pins, and D_2 is the distance between the shorting wall and the shorting pins. Figure 7 shows the parametric deviation of spacing between two shorting pins S_1 and their return loss characteristics. Holding the other parameters as constant and varying the S_1 from 14 to 26 mm with different step size, it is obvious that as the distance between shorting pins increases, there is an increase in the frequency shift and their return loss value. Similarly, if the distance between the shorting wall and shorting pin was increased, then there is a shift in frequency from 3.18 GHz at $D_2 = 11.825$ mm to 3.6 GHz at $D_2 = 15.825$ mm. It can be summarized that, as the distance increases, the frequency shift was increased.

4.4 *Effect of Position of Slot D_3*

The spacing D_3 is the distance between shorting wall and narrow slot. Figure 8 displays the parametric study of their return loss characteristics by changing the distance (D_3). Keeping all the remaining parameters as constant and varying the

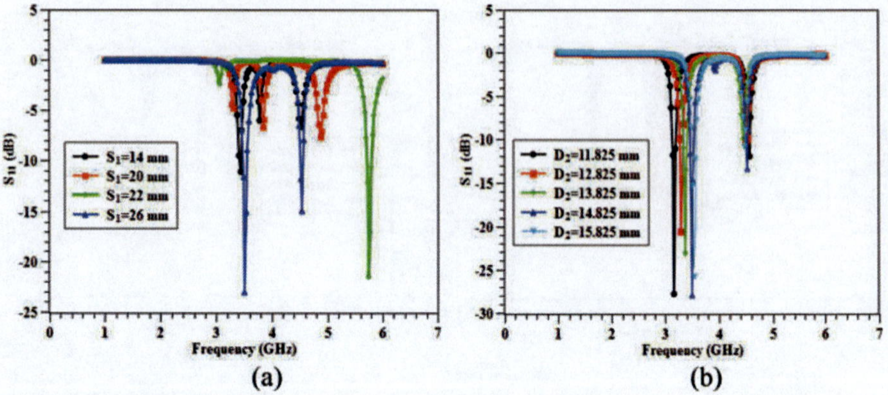

Fig. 7 (a) Effect of position between two shorting pins. (b) Effect of distance between shorting pin and shorting wall

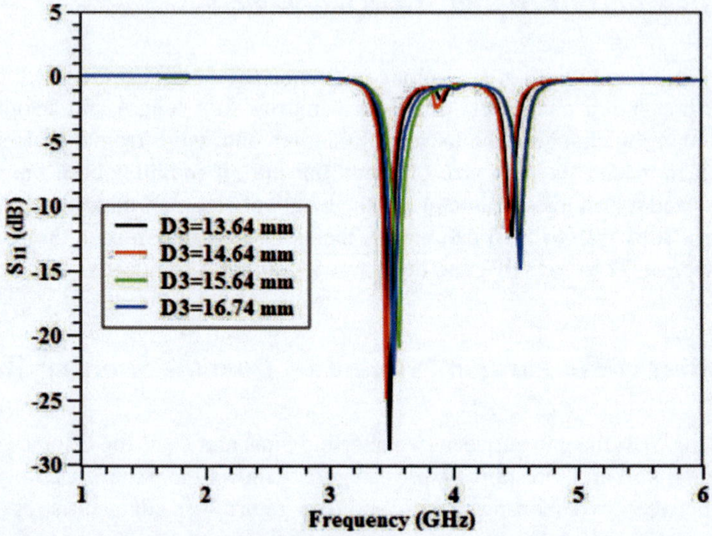

Fig. 8 Effect of distance between shorting wall and slot

D_3 from 13.64 to 16.74 mm in steps of 1 mm changes the return loss and shift in their frequency. It can be concluded that, at the distance between shorting wall and narrow slot was increased from 13.64 to 16.74 mm, the return loss increased from −29 to −23 dB.

Figure 9 shows peak gain of the shorted patch antenna over the operating frequency. It is measured for $\theta = 0°$ and $\phi = 90°$. It can be determined that the proposed method exhibits peak gain of 6.45 dBi compared with the conventional patch antenna which has the peak gain of 5.72 dBi.

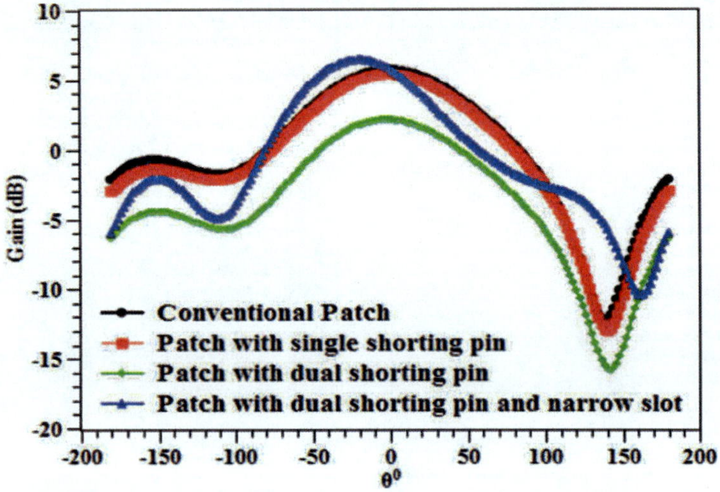

Fig. 9 Gain (dB) versus θ (°)

4.5 *Current Distribution and Radiation Pattern*

Figure 10 illustrates the simulated current distribution in PIFA loaded with their shorting pins and patch with narrow slot. At the lower operating frequency, the minimum values of electric field intensities are mainly focused at the corner of the shorted wall. Hence, the lower operating mode is excited at $f_{0,1/2}$. At the same time, $f_{0,3/2}$ is excited by loading the couple of shorting pins under the radiating element, and it is shown in blue color in Fig. 10d. The simulated radiation patterns for PIFA are illustrated in Fig. 11. From the given figure, it is witnessed that the radiation pattern for each resonant frequency was omnidirectional.

Figure 10a, b shows the *E*-field and *H*-field pattern of conventional patch antenna and patch under single shorting pin. The radiation characteristics of conventional PIFA lead the modular gain variation given with ratio of 7:1 compared with the sorting pin. Considering the effect of inductive loss introduced by the shorting pin reduces the gain to −6 dB as compared with the conventional patch gain of −46 dB. Comparing the 4.81 GHz of conventional patch and 4.71 GHz of shorting pin patch gives the inductive loss parameter with less effect in high-frequency band and dramatic effect at lower-frequency band.

Figure 10c, d shows the *E*-field and *H*-field pattern of patch with dual pin and patch with dual pin and narrow slot. The radiation characteristics of dual pin with narrow slot lead the modular gain variation given with ratio of 5:4 compared with dual pin patch. Considering the narrow slot in patch antenna minimizes the effect of inductive loss that occurs by the shorting pin under the radiating patch. Hence the maximum gain increased to −46 dB as compared with dual pin gain of −38 dB, and comparing the 4.49 GHz of dual pin patch antenna and 4.89 GHz of dual pin

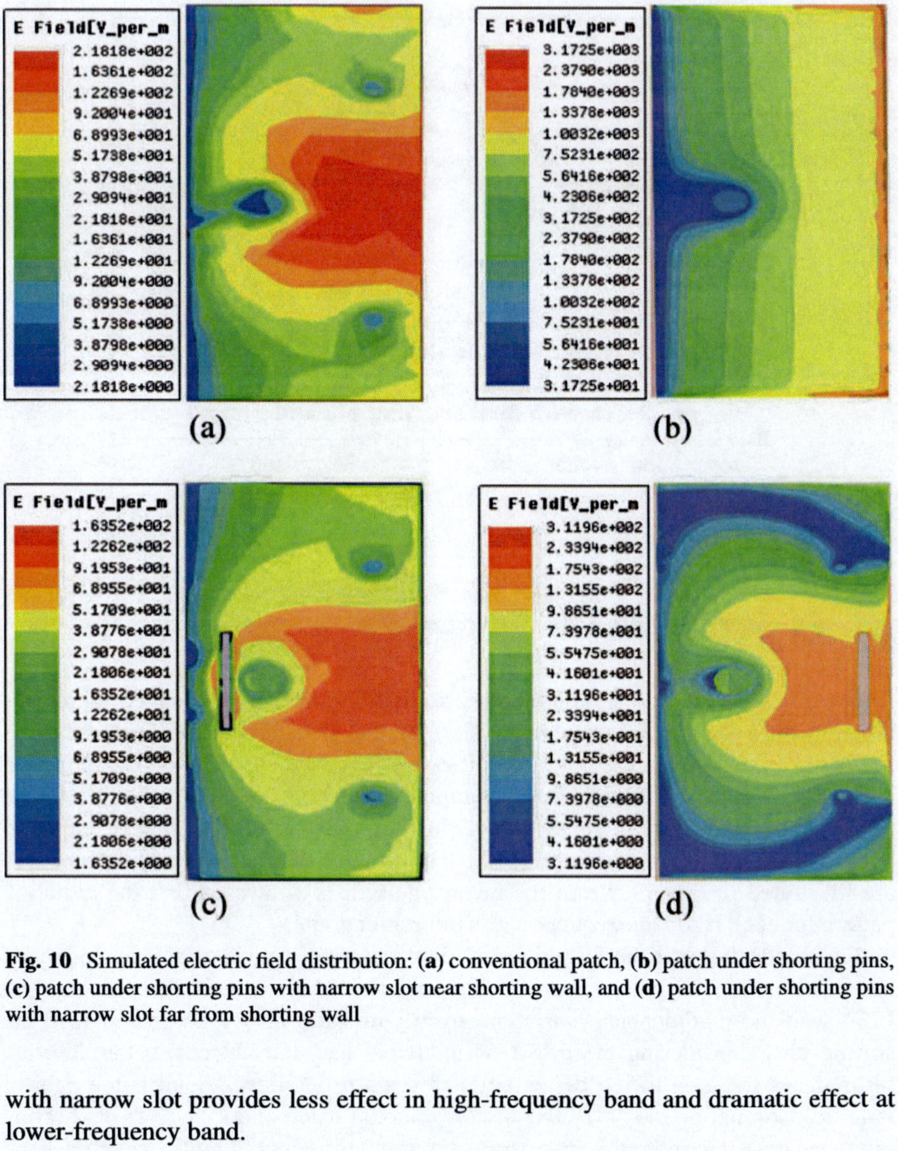

Fig. 10 Simulated electric field distribution: (**a**) conventional patch, (**b**) patch under shorting pins, (**c**) patch under shorting pins with narrow slot near shorting wall, and (**d**) patch under shorting pins with narrow slot far from shorting wall

with narrow slot provides less effect in high-frequency band and dramatic effect at lower-frequency band.

5 Conclusion

This paper is proposed to boost the gain and bandwidth of a low-profile PIFA to work in dual resonant frequencies of $f_{0,1/2}$ and $f_{2,1/2}$. Initially, the design starts to eliminate the even-order modes in order to obtain the uneven-order modes, and it is obtained by placing the shorting wall at one side of the radiating patch. Secondly, a

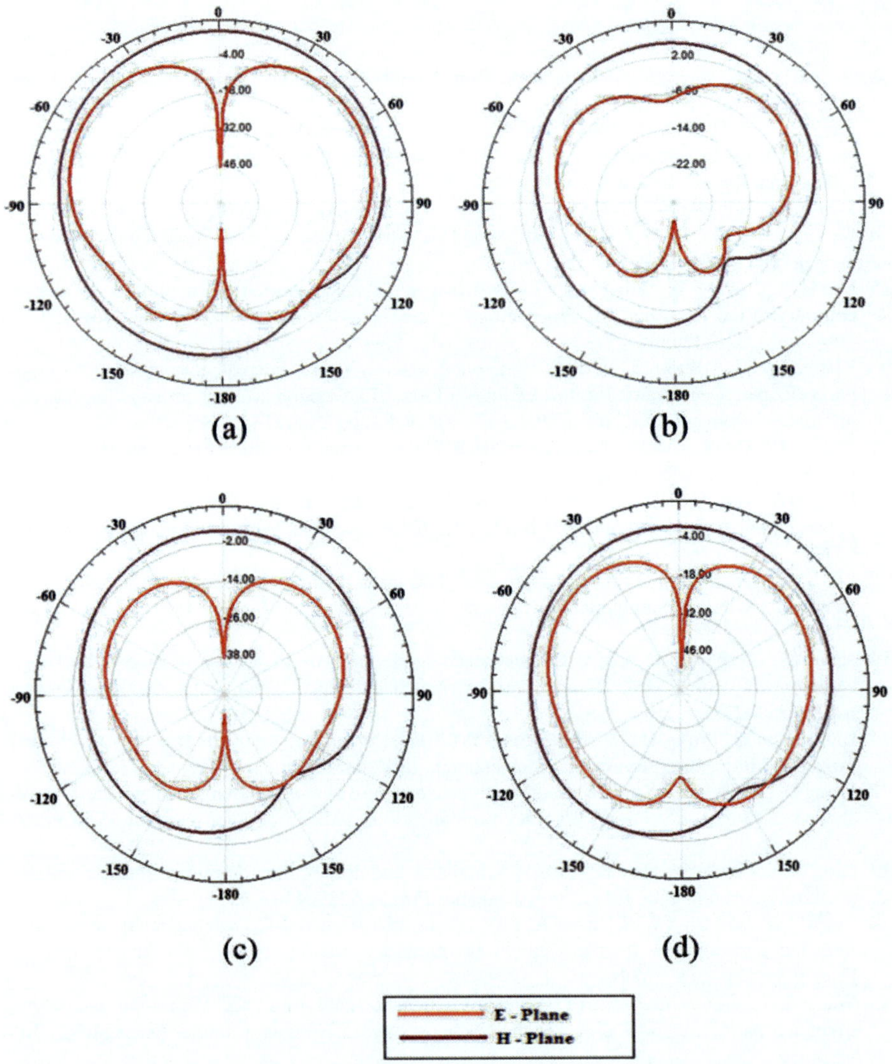

Fig. 11 *E–H* plane pattern: (**a**) conventional PIFA, (**b**) single shorting pin, (**c**) dual shorting pin, and (**d**) dual shorting pin with narrow slot

couple of conducting pins are fixed under the radiating patch in directive to transport the lower resonant frequency of $f_{0,1/2}$. Finally, a small rectangular slot is created on the radiating patch element to decrease the corresponding inductance initiated by the feeding probes and a couple of shorting pins. The results show that $f_{0,1/2}$ and $f_{2,1/2}$ are moved close to each other by placing a couple of shorting pins under the radiating element. By reallocating the lower and higher resonant frequency from 2.48 to 3.49 GHz and 4.81 to 4.49 GHz, bandwidth was improved under these two resonant modes from 14.9 to 76.8 MHz and gain from 5.72 to 6.46 dB.

References

1. Row J-S, Liou Y-Y (2006) Broadband short-circuited triangular patch antenna. IEEE Trans Antennas Propag 54(7):2137–2141
2. Lau KL, Li P, Luk KM (2004) A wideband and dual-frequency shorted-patch antenna with compact size. In: Proceedings of the IEEE antennas and propagation society international symposium digest, Jun 2004, vol 1, pp 249–252
3. James JR, Hall PS (1989) Handbook of microstrip antennas. Peter Peregrinus, London
4. Ge L, Luk KM (2012) A wideband magneto-electric dipole antenna. IEEE Trans Antennas Propag 60(11):4987–4991
5. Lee G-Y, Chiou T-W, Wong K-L (2001) Broadband stacked shorted patch antenna for mobile communication handsets. In: Proceedings of the Asia–Pacific microwave conference, Dec 2001, vol 1, pp 232–235
6. Nishiyama E, Aikawa M (2004) Wide-band and high-gain microstrip antenna with thick parasitic patch substrate. In: Proceedings of the IEEE antennas and propagation society international symposium (APSURSI), Jun 2004, vol 1, pp 273–276
7. Zhang ZY, Fu G, Zuo SL, Gong SX (2010) Wideband unidirectional patch antenna with Γ-shaped strip feed. Electron Lett 46(1):24–26
8. Lin QW, Wong H, Zhang XY, Lai HW (2014) Printed meandering probe-fed circularly polarized patch antenna with wide bandwidth. IEEE Antennas Wireless Propag Lett 13:654–657
9. Liu S, Wu W, Fang DG (2016) Single-feed dual-layer dual-band E-shaped and U-slot patch antenna for wireless communication application. IEEE Antennas Wireless Propag Lett 15:468–471
10. Khan M, Chatterjee D (2016) Characteristic mode analysis of a class of empirical design techniques for probe-fed, U-slot microstrip patch antennas. IEEE Trans Antennas Propag 64(7):2758–2770
11. Liu J, Xue Q, Wong H, Lai HW, Long Y (2013) Design and analysis of a low-profile and broadband microstrip monopolar patch antenna. IEEE Trans Antennas Propag 61(1):11–18
12. Wong H, So KK, Gao X (2016) Bandwidth enhancement of a monopolar patch antenna with V-shaped slot for car-to-car and WLAN communications. IEEE Trans Veh Technol 65(3):1130–1136
13. Pan YM, Zheng SY, Hu BJ (2014) Wideband and low-profile omnidirectional circularly polarized patch antenna. IEEE Trans Antennas Propag 62(8):4347–4351
14. Liu NW, Zhu L, Choi WW, Zhang JD (2016) A novel differential-fed patch antenna on stepped-impedance resonator with enhanced bandwidth under dual-resonance. IEEE Trans Antennas Propag 64(11):4618–4625
15. Wang J, Liu Q, Zhu L (2017) Bandwidth enhancement of a differential-fed equilateral triangular patch antenna via loading of shorting posts. IEEE Trans Antennas Propag 65(1):36–43
16. Da Xu K, Xu H, Liu Y, Li J, Liu QH (2018) Microstrip patch antennas with multiple parasitic patches and shorting vias for bandwidth enhancement. IEEE Access 6:11624–11633
17. Ge L, Gao S, Li Y, Qin W, Wang J (2019) A low-profile dual-band antenna with different polarization and radiation properties over two bands for vehicular communications. IEEE Trans Veh Technol 68(1):1004–1008
18. Mittal D, Dhillon AS, Nag A, Bargota R (2019) High gain and highly directive microstrip patch antenna for radar and satellite communication. In: Information and communication technology for intelligent systems. Springer, Singapore, pp 437–446
19. Prabhu T, Pandian SC, Suganya E (2019) Contact feeding techniques of rectangular microstrip patch antenna for 5 GHz Wi-Fi. In: 2019 5th International conference on advanced computing & communication systems (ICACCS), Mar 2019. IEEE, pp 1123–1127

Fog Computing for Smart Grid Transition: Requirements, Prospects, Status Quos, and Challenges

Md. Muzakkir Hussain, Mohammad Saad Alam, and M. M. Sufyan Beg

1 Introduction

The dawn of the smart grid (SG) has acquired the global consensus because of the intrinsic shortcomings associated with the century-old hierarchical power grid [1]. The shortcomings arise due to primitive generation methodologies, irregular generation-consumption profiles, underutilization of infrastructure resources, security breaches, global climatic concerns, etc. [2, 3]. The SG aims to resolve the voltage sags, overloads, blackouts, and brownouts caused due to erratic nature of power consumption and demand, rise in complexity of power system networks triggered by varying modalities such as distributed generation, dynamic energy management (DEM), electric vehicles, micro-nano grid, etc. [4]. The SG congregates electrical and communication network elements to enable noble and bidirectional energy cum data flows across the whole infrastructure [3, 5, 6].

The future SG operation is envisioned to be more data reliant than the electrical power [7–9]. The self-governing communicate interact and operate (CIO) protocols established by geo-distributed intelligent network nodes in a data-aware SG architecture will ensure hassle-free back-and-forth power delivery to and from the SG stakeholders. The entities such as SCADA systems, smart meters from advanced metering infrastructure (AMI), roadside units (RSU) and on-board units (OBU) from electrically run transportation telematics and miscellaneous sensors, etc.

Md. Muzakkir Hussain (✉) · M. M. Sufyan Beg
Department of Computer Engineering, ZHCET, AMU, Aligarh, India
e-mail: md.muzakkirhussain@zhcet.ac.in; mmsbeg@cs.berkeley.edu

M. Saad Alam
Department of Electrical Engineering, ZHCET, AMU, Aligarh, India
e-mail: saad.alam@zhcet.ac.in

A. Haldorai et al. (eds.), *2nd EAI International Conference on Big Data Innovation for Sustainable Cognitive Computing*, EAI/Springer Innovations in Communication and Computing, https://doi.org/10.1007/978-3-030-47560-4_5

dispersed across multidimensional networks such as home area networks (HAN), metropolitan area networks (MAN), and transport-oriented cities (TOC) form the prime source of data generation and consumption in a typical SG infrastructure [10–12].

A smart SG architecture is equipped with facilities for real-time, duplex communication between producers and consumers, having software modules to control and manage the power usage at both ends [4, 5]. Meanwhile, the current landscape of IoT penetration into smart grid architectures alters the data generation and consumption profiles [13]. The components of SG now called to be "things" will produce galactic mass of data while in execution, thus urging for robust storage and compute framework that can process and serve the service requests of SG stakeholders.

Due to the primitive aspect of the foundation percentage of the IoT endpoint in SG, i.e., the implementation of the necessary storage and compute resources that is never guaranteed each time, the outer agent has to consider the analytical and computation obligations. The processing and storage load in a normal SG is leveraged from massive mobile and static nodes that rotate over an extensive geographical area. This form of heterogeneity in the information architecture of an intelligent grid envisioned the utility of novel computing tech advancement to effectively mitigate the potential problems of various levels of processing and computation. Instead of depending on the master and slave computation framework as a legendary framework, the present idea is to focus on the information center phase analytics that operate under the client and server parameter [1, 10, 11].

The main aim of attaining a consensus on the segment to incorporate storage or computation resources still remains to be a pending concern for industries, R&Ds, academia, and legal entities [12, 14]. Cloud computation remains to be a novel tech advancement meant to fund SG due to its capability to avail more demand and consistency of network accessibility, which control the collective computing resources. All these are released and provisioned with less management services and efforts from providers [6, 15, 16]. The cloud services are liberated through many IoT devices by draining their batteries but present a lot of pay-per-use resources via virtualization [15]. Nonetheless, the changing service modalities based on cloud computing parameters fail to attain the vision of the SG necessities [1, 2]. The present cloud computing parameters' purpose is to introduce various proponents due to its notable failures in establishing multipurpose and common platform potential to promise novel remedies to the anticipated crucial necessities of the SG in the IoT ecosystem.

The main driving force for the emergence of FOG (From cOre to edGe) computing model (FC) is to abridge the computational hysteresis prevalent in cloud computing where the elastic cloud resources are extended to the end points of the network [17, 18]. In the SG ecosystem, fog computing has been noted as a key architectural element used to federate the associated processing where networking-specific logics are embedded in the edge frameworks and wireless clouds. Moreover, these are federated in the intermediary network infrastructure elements like portable devices, gateways, smart meters, RSUs and OBUs at electric vehicular networks, wireless sensors at smart homes and micro-nano grids, and miscellaneous IoT devices [19]. Brute force analysis of the requirement specification plans (RSPs) of smart grid architectures reveals that the mere fact of migrating and executing everything to the mega data centers creates the prime unfeasibility concerns. Thus the notion is to develop platforms for IoT-aware smart grid architectures where significant proportion of compute and storage activities will be offloaded to geo-distributed nodes named as fog nodes.

Based on the idea that the energy framework ecosystem is never thrilled to formulate the network infrastructure or incorporate its self-made computing system, there is need for it to operate with the generally acknowledged protocols and economically appreciated software and hardware platforms. This research purpose is to investigate the degree to which cloud computing facilities can effectively attain the vision of crucial necessities of the SG environment and the kind of subdomains and networking services for the fog-centered computing archetypes.

Obviously, rigorous research and investment effort are required to roll out the century-old legacy power grid with a robust and reliable SG and the objective of this research is to assess how far the current cloud computing platforms are equipped to meet the needs of this transition and which attributes of current computing industry need to be augmented or revised using fog approaches.

The contribution of the paper can be outlined as follows:

1. Analyzed the existing cloud infrastructure to see if they respond to SG computing requirements and in which aspects they need fog reforms
2. Demystified the fact of how fog (FC) paradigms can serve as an ally to cloud platforms and assess how far such a noble mix of both computation models will successfully satisfy the high assurance and mission-critical computing needs of SG
3. Outlined the significant research and development (R&D) opportunities generated along with the adoption challenges encountered towards realizing the commercial viability and optimal implementation of any fog model proposed for SG

2 Cloud-Fog Computing for Computationally Robust Smart Grid: A Synoptic Overview

The fifth generation completely revolutionizes the dynamism transmission, generation, and resourceful landscapes of the energy framework. The four vital SG elements are practices over the home area network (HAN), the neighborhood area network (NAN), and the wide area network (WAN). The HAN is the part that is closest to the ground, which contains the intelligent devices such as the TV, air conditioners, washing machines, ovens, and refrigerators. In addition to that, there are appliances such as electric motors and various forms of renewable sources of energy like the solar panels. The HAN is formed within its local units, found in the company plants, and in commercial plants to connect these electrical appliances with the intelligent meter. As such, they are capable of managing the demands of consumers through various subdomains. The NANs link up with the intelligent metered appliances over the multiple HAN, hence effectively supporting the data transfer through the distributed fields and substation appliances for the power distribution frameworks. The WANs, considered as the premium layer, act as the foundation of communication between aggregate and gateway elements. The information transferred from the NAN data collectors and nodes are thus combined in the WAN layer. As a result, this facilitates effective synchronization within the data transmission frameworks, generation structures, and renewable power sources, including the control center.

Let us consider smart homes as a representative SG use case to show how large-scale computing must play a key role in the smart power grid. The IoT technologies will result in high adoption of SG-powered home appliances such as smart heating/refrigeration system, smart infotainment services, home security subsystem, and smart emergency response devices like lighting control, fire alarms, and temperature-monitoring devices [20, 21]. The integration of IoT services will enhance transparency in the infrastructure by tracking the activities of the smart city residents. These smart homes will be leveraged with a range of AMI and monitoring devices ensuring fine-tuning of consumption patterns with power tariffs and load surges on the power grid [21].

The IoT sensors and actuators dispersed across the smart HANs sense the data associated to that environment and offload to the controlling module operated by householders [22]. Often the smart homes are also equipped with utilities that assimilate the surveillance data in order to predict the future occurrence of events, thereby preparing the householders to behave according to the contingencies. For instance, a geyser might be on when power is cheap but the water is allowed to cool when hot water is unlikely to be needed. An air conditioning, washing machine system, or even smart TVs might time themselves to match use patterns, power costs, and overall grid state [23]. In fact, there seems to be a one-one mapping between a SCADA system and a smart home, having the realization that the former directly reaches into the smart homes. The smart homes or customers may also form communities through neighbor area network (NAN) to have access to shared resources and response intelligently to the community-based cooperative energy consumption schemes [24].

The scope of this chapter is to summarize the status of cloud-based SG solutions coupled with application domains and benefits of fog computing for successful SG rollout. In this section, we analyze the current computing requirements of SG in six key metrics, namely decentralization, scalability, consistency, latency, privacy and security, and availability and reliability. Correspondingly, we also assess the status of cloud-based solutions and discuss the suitability of fog-based solutions while satisfying those requirements.

2.1 Decentralization

The generation of data and nodes in the SG structure has been distributed sparsely. On top of the centralized controls, the actuator and sensor nodes are incorporated in the intelligent homes such as the subsystems require more geo-generated competency. The data transparency in the distributed domains requires a more developed SCADA framework, which is guaranteed to accomplish a national form of data visibility.

The platform as a service (PaaS) resources can be hired to install the SCADA systems, a significant portion of which can be shared across power providers and distributed generation sources. The data center-based SG analytics will maintain transparency at varying levels of granularity. The web applications running in software as a service (SaaS) mode will allow authenticated web access to track the status of IoT end points. Collaboration of cloud computing technologies leveraged with data analytics modules into electrified transportation (vehicle-to-grid mode) use cases will ensure robustness and resiliency in the penetration of EV fleet of any size. The current centralized cloud deployments fail to capture the context of the decentralized SG computing. Since the protocols for visibility delineations are defined by third-party cloud vendors, discrepancies may arise duo to biased favors. As an illustrative example, let us take the V2G scenario where the EVs participate in the energy market acting as prosumers (both producer and consumer). The EVs will discharge its power into the energy market following an incentive policy. Since the V2G integration and recommendation policy is done via aggregators, there may be the case that the cloud market policy is more inclined to aggregator's payoff and the xEV customers may be deprived of appropriate benefits. Moreover, in data center-based centralized control policies, the degree of transparency enforcements is still in the hand of business giants. Under such circumstances there may be a slender abuse of decentralization.

The fog computing platforms will potentially realize both situational and context awareness via the smart application of the concept in the fog computing node (FCN). The energy-producing bodies will make use of the synchronized information retrieved from the decentralized SCADA elements like the phasor measurement unit (PMU) and the phasor data concentrator manufactured in the FCN. As such, this makes it possible to obtain a considerable operational form of visibility of the energy grid dynamics, in preparing to effectively deal with, smartly, the problems leading to brownouts and blackouts. Similar to the V2G, the EV end users in the environment depend on localized computation done in connection to the FCN. Therefore, EV shall depend on the private power of the market, but is based on the output in the associated algorithms.

2.2 Scalability

The IoT deployments unify the multitude of such controllable entities into a single pulpit all demanding computational scalability as an indispensable necessity. These smart homes may behave as independent candidates for intelligent SCADA control. The EVs may also be coordinated via home energy management systems (HEMS) and micro-grid integration, indicating computational scalability concerns. The IoT-equipped autonomous vehicles must tap to the dynamic data generated by SG utilities, viz. state of charge (SOC) and load predictions, tariff structures, and miscellaneous attributes that ensure grid stability. All such instances create potential thrusts for scalability enforcements and demand a paradigm shift into new computational paradigms.

The data center-based computational transformation has resulted in the displacement of traditional SCADA-based state estimation paradigms. The traditional SCADA-like computational platforms derelict because of the scalability problem associated with SG components when deployed on a large scale. The cloud system is bestowed with large numbers of lightweight, inexpensive servers, thus becoming irreversibly dominant to support scalable analytical solutions from large numbers of sensors, actuators, customers, and other SG stakeholder's data. One cloud computing information center can possibly be composed of computing and storage capability, which can be massive compared to the global super-computing utilities linked together. The incentivized SG ecosystem is meant to transform the extreme-performance computing application of the data centers. As time goes, the cloud PaaS services promise a counter and incentive feature for companies that have deployed the high-performance computing frameworks.

Moreover the horizontal as well as vertical extendibility support provided by cloud platforms enables the SG utilities to expeditiously respond to any market or regulatory changes and to the state-of-the-art products and services.

Talking about scalable-consistency-preserving SG requirements, the current cloud vendors seem unilaterally focused towards scalability challenge, and are deploying massive but weak storage and processing configurations that often "embrace inconsistency." For high-assurance applications they seem to be annoyed by scalability-consistency dilemma. The business behavior is more dedicated to the motto "serve more customers" rather than "serve more critical customers." The cloud utilities are tuned with stale data to respond to user's vectored requests and the later required to deal with this. Fog models leverage massive scalability support from multi-hypervisor virtualized systems with bandwidth-optimized paths. Fog architectures are equipped with fog nodes (FNs) that depend on three vital tech advancements used in effective and scalable virtualization of the vital and resourceful categories. In that case, it is possible to completely manifest the SG distribution architecture. These include the following:

(a) Compute: To effectively virtualize both the I/O and computing resources, there is a need to facilitate novel selection of computation resources such as hypervisors.

(b) Network: To control network resources identified above, it is important to embrace robust and viable networking virtualization infrastructure like the software defined network (SDN) and the network function virtualization (NFV) methods.
(c) Storage: This aspect denotes a virtual file system (VFS) and the virtual object and block store.

2.3 *Consistency*

To understand the consistency need of SG use cases, consider an EV fleet where multiple vehicles are concurrently communicating to SCADA system; even if the former are under distributed control and the communication may also be across multiple networks, they all should receive the right control instructions. Minute deviation from the communication synchrony may lead to catastrophic deformation in the fleet dynamics. Ubiquitous network access provided by cloud platforms guarantees the entry of use to the cloud service and procuring of interpretability over the SG framework. A similar information transfer platform offered by the cloud servers assures information flow in the SG, hence evading the application of a lot of Middleware software and interfaces used to access information from the SG framework. Information consistency can also be generated based on data standardization formats on a single platform.

However, often cloud giants employ the embrace of the consistency availability and partitioning (CAP) [3] to justify their consistency-scalability trade-offs. The SG applications feed into a service spectrum supporting a wide degree of shared access; thus poor scaling of consistency is not feasible. Though the CAP theorem can also be formalized under weaker assumptions, the clouds make stronger assumptions in practice and cite such folk logic for offering weak consistency guarantees, although they consider strict consistency assurances for their own needs. Based on the initiation of the fog later, the problem of inconsistency in the cloud architecture can therefore be mitigated to a significant extent. Operating applications and systems can effectively support fog computing based on the application and device-level structures. The ecosystem services will obviously have a vital role to play in attaining an off-line computation. As such, architecture can be considered as an option in many cases, whereas it remains to be complementary to one another in many cases. Based on the necessities and setting, a lot of modification can be initiated.

2.4 *Real-Time Analytics*

Based on the data latency hierarchy requirements, the SG applications can be grouped into three categories with loosely defined boundaries. The data latency hierarchy of typical SG applications is shown in Fig. 1. Installation of transmission

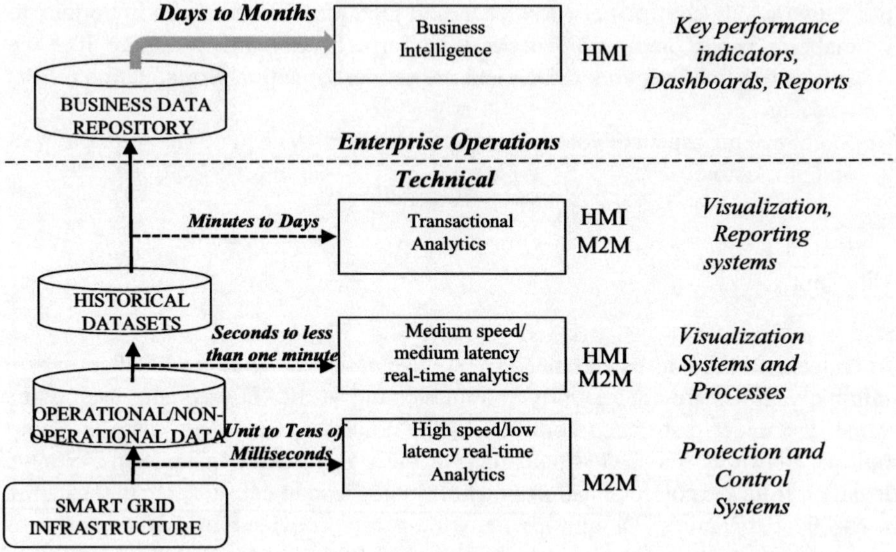

Fig. 1 Latency hierarchy of typical SG applications

infrastructure, power delivery road maps, etc. come under group A type applications with relaxed timing requirements. For such applications current computing configurations guarantee service-level agreements (SLA) because of informal service constraints. Group B applications are those that need super-paced communication and transport channels such as circulating the smart meter data to regulate SCADA control. Such applications can tolerate a delay of only a few tens of microseconds, caused either due to node failure or connectivity disruption. Such SG solutions demand real-time response even in the presence of failures. The third class C includes mission-critical SG applications that require high assurance, stringent privacy and security enforcements, robust access control, and consistent behavior across the nodes of action. The applications acting on real-time data may produce glitches if exposed to stale data. The SCADA framework is implemented in the modern information stream driving SG. This framework is timed since it can malfunction during the process of operation over the TCP and IP protocols, which limits the sustainability of the all-inclusive TCP flow controls.

To obtain a comprehensive and actual-time data concerning the condition of devices, the present SG infrastructure are planned with deployment of a large number of sensors and smart meters in various points of the grid. For real-time monitoring of SG applications the cloud data centers obtain two-way communication and relay data from wireless sensor networks (WSN), and disseminate it for proactive diagnosis and timely response to any erroneous context that can lead to transient faults or blackouts in worst case.

2.5 *Privacy and Security*

The SG utilities when leveraged with IoT end points need robust protective mechanisms that ensure restricted and entrusted access to critical data. The SG confidential data might be of interest to the criminals or manipulating entities seeking edge to energy trade, thus exposing the system to cyberattacks. The denial-of-service (DoS) or distributed denial-of-service (DDoS) attacks on AMI or vehicular data may cause severe vulnerabilities such as bandwidth drainage, excessive CPU utilization, irregular memory surges, and halting of the client's or host's operations [25]. The designed computing architectures should be leveraged with robust aggregation algorithms that can guarantee the privacy and data anonymization of respective stakeholders and motivate consumer's participation [15, 26].

In the general setting of the SG, *application privacy here* implies to the stability and safety of the energy grid instead of safeguarding it against the potential malice that falls under the umbrella of privacy. Malignancy of unpremeditated defense of the CAP theory is to enhance manifestation in the present segment of intelligent grid cloud privacy. In the present deigns, all the information are considered to fit in the cloud storage featured with a significant number of storage elements and servers composed of vertical and horizontal elasticity. Nonetheless, the present cloud-centered privacy and security enforcement are significantly erratic. Sometimes, cloud experts can become evil whenever they decide it. In the competitive collaborative cloud SG ecosystem, the rivals can possibly tamper with the privacy of information, while there is a surge of competition over capturing of power infrastructures, there is possibility of privacy breach, data commingling, Denial-of-Service (DoS) attack, eavesdropping, etc, leading to Cyber-Physical Warfare (CPW) among nations. These platforms can sometimes guarantee that information centers can be on and the network application can keep on running. However, when this data is evaluated deeply, they can be considered to be far from the typical information items and single computations.

Gartner argues that cloud systems are fraught with a lot of privacy issues and recommends that SG, like clients, has to pose difficult specifications and questions to cloud service providers. Therefore, they have to consider a certified privacy evaluation criterion from a third party, which is neutral to execute the commitments [21].

2.6 *Availability Reliability and High-Assurance Computing*

A subset of SG applications need to have hardware and software solutions that are "everytime on." Any interruption in computing services may cause increased costs and many a times it may lead to loss of consumer confidence. Because cloud computing by its nature relies primarily on the internet connectivity, the SG utility vendors interested in starting or expanding their business strategies with current cloud platforms must have rigorous IT consultations for showing them ways to schedule network resources such as bandwidth levels which will suffice their mobility and availability constraints [6, 14]. While conducting technical operations in such remote data centers the control of SG stakeholders over the infrastructure will be seized. Thus the manageability is not as robust as when they are used to with SCADA workstations and other controllers.

3 Adoption Challenges and Future Research Opportunities

In this section we inspect the chances facilitated by the fog-based SG deployment. Considerably, we have highlighted the problems and study thrusts brought out towards the viability of fog computing, which enhances effective SG change.

3.1 *Opportunities*

Compute and Connectivity Convergence

Fog computing facilitates the base station (BS) of the mobile network especially the F-RAN to incorporate the ad hoc and the integrated FCN to fully leverage the SG information with computation and temporary storage services. The FCN embedding in BS(s) allows the latter to process computation and communication workload, thereby bringing merits as a form of performance for the purpose of application in diversified applications. Conjunction of the computation and information transfer resources will therefore develop the performance of a lot of SG applications like:

(a) Actionable real-time intelligence in micro-grid systems such as wind farm and solar panels
(b) Real-time outage management in generation-distribution subsystems
(c) Dynamically optimizing the wind farm operation through forecasting at finer granularities (once every 5 min or less)

The connectively and compute convergence consent more to the prevalent SG network architectures, which effectively make use of the context data of communication to effectively undertake the anticipated categorization aspect on the significant amount of achieved information by the IoT end point.

Flexibility in Application Development

The developers facilitating the development of software are vital for the success of the fog computing and its implementation. These developers can effectively develop and design a state-of-earth service or application that can reap off the benefits or contextual data in the fog radio network (F-RANs) certainly in the fog gateway. The developers of application can effectively formulate their applications to be significantly flexible in the fog ecosystem through the application of programming languages, software development kits (SDKs), and open standards. The applications found in the cloud environments are fundamentally found in the servers. Nonetheless, as for the fog paradigm, applications have to possibly enhance the migration of things from cellular devices to the fog nodes and servers to enhance execution. In that regard, the developers of applications have to make the applications by practically separating the relay-sensitive compute and intensive segments from the delay-tolerant segments of application. Resultantly, this forms more room for application designers to reap off the benefits from a certain fog demand based on their share of success in the emergent paradigm.

New Revenue Stream for Service Providers

The SG software developers can typically avail novel services for cellular enterprises and application meant to leverage users to advance productivity. This context-aware localized network can effectively present a novel category of services, which potentially boost the service feature promised by the relevant SG stakeholders. The initiation of novel services can complement new streams of revenue for the fog service designers and vendors. The deployment of services and applications in the HAN will possibly not advance the end-user application performance, but will lead to shrinkage of the signal traffic volume of the founding network. Resultantly, it limits the costs of operation of the SG service provider.

Network Equipment Vendors

The network elements like the fog gateway and access point require significant energy meant to involve the end user with demanded resources. The network equipment vendor has a significant demand for the processing of networking elements as service provider might possibly have replaced the present base station and access points with computing resources. The fog platform is concerned with not only novel streams of application designers or network services, but also networking equipment and vendors that present products with unique capabilities and features. Moreover, the introduction of unique features will possibly facilitate the creation of novel streams for SG software developers.

3.2 *Challenges*

Compatibility, Scalability, and Complexity

The contemporary IoT-centered SG application is diversified based on interoperability, reliability, security, and scalability. Configuration, location, and server functionality of the diversified FNs remain to be major concerns. Choosing the novel IoT facilities and the fog element is meant to create an optimum application workflow while limiting nonfunctional necessities like networking latency, QoS, and security, hence creating nascent research thrusts across the computational domain. Since numerous IoT vendors and manufacturers are involved in developing heterogeneous sensors and smart devices, it is increasingly complicated to reach the consensus on how to select optimal components customized to SG hardware configuration and customize necessities. There might be SG applications, which require specific hardware and protocols and might even operate under high-security-constrained environments. An efficient orchestration framework can cater such disruptions in functionalities and requirements and successfully manage large and dynamic workflows [27].

Security

Security is among the prime issues for a typical data-driven SG platform, as the transportation utility vendors may agonize for the repercussions if the privacy of entrusted cloud data is compromised. Due to the dynamic nature of a transportation and SG infrastructure, it becomes nearly infeasible to create coherent cross-cloud trust relationships. Further, existence of complex relationships and dependencies among varying range of stakeholders in contemporary smart cities hinder the compatibility and cost-effectiveness of data clouds employed in transport-oriented cities (TOCs). Global security standards are essential to cope up with such privacy and flexibility concerns.

Decisions regarding selective migration of information hosted in private clouds, FNs, and vehicular cloudlets to the public storage space require rigor research. Robust and fine-grained authorization protocols and access grants should be defined to ensure multiple accesses to federated cloud and federated fog repositories.

Due to the fact that IoT developers and SG prefer use cases, a certain service is used to create sensors, computer devices, or chips and hence deployment in different geographical areas. As such, this results in a boosted attack vector of the associated objects. The samples of attack vectors can be categorized as human-led sabotage of networking infrastructure, dangerous programs affecting the leakage of data, and even physical accessibility to networking devices [28]. The novel security and risk evaluation processes are required to categorically and vibrantly analyze the measure and security risks, as well as analyze the privacy concerns of the vibrant IoT-based applications that are composed to become more crucial for securing information processing and placement. IoT-integrated appliances and devices facilitating fog support like the switches, base stations, and router are utilized publicly to access computing edge nodes. The problems connected by the private and public vendors can be implemented in the devices requiring more advance articulation. Moreover, the purposed aim of devices, such as internet routers used to handle networking traffic, cannot be relied to compromise based on fog nodes. The fogs can be modelled as multiple tenants whenever the stringent privacy protocols are considered obligatory.

Performance and Reliability

With numerous fog utility offerings available on the market with varying pricing schemes, decision on selecting one that is commercially optimal to SG entities needs to be standardized. A budding informatics thrust is to evaluate the complexity and financial viability of fog service deployments in price-diverse environments. The infrastructure assembles multiple cloud and fog genres into a common platform; thus uncertainties in pricing models are obvious. The stakeholders, if aware of the future service tariffs and incentives, will allow them to ponder for the optimum. The incentives of V2F like use cases are primarily dedicated to promote the development of intelligent vehicular services and offer a range of anxiety-free drives to the naïve xEV users. The SG infrastructure assembles the distributed cloud and fog platforms to co-work with each other for smooth and reliable operation of its entities. However, maintaining an optimal balance in the distribution of data, control, and computation among the dedicated cloud-cloudlets-FNs determines the performance of the whole system. Commercial realization of the notion of cloud of things (CoT) and fog of things (FoT) from billions of sensors and low-power devices in a sensor network and connectivity with the data centers demands reliable and permanent sources of energy. Efficient fabrication techniques can enable the sensors to generate onsite power from renewables and environment.

For V2F-like infrastructure that employ dynamic and ad hoc fog nodes, intermittent vehicular networking will hamper the service quality and nonfunctional QoS. The mesh created by seamless communication among cloud-cloudlet utilities will create galactic volumes of information to flow across the interfaces and data centers; thus uncertain network and communication failure will adversely affect the execution of backend SG infrastructure. Intelligent controllers and gateways coupled with mobile networking paradigms can manage the connectivity control of distributed and networked cloud and fog resources in SH-based TOC cyber infrastructure (Fig. 2).

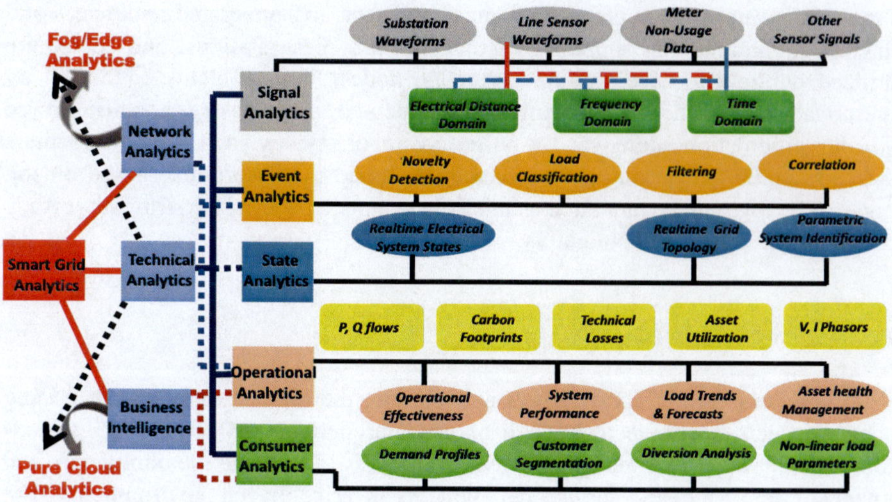

Fig. 2 Landscape of analytics services that can be supported over fog-based smart grid architectures

4 Conclusion

This research has evaluated the present status of cloud computing services in attaining mission-critical computing demands of SG, and side by side we identified the requirements that could be fulfilled, if carried out through fog computing. The work comprehends the opportunities and prospects of fog-based SG analytics. Finally, the significant adoption challenges encountered towards fog deployment are outlined along with the future research avenues that will prove to be productive while going for fog-based SG computing.

References

1. Saleem Y, Crespi N, Rehmani MH, Copeland R (2019) Internet of Things-aided smart grid: technologies, architectures, applications, prototypes, and future research directions, pp 1–30
2. Birman K, Ganesh L, van Renesse R (2010) Running smart grid control software on cloud computing architectures 1. Introduction: the evolving power grid. In: Workshop on computational needs for the next generation electric grid, pp 1–28
3. Hussain MM, Beg MMS (2019) Fog computing for Internet of Things (IoT)-aided smart grid architectures. Big Data Cognit Comput 3(8):1–29
4. Markovic DS, Zivkovic D, Branovic I, Popovic R, Cvetkovic D (2013) Smart power grid and cloud computing. Renew Sustain Energy Rev 24:566–577
5. Gao J, Xiao Y, Liu J, Liang W, Chen CLP (2012) A survey of communication/networking in smart grids. Futur Gener Comput Syst 28(2):391–404
6. Al-Ali AR, Aburukba R (2015) Role of Internet of Things in the smart grid technology. J Comput Commun Technol 3(3):229–233

7. Chan CC, Tu D, Jian L (2014) Smart charging of electric vehicles—integration of energy and information. IET Electr Syst Transp 4(4):89–96
8. Hussain MM, Alam MS, Beg MMS, Laskar SH (2019) Big data analytics platforms for electric vehicle integration in transport oriented smart cities. Int J Dig Crime For 11(3):2
9. Lv Z et al (2017) Next-generation big data analytics: state of the art, challenges, and future. IEEE Trans Ind Inform 13(4):1891–1899
10. Rehmani MH, Erol Kantarci M, Rachedi A, Radenkovic M, Reisslein M (2015) IEEE access special section editorial smart grids: a hub of interdisciplinary research. IEEE Access 3:3114–3118
11. Xu S, Qian Y, Hu RQ (2015) On reliability of smart grid neighborhood area networks. IEEE Access 3:2352–2365
12. Yu L, Jiang T, Zou Y (2017) Fog-assisted operational cost reduction for cloud data centers. IEEE Access 5:1–8
13. Hoang DT, Wang P, Niyato D, Hossain E (2017) Charging and discharging of plug-in electric vehicles (PEVs) in vehicle-to-grid (V2G) systems: a cyber insurance-based model. IEEE Access 5:732–754
14. Osanaiye O, Chen S, Yan Z, Lu R, Choo KKR, Dlodlo M (2017) From cloud to fog computing: a review and a conceptual live VM migration framework. IEEE Access 5:8284–8300
15. Bera S, Misra S, Rodrigues JJPC (2015) Cloud computing applications for smart grid: a survey. IEEE Trans Parallel Distrib Syst 26(5):1477–1494
16. Cisco Systems (2016) Fog computing and the Internet of Things: extend the cloud to where the things are, p 6. www.cisco.com
17. Sarkar S, Chatterjee S, Misra S (2018) Assessment of the suitability of fog computing in the context of Internet of Things. IEEE Trans Cloud Comput 6(1):46–59
18. Misra S, Sarkar S (2016) Theoretical modelling of fog computing: a green computing paradigm to support IoT applications. IET Networks 5(2):23–29
19. Hussain MM, Alam MS, Beg MMS (2019) Fog computing based big data analytics in cyber-physical systems—a smart grid case study. In: Di Martino B, Yang LT, Zhang Q (eds) Smart data: state-of-the-art and perspectives in computing and applications. CRC Press, Taylor & Francis, Boca Raton, FL, pp 289–317
20. Zanella A, Bui N, Castellani A, Vangelista L, Zorzi M (2014) Internet of things for smart cities. IEEE Internet Things J 1(1):22–32
21. Ullah R, Faheem Y, Kim BS (2017) Energy and congestion-aware routing metric for smart grid AMI networks in smart city. IEEE Access 5:13799–13810
22. Scarpiniti M (2017) Fog of everything: energy-efficient networked computing architectures, research challenges, and a case study fog of everything: energy-efficient networked computing architectures, research challenges, and a case study, May 2017
23. Hussain MM, Alam MS, Beg MMS (2019) Feasibility of fog computing in smart grid architectures. In: Proceedings of 2nd International conference on communication, computing and networking. Springer, Singapore, pp. 999–1010
24. Hussain MM, Alam MS, Beg MMS, Malik H (2017) A risk averse business model for smart charging of electric vehicles. In: Smart innovation systems and technologies (SIST) series, vol 79. Springer, Berlin
25. Asri S, Pranggono B (2015) Impact of distributed denial-of-service attack on advanced metering infrastructure. Wirel Pers Commun 83(3):2211–2223
26. Diamantoulakis PD, Kapinas VM, Karagiannidis GK (2015) Big data analytics for dynamic energy management in smart grids. Big Data Res 2(3):94–101
27. Wen Z, Yang R, Garraghan P, Lin T, Xu J, Rovatsos M (2017) Fog orchestration for internet of things services. IEEE Internet Comput 21(2):16–24
28. Hussain MM, Alam MS, Sufyan Beg MM, Krishnamurthy M, Ali QM (2018) Computing platforms for big data analytics in electric vehicle infrastructures. In: The 4th international conference on big data computing and communications (BigCom-2018), 7–9 Aug 2018, Illinois Institute of Technology, Chicago, IL, USA

Multispectral Data Processing for Agricultural Applications Using Deep Learning Classification Methods

Anuj Rapaka and Arulmurugan Ramu

1 Introduction

Agricultural multispectral satellite imaging devices enable the farmer to handle plants, land, fertilization and drainage more efficiently. Either the farmer or the wider atmosphere benefits enormously by reducing the use of sprays, fertilizers and water wastage while at the same moment improving crop yields. Multispectral remote imaging technology uses green, red, red-edge or near-infrared wavebands to Record both noticeable unseen crop and also vegetation pictures. The multispectral pictures incorporate the information into significant data with dedicated agricultural software. This land telemetry, soil as well as plant information allows the farmer to track, schedule as well as handle the farm to save time and cash more efficiently, together with decreasing pesticide usage. The foundations of multispectral imaging technique, reflection, wavebands as well as vegetation indicators such as NDVI and NDRE are explained in this study. All this data provides the farmer with tremendous ideas into soil as well as plant health [1].

Multispectral. First let's review "RGB," which means "red, green, blue" because for most photo is a new format. RGB databases have three data levels, one corresponding to the image's reds, one corresponding to the greens and only one corresponding to the blues. In many other phrases, only visible light is recorded in RGB pictures. If we look at the same Google Earth pictures, you can't see the additional bands because we don't have links to the initial documents. The non-visible frequencies are essential as they can be used to identify various rates of

A. Rapaka (✉) · A. Ramu
Department of Computer Science and Engineering, Presidency University, Bengaluru, Karnataka, India
e-mail: arulmurugan@presidencyuniversity.in

A. Haldorai et al. (eds.), *2nd EAI International Conference on Big Data Innovation for Sustainable Cognitive Computing*, EAI/Springer Innovations in Communication and Computing, https://doi.org/10.1007/978-3-030-47560-4_6

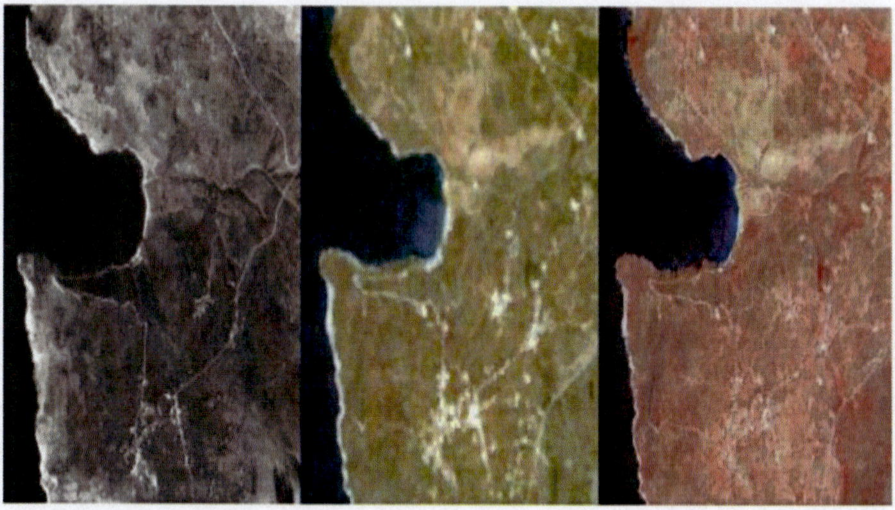

Fig. 1 Multispectral satellite image

development in vegetation. Chlorophyll in trees provides some kind of light that even the human eye cannot see but can also be captured by dedicated sensors. These additional groups open a whole universe of prospective analyses that can sense small distinctions along a landscape in vegetation development. They can also be used in order to define archaeological underground characteristics or ancestral rivers. It can be combined with a greater-resolution, panchromatic picture when you buy a multispectral image. This enables the customer to merge the two in a method called pan sharpening that fuses the colour fields with the black-and-white imagery of high resolution [2].

The next phase is Fig. 1 the information collection for temporal resolution we use for our venture. Wavelength may differ considerably based on the picture drawn and also the product's price range. It is generally evaluated in metres relative to aerial photos or chart size, which is generally provided as a scale, such as 1:5000. A 2 m range implies we're not likely to see something lower than that. If we are searching for pillars or terraces, a larger-resolution image must be considered [3, 4].

2 Literature Review

In this chapter, we address certain hyperspectral information features 43 that challenge the ranking phase. In the curse of dimensionality in [5–7], the scientists recorded several differentiated geometric, statistical and also asymptotic characteristics of high-dimensional information through certain laboratory instances.

A high-dimensional storage is nearly vacant, meaning that multivariate IR data is generally in a lower-dimensional framework. In certain phrases, in addition of category separability, high-dimensional information could be mapped into a reduced hyperspace without compromising significant information [1].

Gaussian-dispersed information tend to focus in the tails, while consistently distributed information tend to focus in the angles, making it much harder for both models to estimate the thickness of high-dimensional information.

Fukunaga [8] reveals a relationship between both the amount of teaching samples needed and the amount of measurements for distinct classifier kinds. The amount of practice samples needed is sequentially connected to the dimensionality of linear classifications as well as quadratic classifier dimensionality squares (e.g. Gaussian MLC [8]).

In [9], Landgrebe reveals that in addition to anticipated identification precision, too many spectral groups could be undesirable. When dimensionality (the amount of bands) rises, a higher-dimensional collection of statistical data should be calculated with a continuous amount of training samples. In other phrases, while greater spatial sizes boost class separability, statistical assessment precision reduces. This results in a decline in ranking accuracy beyond an amount of groups. These issues are linked to the so-called dimensionality curse for ranking purposes. It is anticipated that even more information will be required to identify more courses with greater precision as dimensionality rises. At the same moment, the above features show that standard methods created for multispectral information might not be appropriate for hyperspectral information ranking. The above-mentioned problems associated with the data's high-dimensional design have a drastic impact on the methods of controlled classification [10]. To create an accurate estimate, these methods require a big amount of training samples (which in reality is almost difficult to achieve). If dimensionality rises, this issue is still more serious. Thus, when only a limited number of practice samples are accessible, ranking methods created on multispectral information should be able to manage high-dimensional information.

2.1 Uncertainties

Uncertainties produced at various phases of the information procurement and also evaluation process could have a dramatic impact on the accuracy of measurement as well as the performance of the initial test map [11–14]. There are also many explanations for these uncertainties, along with weather circumstances at the moment of data capture, information restriction in aspects of radiometric as well as spatial resolution and multiple image mosaicking, among many others. Registration of images and geometric rectification trigger confusion of location. Furthermore/Moreover, algorithmic mistakes can contribute to radiometric uncertainties when calibrating whether environmental or spatial effects [15].

It was stated in that the accuracy of ranking is the consequence of a trade-off between two elements. The first relates to the effect of border pixels on the outcomes of ranking. In this scenario, the amount of pixels dropping on the boundary of various objects will reduce as the spatial resolution becomes finer. The second element relates to the enhanced colour variability connected with better spatial resolution of distinct soil parts. The presence of several blended pixels between distinct land-cover groups is the primary cause of confusion when dealing with optical information of small or intermediate spatial resolution and also can dramatically impact ranking outcomes. Fine geographic measurement could provide comprehensive data on distinct soil cover form and composition. Such data could also be supplied to the ranking scheme in order to further enhance the precision of ranking measurements and enhance the performance of range charts. Concern of temporal data in the ranking scheme is a dynamic study subject in the hyperspectral society and has already been studied in several publications such as [1, 17–21]. As stated above, considering temporal data in the conceptual model is beyond the range of this job, which relies on monitored spectral classifiers. Using high-resolution hyperspectral pictures, though, presents many fresh issues, particularly those induced by darkness, resulting in elevated colour differences within that land-cover category. These disadvantages may decrease the precision of ranking if classifiers are unable to manage these impacts effectively [22].

3 Existing Methods

3.1 Random Forests (RFs)

First launched, RFs are a common ranking as well as regression sample technique. This classifier was commonly used in combination with hyperspectral information as it assumes no fundamental allocation of likelihood for entry information. In addition, if there is no equilibrium among linear and amount of accessible teaching samples, it will provide a nice identification outcome in terms of effectiveness in an ill-posed scenario. Rotation forest is suggested depending on the concept of RFs to concurrently promote participant diversity as well as personal precision within a classifier set. See [1, 23] for a comprehensive explanation of the strategy.

3.2 Sparse Representation Classifiers

The use of scarce image classifiers (SRCs) to dictionary-based conceptual models [9, 10] was also a significant innovation. In this scenario, a sparse linear combination of samples (atoms) from either a dictionary represents an input signal, in which the coaching information is usually used as the dictionary. SRCs' primary benefit is that they prevent the intensive practice operation usually conducted by a monitored classifier, and the evaluation is conducted straight on the dictionary. Some scientists have also created discriminative and compact category dictionaries to enhance ranking efficiency, given the accessibility of adequate training data.

3.3 Deep Learning

Deep learning is something of a multi-layered neural network, usually greater than three levels, that attempts to know entry information characteristics hierarchically. Deep learning is a rapidly growing subject that has proved invaluable in many fields of studies, along with computer vision and handling of natural language. In the context of RS, some deep designs were suggested for the production and also classification of hyperspectral information features. For hyperspectral data classification, the stacked auto encoder (SAE) and also the auto encoder (AE) with sparse restriction were suggested. Later, some other profound model was suggested for the ranking of hyperspectral information, i.e. the Deep Belief Network (DBN). Recently, for RS image analysis, an unmonitored convolutionary wireless network (CNN) was suggested using selective layer-wise unsupervised learning to formulate a profound CNN model.

3.4 Hyperspectral Method

Both RGB and RE graphics were presented to a handling level that allowed the computing of ortho-picture mosaics. They are subsequently segmented separately into 512 × 512 pixel parts with a 64 pixel path used to coach a CNN. Using Keras with TensorFlow as backend described in Python 2.7, the network was introduced. All tests were conducted with such an Intel Xeon CPU E5-2650, 64 GB RAM and a NVIDIA GTX-1080Ti GPU (11 GB RAM) used only to boost the CNN on the workspace. The CNN was taught with instruction array (Fig. 2, remaining) from baseline, which was divided evenly into 80% (12,109 bugs) for instruction and 20% (3002 bugs) for experimentation. A research was conducted using the price feature of Dice coefficient, with both the Adam optimizer in 3 runs of 128 epochs. The teaching pace for each phase was 1E-4, 1E-5 and also 1E-6. Recognition utilizing 153 Dice coefficient reached 0.9953 and is also represented/illustrated graphically (Fig. 2, correct).

The qualified system was then implemented of the same variables (spots as well as step) to a separate vineyard region for verification. Taking into account that all pixels with such a significance above null belong to the vineyard plan, the subsequent picture from consideration was binarized. A number of morphological activities were then implemented in order to extract probable outliers to provide a clearer knowledge of the region being identified. In the previous sequence, the morphological operations were implemented: enter, near, and extract items below 2000 pixels corresponding to prospective outliers. A pixel-wise comparative with

Fig. 2 Existed methods

a guide differentiated mask was conducted for verification reasons, which amounts to ground-based reality. With 85.07% precision, the outcome for accurate identification was accomplished, preceded by 14.64% for identification and 0.29% for identification. Figure 2 presents a graphic understanding of the outcomes acquired. A definite absence of identification is visible in fields with a powerful cloud presence and a lower density of inter-row vegetation, this absence of identification being linked to the occurrence of such circumstances in the information collection used for teaching.

4 Proposed Method (Max Pooling Deep Learning (MPDL))

Here we proposed a multispectral-based agriculture monitoring system, using the max pooling deep learning (MPDL) method. The above-mentioned methods have problems like efficiency and accuracy. Problems with existed methods are clear visionary, disease finding, future extraction, etc. Multispectral pictures comprise hundreds or thousands of rows as compared to hyperspectral pictures, typically between 5 and 12 bands [10]. This could be obtained from satellites in this study using Manu's Disease Dataset, and the large amount of spectral bands triggered the development of a third dimension to the temporal picture arising in a 3D data cube. The special features of hyperspectral pictures elevated dimensionality. Distinctive brightness models or spectral band tags occur in every group. The wealthy colour data accessible in the hundreds (ultimately thousands) of tight channels may enable the precise discrimination of various materials [3]. This reality allows useful multispectral information. At the same moment, the greatest task to handling and evaluating is the greatest benefit of multispectral information, the elevated dimensional dimensionality and transparency, resulting from the need to establish computationally effective algorithms. Indeed, normal parametric classifiers will

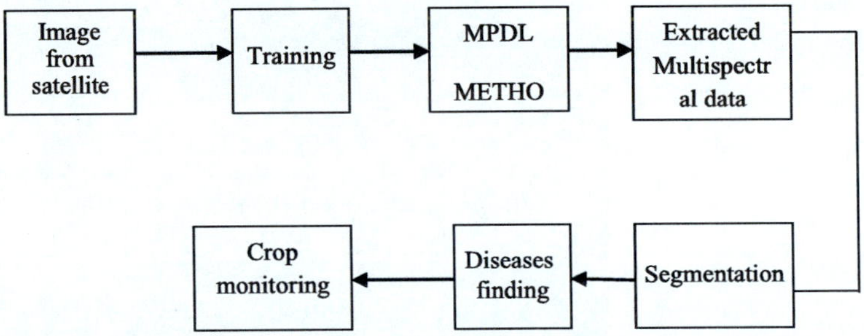

Fig. 3 Proposed block MPDL diagram

no longer be enough. ML seems to be a more appropriate approach [11]. To maintain the multi-resolution property at each scale, these filters should be adjusted accordingly. The up-sampling can be defined as follows:

$$X[k] \uparrow 2 = \begin{cases} x\left[\frac{k}{2}\right] & \text{if } k \text{ is even} \\ 0 & k \text{ otherwise} \end{cases} \quad (1)$$

Use max pooling around dimensionality of the matrices rather than changing the distance of the step. Pooling aims at a region that can be assumed to be 2 × 2 and only retains the greatest or median importance. A 2 × 2 matrix depicting pooling is shown in the preceding image: a pooling region still has the same length step as the pool size. This prevents overlap. Here's a comparatively small depiction of max pooling deep learning (MPDL): you can see the entry picture is exposed to different convolutions and pooling stages with ReLU activations between them finally reaching a traditionally fully integrated network. Although not seen in the graph, the fully integrated network eventually predicts the category and is also shown in Fig. 3.

As in most MPDLs, there will be numerous Fourier transforms at each level in this technique. We watch ten here, which are shown as lines. Each one of these ten convolutions has its own matrices for each row so that at each phase, distinct convolutions could be studied. The entirely linked strata on the left will decide which curves better define the vehicle or truck and so on as displayed in Fig. 4.

4.1 Algorithms and Mathematical Analysis

By splitting the entry into triangular pooling areas and calculating the peak from each region, a max pooling matrix conducts down-sampling. Layer = maxPooling2dLayer(poolSize, Name, Value) establishes the mandatory characteristics of Stride, Name and has unpooling outputs using combinations of name-value. Use

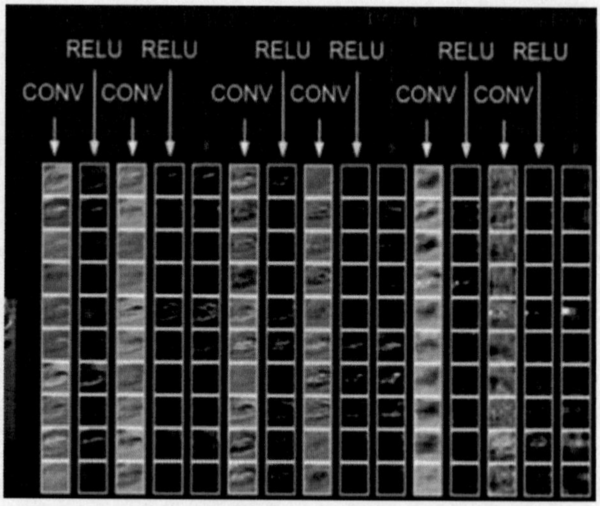

Fig. 4 Pool position

Fig. 5 Maximum poling

12	20	30	0
8	12	2	0
34	70	37	4
112	100	25	12

2×2 Max-Pool →

20	30
112	37

the name-value pair contention "Padding" to indicate entry padding. For example, max Pooling2dLayer(2,′Stride′,3) creates a pooling layer of [2 2] and [3 3] pool size. Multiple name-value combinations can be specified. Enter each title of the estate in single quotes as shown in Fig. 5.

$$\sum_{i=1}^{N} \sigma\,(x - i + 0.5) \approx \log\left(1 + e^{x}\right) \quad (2)$$

where $X = \mathrm{VW}^{\mathrm{T}} + b$.

The most commonly utilized technique in pest as well as disease management is to pour pesticides evenly over most of the crop region, both in open-air and in also greenhouse circumstances. So we are applying the maximum pooling on this satellite image using deep learning method such that calculating (Eq. 1) to be effective parameters like, accuracy and efficiency, this approach requires significant amounts of pesticides which results in a high financial and significant environmental cost. DL is used as portion of overall agricultural accuracy leadership, where the entry of agrochemicals is aimed in aspects of moment, location and crops impacted.

5 Results

Figure 6 shows the target image. This is obtained from the satellite system. SS system is at out of earth surface. These are continuously monitoring the crops. Satellite systems have HD cameras, and their lens are more efficient than the optical receivers. But the image quality is less, because of distance.

Figure 7 is a comparison of foreground images. These images are background-related data. Here we got binary image information from digital one. Labelled images are processed data from machine learning algorithms.

Figure 8 explains the colour model of our selected crop field. Here red, blue, green, pink and yellow colours represent different types of message indications.

Figure 9 shows the message window which is obtained from run time.

Figure 10 shows the distorted image vs the original image. Distorted image is less efficient than the original image. Here the original image has all the information.

Figures 11 and 12 explain the field selection from the selected graph and the obtained message is shown in the message window.

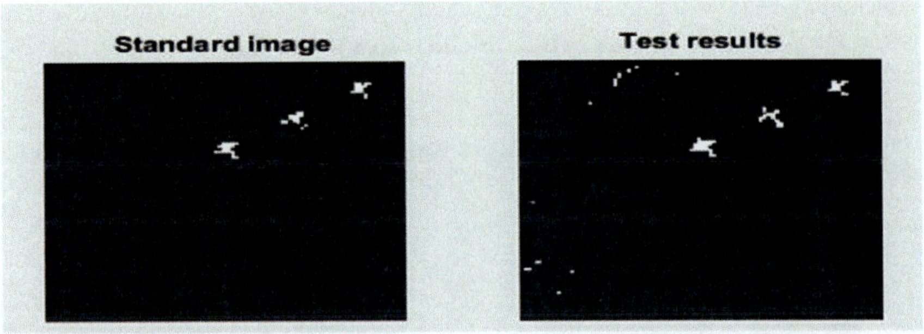

Fig. 6 Satellite image

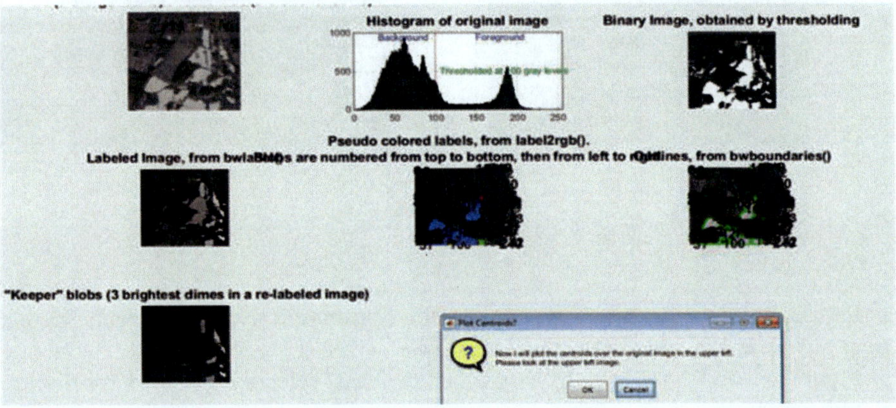

Fig. 7 Histogram comparison with satellite image

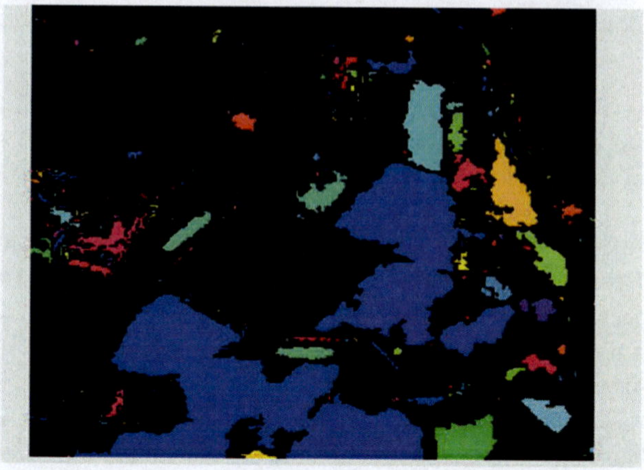

Fig. 8 Colour model of crop

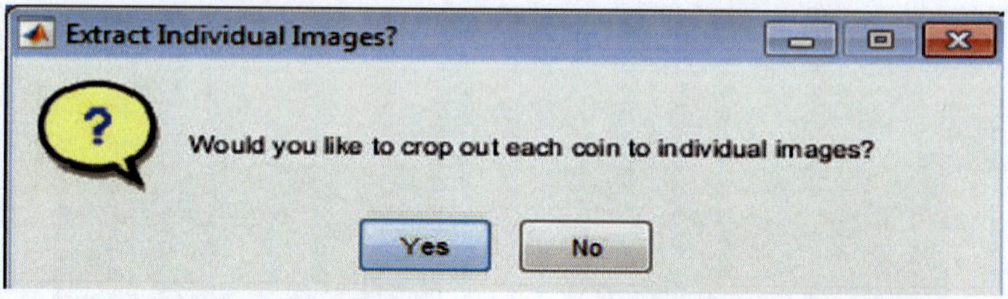

Fig. 9 Message window

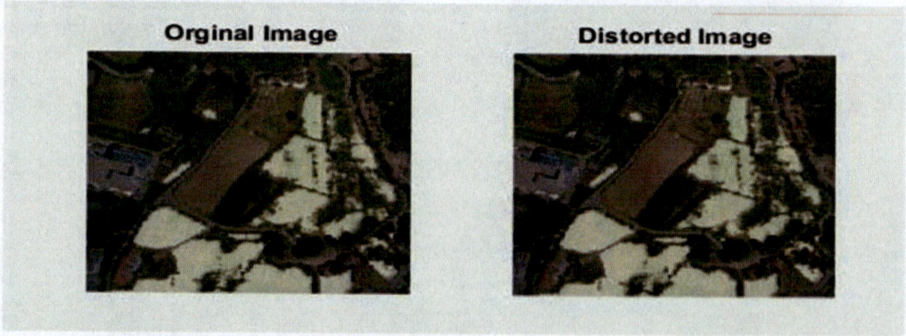

Fig. 10 Comparison of distorted and original image

Figures 13 and 14 exhibits the comparison of filtration from coefficients. Here a clear picture is available in the third window.

Figures 15 and 16 shows the segmented data and selection of slices for disease classification.

Fig. 11 Selected region

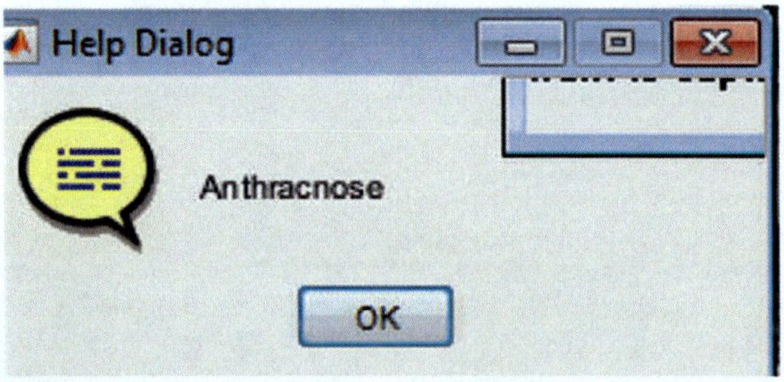

Fig. 12 Message window

Figure 17 exhibits the contrast adjustment for a better quality image, such that we get good results at the final stage. Such that we get good results at final stage.

Figures 18 and 19 are window boxes for information sharing. We click the buttons to get the results simultaneously.

5.1 *Work Flow*

Figure 20 shows the original and tracking data. Here recovery image analysis is also done.

A satellite picture with MPDL production phase is centred on DL that allowed the computing of picture mosaics (Fig. 21). They were subsequently divided separately into sheets of 256 × 256 pixels with a step of 64 pixels used to coach a DL. The network has been introduced using the backend of MPDL flow, specified with DL in the MATLAB tool cabinet. All tests were conducted on Laptop i3, 8 GB RAM and a GPU NVIDIA GTX-1080Ti (11 GB storage) used to extend the DL. The MPDL was taught from scratch with instruction collection (Fig. 22 shown above),

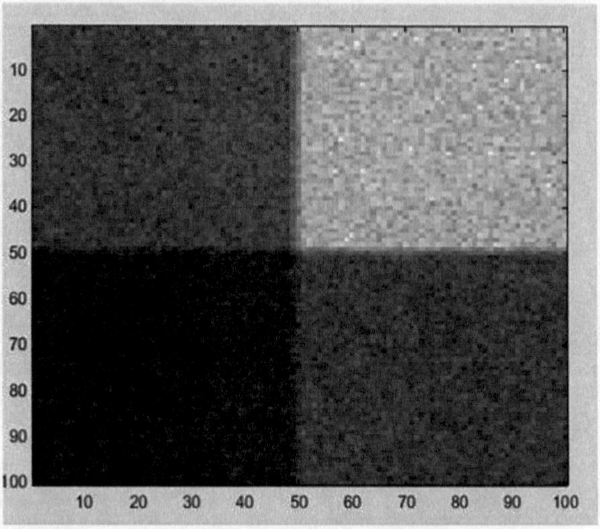

Fig. 13 Filtered image

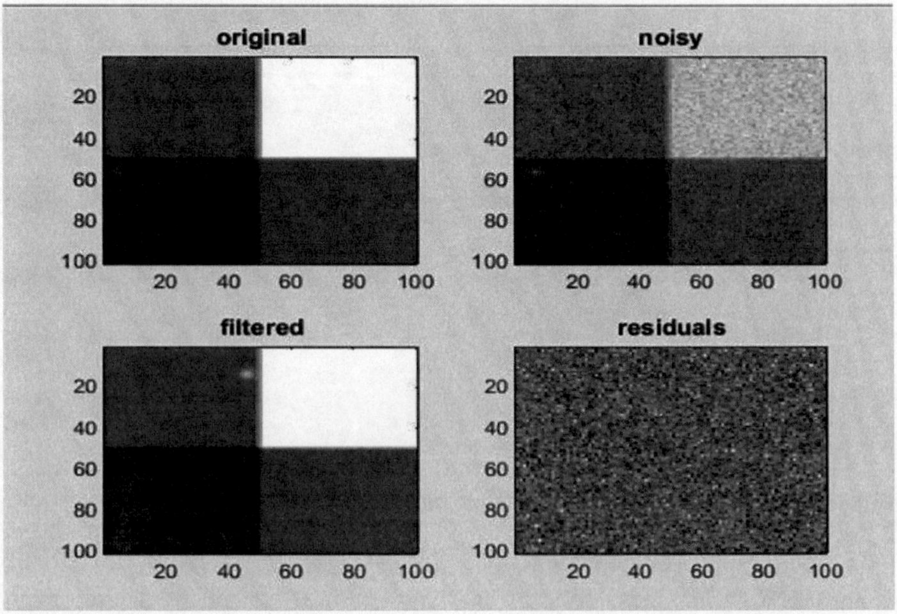

Fig. 14 Final filtered image

Fig. 15 Segmented data

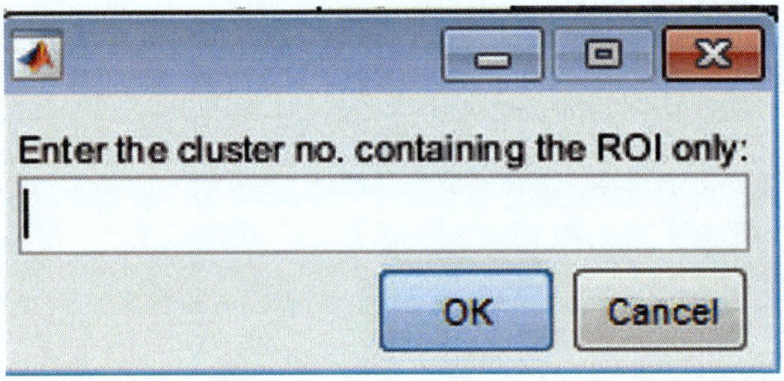

Fig. 16 Slice selection

Fig. 17 Contrasted image

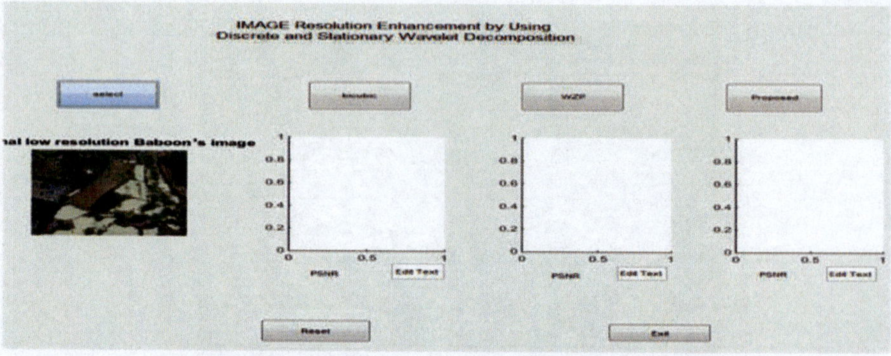

Fig. 18 Image selection

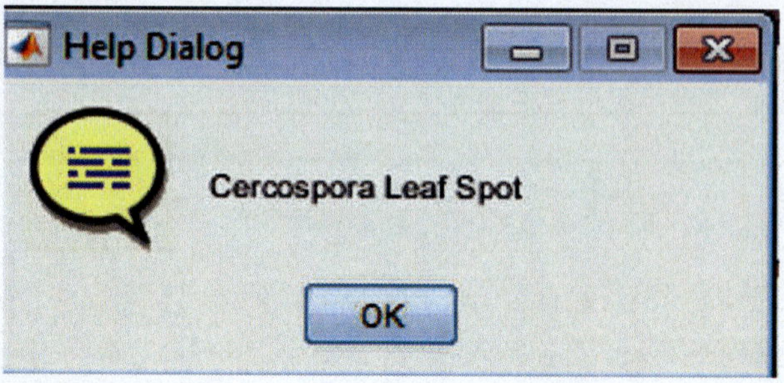

Fig. 19 Message windows

which was divided evenly for practice in 92% (14,109 bugs) and screening in 18% (3002 bugs). The learning was conducted using the cost function of Dice coefficient, with the Adam optimizer in 4 runs of 138 epochs. The teaching level in each phase was 1E-6, 1E-7 and 1E-8. Validation using Dice coefficient reached 0.9987 and is displayed graphically (shown in the workflow chart Fig. 2) (Table 1).

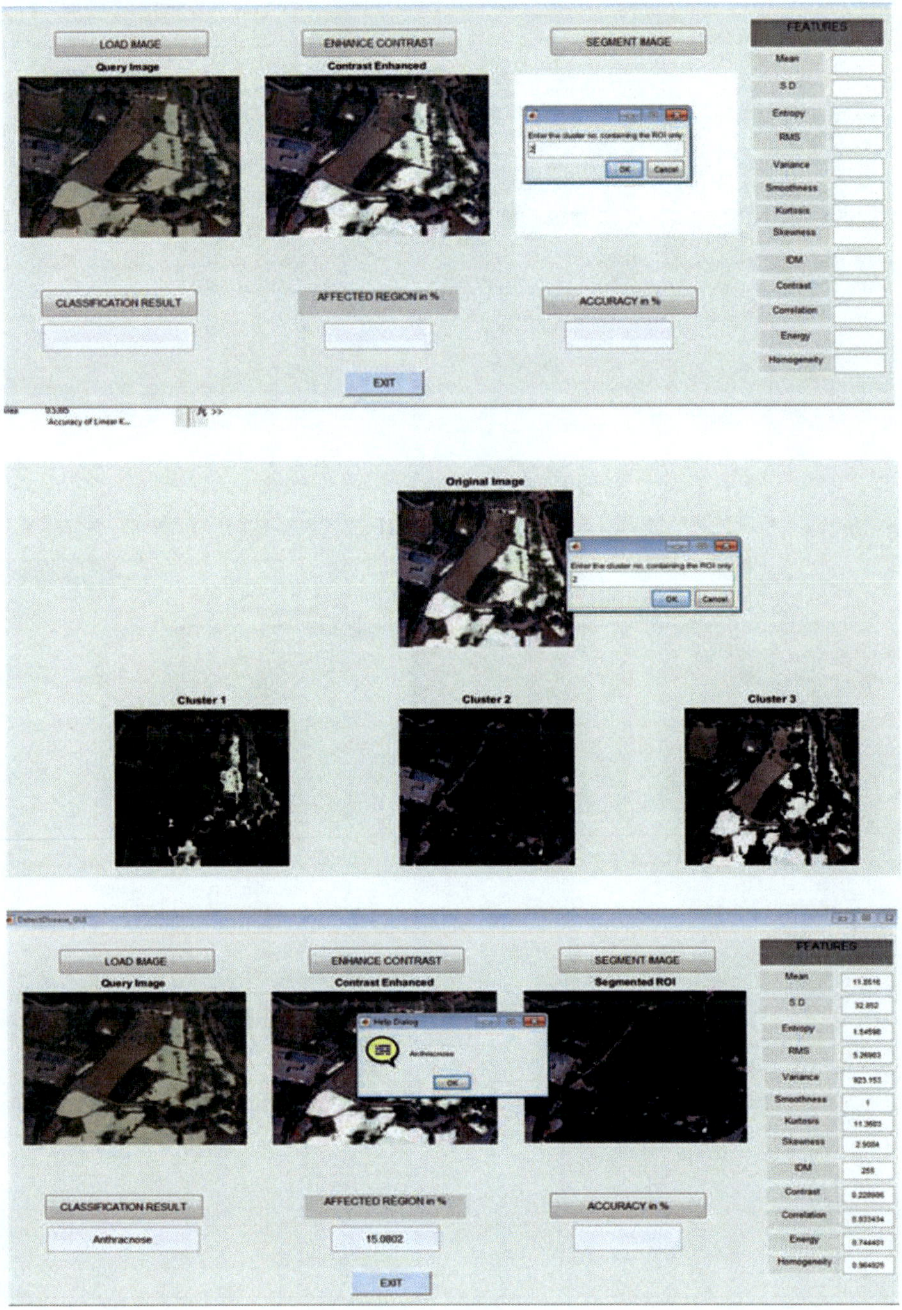

Fig. 20 Comparison

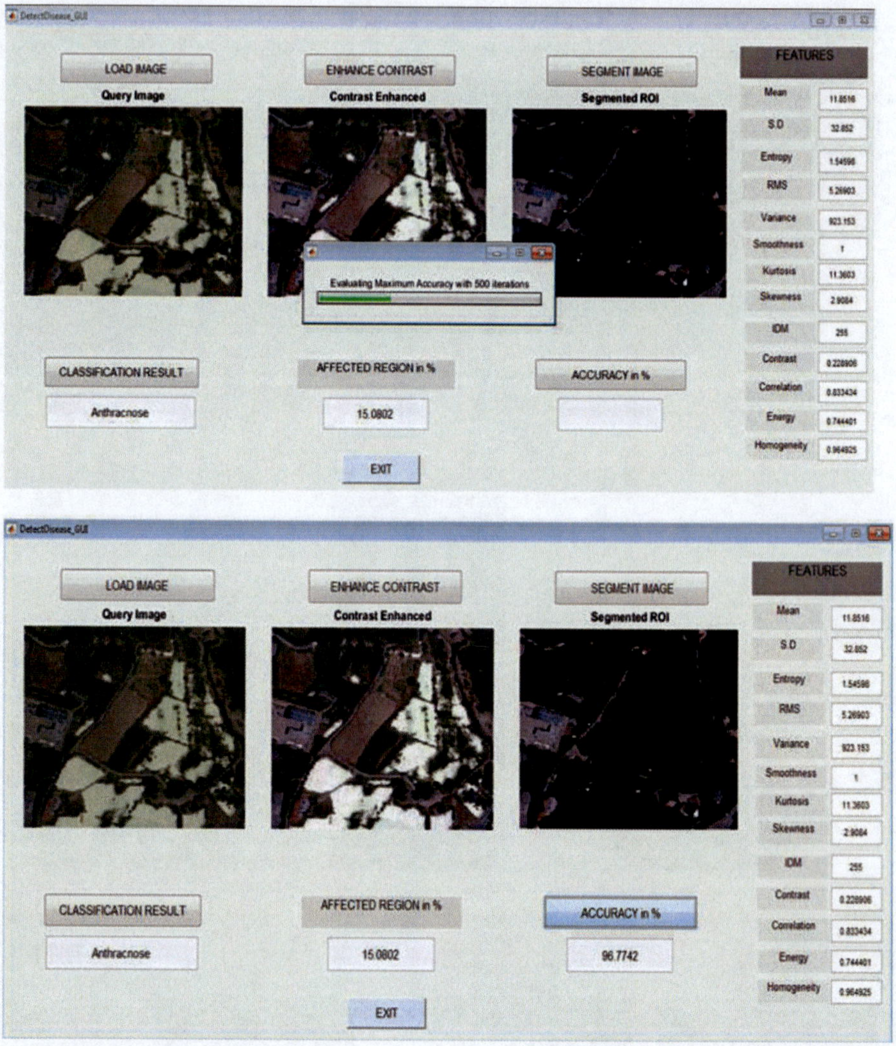

Fig. 20 (continued)

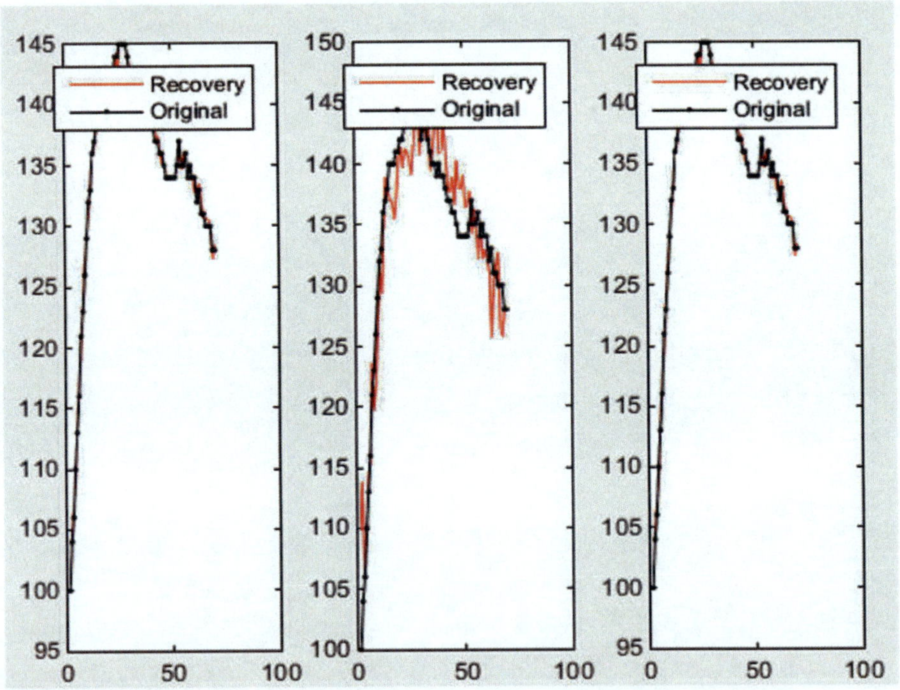

Fig. 20 (continued)

Fig. 21 Deep-scanned image from database

6 Conclusion

Together with DL, multispectral remote sensing technology can give some advantages to the farming sector. It is a non-invasive, less laborious, precise technique that can decrease expenses and activities on the ground. Not many research have been discovered, particularly in agriculture, that applied ML in remote sensing apps using

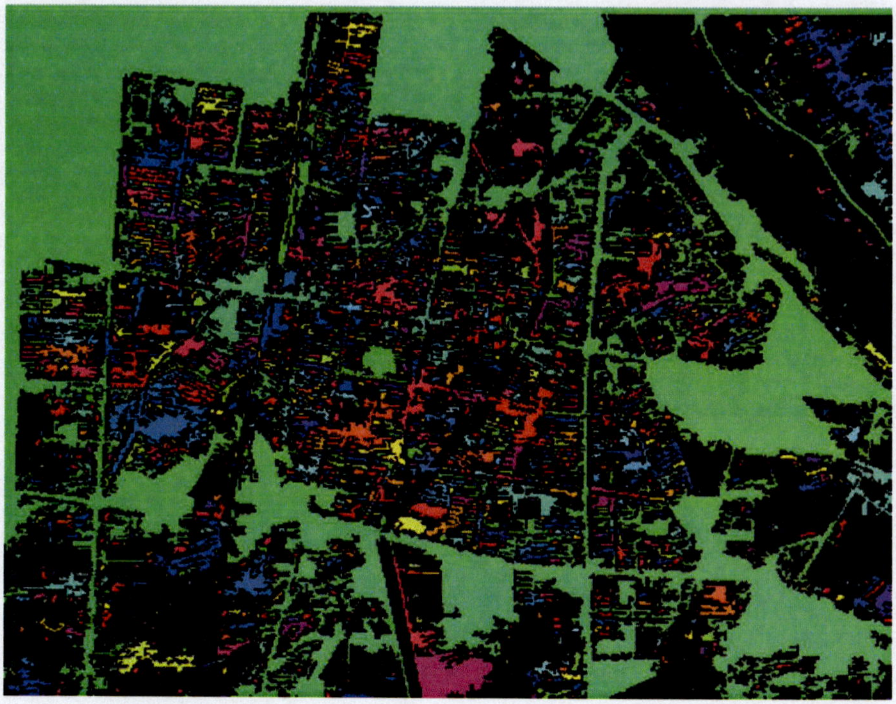

Fig. 22 High-resolution images with training methods

Table 1 Parameter analysis table

S. no.	Parameter	Conventional methods	Hyperspectral data processing	Multispectral MPDL	% Achievement
1	Randomness	78%	80%	92%	12% increases
2	Training	22%	20%	18%	2% increases
3	Cycles	3	3	4	1% increases
4	Epochs	129	132	138	6% increases
5	Accuracy	72%	83%	92.86%	9.86% increases
5	Efficiency	82%	88%	90.86%	2.86% increases

multispectral information. As for the outcomes obtained, the excellent identification precision and knowledge gained from these preliminary studies motivated the implementation of this strategy to multispectral information handling in the close future, motivated by the powerful belief that DL-based selection has great capacity for high-dimensional information, which both literature and our tests prove to be achieved. Finally, there is an increase of 9.86% and 2.86% in accuracy and efficiency, respectively, shown in Figure. Figs. 23 and 24, obtained precision and effectiveness.

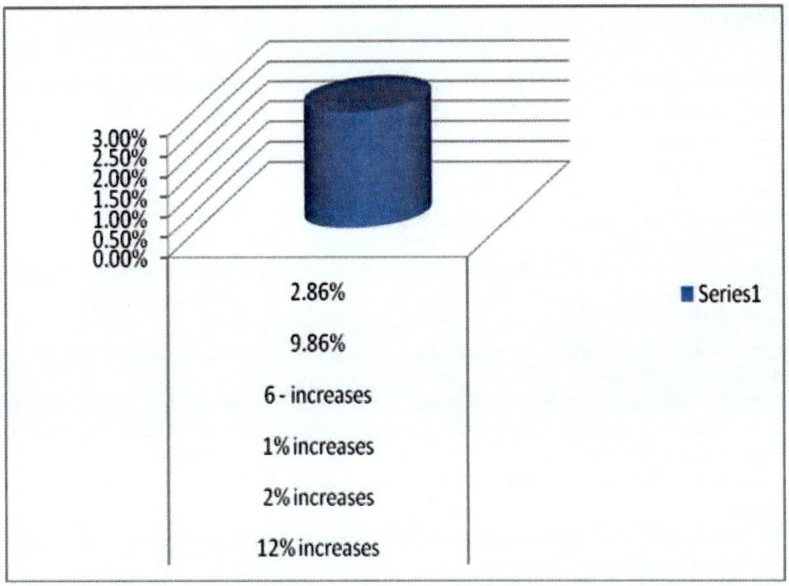

Fig. 23 Comparisons of all parameters

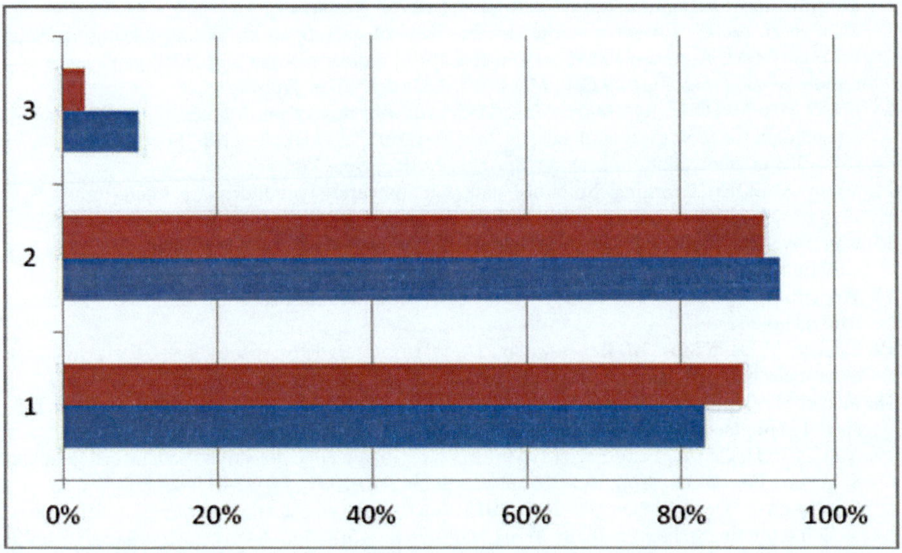

Fig. 24 Comparisons of accuracy and efficiency

References

1. Benediktsson JA, Ghamisi P (2015) Spectral-spatial classification of hyperspectral remote sensing images. Artech House, Norwood, MA
2. Ghamisi P, Benediktsson JA (2015) Feature selection based on hybridization of genetic algorithm and particle swarm optimization. IEEE Geosci Remote Sens Lett 12(2):309–313
3. Ghamisi P, Ali AR, Couceiro M, Benediktsson J (2015) A novel evolutionary swarm fuzzy clustering approach for hyperspectral imagery. IEEE J Sel Top Appl Earth Observ Remote Sens 8(6):2447–2456
4. Jain AK, Duin RP, Mao J (2000) Statistical pattern recognition: a review. IEEE Trans Pattern Anal Mach Intell 22(1):4–37
5. Scott DW (1992) Multivariate density estimation. Wiley, New York
6. Wegman EJ (1990) Hyperdimensional data analysis using parallel coordinates. J Am Stat Assoc 85(411):664–675
7. Jimenez L, Landgrebe D (1998) Supervised classification in high dimensional space: geometrical, statistical, and asymptotical properties of multivariate data. IEEE Trans Syst Man Cybern A Syst Humans 28(1):39–54
8. Fukunaga K (1990) Introduction to statistical pattern recognition, 2nd edn. Academic, San Diego, CA
9. Landgrebe DA (2003) Signal theory methods in multispectral remote sensing. Wiley, Hoboken, NJ
10. Qian Y, Yao F, Jia S (2009) Band selection for hyperspectral imagery using affinity propagation. IET Comput Vis 3(4):213. https://doi.org/10.1049/iet-cvi.2009.0034
11. Canters F (1997) Evaluating the uncertainty of area estimates derived from fuzzy landcover classification. Photogrammetric Eng Remote Sens 63:403–414
12. Dungan JL (2002) Toward a comprehensive view of uncertainty in remote sensing analysis. In: Foody GM, Atkinson PM (eds) Uncertainty in remote sensing and IEEE geoscience and remote sensing magazine 29 GIS, Mar 2017, 2nd edn. Wiley, Hoboken, NJ
13. Friedl MA, McGwire KC, McIver DK (2001) An overview of uncertainty in optical remotely sensed data for ecological applications. In: Hunsaker CT, Goodchild MF, Friedl MA, Case TJ (eds) Spatial uncertainty in ecology. Springer-Verlag, New York
14. Wang X (2015) Learning from big data with uncertainty editorial. J Intell Fuzzy Syst 28(5):2329–2330
15. Lu D, Weng Q (2007) A survey of image classification methods and techniques for improving classification performance. Int J Remote Sens 28(5):823–870
16. Woodcock CE, Strahler AH (1987) The factor of scale in remote sensing. Remote Sens Environ 21(3):311–332
17. Ghamisi P, Dalla Mura M, Benediktsson JA (2015) A survey on spectral–spatial classification techniques based on attribute profiles. IEEE Trans Geosci Remote Sens 53(5):2335–2353
18. Fauvel M, Tarabalka Y, Benediktsson JA, Chanussot J, Tilton JC (2013) Advances in spectral-spatial classification of hyperspectral images. Proc IEEE 101(3):652–675
19. Xu C, Liu H, Cao W, Feng J (2012) Multispectral image edge detection via Clifford gradient. Sci China Inform Sci 55(2):260–269. https://doi.org/10.1007/s11432-011-4540-0
20. Su Z, Luo X, Deng Z, Liang Y, Ji Z (2013) Edge preserving texture suppression filter based on joint filtering schemes. IEEE Trans Multimedia 15(3):535–548. https://doi.org/10.1109/TMM.2012.2237025
21. Zhu Z, Jia S, He S, Sun Y, Ji Z, Shen L (2015) Three-dimensional Gabor feature extraction for hyperspectral imagery classification using a memetic framework. Inf Sci 298:274–287
22. Cushnie JL (1987) The interactive effect of spatial resolution and degree of internal variability within land-cover types on classification accuracies. Int J Remote Sens 8(1):15–29
23. Wang W, Zhang Y, Li Y, Zhang X (2006) The global fuzzy c-means clustering algorithm. In: Intelligent control and automation, Jun 2006, vol 1, pp 3604–3607

Lecture Video Summarization Using Subtitles

Ramamohan Kashyap Abhilash, Choudhary Anurag, Vaka Avinash, and D. Uma

1 Introduction

The amount of influence that videos have on one's daily lives is undeniable. Online video sharing sites have millions of monthly audience numbers. With videos continuing to gain popularity, this familiar and widespread platform appears only natural to extend into the educational setting, and this has already started to occur with the appearance of so many online e-Learning platforms. Certain colleges and universities also upload videos of lectures online so that many people have access to a better quality of teaching.

Students today use educational videos as a tool to learn anything and everything. Because of the availability of educational videos, abstract topics that once seemed difficult to teach and learn are now more accessible and understandable. Studies have shown that using short video clips can make knowledge intake and memory recall more efficient. The visual and auditory characteristics of a video appeal to a wide audience and allow the users to learn information in a way that is easier for them.

Using videos for teaching and learning will benefit students, teachers, and their respective colleges or universities as well. Institutions nowadays are facing the challenge of meeting the growing demand for quality teaching. The number of learning portals with video content is also growing fast. Online courses are being introduced newly to improve accessibility to the resources. Since the videos are going to be available on the internet, this enables them to be shared at all times

R. K. Abhilash · C. Anurag · V. Avinash · D. Uma (✉)
PES University, Bengaluru, Karnataka, India
e-mail: umaprabha@pes.edu

A. Haldorai et al. (eds.), *2nd EAI International Conference on Big Data Innovation for Sustainable Cognitive Computing*, EAI/Springer Innovations in Communication and Computing, https://doi.org/10.1007/978-3-030-47560-4_7

around the world by different types of users. By using videos to reach their wide audience, educational institutions can gain great autonomy.

Video summarization helps people to go through a long lecture without wasting much time and also makes sure no important information is lost. Our target audience will be students and teachers. It will massively help students especially during their exams, where the time to study is limited and the content to cover is too large. It can play an important role for teachers as well to keep themselves updated with the current technology/information without spending too much time.

Long lecture videos captured are typically linked to some of the most important information, but often the least frequently watched. The time it takes to retrieve and watch sections initially may be daunting. This problem can be tackled by summarizing the video without losing important information. Video summarization can be done by outlining the main points, including the issue, solution, or any other things discussed. It should not contain all the details presented in the video, however, but only the most important ones.

One of the ways to summarize the video is to extract the transcript of the video and use text summarization techniques. Text summarization can be widely divided into two categories—extractive summarization and abstractive summarization. Extractive summarization methods are based on extracting multiple parts from a piece of text, such as phrases and sentences, and stacking them together to create a summary. Identifying the right sentences for summarization in an extractive method is therefore of utmost importance. Abstractive Summarization methods generate an entirely new summary using advanced Natural Language Processing techniques. There may not even be parts of this summary in the original text.

Some of the latest techniques being used in the field of extractive text summarization include machine learning and deep learning. In these approaches, a machine is trained to identify and assign importance to different sentences based on its context. One of these deep learning methods is to use a Convolutional Neural Network to identify similarity between sentences and select only those sentences to the summary which are least similar to one another. Apart from the deep learning methods, fuzzy logic and rule based models are also being used in order to classify and summarize documents in virtual learning environments. In the fuzzy model, features are extracted from the textual data and used to evaluate sentences and documents. This fuzzy model is used to deal with the imprecise and uncertainty of feature weight.

The proposed method utilizes extractive method for summarization of video transcripts. While Sect. 2 provides information on the work done in this field and the inferences used from previous work, Sect. 3 gives a short description of the methodology discussed in this work. Section 4 shows the detailed methodology used with all the involved modules. It describes the data used, preprocessing done, and the data flow through the system. Finally the results are presented with a comparison to other existing methods and conclude the study. The paper also talks about the scope of the solution and future work that can be done.

2 Related Work

The main problem with the summarization task, either with text or video data is to have continuity in the summarized data that is presented at the end after summarizing. Reference [1] tries to solve this by using a support vector machine. The importance of each statement is identified using the Support Vector Machine and dealt by having a relationship among the extracted sentences using dynamic and differentiation features. So the decision of the present sentence is based on the previous sentence. It is a good approach to gain continuity in the summarized data. This approach obtained a ROGUE-4 score of 0.69 evaluated on 8 Japanese lectures.

Coming to video summarization techniques in particular, most of the literature uses the video content to summarize the videos. Reference [2] uses middle-level features like the extracted text and figures instead of low-level features like edges and colors of the video frames. It addresses many real-time video capturing problems like the camera movements, non-uniformity of the board, and occlusions. The K-th Hausdorff distance is used as the distance measure and connected component decomposition to extract the relevant frames. This method was evaluated on 4 instructional videos and it performed better than the Dynamic Clustering and Tolerance Based methods consistently. This demonstrated that the algorithm is highly effective in extracting and summarizing the visual content.

Along the same lines, [3] and [4] use video frames as a key component to summarize video data, and analyze the video statically rather than playing the video as a whole. Reference [3] uses moving object detection and tracking to produce a summary for video surveillance applications, as these applications generate large amounts of video data, most of which are deemed redundant. Various object detection and optical flow techniques are used to gain information about each of the frames of the video. This data was processed by summarization techniques and utilized to provide a video summary. The techniques used in this include context-aware video summarization (CAVS), which is a framework developed by Shu et al. [5]. This method is capable of capturing the important portions of video by data on specific regional motion regions as well as interactions between these motion regions. The theoretical model is capable of finding new events and various associations of events. Another technique used in this is geographic region of interest (GROI) summarization, which is yet another framework used to produce a summary automatically from multiple geo-referenced user-generated videos in [6]. These were evaluated using metrics such as precision, recall, average precision, and average redundancy rate.

The approach followed by Mundur et al. [4] is to use a clustering algorithm in order to extract a set of key frames which may not preserve temporal order of the video, but instead reduces overall redundancy of the video summary. This is in contrast to shot detection based key frame selection, used in [7] and [8], which yields a summary that maintains the temporal order, but at the cost of increased redundancy. Delaunay clustering is a fully automatic technique which requires no user input or parameters for its operation, therefore the result remains independent of external factors and only depends on the content of the video frames. The main goal is to represent individual frames as data points in generating a Delaunay Triangulation (DT) diagram. Using DT, the inter-frame similarity relationship is mapped to the spatial proximity relationship among data points. This mapping is then used to identify clusters using the mean length of edges between the points. Key frames are identified from cluster based approach on their distance to the center of the cluster. In order to evaluate the summary generated, metrics such as significance factor, which depends on the size of the cluster from which the key frame is chosen, compression factor, and overlap factor are used.

In [9], different image and text processing techniques are used to get a preview of the lecture materials. The process of previewing is customized with the option of time to the user. The summarization has been trained with the answers to various quizzes, tests, and examinations. It has been very successful in this process as the scores of 372 students who used it showed improvement in their scores after using the preview of the lectures before exams even without spending much time. One of the inputs—the preview time of each slide—was taken from the professors as well and used it while summarizing at the Kyushu University.

A deep learning method using feed-forward neural networks for single document summarization was suggested in [10]. It is a fully data-driven approach trained and evaluated on the standard DUC 2002 data-set. The results are comparable to other state-of-the-art methodologies. The proposed model is scalable and is able to produce the summary of arbitrarily sized documents by breaking the original document into fixed sized parts and then feeding it recursively to the network. It has managed to get a ROGUE-1 score of 0.551 for the data-set.

The notable feature used in [11] is that they have considered the assumptions a human would make while summarizing any data. It uses a genetic algorithm for summarizing a single piece of text automatically. A special emphasis is given to the fitness function over the rest of the steps. Multiple steps like chromosome encoding, parent selection, crossovers, and mutations are used in the process. The genetic algorithm used will allow the user to select the length of the summarized text he/she desires. This work was evaluated over 567 documents of the DUC collection and achieved a ROGUE F-Measure of 0.48. The evaluation is done with the use of the original text and not a reference summary.

The authors of paper [12] have suggested a new experimental technique for lecture video summarization. This technique constructs a graph out of the spoken lecture using the key terms and phrases present in it, using the Probabilistic LSA algorithm. Each sentence of the lecture is assumed to be a node and edges between nodes are assigned a weight based on the topical similarity of nodes. Using this, the main aim which they are trying to achieve is to consider sentences globally, rather than individually. Key terms are defined as the single terms used in the documents carrying core concepts of the content, and key phrases are multiple terms which are used together. They have used lectures for a course offered in National Taiwan University for evaluating and the best results they have obtained is an F-measure of 0.57, with ROGUE-3 as the evaluation standard.

The work that is closest to what has been done is in [13]. In this, the subtitles are used to summarize the lecture videos automatically. A textual summary of the class lecture is produced using the Term Frequency-Inverse Document Frequency norm. The text data is then reduced by 44–45% of the original source and evaluated the results using the lecture videos available on online platforms like NPTEL, Coursera, etc. The present work described in this paper extends this work to achieve better accuracy. The belief is that more value can be added to the idea. This also gets the summarizing process costs to a record low as it does not need to deal with the video content anymore while summarizing.

3 Proposed Approach

As seen in the related work section, most of the models proposed for video summarization concentrate on extracting the content of the frames from the video, and then these frames would be combined together to shorten the video. In lecture videos, to reduce the redundancy contents from the chalkboard or the slide-set being presented by the speaker are processed and important parts are extracted. But this will not be of as much use because the slide sets can be shared with the students anyways. As the attention span of students is less, it is important to capture what the lecturer is speaking and present in a shorter format. So, unlike most of the video summarization work that has been done before, this paper uses the subtitles from the video to do text summarization first and then get the summary of the video. The details of each phase are mentioned in the next section. Adding on to [12], punctuations are used to tokenize the subtitles into sentences. However, subtitles for lecture videos are not readily available, and more often than not textual subtitles have to be extracted from the audio of lecturer. Another point to be noted is that audio-to-text techniques cannot accurately provide punctuations for the text. Hence, the result is a stream of textual data which becomes the generated subtitles for the video. In order to overcome this problem, fixed length for all the sentences was assumed, and tokenized the subtitles accordingly. In the result section, it can be observed that the summarized video obtained from the aforementioned method has a much lesser ROGUE score than the summarized video which is formed by using subtitles with punctuations. Hence it can be proved that summarization can be very effective when punctuations are part of the subtitles. This has given better accuracy than the previously existing models and approaches.

4 Methodology

Figure 1 depicts the different modules of the system and how they interact with each other. It also shows the flow of data from the time the user gives the input to the time summarized video is given as output. The working of the system is divided into four

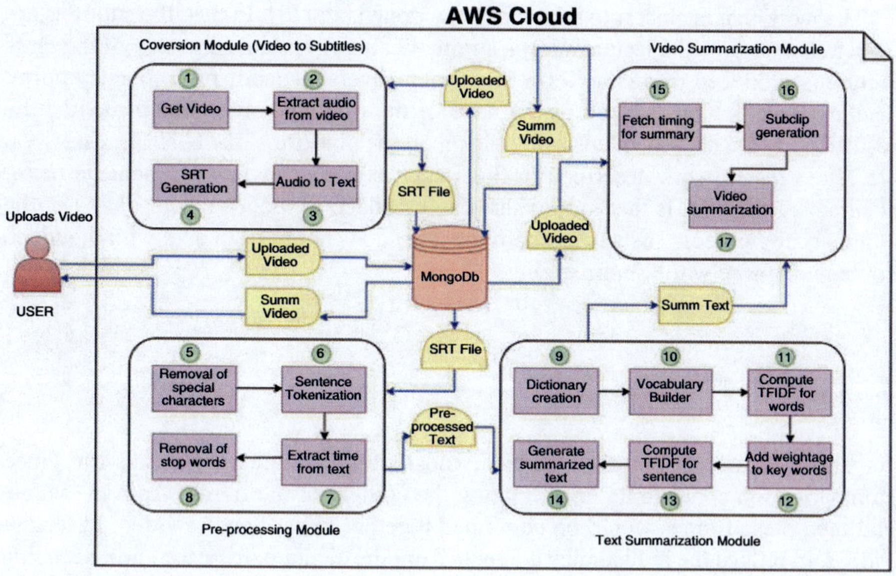

Fig. 1 System architecture

modules. They are conversion module, preprocessing module, text summarization module, and video summarization module.

The conversion module takes in the video as input from the user and gives subtitles as the output. The preprocessing module removes the special characters of the subtitle ripped file and then sentences are tokenized. After that, time is extracted from these ripped files. Finally the stop words and white spaces are removed. The text summarization module then builds the vocabulary set and computes the TFIDF (Term Frequency-Inverse Document Frequency) scores for all the words. Weightage is added to keywords and then TFIDF scores for sentences are computed. Based on these scores, a summary text is generated. The last module is the video summarization module which fetches timing for the summary, generates the subclips, and forms the final video.

Conversion module is mainly used for the generation of the subtitle file. The video will be taken from the user and the audio is extracted from the video. This audio is processed and the speech is converted into text. For the above process, Google's speech to text conversion module is used for the same, but this does not produce punctuations in the subtitles. So, while converting, the time gaps between two utterances from the speaker are noted down to punctuate the sentences. These punctuations show to be effective in the process of summarization. This module gives a subtitle file as the output.

The preprocessing module uses the subtitle file given as an output from the conversion module and processes the text so as to prepare it for text summarization. The preprocessing includes converting all the words into lower case, removal of special characters, white spaces, and stop words. Stemming and lemmatization of words will be done whenever required. These subtitles are then tokenized to form sentences. This module also extracts the time stamps from the subtitles.

The text summarization module is one of the most important modules of the system. It uses the output of the text preprocessing module as its input. A dictionary of words and timings is created for future use. The vocabulary set is generated and vectors from the subtitles are formed. TFIDF algorithm is used here to summarize the text data. This algorithm has two main terms to it: term frequency (TF) and inverse document frequency (IDF). The term frequency is the number of times a word occurs in the present document. Inverse document frequency is the inverse of the number of documents that a term occurs in. Weightage is given to terms using the TF factor and IDF factor. The keywords are identified and their scores are doubled here so as to ensure that the relevant terms get higher scores. If the IDF factor is not considered, then the stop words would show up to be as the most important words in the document. The subtitle lines which contained the most weighted keywords are then included in the textual summary, and the timestamps of these lines are saved for further use.

This method is used by most of the search engines to find the relevant results for the search term given by the users. First, the TFIDF scores are calculated for each of the terms in the document. To get the score for each sentence, the scores of all the words are summed up and normalized by the length of the sentence. This normalization ensures that there is no unfair advantage to the longer sentences. These are the sentences that make up the textual summary for the original video. It generates the sequences of text data with their corresponding scores. The sentences that have scores higher than the average are retained and the rest are discarded. These sentences that are remaining will be mapped to the timings when they occur in the original video and ordered according to the time of the appearance in the video.

The video summarization module uses the results of the text summarization module as an input. It also uses the timings extracted in the text preprocessing module so as to choose only those timestamps for the sentences which are present in the summary. Using the list of timestamps hence formed, small subclips of the original video are created. These subclips are then concatenated together to form the summarized video. For the above process, Moviepy library was used for manipulating and forming the final summarized video. This will be sent back to the user for viewing.

5 Evaluation and Results

For evaluating video summarization, duration of the summarized and original videos is compared. The difference can be observed between using the punctuations and not using them. The results for 4 such NPTEL lectures and the duration of the original and summarized videos with and without punctuations can be seen in Table 1.

Table 1 Durations of original and summarized videos with and without punctuations

Video name	Duration of original video	Duration of summarized video with punctuations	Duration of summarized video without punctuations
Introduction to OS	19:09	8:08	5:41
Introduction to ML	15:27	8:30	4:09
Relational databases	24:51	12:13	7:08
Information security	15:42	7:21	4:19

Table 2 Results for videos whose subtitles have punctuations

	ROGUE-1			ROGUE-2		
Video name	*F*-measure	Precision	Recall	*F*-measure	Precision	Recall
Introduction to OS	0.82	0.74	0.91	0.65	0.53	0.84
Introduction to ML	0.78	0.86	0.72	0.67	0.74	0.62
Relational databases	0.86	0.85	0.87	0.73	0.68	0.78
Information security	0.76	0.84	0.71	0.65	0.72	0.61
Average	0.805	0.822	0.802	0.675	0.667	0.712

Table 3 Results for videos whose subtitles do not have punctuations

	ROGUE-1			ROGUE-2		
Video name	*F*-measure	Precision	Recall	*F*-measure	Precision	Recall
Introduction to OS	0.48	0.46	0.5	0.23	0.18	0.34
Introduction to ML	0.36	0.38	0.35	0.12	0.094	0.18
Relational databases	0.51	0.48	0.53	0.26	0.21	0.38
Information security	0.35	0.36	0.34	0.11	0.089	0.15
Average	0.425	0.42	0.43	0.18	0.143	0.262

The summaries generated were also evaluated using the Recall-Oriented Understudy for Gisting Evaluation (ROUGE) toolkit with the reference being manually generated summaries for the videos. ROUGE compares an automatically produced summary against a reference (generally human-produced). The F-measure, precision, and recall calculated using the ROGUE toolkit are recorded in Tables 2 and 3. Table 2 contains the results for videos whose subtitles have punctuations and Table 3 contains results for videos whose subtitles do not have punctuations.

From the results presented, a clear improvement can be seen in the scores while using punctuations.

6 Conclusion

The proposed solution for video summarization proves to be efficient while using punctuations from the subtitles. The system summarizes the videos based on the subtitles generated from the videos. The subtitles are then preprocessed and converted to textual data. The data is then cut into sentences using the punctuations. These sentences are concatenated to form textual data and that is summarized using the TFIDF algorithm. The keywords are identified and given more importance. The important sentences have been identified and they are mapped to the timestamps from the original video. The original video is clipped using only the timestamp generated to produce the summarized video. The results have been evaluated using ROGUE toolkit to get the average precision score of 0.822, recall score of 0.802, and F-measure score of 0.805.

7 Future Enhancements

The generation of subtitles using Google speech to text converter does not include punctuations in the text. The video summarization will be more meaningful for the users if the sentences can be broken wherever punctuations are used. This can be done by looking at the length of pauses and classifying these into different punctuation marks like full stops and commas. It can be achieved by a binary classification of these pauses and identify the points where the video can be broken. This will give a smooth transition of video frames and the summarized video will be more meaningful.

Recurrent neural networks can be used to have a sequence to sequence encoder-decoder connections which can learn from large data-sets of auto summarization and predict the summary. The summarized video will be cut based on the summarized text. If the summarized text is enough for the user, this can be complemented with a text summary that can be done using abstractive summarization. Capsule network is one of the latest technologies that are being used to summarize text and can be used too.

References

1. Fujii Y, Yamamoto K, Kitaoka N, Nakagawa S (2008) Class lecture summarization taking into account consecutiveness of important sentences. In: Inter-National Speech Communication Association
2. Choudary C, Liu T (2013) Summarization of visual content in instructional videos. IEEE Trans MultiMedia 9:1443–1455
3. Senthil Murugan A, Suganya Devi K, Sivaranjani1 A, Srinivasan P (2018) A study on various methods used for video summarization and moving object detection for video surveillance applications. Multimed Tools Appl 77(18):73–90
4. Mundur P, Rao Y, Yesha Y (2006) Keyframe-based video summarization using Delaunay clustering. Int J Digit Libr 6:219
5. Shu Z, Yingying Z, Roy-Chowdhury AK (2016) Context-aware surveillance video summarization. IEEE Trans Image Process 25(11):5469–5478
6. Ying Z, Roger Z (2015) Efficient summarization from multiple georeferenced user-generated videos. IEEE Trans Multimedia 2:1–30

7. Cernekova Z, Pitas I, Nikou C (2005) Information theory-based shot cut/fade detection and video summarization. IEEE Trans Circuits Syst Video Technol 16(1):82–91
8. Dirfau F (2000) Key frame selection to represent a video. In: International conference on image processing, vol 2. IEEE, Piscataway, pp 275–278
9. Shimada A, Okubo F, Yin C, Ogata H (2018) Automatic summarization of lecture slides for enhanced student preview. IEEE Trans Learn Technol 11:165–178
10. Aakash S, Yadav A, Gahlot A (2018) Extractive text summarization using neural networks. Preprint. arXiv:1802.10137
11. García-Hernández RA, Ledeneva Y (2013) Single extractive text summarization based on a genetic algorithm. In: MCPR. Springer, Berlin
12. Chen Y-N, Huang Y, Yeh C-F, Lee L-S (2011) Spoken lecture summarization by random walk over a graph constructed with automatically extracted key terms In: INTERSPEECH 2011, 12th annual conference of the International Speech Communication Association, Florence, 27–31 August 2011
13. Garg S (2017) Automatic text summarization of video lectures using subtitles. In: Developments in intelligent computing, communication and devices. Springer, Singapore, pp 45–52

Retinal Image Enhancement by Intensity Index Based Histogram Equalization for Diabetic Retinopathy Screening

Arun Pradeep, X. Felix Joseph, and K. A. Sreeja

1 Introduction

Retinal exudates that can be visually identified as yellow flecks in fundus images and is considered one of the symptoms arised due to Diabetic Retinopathy. These are mainly due to leakage of lipids in the eyes from the damaged capillaries as shown in Fig. 1. Diagnosis done at an earlier stage can control the degree of impairment caused by leakage of lipids that can ultimately lead to loss of eyesight. Patient friendly studies are centered on the accuracy of exudate detection from RGB fundus images with the help of machine learning.

These images are captured using a fundus camera which may contain effects of noise and uneven illumination and contrast. In order to filter out these undesired effects, literature suggests that pre-processing and image enhancement should be more focused before image segmentation and classification. The study presented by [1] identifies retinal exudates established on spider monkey optimization using an SMO-GBM classifier. Likewise, the image enhancement was done using contourlet transform. The method proposed in [2] uses classification based on Top-k loss method instead of Class Balance Entropy (CBCE) to reduce misclassification in exudate detection.

A. Pradeep (✉)
Noorul Islam University, Thucklay, Kanyakumari, India

X. F. Joseph
Bule Hora University, Hagere Maryam, Ethiopia

K. A. Sreeja
SCMS School of Engineering and Technology, Ernakulam, India

A. Haldorai et al. (eds.), *2nd EAI International Conference on Big Data Innovation for Sustainable Cognitive Computing*, EAI/Springer Innovations in Communication and Computing, https://doi.org/10.1007/978-3-030-47560-4_8

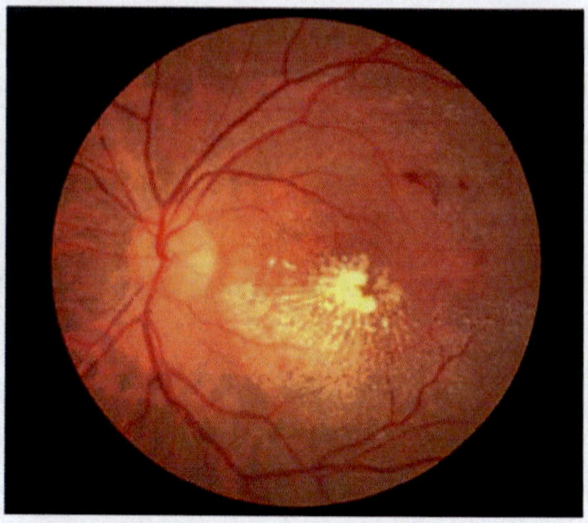

Fig. 1 Fundus image of diabetic eye

The analysis in [3] proposes that Convolutional Neural Network (CNN) can be utilized for a deep learning technique, for exudate detection, but the efficiency is deprived when compared with Residual Network and Discriminative Restricted Boltzmann machines. The color space used in our work is HSI instead of RGB, which gives more attenuation to noise. This method is reiterated by the work suggested by Khojasteh et al. [4] for exudate detection. Holistic texture features of fundus images were extracted and trained to four different classifiers in the study [5] conducted on a public database. Classification of hard exudates from soft exudate using fuzzy logic was the area of interest in [6]. Segmentation of exudates using dynamic decision thresholding was the focus of study in [7]. Their results were validated using lesion and image based evaluation criteria. Circular Hough Transform(CHT) and CNN based detection of exudates were suggested in [8]. A reduced pre-processing strategy for exudate based macular edema recognition using deep residual network was put forward in [9]. Multilayer perceptron based supervised learning is studied in [10] to identify exudate pixels. Further segmentation was done using unsupervised learning with the help of iterative Graph Cut (GC). The entire image is segmented into a series of super pixels in [11] which are considered as candidate pixels.

Also each candidate is characterized by multi-channel intensity features and contextual features. The study in [12] using a neighborhood estimator presents detection of blood vessels followed by segmentation with the help of in-painting the exudates with the help of this estimator. A new approach called voxel classification by a strategy based on layer dependent stratified sampling on OCT image was introduced in [13]. Grayscale morphology based segmentation of exudates was presented in [14] where the candidate pixels' shape was determined with the help of Markovian Segmentation model. Another method using Partial Least Squares (PLS) for detection of exudates is studied in [15]. An image segmentation based high level entity known as splat is used to identify retinal hemorrhages in [16] where pixels sharing similar properties are grouped together to form non-overlapping splats and the features are extracted and classified using supervised learning. The research

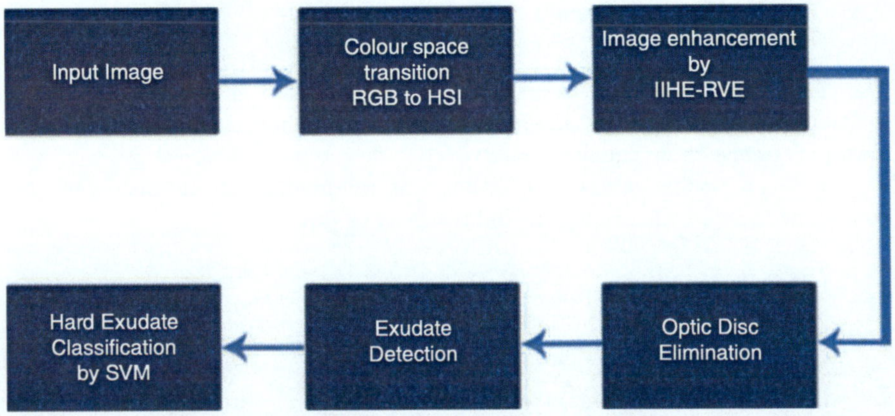

Fig. 2 Depiction of total work flow

study presented in this paper is a modification of our existing algorithm presented in [17]. The method associates both the principles of mathematical morphology operation for detection of exudates and classification and extraction of exudates using a trained classifier. Before the mathematical binary operation, initial pre-processing is done to enhance the fundus image where an algorithm called Intensity Index based Histogram Equalization Technique for retinal vessel enhancement (IIHE-RVE) is proposed. The algorithm of the total work is depicted in Fig. 2.

2 Methodology

Color plane transition from RGB to HSI is performed because the Optic Disc (OD) as well as exudates have analogous brightness characteristics. Many of the imperfections caused by noise and texture in the image can also be reduced by transition to HSI plane [18]. Median filter is applied to reduce the noise in the intensity band of the image. A novel method called Intensity Index based Histogram Equalization Technique for retinal vessel enhancement (IIHE-RVE) is applied to enhance the contrast of the noise free image.

IIHE-RVE is based on the estimation of under radiance of the image which is more effective than the existing Contrast limited adaptive Histogram equalization (CLAHE) algorithm or any other Gaussian function equalization algorithms. The following step is involved in the removal of Optic Disc (OD). It is assumed that the OD exist as the largest bright circular shape component in the image. Finally, exudates are classified into hard and soft exudates using a supervised classifier. Clinical images as well as images from publically available database are validated for the proposed algorithm.

2.1 Image Enhancement

The pre-processing steps involved in this work are shown in Fig. 3. RGB to HSI transition is followed by median filtering and contrast enhancement is done the new technique of histogram equalization.

Applying a tunable parameter ξ, histograms are divided into sub-histograms by computing the split value using the following set of Eqs. 1 and 2

$$\alpha_c(i) = \frac{\phi(i)}{\epsilon} \quad \text{for} \quad 0 \le i \le I - 1 \tag{1}$$

$$\Gamma(k) = \sum_{i=0}^{k} \alpha_c(i) \quad \text{for} \quad 0 \le k \le I - 1 \tag{2}$$

where ϕ denotes the histogram of the image, i represents the intensity value, ϵ represents the pixel numbers for the whole image, and I signifies the total brightness levels in numbers. The parameters Γ and α_c give accumulated normalized histogram count and normalized histogram count, respectively, for the given image. The controlling parameter Γ_p is found by Eq. 3.

$$\sum_{j=0}^{\Gamma_p} \Gamma(j) \approx \xi \quad \text{for any } 0.1 \le \xi \le 0.9 \tag{3}$$

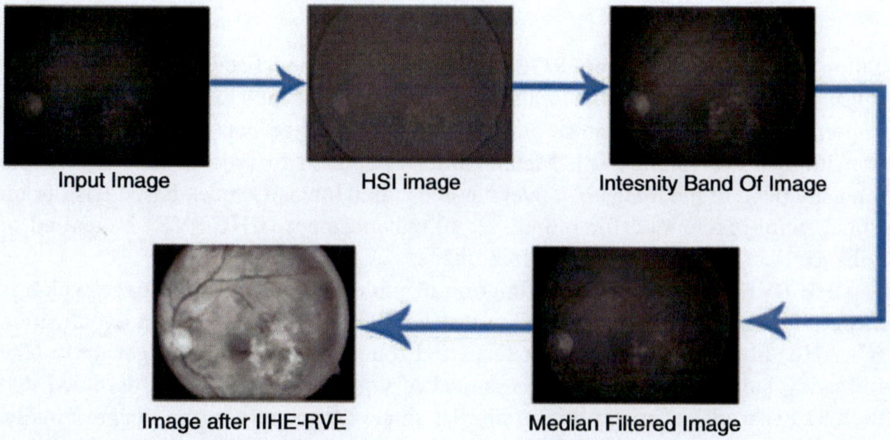

Fig. 3 Steps involved in pre-processing

The split value S_v is found from Eq. 4.

$$S_v = (I - 1) - \Gamma_p - 1 \tag{4}$$

The value of tunable parameter ξ is inversely proportional to enhancement level of the image. Also when ξ increases the value of Γ_p also increases. For a certain low value of ξ, we can acquire a first sub-histogram and for another high value of ξ we can acquire a second sub-histogram. The first and second sub-histograms are equalized specifically. Due to the extendedness of these histograms, the range of pixels having lower intensity can be mapped to a range of higher intensity. Whereas, in the second sub-histogram, the range is less and contains only larger intensity range pixels. Because of this small range, the larger intensity pixels are saved from over enhancement.

2.2 *Intensity Index Based Histogram Equalization Technique for Retinal Vessel Enhancement (IIHE-RVE)*

According to the algorithm, when the two histograms are obtained, successive integration based on difference of intensity parameters obtained from the iteratively enhanced images is performed. Integration is continued till the absolute difference between the intensity values ω_1 and ω_2 obtained from Eq. 5, for the given image and the equalized image is lower than error referred to as threshold, e. Here, the value of e is taken as 0.002.

Algorithm 1 IIHE algorithm

1: Compute histogram ϕ for image f.
2: Compute intensity values for input image from Eq. 5 for $I = 256$.

$$\omega_1 = \frac{\sum_{i=0}^{I-1} \phi(i) \cdot i}{I \cdot \sum_{i=0}^{I-1} \phi(i)} \tag{5}$$

3: S_v which is the split value is calculated from Eq. 4.
4: Separate the histogram into sub-histograms ϕ_l from radiance range 0 to S_v and ϕ_u from S_{v+1} to $I - 1$.
5: Equalize histograms ϕ_l and ϕ_u in the respective intensity range.
6: Reiterate Step 2 to find the intensity values ω_2 of equalized image.
7: Repeat steps 1–6 until $|\omega_1 - \omega_2| \leq e$.
8: Integrate ϕ_l and ϕ_u to re-establish histogram ϕ.

2.3 *Optic Disc Elimination*

The exudates have similar intensity values as that of optic disc. Opening and Closing are the two binary operations used for detection of OD in retinal image. The shape of the OD is obtained from the image I by employing the mathematical closing operation. Using a threshold operation, the suitable binary image is produced.

The binary image Ω contains various connected components known as C_i which is based on Eq. 6.

$$\Omega = \bigcup_{k \in m} C_k, C_i \bigcap C_j = 0, \quad \forall \quad i, j \in m, \quad i \neq j \tag{6}$$

where m varies from 1 to k, k symbolizes the connected components. The disc shape structure when compared to the background pixels are the components of C_i. This includes the OD also. Hence, an effective separation of OD from other structures is established. Now, R_i becomes the greatest component that is connected in C_i . The conciseness of R_i is calculated using Eq. 7:

$$C(R_i) = 4\pi \frac{A(R_i)}{P^2(R_i)} \tag{7}$$

In this equation, $A(R_i)$ signifies pixels' number in the ith region and $P(R_i)$ represents the pixels in region (R_i). Another threshold method is obtained from P-tile method [19] and Nilback's method[20, 21] in order to obtain the binary image.

The weight factor chosen is 1.3 based on previous conclusions in our method [17]. In order to delineate the OD on the retinal image, Circular Hough Transformation (CHT) is employed as studied in[22]. The OD elimination is depicted in Fig. 4.

2.4 *Detection of Exudates*

After optic disc elimination, exudate pixels are identified. Using binary closing operation a 16-pixel radius, flat disc shaped structuring element is utilized and the exudates pixels are directly identified. Binary closing operation follows this threshold operation. The blood vessels have a contrast component which is similar to the contrast component applied in this operation. Hence the image's Standard Deviation (SD) is calculated using Eq. 8.

$$I_3(x) = \frac{1}{N-1} \sum_{i \in W(x)} \left(I_2(i) - \overline{I_3(x)}\right)^2 \tag{8}$$

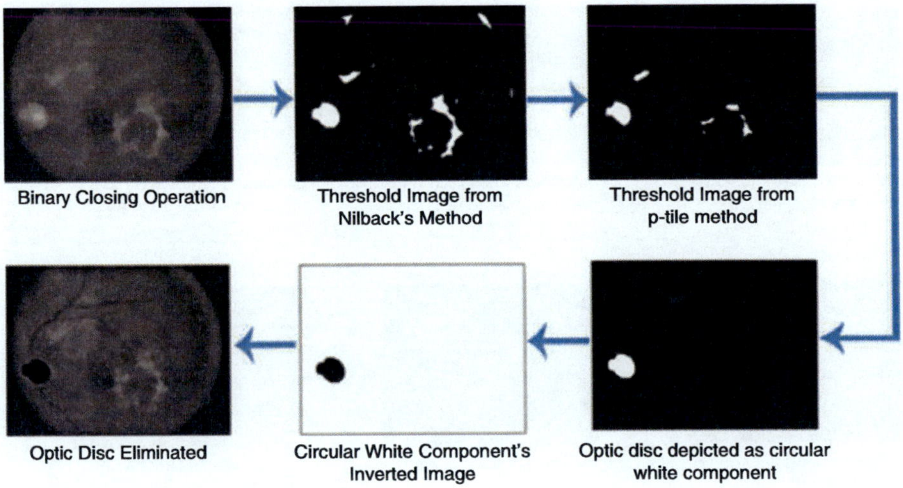

Fig. 4 Steps involved in OD elimination

In this above equation, $W(x)$ symbolizes available pixels available for a sub-window, N symbolizes pixels available in $W(x)$ and $\overline{I_3(x)}$ give the average value for the image $I_3(x)$ where the local contrast image is symbolized by I_3. Using a method called Triangle based threshold [23], the bright regions can be precisely detected and the components can be differentiated. Followed by identification of the high intensity regions, unwanted pixels on the image are eliminated using binary operation called dilation. This method is followed by a flood fill operation that is done on holes so as to regenerate the image. Next, the final step involved in exudate detection is difference image acquisition between from the output image from the threshold image, which is nothing but the brightness based image. As a result, the difference image is superimposed on the original image in order to extract exudate features from the pixels. The whole process of detection of exudates is illustrated in Fig. 5.

2.5 *Hard Exudate Classification*

The final operation which is the classification of hard exudates from the exudate pixels comprises of a valuation using the features that are usually employed by ophthalmologists to visually distinguish hard exudates. The same features are employed as SVM Classifier's input. The set of features is mentioned in Table 1.

Compared with features published in algorithms [24–26], the above eight features were measured important to decrease processing time by not compromising the efficiency for hard exudate classification. The features mentioned in Table 1 are given to the input of an SVM classifier where the output shows the classification

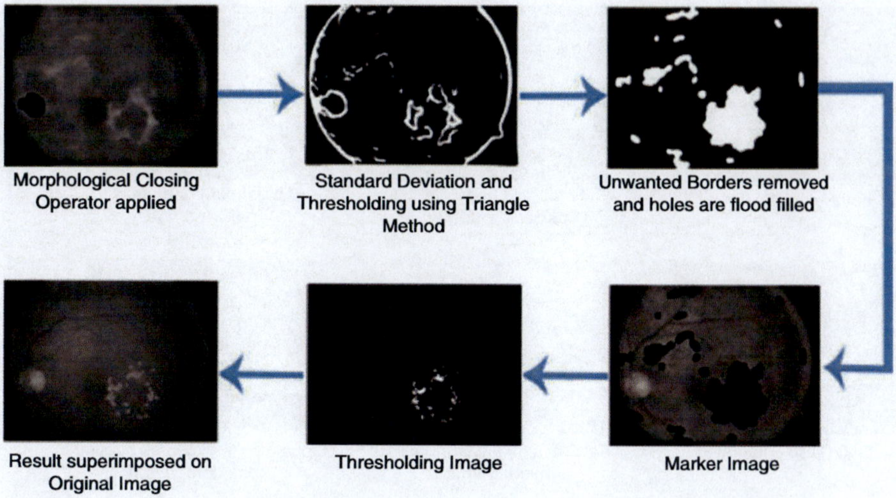

Fig. 5 Steps involved in exudate detection

Table 1 Feature sets for hard exudate classification

Feature sl. no.	Feature type	Description
f_1	Green intensity of mean channel	The green channel image is applied with a 3 × 3 size Mean filter in order to find each pixels's gray scale intensity
f_2	Gray intensity	Pixel's gray scale value
f_3 f_4 f_5	Mean saturation, mean hue and mean intensity of HSI color model	A 3 × 3 size Mean filter is respectively applied to the images I_h, I_s, I_i. Now, f_4 and f_5 refers to saturation and brightness as exudates can be seen as bright lesions
f_6	Energy	Intensity square of pixels and its summation
f_7	Standard deviation	SD is performed and the foreground regions are preserved in the image which have characteristics similar to structuring element
f_8	Mean gradient magnitude	The edge pixels' intensity in terms of directional change in magnitude

results in the form of a binary matrix. SVM is applied over Radial Basis Function (RBF) kernel. The evaluation using cross validation was performed using the gold standard images obtained from Dr. Bejan Singh Eye Hospital and selected by an expert. A total of 72 images were selected from the gold standard for training. The pixels were categorized as non-exudate regions and exudate regions. The cross validation was performed in ten folds to check the SVM classifier's efficiency.

The database images from DIARETDB1 were selected and split arbitrarily into ten subsets (ten folds) which were mutually exclusive and has exudate connected components. They are $B_1, B_2, B_3 \ldots B_{10}$ that have same size. Sixty-seven images were trained on the classifier from the gold standard and the remaining 5 were employed for testing. The output obtained was a binary matrix. And for cross validation the process was repeated ten times with each subset. Thus every pixel provided a vector set containing all the features mentioned in Table 1 as:

$$a_i = (f_1, f_2, f_3 \ldots f_8) \tag{9}$$

Another entity b_j is defined as a flag to define the category which is represented as

$$b_j = \begin{cases} -1 & a_i \in A \\ +1 & a_i \in B \end{cases} \tag{10}$$

where $j \subset \{1, 2, 3 \ldots W\}$ denotes the dimensions of the vector set sample. The hard exudate region is represented by A and the non-hard exudate region is represented by B.

The SVM classifier was trained using the sample set (a_i, b_j). The value of W is chosen as 4200, which means 4200 pixels in 67 samples were categorized by the expert.

2.6 Evaluation Parameters

In this research work, the database candidate subset is considered as $\{B_1, B_2, B_3, \ldots B_N\}$ and gold standard subset is $\{T_1, T_2, T_3, \ldots T_M\}$.

The equation for a pixel to be True Positive (TP) is given in Eq. 11

$$\{B \cap T\} \cup \left\{B_i | \frac{|B_i \cap T|}{B_i} > \sigma\right\} \cup \left\{T_j | \frac{|T_j \cap B|}{T_j} > \sigma\right\} \tag{11}$$

In this research work the σ value is fixed at 0.2 which has a global range of {0,1}. The equation for a pixel to be False Positive (FP) is given in Eq. 12.

$$\{B_i | B_i \cap T = \phi\} \cup \left\{B_i \cap \overline{T} | \frac{|B_i \cap T|}{|B_i|} \leq \sigma\right\} \tag{12}$$

The equation for a pixel to be False Negative (FN) is given in Eq. 13

$$\{T_j | T_j \cap B = \phi\} \cup \left\{T_j \cap \overline{B} | \frac{|T_j \cap B|}{|T_j|} \leq \sigma\right\} \tag{13}$$

Finally, all the remaining pixels can be referred to True Negatives (TN).

3 Results and Discussions

There are mainly two sources for the fundus image acquisition. Dr. Bejan Singh Eye Hospital provided with the clinical image which were captured by "Remidio Non-Mydriatic Fundus On Phone (FOP-NM10)" [27] Fundus camera having a Field-Of-View: 40°, having 100–400 ISO range and has a 33 mm working distance. The public database DIARETDB1 was utilized for images required for validation. Table 2 shows the observations of 30 images that were validated.

Since there is an asymmetry between the classes of TP, FN, and FP when compared with TN, by computing just the Area Under Curve (AUC) of Receiver operator

Table 2 Performance matrix evaluated for 30 fundus images

	TP	FP	FN	TN	Accuracy	Sensitivity	Specificity	PPV	*F*-score
Image 1	349	78	35	431,651	99.97%	90.89%	99.98%	81.73%	86.07%
Image 2	372	106	35	431,487	99.97%	91.40%	99.98%	77.82%	84.07%
Image 3	6835	83	52	419,183	99.97%	99.24%	99.98%	98.80%	99.02%
Image 4	54	89	30	431,946	99.97%	64.29%	99.98%	37.76%	47.58%
Image 5	321	34	23	431,630	99.99%	93.31%	99.99%	90.42%	91.85%
Image 6	1488	31	80	429,122	99.97%	94.90%	99.99%	97.96%	96.40%
Image 7	409	26	37	431,420	99.99%	91.70%	99.99%	94.02%	92.85%
Image 8	964	40	54	430,947	99.98%	94.70%	99.99%	96.02%	95.35%
Image 9	6543	56	67	422,555	99.97%	98.99%	99.99%	99.15%	99.07%
Image 10	811	80	78	430,774	99.96%	91.23%	99.98%	91.02%	91.12%
Image 11	1166	49	52	430,535	99.98%	95.73%	99.99%	95.97%	95.85%
Image 12	3522	39	40	427,474	99.98%	98.88%	99.99%	98.90%	98.89%
Image 13	818	30	67	430,259	99.98%	92.43%	99.99%	96.46%	94.40%
Image 14	435	88	23	431,328	99.97%	94.98%	99.98%	83.17%	88.69%
Image 15	1536	40	57	428,684	99.98%	96.42%	99.99%	97.46%	96.94%
Image 16	623	56	35	431,002	99.98%	94.68%	99.99%	91.75%	93.19%
Image 17	3421	38	22	427,567	99.99%	99.36%	99.99%	98.90%	99.13%
Image 18	4090	49	25	427,468	99.98%	99.39%	99.99%	98.82%	99.10%
Image 19	233	39	55	431,731	99.98%	80.90%	99.99%	85.66%	83.21%
Image 20	785	30	22	431,053	99.99%	97.27%	99.99%	96.32%	96.79%
Image 21	327	88	15	431,563	99.98%	95.61%	99.98%	78.80%	86.39%
Image 22	1053	33	24	430,947	99.99%	97.77%	99.99%	96.96%	97.36%
Image 23	188	70	22	431,441	99.98%	89.52%	99.98%	72.87%	80.34%
Image 24	2213	44	21	429,216	99.98%	99.06%	99.99%	98.05%	98.55%
Image 25	964	33	37	430,750	99.98%	96.30%	99.99%	96.69%	96.50%
Image 26	521	25	6	431,650	99.99%	98.86%	99.99%	95.42%	97.11%
Image 27	848	90	5	429,132	99.98%	99.41%	99.98%	90.41%	94.70%
Image 28	904	24	56	431,480	99.98%	94.17%	99.99%	97.41%	95.76%
Image 29	842	99	34	430,927	99.97%	96.12%	99.98%	89.48%	92.68%
Image 30	4543	35	68	422,565	99.98%	98.53%	99.99%	99.24%	98.88%

characteristic (ROC) is not appropriate. So 5 different evaluation parameters are taken into consideration. They are

$$\text{accuracy} = \frac{\text{TN} + \text{TP}}{\text{TP} + \text{FP} + \text{TN} + \text{FN}} \tag{14}$$

$$\text{sensitivity} = \frac{\text{TP}}{\text{TP} + \text{FN}} \tag{15}$$

$$\text{specificity} = \frac{\text{TN}}{\text{TN} + \text{FP}} \tag{16}$$

$$\text{Positive Prediction Value (PPV)} = \frac{\text{TP}}{\text{TP} + \text{FP}} \tag{17}$$

$$F \text{ score} = 2 \times \frac{\text{sensitivity} \times \text{PPV}}{\text{sensitivity} + \text{PPV}} \tag{18}$$

The table shows good results with respect to the average sensitivity, specificity as well as accuracy having a value of 87%, 98%, and 98.7%, respectively. The F-score as well as the precision calculated were far higher than other works published in the literature in [28, 29] that is F-score = 89.91% and precision = 88.10%. Table 3 shows a comparative study with algorithms that were already published and it can be inferred that accuracy as well as specificity of this research work is greater than the other methods in literature.

Table 4 gives a comparison of the improved method of image enhancement that is IIHE-RVE with our previous method—contrast limited adaptive histogram equalization (CLAHE) which shows a reasonable increase in the value of specificity, PPV, and F-score.

Table 3 Comparison with existing algorithms

Methodology	Sensitivity	Specificity	Accuracy
Chen et al. [29]	83	75	79
Travieso et al. [30]	91.67	92.68	92.13
Barman et al. [31]	92.42	81.25	87.72
Proposed method	87.90	99.97	99.92
A Hajdu et al. [26]	92	68	82
R Sinha et al. [25]	96.54	93.15	N.A.
Pourreza et al. [28]	86.01	99.93	N.A.

Table 4 Performance matrix of 30 images evaluated

Methodology	Sensitivity	Specificity	Accuracy	PPV	F-score
CLAHE [17]	99.81%	80.06%	99.96%	88.03%	81.90%
IIHE-RVE	99.92%	87.90%	99.97%	89.91%	88.10%

4 Conclusion

The proposed work is a novel technique to detect exudates using morphological operation. The new enhancement method IIHE-RVE was used to increase the sensitivity of our existing algorithm that originally involved enhancement using CLAHE. A considerable increase in specificity indicates that the algorithm is more accurate while considering low intensity images. Using the same feature set to the classifier, the score of evaluation parameters could be increased by changing the enhancement technique. Further studies can be implicated to increase the PPV and F-score of this algorithm.

References

1. Badgujar RD, Deore PJ (2019) Hybrid nature inspired SMO-GBM classifier for exudate classification on fundus retinal images. Innov Res BioMed Eng 40(2):69–77
2. Guo S, Wang K, Kang H, Liu T, Gao Y, Li T (2019) Bin loss for hard exudates segmentation in fundus images. Neurocomputing 392:314–324
3. Khojasteh P et al (2019) Exudate detection in fundus images using deeply-learnable features. Comput Biol Med 104:62–69
4. Khojasteh P, Aliahmad B, Kumar DK (2019) A novel color space of fundus images for automatic exudates detection. Biomed Signal Process Control 49:240–249
5. Frazao LB, Theera-Umpon N, Auephanwiriyakul S (2019) Diagnosis of diabetic retinopathy based on holistic texture and local retinal features. Inf Sci (NY) 475:44–66
6. Kumar RS, Karthikamani R, Vinodhini S (2018) Mathematical morphology for recognition of hard exudates from diabetic retinopathy images. Int J Recent Technol Eng 7(4S):367–370
7. Kaur J, Mittal D (2018) A generalized method for the segmentation of exudates from pathological retinal fundus images. Biocybern Biomed Eng 38(1):27–53
8. Adem K (2018) Exudate detection for diabetic retinopathy with circular Hough transformation and convolutional neural networks. Expert Syst Appl 114:289–295
9. Mo J, Zhang L, Feng Y (2018) Exudate-based diabetic macular edema recognition in retinal images using cascaded deep residual networks. Neurocomputing 290:161–171
10. Kusakunniran W, Wu Q, Ritthipravat P, Zhang J (2018) Hard exudates segmentation based on learned initial seeds and iterative graph cut. Comput Methods Programs Biomed 158:173–183
11. Zhou W, Wu C, Yi Y, Du W (2017) Automatic detection of exudates in digital color fundus images using superpixel multi-feature classification. IEEE Access 5:17077–17088
12. Annunziata R, Garzelli A, Ballerini L, Mecocci A, Trucco E (2016) Leveraging multiscale Hessian-based enhancement with a novel exudate inpainting technique for retinal vessel segmentation. IEEE J Biomed Health Inform 20(4):1129–1138
13. Xu X, Lee K, Zhang L, Sonka M, Abramoff MD (2015) Stratified sampling voxel classification for segmentation of intraretinal and subretinal fluid in longitudinal clinical OCT data. IEEE Trans Med Imaging 34(7):1616–1623
14. Harangi B, Hajdu A (2014) Detection of exudates in fundus images using a Markovian segmentation model. In: 36th annual international conference of the IEEE Engineering in Medicine and Biology Society, 2014, vol 2014, pp 130–133
15. Agurto C et al (2014) A multiscale optimization approach to detect exudates in the macula. IEEE J Biomed Health Inform 18(4):1328–1336
16. Sreeja KA, Kumar SS (2019) Comparison of classifier strength for detection of retinal hemorrhages. Int J Innov Technol Exploring Eng 8(6S3):688–693

17. Pradeep A, Joseph XF (2019) Retinal exudate detection using binary operation and hard exudate classification using support vector machine. Int J Innov Technol Exploring Eng 8(9):149–154
18. Arpit S, Singh M (2011) Speckle noise removal and edge detection using mathematical morphology. Int J Soft Comput Eng 1(5):146–149
19. Taghizadeh M, Mahzoun MR (2011) Bidirectional image thresholding algorithm using combined edge detection and P-tile algorithms. J Math Comput Sci 02(02):255–261
20. Rais NB, Hanif MS, Taj IA (2004) Adaptive thresholding technique for document image analysis. In: 8th international multitopic conference, 2004. Proceedings of INMIC 2004, pp 61–66
21. Leedham G, Chen Y, Takru K, Tan JHN, Mian L (2003) Comparison of some thresholding algorithms for text/background segmentation in difficult document images. In: Seventh international conference on document analysis and recognition, 2003. Proceedings, vol 1, pp 859–864
22. Long S, Huang X, Chen Z, Pardhan S, Zheng D (2019) Automatic detection of hard exudates in color retinal images using dynamic threshold and SVM classification: algorithm development and evaluation. Biomed Res Int 2019:1–13
23. Baisantry M, Negi DS, Manocha OP (2012) Change vector analysis using enhanced PCA and inverse triangular function-based thresholding. Def Sci J 62:236–242
24. Akram MU, Tariq A, Khan SA, Javed MY (2014) Automated detection of exudates and macula for grading of diabetic macular edema. Comput Methods Programs Biomed 114(2):141–152
25. Haloi M, Dandapat S, Sinha R (2015) A Gaussian scale space approach for exudates detection, classification and severity prediction. In: ICIP, May 2015
26. Harangi B, Hajdu A (2014) Automatic exudate detection by fusing multiple active contours and regionwise classification. Comput Biol Med 54:156–171
27. Remidio Non-Mydriatic Fundus On Phone (FOP-NM10)
28. Imani E, Pourreza H-R (2016) A novel method for retinal exudate segmentation using signal separation algorithm. Comput Methods Programs Biomed 133:195–205
29. Liu Q, Chen J, Ke W, Yue K, Chen Z, Zhao G (2017) A location-to-segmentation strategy for automatic exudate segmentation in colour retinal fundus images. Comput Med Imaging Graph 55:78–86
30. Rekhi RS, Issac A, Dutta MK, Travieso CM (2017) Automated classification of exudates from digital fundus images. In: 2017 international conference and workshop on bioinspired intelligence (IWOBI), 2017, pp 1–6
31. Fraz MM, Jahangir W, Zahid S, Hamayun MM, Barman SA (2017) Multiscale segmentation of exudates in retinal images using contextual cues and ensemble classification. Biomed Signal Process Control 35:50–62

A PGF-Mediated Social Media Approach for Cyberpolicing India Abusers in Telangana Academics

S. Ravi Kumar, K. Chandra Sekharaiah, and Y. K. Sundara Krishna

1 Introduction

The Government of Telangana (GoT) was created on 2 June 2014. But online cybercriminal fake website JNTUHJAC was formed before the creation of GoT in a JNTUH academic environment. It has been working 2011 onwards and encouraging students to join in the fake website.

The students and faculty members who unknowingly joined in this fake website are approximately 2000 members. We collected the evidence from a Wayback Machine web crawler tool [1]. It is an Internet archive tool which captured the data from the 2011 onward. It also committed multiple cybercrimes of IT Act, 2008, under the Identity Theft 88-A, cheating the nation, the State Emblem of India Act 2005 (Prohibition of Improper Use). Figure 1 depicts the home page of the JNTUHJAC [1–16] captured from Wayback Machine web crawler tool [1].

S. Ravi Kumar (✉)
Krishna University, Machilipatnam, AP, India

K. Chandra Sekharaiah
Department of Computer Science and Engineering, JNTUH University, Hyderabad, Telangana, India

Y. K. Sundara Krishna
Department of Computer Science and Engineering, Krishna University, Machilipatnam, AP, India

A. Haldorai et al. (eds.), *2nd EAI International Conference on Big Data Innovation for Sustainable Cognitive Computing*, EAI/Springer Innovations in Communication and Computing, https://doi.org/10.1007/978-3-030-47560-4_9

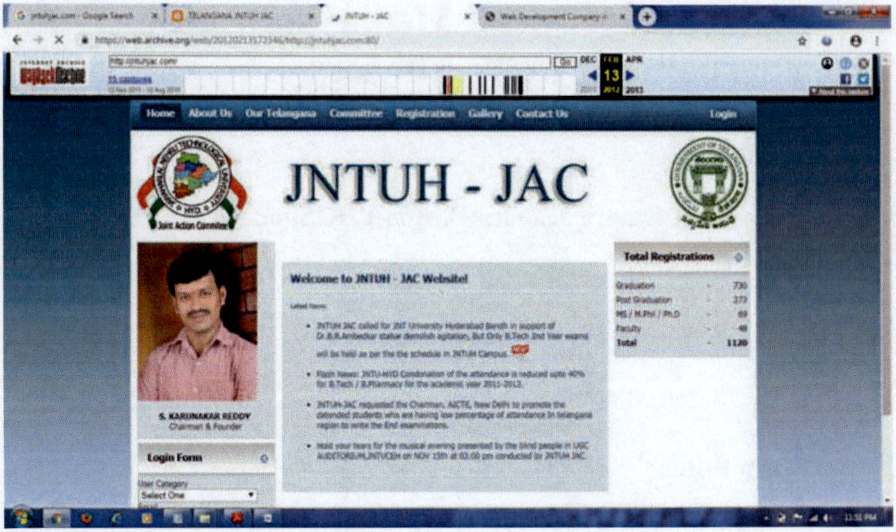

Fig. 1 Screenshot of homepage of jntuhjac.com website as of 13 February 2012 [1]

2 Related Work

The literature review conducted in our research work shows as follows. The initial work is initiated in continuing the case study with respect to the issues related to the cybercrime. The research work focuses on creating the awareness through the remedial forum of cybercrime of JNTUHJAC.

The RTI Act is a road map to the remedial solution for the cybercrime. We presented the RTI Act to various government organizations in India and gathered information related to the cybercrime, on whether or not JNTUHJAC received any funds from any of the organizations.

This paper deals with the social media approach for cyberpolicing in the academic environment. We approached different social media platforms like Telegram, WhatsApp, Twitter, Facebook, etc. I am also creating a PGF (People's Governance Forum) for creating awareness on the cybercrime. In this chapter, I represent a research methodology to come up with positive results in order to get an awareness of national spirit, Mother India Consciousness (MIC).

3 Research Methodology

As part of research work, we took 18 masters in technology students and research scholars for sampling to get responses in handling cybercrimes committed to the academic environment of JNTUH. This work is done by

Fig. 2 Facebook: People's Governance Forum

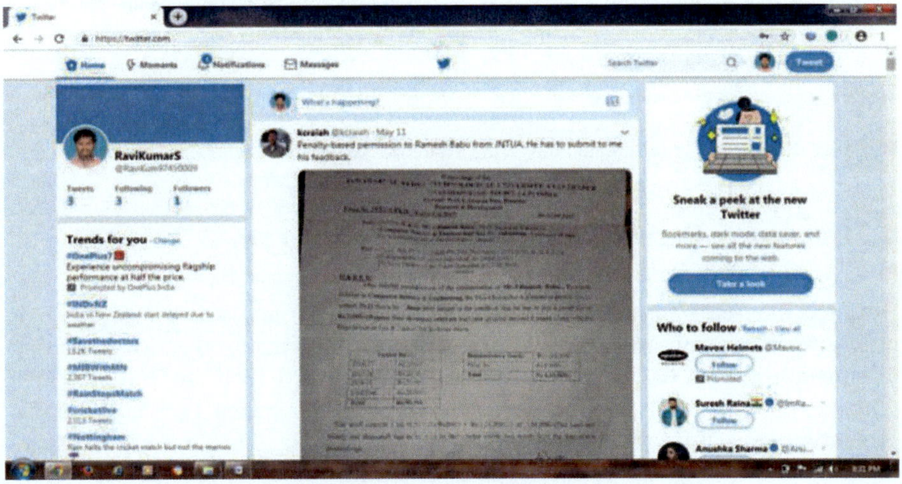

Fig. 3 Twitter: People's Governance Forum

using the six social media platforms such as Facebook, Twitter, WhatsApp, Telegram, etc. The main aim of conducting the case study is to achieve social media intelligence (SMI) for the JNTUH students in an academic environment. In the cybercriminal website, 2000 students were registered, knowingly or unknowingly. The following are the some of the social media, screenshots of remedial solution to create awareness to the students (Figs. 2 and 3).

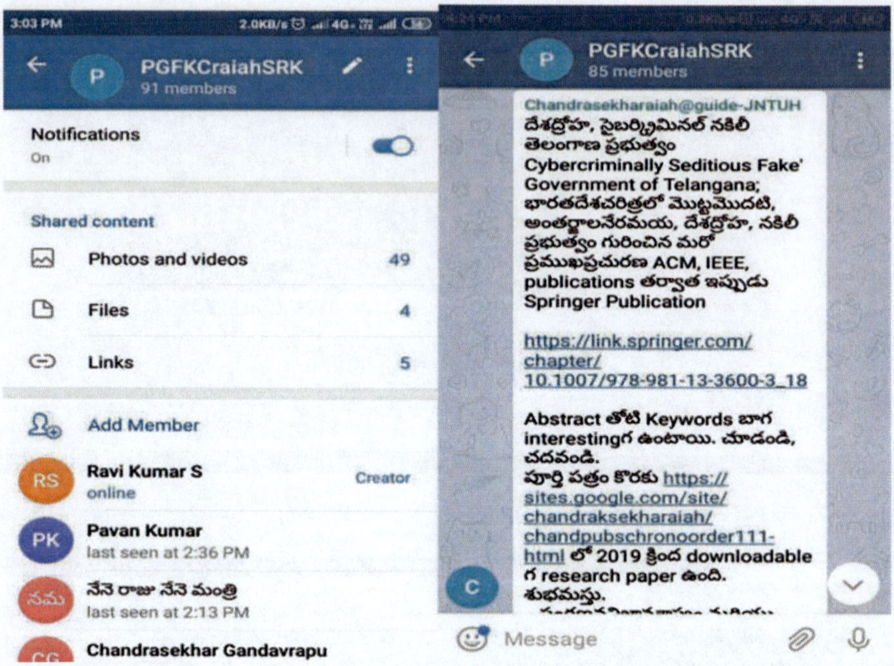

Fig. 4 PGF Telegram group created for giving awareness to the students of JNTUH

4 Difference Between WhatsApp and Telegram

I created the Telegram PGF group (PGFKCraiahSRK) and was joined approximately by more than 90 persons. The Telegram app's main objective is sharing information related to the cybercrime. On this app, M.Tech students, research scholars, police officials, etc. participated. Here M.Tech students posted about the cybercrime-related information. The telegram group (Fig. 4) shows members and masters in technology students, who participated with their message posted. Figure 5 depicts the stickers pasted on the JNTUH class doors and circulated in the telegram app.

A research scholar (Punitha) created the WhatsApp PGF group on 25 December 2018 and shared information about JNTUHJAC. In this group PGF members participate and share cybercrime-related information. Figure 6 depicts the WhatsApp PGF members circulating JNTUHJAC information within the group.

5 Conclusions and Future Work

We used social media analysis to cybercrime and its impact in the academic environment. The cybercrime information shared in the social media like Telegram,

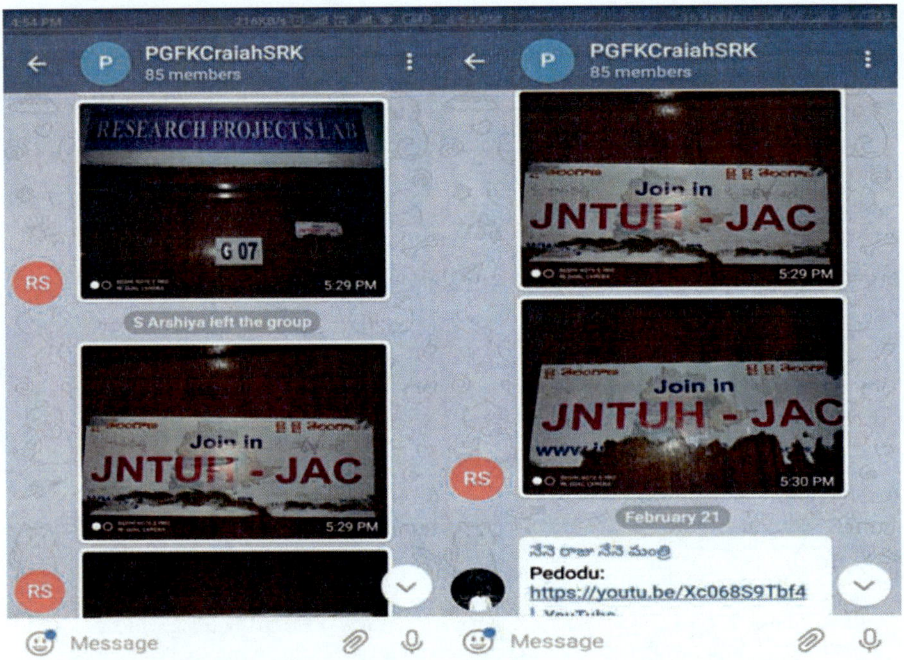

Fig. 5 Cybercrime sticker on JNTUH classroom doors circulated in Telegram group

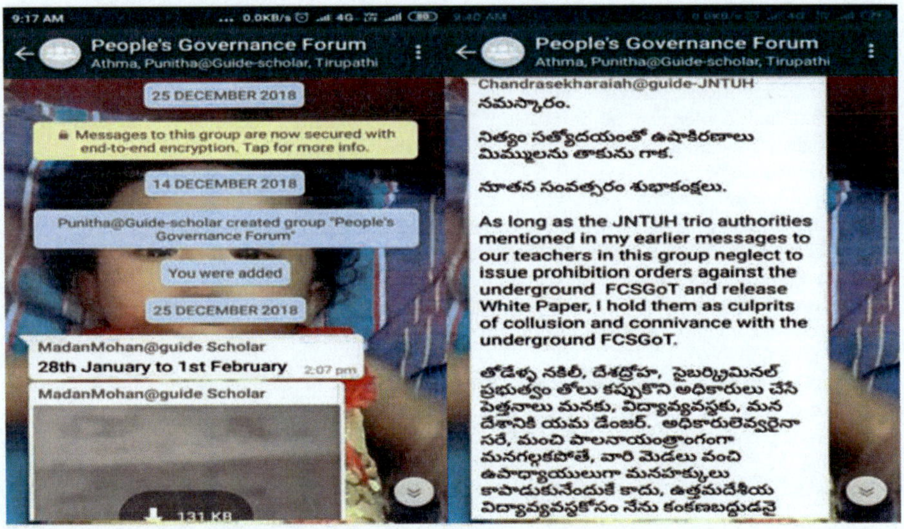

Fig. 6 PGF WhatsApp group created for giving awareness to the students of JNTUH

Facebook, Twitter, etc. We should abolish these types of cybercrimes in smart cities. To reduce cybercrimes, we are creating awareness to the people and also in the academic environment. For the remedial solution, we implemented the PGF (People's Governance Forum) website. It is a remedial forum which gives awareness to the students, faculties, etc. We have to provide a better cybercrime-free society for the next young generation. So that, Mother India Consciousness (MIC) is now reflected through social media is free from cybercrimes. The case study gives the remedial solution to reduce the cybercrime in the society. In the future this case study data are very essential in preventing similar issues such as theft, hacking and other similar kind of cybercrimes, etc.

References

1. https://archive.org/web/
2. Ravi Kumar S, Chandra Sekharaih K, Sundara Krishna K (2018) Cybercrimes—trends and challenges in achieving swachh and digital India using a public cloud: a case study. In Proceedings of the national conference on role of law enforcement authorities, government in upholding justice, SOL, Pondicherry, India, 2–3 Mar 2018, p 149
3. Ravi Kumar S, Chandra Sekharaih K, Sundara Krishna YK (2018) Impact of the RTI act within a public authority organization towards employee-employer engagement: a case study. In: Proceedings of the international conference on science and technology at MRIT, Telangana, India, 19–20 Jan 2018
4. Srihari Rao N, Chandra Sekharaiah K, Ananda Rao A (2018) An approach to distinguish the conditions of flash crowd versus DDoS attacks and to remedy a cyber crime. In: Proceedings of the national conference on role of law enforcement authorities and government in upholding justice, SOL, Pondicherry, India, 2–3 Mar 2018, p 146
5. Tirupathi Kumar B, Chandra Sekharaiah K, Suresh Babu D (2016) Towards national integration by analyzing a case study of cybercrimes. In: ICTCS'16: Proceedings of the ACM second international conference on information and communication technology for competitive strategies, Udaipur, Rajasthan, India, 4–5 Mar 2016, Article no. 79
6. Gouri Shankar M, Usha Gayatri P, Niraja S, Chandra Sekharaiah K (2017) Dealing with Indian Jurisprudence by analyzing the web mining results of a case of cybercrimes. In: Proceedings of international conference on communication and networks. Advances in intelligent systems and computing, vol 508. Springer, Singapore, pp 655–665
7. Usha Gayatri P, Neeraja S, Leela Poornima C, Chandra Sekharaiah K, Yuvaraj M (2014) Exploring cyber intelligence alternatives for countering cyber crime. In: International conference on computing for sustainable global development (INDIACom'14). BVICAM, New Delhi, pp 900–902
8. Usha Gayatri P, Chandra Sekharaiah K (2017) A case study of multiple cybercrimes against the Union of India. In: National conference on innovations in science and technology (NCIST'17), Manipur University, Imphal, 20–21 Mar 2017; Int J Comput Math Sci (IJCMS) 6(3). ISSN: 2347-8527
9. Tirupathi Kumar B, Chandra Sekharaiah K, Mounitha P (2015) A case study of web content mining in handling cybercrime. In: Second international conference on science, technology and management, University of Delhi, Conference Center, New Delhi, India, 27 Sept 2015; Int J Adv Res Sci Eng 4(1)
10. Madan Mohan K, Chandra Sekharaiah K (2017) A case study of ICT solutions against ICT abuse: an RTI act 2005 success story. In: National seminar on science and technology for national development, Manipur University, Imphal, Mar 2017

11. Usha Gayatri P, Chandra Sekharaiah K (2013) Encasing the baneful side of internet. In: National conference on computer science & security (COCSS-2013), 5–6 Apr 2013, Sardar Vallabhbhai Patel Institute of Technology, Vasad, Gujarat, India
12. Usha Gayatri P, Chandra Sekharaiah K, Exploring cyber intelligence alternatives for countering cyber crime: a continuing case study for the nation. In: Proceedings of the international conference at Bharati Vidyapeeth's Institute of Computer Applications and Management (BVICAM), New Delhi, India
13. Usha Gayatri P, Chandra Sekharaiah K, Premchand P (2018) Analytics of a judicial case study of multiple cybercrimes against the Union of India. In: Proceedings of the national conference on role of law enforcement authorities and government in upholding justice, SOL, Pondicherry, India, 2–3 Mar 2018, pp 151
14. Santhoshi N, Chandra Sekharaiah K, Madan Mohan K, Nagalaxmi (2018) ICT based social policy for swatch digital India. In: Proceedings of the national conference on role of law enforcement authorities and government in upholding justice, SOL, Pondicherry, India, 2–3 Mar 2018, pp 151–152
15. https://sites.google.com/view/pgfsrk
16. https://sites.google.com/site/sekharaiahk/apeoples-governanceforumwebpage

Quantitative Performance Analysis of Hybrid Mesh Segmentation

Vaibhav J. Hase, Yogesh J. Bhalerao, Mahesh P. Nagarkar, and Sandip N. Jadhav

1 Introduction

In today's industrial perspective, automation of design and manufacturing activities poses many difficulties in seamless CAD-CAM integration. Almost all commercial CAD-CAM systems used their proprietary file formats to store and retrieval of feature data, which lead to interoperability and CAD-CAM integration problem.

Feature Recognition (FR) provides a communication medium between CAD and manufacturing application. FR manipulates geometrical data seamlessly from the CAD system to a CAM system or vice versa. FR is the first stage for seamless CAD-CAM integration. FR makes smart solid out of dumb solid. It acts as a bridge

V. J. Hase (✉)
Amrutvahini College of Engineering, Department of Mechanical Engineering, Savitribai Phule Pune University, Sangamner, India
e-mail: vaibhav.hase@avcoe.org

Y. J. Bhalerao
Engineering, Faculty of Science, University of East Anglia, Norwich, UK

School of Mechanical Engineering, MIT Academy of Engineering, Alandi, Pune, India
e-mail: Y.bhalerao@uea.ac.uk;
http://www.yogeshbhalerao.com

M. P. Nagarkar
SCSM College of Engineering, Department of Mechanical Engineering, Savitribai Phule Pune University, Ahmednagar, India

S. N. Jadhav
Centre for Computational Technologies (CCTech), Pune, India
e-mail: sandip@cctech.co.in

A. Haldorai et al. (eds.), *2nd EAI International Conference on Big Data Innovation for Sustainable Cognitive Computing*, EAI/Springer Innovations in Communication and Computing, https://doi.org/10.1007/978-3-030-47560-4_10

between CAD-CAM. However, FR technology is not mature enough, and platform dependent.

Standard Triangulated Language (STL) format is not much explored for CAM due to lack of feature recognition interface. Thus, a practical approach for FR from the CAD model is required, to work as an interface between CAD and CAM. Features from Computer-Aided Design (CAD) mesh models (CMM) can be employed to enhance mesh model, mesh simplification, and Finite Element Analysis (FEA).

In many real-world mechanical engineering parts, it is essential to recognize machining features such as complex interacting holes and blends which constitute a significant percentage of features in CMM [1]. The last four decades have witnessed significant research work in FR from B-rep models, while innovative manufacturing and design systems are mesh-based. Therefore, there is an urgent need to make smart solid out of dumb mesh model. Hence, this research work primarily aims at addressing this issue. The focus of the proposed work is to extract complex interacting features along with blends from the STL model. Furthermore, developing a fully automatic FR system from the CAD Mesh Model (CMM) is a challenging task.

Several mesh segmentation algorithms are available in the literature. As Scan Derived Mesh (SDM) have uniform tessellation throughout the mesh, segmentation methods of SDM cannot be applied to CMM [2]. The commonly utilized mesh attributes for mesh segmentation are curvature, convexity, dihedral angle, geodesic distance, etc.

Mesh segmentation is the most supported methodology for FR [3]. An elegant, unique, platform-independent hybrid mesh segmentation method has been developed, for the extraction of features from the CMM. The proposed method extracts intersecting features along with their parameters built over the segmentation workflow. The HMS algorithm also detects intersecting features and separates them.

This paper presents a comprehensive performance analysis of Hybrid Mesh Segmentation (HMS) algorithm quantitatively. The performance of the proposed algorithm is evaluated by comparing the proposed algorithm with the recently developed state-of-the-art algorithms like Attene et al. [4], RANSAC [5], Li et al. [6], Yan et al. [7], Adhikary and Gurumoorthy [8], and Le and Duan [9] in terms of coverage, time complexity, and accuracy. An important contribution of this proposed HMS is that it clusters facets using “facet area” as a novel mesh attribute. The method does not require to set any critical parameters for segmentation.

The rest of the paper is structured as follows: Sect. 2 provides a comprehensive review of relevant literature; Sect. 3 illustrates a proposed Hybrid mesh segmentation algorithm. Section 4 deals with the quantitative performance analysis of HMS. Discussion based on results is provided in Sect. 5. Section 6 present conclusion and future scope.

2 Literature Review

Mesh segmentation partitions the input CMM into "meaningful" regions [7]. Over past four decades, several researches [3, 9–15] have comprehensively summarized mesh segmentation methods with their strengths and weaknesses. Mesh attributes play a crucial role in success of segmentation results. Mesh segmentation is the most preferred approach for FR [3].

STL models of mechanical parts have sparse and dense triangles (see Figs. 1a and 2a). Flat surfaces have sparse and big triangles. Highly curved surfaces have dense and small triangles. Ruled surfaces have triangles of a small base [16, 17]. Because of such a diverse variety of triangles in the STL model, computing the principal curvatures accurately is a tough task for coarse meshes [18]. Angelo et al. [19] presented a segmentation method to extract blend features using principal curvatures from a tessellated model. Features detected are fillets, rounds, and grooves. Conical surfaces were not detected.

Several scientists have tried to estimate curvature along the boundary [20]. Nevertheless, only the curvature knowledge ("Gaussian curvature" and "absolute mean curvature") is not sufficient to identify sphere or cylinder alone. The curvature is also strongly influenced by non-uniformly and sparsely distributed facets [21]. It takes a lot of time to measure the curvature. Many mesh segmentation methods,

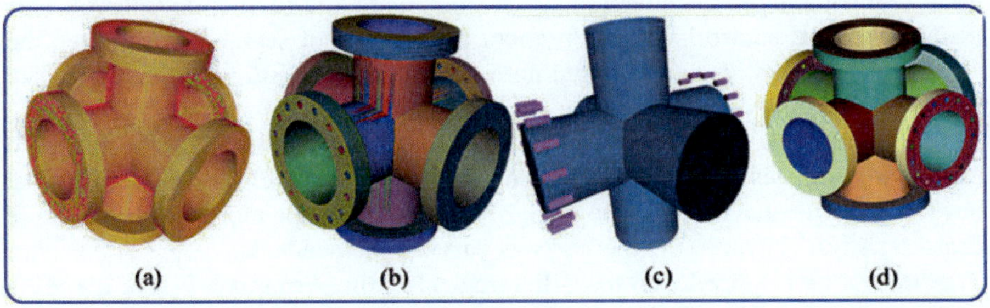

Fig. 1 Illustration of benchmark test cases failures. **(a)** Input CMM; **(b)** Attene et al. [4]; **(c)** Muraleedharan et al. [32]; **(d)** Output

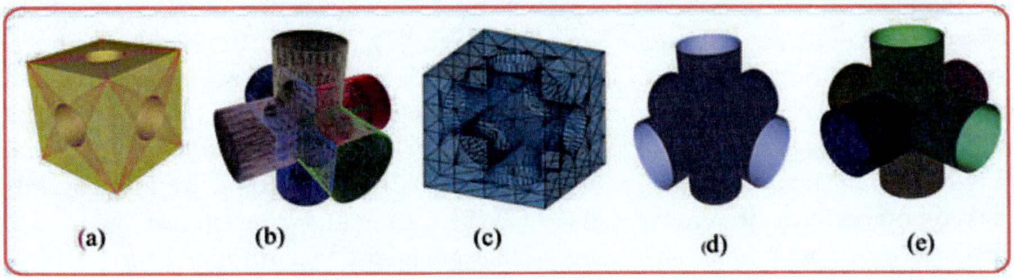

Fig. 2 Illustration of benchmark test cases failures. **(a)** Input CMM; **(b)** Attene et al. [4]; **(c)** Adhikary and Gurumoorthy [8]; **(d)** Muraleedharan et al. [32]; **(e)** Output

when computing curvature, set a local threshold. It is difficult to establish a single global threshold [22–25]. Sunil and Pande [16] proposed a hybrid region-based segmentation system based on shape properties to identify free-form features from a tessellated sheet metal parts. This technique, however, can only be used to identify features that have a limited set of shape properties. When detecting complex parts, this method requires a user's interaction. The method did not identify blends.

Research shows that triangle shape has more influence on discrete curvature than triangle size does [26]. There are very few researchers who use facets distribution properties of the STL for FR [27]. Hardly any of the algorithms took benefit of intrinsic surface properties of the facets distribution. This could be the first attempt, to take advantage of facets distribution for PM segmentation, and to extract various types of geometric primitives. In this research work, an elegant, unique, platform-independent HMS (vertex-based + facet-based +Artificial Neural Network (ANN) + and Rule-based techniques) is proposed and implemented to partition CMM using "facet area," avoiding tedious curvature estimation [28].

As the focus of this research paper is to evaluate the performance of HMS by comparing with existing and recent state-of-the-art approaches, here, we limit our review of those approaches only.

Katz and Tal [29] proposed a "Hierarchical Mesh Decomposition using fuzzy clustering and cuts (HMD) method based on a fuzzy K-means iterative clustering algorithm." However, the iterative clustering technique cannot be applied directly to segment mechanical parts [9]. Further, Katz et al. [30] presented a hierarchical mesh segmentation method using Feature Point and Core Extraction (FPCA). This method did not require information about the number of segments. However, the algorithm was iterative and reiterate until getting characteristic feature points like high convexities or concavities.

Mortara et al. [31] developed "Multi-Scale mesh Analysis by using the paradigm of Blowing Bubbles (MSABB)." They segmented shape into clusters of vertices that have a uniform behavior from the point of view of the shape morphological feature characteristics. However, the method was curvature dependent.

Attene et al. [4] developed the "Hierarchical Fitting Primitives" (HFP), a mesh segmentation framework that involves visual inspection along with a number of clusters as an input parameter. However, it is difficult to know a number of clusters before FR. Figures 1b and 2b illustrate the failure cases of Attene et al.[4].

Schnabel et al. [5] developed a system for the identification of basic primitives based on "RANSAC (RANdom Sample Consensus)." This method over- or under-segments the model. Li et al. [6] invented the "GlobFit" method, which is a modified version of "RANSAC" [5] approach. Instead of segmentation, this approach is a primitive fitting. For primitive extraction, they used parallel, orthogonal, equal angle relationships. This method is computationally more expensive and depends heavily on performance from "RANSAC" [5]. Yan et al. [7] developed "geometric distance-based error function" based mesh segmentation algorithm for the CMM or scanned model by fitting a general quadric surfaces. However, the technique was only suitable for the quadric surface. Not appropriate to blend detection.

Shapira et al. [33] presented a Gaussian distribution based part type mesh segmentation using “Shape Diameter Function (SDF).” SDF provides a good distinction between thick and thin parts of the object. They clustered facets using “Gaussian Mixture Model (GMM)” which is sensitive to noise. However, SDF has its limitation on non-cylindrical parts of objects.

Adhikary and Gurumoorthy [8] had developed “Minimum Feature Dimension (MFD)” based free-form volumetric features extraction technique. They identified feature boundary edges from CMM by 2D slicing, without segmentation. The algorithm does not rely on mesh triangle density, and mesh geometrical properties. However, for the test case shown in Fig. 2a, the algorithm was unable to detect and extract features. MBD must be known prior to extraction of the feature. Figure 2c illustrates the failure case of Adhikary and Gurumoorthy [8].

Le and Duan [9] proposed a “dimensional reduction technique” in which profile curve analysis has been carried out. To obtain a profile curve, they transform 3D primitives into 2D. However, the algorithm was dependent on slice thickness, and slicing techniques fail to detect or separate complex interacting features as noted by [8].

Volumetric interacting features were recognized by Muraleedharan et al. [32] using “a random cutting plane technique.” They utilized “Gaussian curvature” for boundary detection and separating the interacting features. However, their algorithm relies on “number of cutting planes” for FR which must be known prior to extraction. The feature must have an inner ring presence, which is the algorithm’s key weakness. If there were no inner rings (for a joint have complex boundary) in a feature, it remained undetected. The failure case of Muraleedharan et al. [32] is illustrated in Figs. 1c and 2d.

HMS extracts and separates intersecting features. Figures 1d and 2e show success of HMS in feature recognition.

3 Hybrid Mesh Segmentation

HMS automatically segment CMM into “meaningful” and distinct, mathematically analyzable analytic regions [7, 21].

3.1 Architectural Framework of Hybrid Mesh Segmentation

Figure 3 illustrates an architectural framework of HMS. It includes three stages viz. preprocessing, mesh segmentation, and iterative region merging.

Preprocessing Topology and facet adjacency are constructed in imported CMM, and automated threshold prediction has been performed.

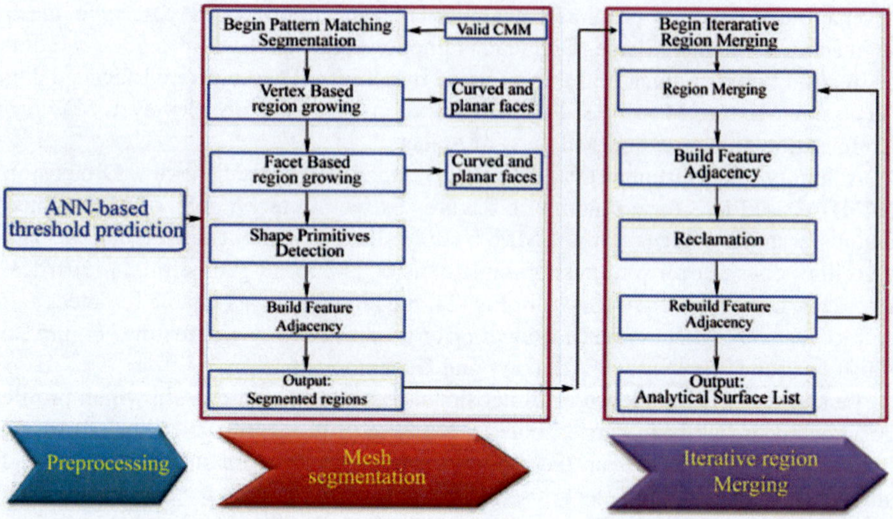

Fig. 3 Illustration of an architectural framework of hybrid mesh segmentation

Input CAD Mesh Model The HMS takes a valid CMM which is free from errors as input in ASCII or Binary format, therefore, there is no need for model healing [16].

Automatic Threshold Prediction Segmentation of CMM leads to under-segmentation or over-segmentation based on input Area Deviation Factor (*Adf*)[12, 32]. Setting the appropriate *Adf* is too complicated for a layman. Therefore, an automatic prediction of *Adf* is of great importance. Hase et al.[34] propose and implement smart prediction of *Adf* using the Artificial Neural Network (ANN). A detailed description is beyond this paper's reach.

Mesh Segmentation Segmenting CMM is difficult by using a stand-alone vertex-based (*VBRG*) or facet-based region growing (*FBRG*) techniques [31]. A promising approach that has become evident is a hybrid (*VBRG* + *FBRG*) one, wherein the advantages of the above approaches are combined.

HMS utilizes the "facet area" as an attribute for segmenting CMM. It combines *VBRG* and *FBRG* algorithms. HMS automatically segments CMM into meaningful analytic surfaces without curvature estimation.

Iterative Region Merging Iterative region merging repeatedly merged over-segmented regions that have similar geometric property to generate a single region. It includes following steps:

Region Merging Region merging merges regions iteratively. It is not enough to combine all regions with a single pass. Two adjacent regions are merged to one, if they satisfy *geometry equality test*. Region adjacency may have changed after merging, so in the next iteration, features that were not eligible for merging in the previous iteration, will be merged.

Reclamation After region merging, small cracks may be found at the region boundaries [35]. To make a watertight model, uncollected facets were reclaimed into the specified surrounding region, based on *reclamation criteria*.

For further implementation details of HMS, please refer to [36, 37].

3.2 Illustrative Examples

Figure 4a–e briefly illustrates all stages in hybrid mesh segmentation for the STL CAD mesh model. Experimental evaluation is carried out on the "Box" model. The part is created in Autodesk™ Inventor™ 2018 and has been exported as a CMM. This part is used to test the efficacy of the proposed algorithm wherein cylindrical features intersect with one another forming complex boundaries at the intersections. Table 1 outlines the parameters for the "Box" model. Table 2 outlines primitives detected before merging and after merging. The system takes 0.31 s for segmentation. Table 3 shows the performance of *HMS* with detailed timings for each step.

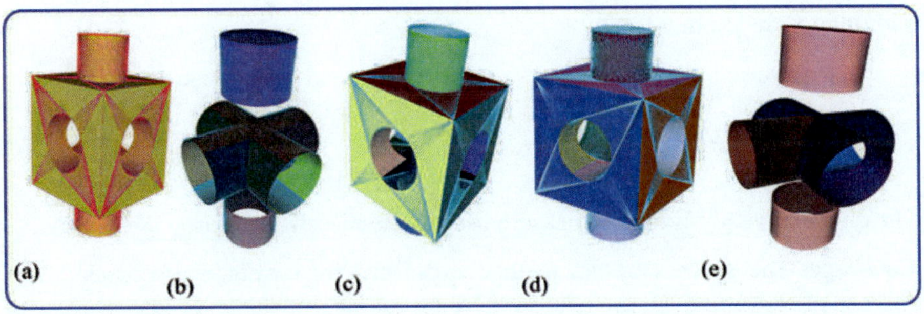

Fig. 4 Illustration of the hybrid mesh segmentation process. (**a**) Input CAD Mesh model. (**b**) Segmentation. (**c**) Region merging. (**d**) Reclamation. (**e**) Region merging after reclamation

Table 1 Outline of the parameters for the Box model

Particulars	Descriptions
Vertex count	1390
Facet count	2788
Area deviation factor	0.75
Sharp edge angle	40°
Dihedral angle	40°
Coverage	100%

Table 2 Illustration of the region merging process

Feature	Before merging	After merging
Planes	8	80
Cylinder	10	6
Overall time elapsed		0.31 s

Table 3 Timings statistics for the "Box" model

Particulars	Time elapsed in seconds
Mesh import and topology generation	0.08
Planar face segmentation	0.043
Curved face segmentation	0.063
Region merging	0.003
Reclamation	0.012
Iterative region merging	0.011
Overall: time elapsed	0.31

4 Quantitative Performance Analysis of HMS

The output of HMS algorithm is quantitatively assessed by performing simulation on benchmark test cases using a computer with windows 8.1 operating system, and with Intel Core i3 processor.

4.1 *Quantitative Performance Measure*

The quantitative performance measure used in the investigation are:

Coverage The success of the method quantified by coverage assessment. The percentage coverage is used as a measure of an indicator of the successful segmentation algorithm. The coverage is calculated as:

$$\text{coverage} = \frac{\text{number of primitives extracted}}{\text{actual number of primitives present}} \tag{1}$$

Absolute Distance Error An absolute error is found by comparing parameters of recovered features with a standard reference.

$$\text{distance error} = |\text{measured distance} - \text{actual distance}| \tag{2}$$

Time The overall time (in seconds) needed for various steps of mesh segmentation are:

Step 1: Mesh import and topology generation
Step 2: Hybrid mesh segmentation

Step 3: Building feature adjacency
Step 4: Region merging
Step 5: Reclamation
Step 6: Iterative region merging
Step 7: Feature recognition

Number of Regions Before/After Region Merging Iterative region merging technique merges a number of regions before merging (N_{Rbrm}) to a single region that has a similar geometric property (N_{Rarm} : number of regions after region merging).

4.2 *Evaluation of Segmentation Algorithms*

The efficacy of the proposed algorithm was tested by comparing with the existing state-of-the-art approaches. Table 4 shows the existing state-of-the-art approaches. The proposed algorithm has been tested on benchmark "Anchor" model with different approaches, as shown in Fig. 5. For methods SDF [33], SSACA [38], and HFP [4], the code was publicly available. For methods BBMSA [31], HMD [29], FPCA [30], and RCPA [32] are taken from [11, 32] as the code was not publically available. The HMS approach extracted all the features with coverage (C) of 100%.

Table 4 Existing state-of-the-art approaches

Authors/Reference	Year	Methodology	Code availability	Abbreviation
Katz and Tal [29]	2003	"Hierarchical mesh decomposition using Fuzzy Clustering"	No	HMD
Mortara et al. [31]	2004	"Blowing Bubbles for Multi-Scale Analysis"	No	BBMSA
Katz et al. [30]	2005	"Mesh segmentation using feature point and core extraction"	No	FPCA
Attene et al.[4]	2006	"Hierarchical mesh segmentation based on fitting primitives"	Yes	HFP
Shapira et al. [33]	2008	"Mesh segmentation using shape diameter function"	Yes	SDF
Kaick et al. [38]	2014	"Shape Segmentation by Approximate Convexity Analysis"	Yes	SSACA
Muraleedharan et al. [32]	2018	"Random cutting plane approach"	No	RCPA
Hase et al. [36]	2019	"Hybrid Mesh segmentation"	Yes	HMS

Fig. 5 Performance comparisons of the proposed method with existing approaches on benchmark "Anchor" model

Table 5 Qualitative performance analysis of existing state-of-the-art approaches

Test cases	F	V	S	Adf	N_{Rbrm}	N_{Rarm}	T
Optics housing	1182	593	0.306	0.75	49	35	0.186
Demo08	2492	1238	0.644	0.75	82	32	0.289
Cami1	944	464	0.246	0.70	36	18	0.18
Caddy02	1644	822	0.43	0.75	30	18	0.268
Gear_38	2696	1340	0.681	0.75	26	23	0.36

Wherein, F: number of facets, V: number of vertex, S: STL size (in MB), Adf: predicted area deviation factor, T: overall timing (in a second)

Comparing with the existing state-of-the-art approaches, the closest one among others is the technique HFP of Attene et al. [4]. Other existing approaches under-segments the model, making feature extraction a difficult task. Table 5 evaluates the time performance of the proposed algorithm for the test cases shown in Fig. 6.

4.3 Comparison with the Recently Developed Algorithm

The experimentation results on various benchmark test cases demonstrate that, HMS approach does not depend on complex attributes, and outperforms the existing state-of-the-art algorithms. Table 6 summarizes the quantitative comparison of HMS for the test cases. We evaluate using the coverage percentage, a number of primitives, and the distance error. As noted by [39], the HMS algorithm yields better results than RANSAC [5] and Attene et al. [4] (see Fig. 7). HMS results are comparable to Le

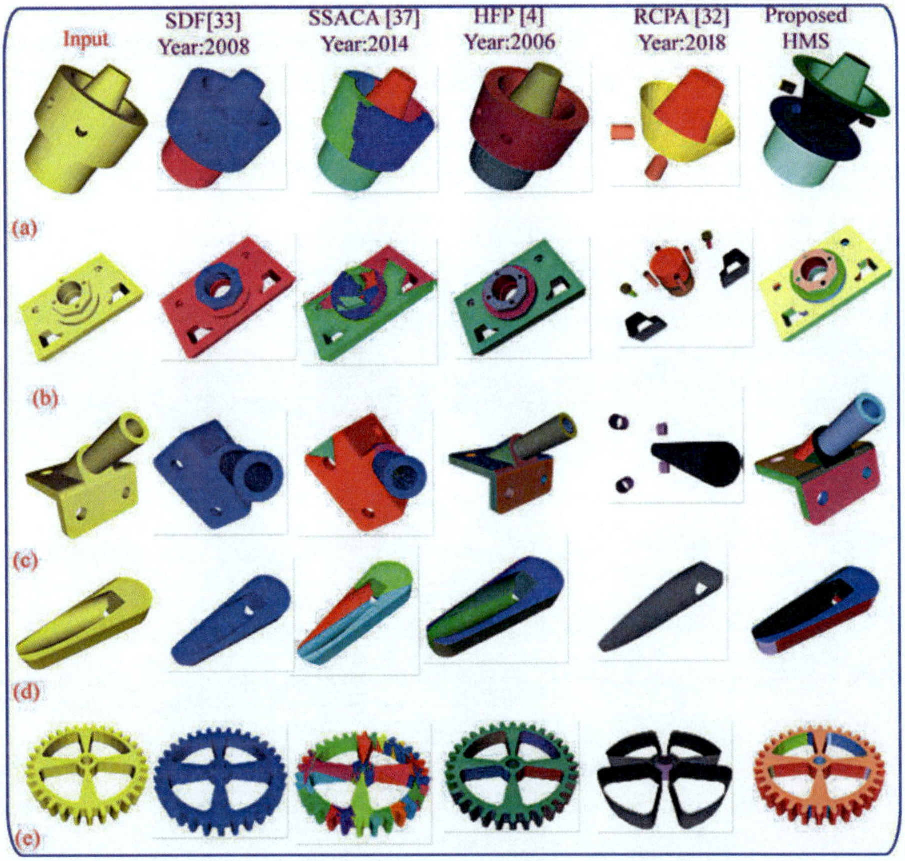

Fig. 6 Comparison of volumetric and surface-based FR with existing approaches. (**a**) Optics housing. (**b**) Demo08. (**c**) Camil. (**d**) Caddy02. (**e**) Gear_38

and Duan [9]. The simulation reveals that HMS achieves a promising performance with coverage of more than 95%.

5 Results and Discussions

5.1 Evaluation of Segmentation Results

As noted by Adhikary and Gurumoorthy [8], it is challenging to extract interacting features. The algorithm developed by Muraleedharan et al. [32] is unable to separate the interacting features as these models do not have a presence of inner rings. Most of the existing algorithms pose difficulties in extracting and separating interacting

Table 6 Quantitative evaluation of primitive quality in Fig. 7

	# of primitives					Coverage (%)					Distance error ($\times 10^{-3}$)				
Model name	I	II	III	IV	V	I	II	III	IV	V	I	II	III	IV	V
Block	14	14	14	9	14	100	99.98	99.98	64.28	98.98	0.04	0.37	0.08	n/a	0.69
Cover rear	45	28	45	45	28	100	87.79	100	100	87.79	0.02	0.11	0.04	n/a	0.15
Stator	12	12	12	6	n/a	100	99.99	100	50	n/a	0.01	0.8	0.47	n/a	n/a

(I) Proposed algorithm (II) RANSAC [5] (III) Le and Duan [9] (IV) Attene et al. [4] (V) GlobFit et al. [6]

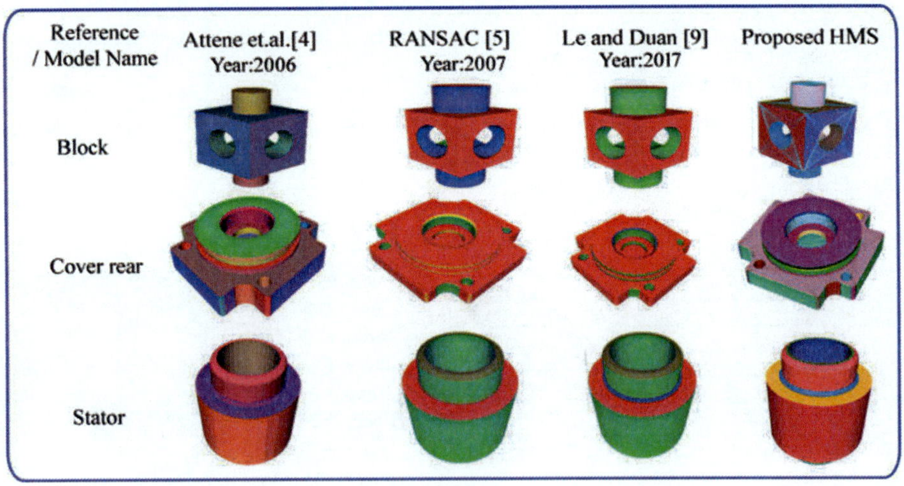

Fig. 7 Comparison with the existing algorithms [39]

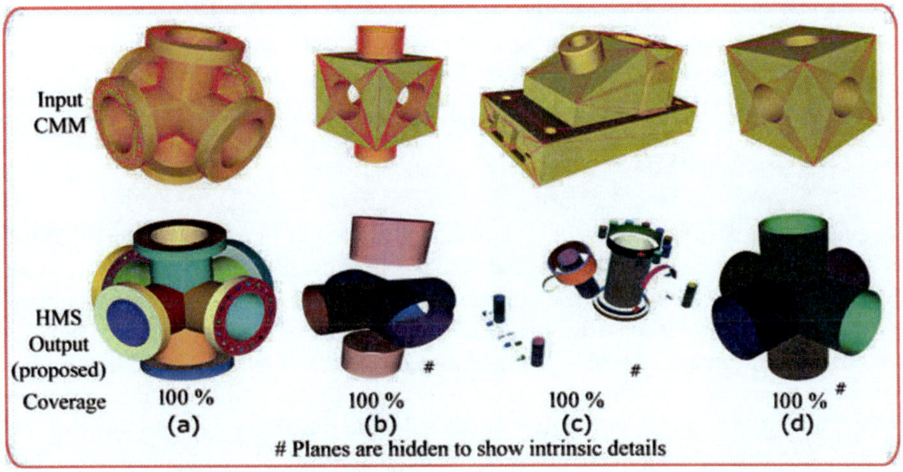

Fig. 8 Experimental results for proposed HMS

features as joints between them have complex boundaries. The HMS algorithm extracts and separates interacting features. The HMS algorithm requires no prior knowledge of attributes like "the number of clusters," "curvature," "the number of cutting planes," "minimum feature dimension," "the orientation of model," and "thickness of the slice" to extract volumetric features. Figure 8a–d briefly illustrates the results of the HMS algorithm for extracting interacting features. Table 7 shows experimental results for different CMM as shown in Fig. 8.

Table 7 Experimental results for different CMM shown in Fig. 8

Test cases	F	V	S	Adf	N_{Rbm}	N_{Rarm}	T	C
Figure 8a	17104	8480	2.931	0.65	537	62	2.52	100
Figure 8b	2788	1390	0.709	0.60	18	14	0.816	100
Figure 8c	7100	3542	1.837	0.75	210	80	0.927	100
Figure 8d	3360	1672	0.875	0.60	117	12	0.375	100

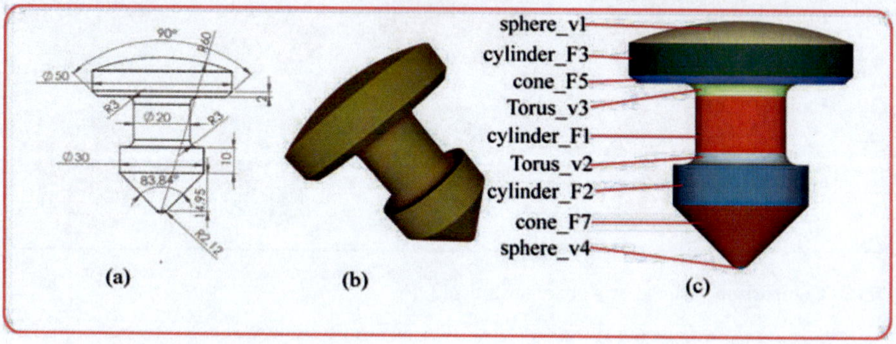

Fig. 9 A quantitative analysis of HMS. (**a**) Input CAD mesh model with parameter. (**b**) Input CAD mesh model. (**c**) Output of HMS

5.2 *Error Analysis*

A quantitative analysis of HMS is performed. Figure 9a illustrates an input CMM with parameters to do quantitative analysis. The primitives (cylinders, spheres, cone, torus etc) parameters of CMM as shown in Fig. 9a has been selected as a standard reference for comparison. After FR, features parameters are recovered.

An absolute error is found by comparing parameters of recovered features with a standard reference. A quantitative error analysis of HMS algorithm in the extraction of the feature parameter shows that recovered values are very close to those of the input CAD model, as shown in Table 8.

5.3 *Mesh Density*

The HMS algorithm has been tested with varying mesh density. As different solid modeler has its own techniques of generating a tessellated model, results in each model have varying mesh density, mesh pattern, and mesh quality. Based on mesh quality, ANN predicts the area deviation factor automatically [34]. Experiments show that the HMS algorithm has no difficulty in extracting the features correctly. The system identifies seven planer, eight cylindrical, one conical, and three torus

Table 8 A quantitative error analysis in the extraction of the feature parameter

Feature	Checked values	Value got from CMM	Value got by mesh FR	Absolute error
Cylinder_F1	Axis	(0,1,0)	(−0.000000, 1.000000, 0.000000)	0
	Radius	10	10	0
	Axis point	(0,0,0)	(0.000000, 0.000000, −0.000000)	0
	End point1	(0, 29.71, 0)	(0.000000, 29.707320, −0.000000)	0
	End point2	(0, 43.71, 0)	(0.000000, 43.707320, 0.000000)	0
Cylinde_F2	Axis	(0,1,0)	(0.000000, 1.000000, 0.000000)	0
	Radius	15	15	0
	Axis point	(0,0,0)	(−0.000000, 0.000000, −0.000000)	0
	End point1	(0, 16.71, 0)	(−0.000000, 16.707320, −0.000000)	0
	End point2	(0, 26.71, 0)	(−0.000000, 26.707320, −0.000000)	0
Cylinder_F3	Axis	(0,1,0)	(−0.000001, 1.000000, 0.000000)	0
	Radius	25	25	0
	Axis point	(0,0,0)	(0.000032, 0.000000, −0.000004)	0
	End point1	(0, 48.71, 0)	(0.000002, 48.707304, −0.000001)	0
	End point2	(0, 56.71, 0)	(−0.000003, 56.707336, 0.000000)	0
Sphere_V1	Center	(0,, 2.16, 0)	(−0.000330, 2.161758, 0.000490)	0
	Radius	60	60.0021	0.0021

(continued)

Table 8 (continued)

Feature	Checked values	Value got from CMM	Value got by mesh FR	Absolute error
Sphere_V4	Center	(0, 3.17, 0)	(0.000003, 3.166820, −0.000007)	0
	Radius	2.12	2.11565	0.00435
Cone_F5	Apex point	(0, 23.71, 0)	(0.000000, 23.707310, −0.000000)	0
	Axis	(0,−1,0)	(0.000000, −1.000000, 0.000000)	0
	Angle	0.79	0.785398	0.004602
	Major center	(0, 48.71, 0)	(−0.000000, 48.707320, −0.000001)	0
	Minor center	(0, 46.71, 0)	(0.000000, 46.707320, −0.000000)	0
Cone_F7	Apex point	(0,0,0)	(−0.000000, 0.000000, 0.000005)	0
	Axis	(0,1,0)	(0.000000, 1.000000, −0.000003)	0
	Angle	0.73	0.731604	0.001604
	Major center	(0, 16.71, 0)	(−0.000000, 16.707317, −0.000000)	0
	Minor center	(0, 1.75, 0)	(−0.000000, 1.753428, 0.000000)	0
Torus_v2	Axis	(0,−1,0)	(−0.000000, −1.000000, −0.000000)	0
	Center	(0, 29.71, 0)	(0.000070, 29.707337, 0.000070)	0
	Major radius	13	12.9999	0.0001
	Minor radius	3	3.00004	0.0004
Torus_v3	Axis	(0,1,0)	(0.000000, 1.000000, 0.000000)	0
	Center	(0,43.71,0)	(−0.002234, 43.707458, 0.000000)	0
	Major radius	13	13.0021	0.0021
	Minor radius	3	2.99983	0.00017

Table 9 Experimental results for "Assy" model with varying mesh density

Test cases	F	V	S	Adf	N_{Rbrm}	N_{Rarm}	T	Solid modeler	C
Assy	5628	2810	1.41	0.8261	43	19	0.605	Creo™ 2.0	100
Assy	7172	3582	1.96	0.75	105	19	0.727	Solidworks™ 2017	100
Assy	8022	4007	2.03	0.8	65	19	0.813	Autodesk™ Inventor™ 2018	100
Assy	34,304	17,148	6.42	0.75	30	19	2.914	Onshape™	100

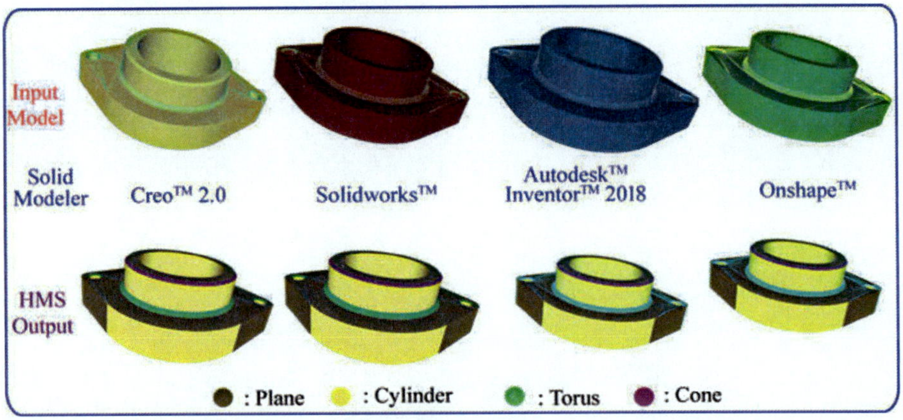

Fig. 10 Experimental results for "Assy" model

surfaces. Also, primitive parameters are estimated accurately. Table 9 shows the details of the "Assy" model shown in Fig. 10.

5.4 *Discussions*

Efficacy Measure of Hybrid Mesh Segmentation The experimentation results on three case studies demonstrate that HMS algorithm extracts and separates interacting features. Results are tabulated in Table 10 and presented in the bar charts, refer Figs. 11, 12, and 13.

From Table 10 and Fig. 11, it is observed that the stand-alone vertex-based region growing (***VBRG***) is able to extract planer (***P***) surface successfully for all three test cases (see Figs. 14b, 17b, 20b) and unable to detect curved (***C***) surface (undetected or miss out surfaces). For the test case "Box," ***VBRG*** detects eight planer surface, and six curved surfaces which are undetected, see Fig. 14d.

For the test case "Stator" ***VBRG*** detects four planer surface, four curved surfaces and four are undetected surfaces, see Fig. 17d. For the test case "Pipe" ***VBRG*** detects 12 planer surface, and 50 curved surfaces which are undetected, see Fig. 20d.

Table 10 Quantitative evaluation of primitive quality for test cases

Model name	F	V	S	Adf	Number of primitives						Coverage (%)			Overall timing (s)		
					VBRG		FBRG		HMS		VBRG	FBRG	HMS	VBRG	FBRG	HMS
					P	C	P	C	P	C						
Box	2788	1390	0.709	0.60	8	0	8	2	8	6	57.14	71.43	100	0.402	5.239	0.368
Stator	2592	1296	0.665	0.75	4	4	4	8	4	8	66.67	100	100	0.37	0.463	0.396
Pipe	17,104	8480	2.87	0.65	12	0	12	50	12	50	19.35	100	100	1.97	2.99	2.52

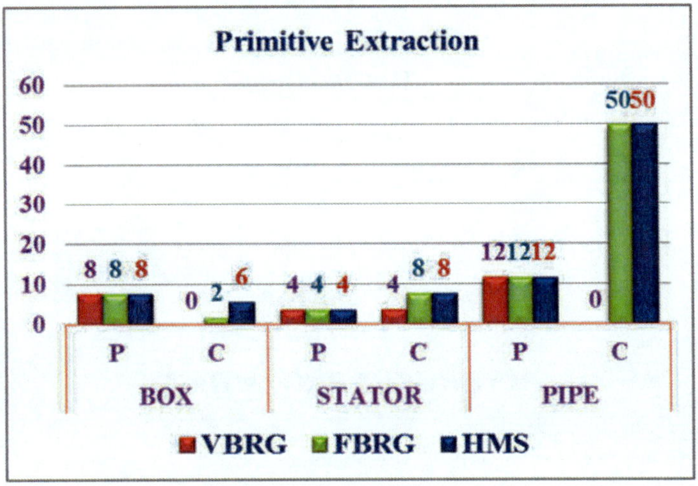

Fig. 11 Performance evaluation of HMS: primitive extraction

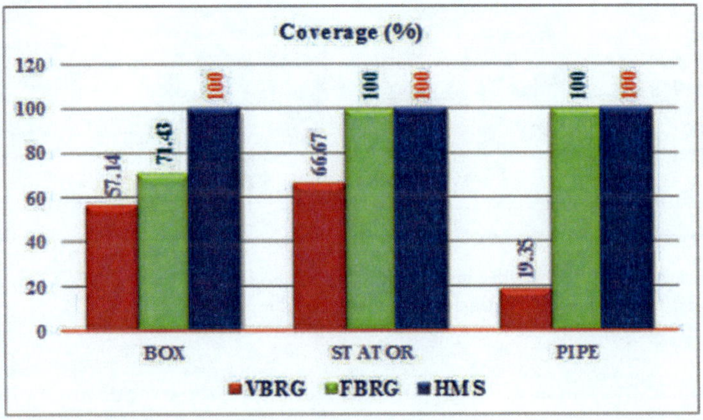

Fig. 12 Performance evaluation of HMS: coverage

The stand-alone facet-based region growing (***FBRG***) is able to extract planer (***P***) and curved (***C***) surface fully or partially for all three test cases. For the test case "Box" ***FBRG*** detects eight planer surface, two curved surfaces, and four undetected curved surfaces, see Fig. 15d. For the test case "Stator" ***FBRG*** detects four planer surface, eight curved surfaces, and zero undetected curved surfaces, see Fig. 18d. For the test case "Pipe" ***FBRG*** detects 12 planer surfaces, 50 curved surfaces, and zero undetected curved surfaces, see Fig. 21d.

To measure the performance of HMS, coverage for ***VBRG***, ***FBRG***, and ***HMS*** is computed. From Table 10 and Fig. 12, it is observed that percentage coverage for all three test cases is 100% for ***HMS***, See Figs. 16b, 19b, 22b.

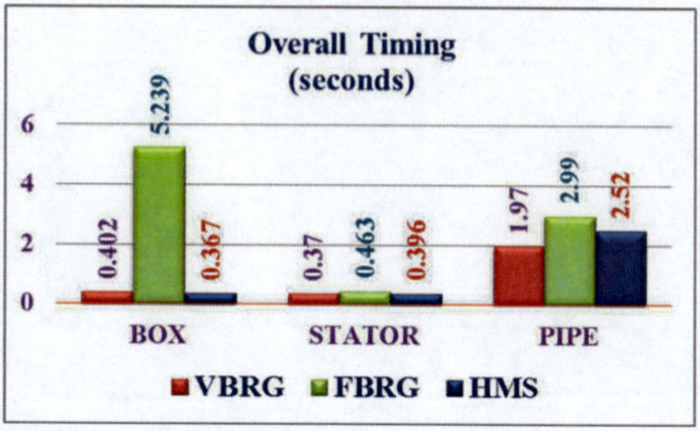

Fig. 13 Performance evaluation of HMS: overall timing

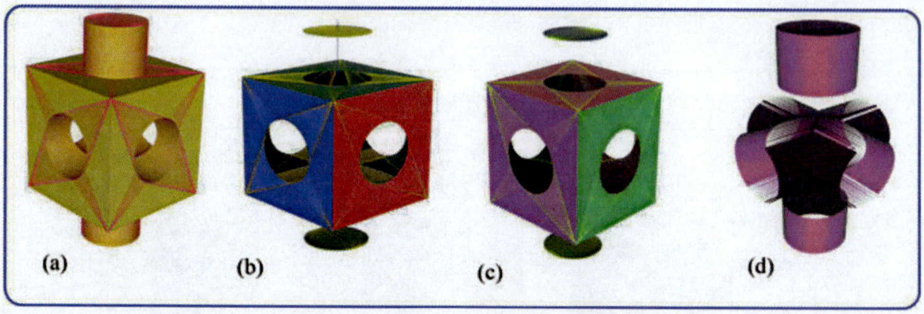

Fig. 14 Vertex-based region growing (VBRG): Box model. (**a**) Input CAD Mesh model. (**b**) Result of VBRG. (**c**) Primitive extracted. (**d**) Undetected regions

To evaluate the efficacy of the proposed technique, the overall timing for ***VBRG***, ***FBRG***, and ***HMS*** is computed.

From Table 10 and Fig. 13, it is observed that the overall timing needed for ***HMS*** is least as compared to stand-alone ***VBRG*** and ***FBRG***. This is due to ***HMS***, wherein an intelligent blending of ***VBRG*** and ***FBRG*** is carried out, See Figs. 16, 19, 22.

Significance of Area Deviation Factor The accuracy and reliability of HMS depend on *Adf*. Inadequate *Adf* leads to under-segmentation or over-segmentation based on input. The small value of $Adf < 0.60$ over-segments the model whereas $Adf > 0.80$ leads to under-segmentation. Iterative region merging technique merges over-segmented regions to the similar adjacent region, resulting in increase of the overall time of FR. Under-segmentation leads to inaccuracy. Artificial neural networks (ANN) based ***Adf*** predictor [34] set optimum ***Adf***, i.e. (greater than 60 and less than 80) which results in better segmentation.

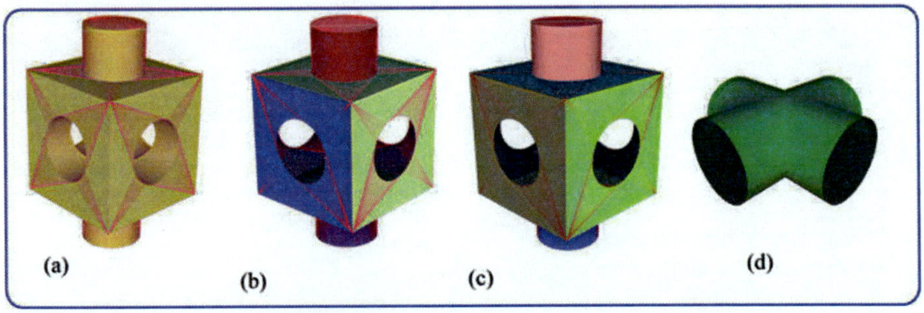

Fig. 15 Facet-based region growing (FBRG): Box model. (**a**) Input CAD Mesh model. (**b**) Result of FBRG. (**c**) Primitive extracted. (**d**) Undetected regions

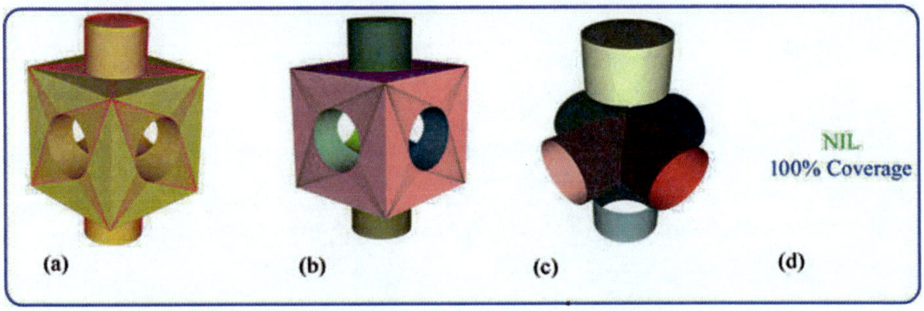

Fig. 16 Proposed hybrid mesh segmentation: Box model. (**a**) Input CAD Mesh model. (**b**) Result of HMS. (**c**) Primitive extracted # planes are hidden. (**d**) Undetected regions

Accuracy of FR The prerequisite step for FR from the CMM is hybrid mesh segmentation. If the algorithm fails at the segmentation stage, it leads to failure of FR. The whole segmentation process must be successfully accomplished for better accuracy of FR.

Interacting Feature Recognition The crucial problem for seamless CAD-CAM integration is interacting feature recognition. As features interact, their topology changes, and makes it challenging to recognize resulting geometry. As previously stated (see Figs. 1 and 2), the existing techniques are unable to separate the interacting features, and HMS algorithm extracts and separates interacting features along with the geometric parameter (Figs. 17, 18, 19, 20, 21, and 22).

Comparison of Techniques The pertinent literature concerns with feature extraction with attributes like "the number of clusters," "curvature," "the number of cutting planes," "minimum feature dimension," "the orientation of model," and "thickness of the slice" to extract features. The HMS technique is independent of these attributes.

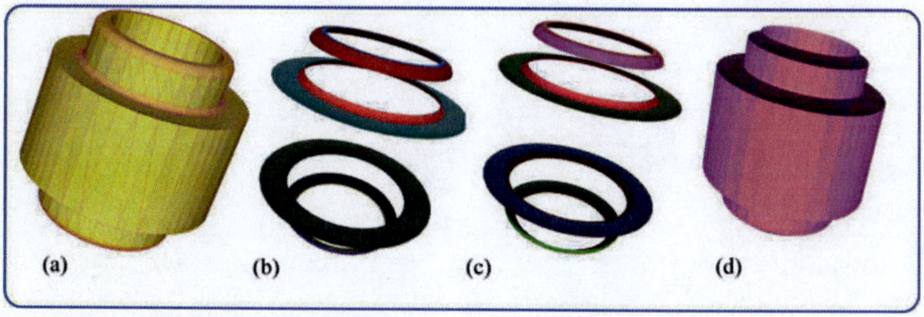

Fig. 17 Vertex-based region growing (VBRG): Stator model. (**a**) Input CAD Mesh model. (**b**) Result of VBRG. (**c**) Primitive extracted. (**d**) Undetected regions

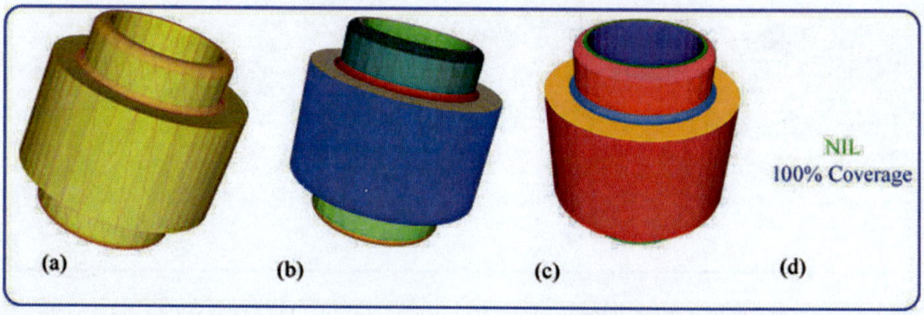

Fig. 18 Facet-based region growing (FBRG): Stator model. (**a**) Input CAD Mesh model. (**b**) Result of FBRG. (**c**) Primitive extracted. (**d**) Undetected regions

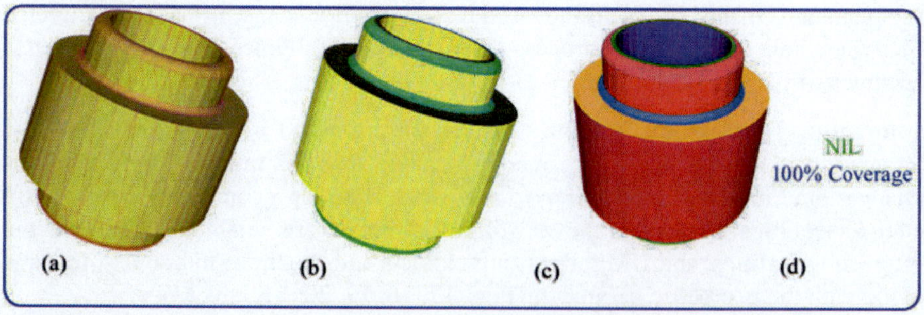

Fig. 19 Proposed hybrid mesh segmentation: Stator model. (**a**) Input CAD Mesh model. (**b**) Result of HMS. (**c**) Primitive extracted. (**d**) Undetected regions

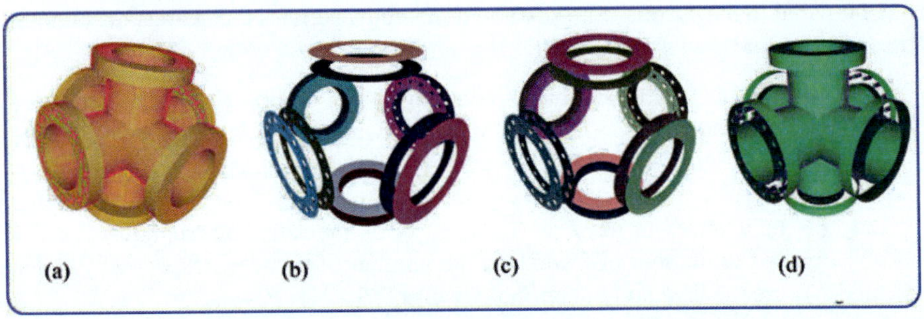

Fig. 20 Vertex-based region growing (VBRG): Pipe model. (**a**) Input CAD Mesh model. (**b**) Result of VBRG. (**c**) Primitive extracted. (**d**) Undetected regions

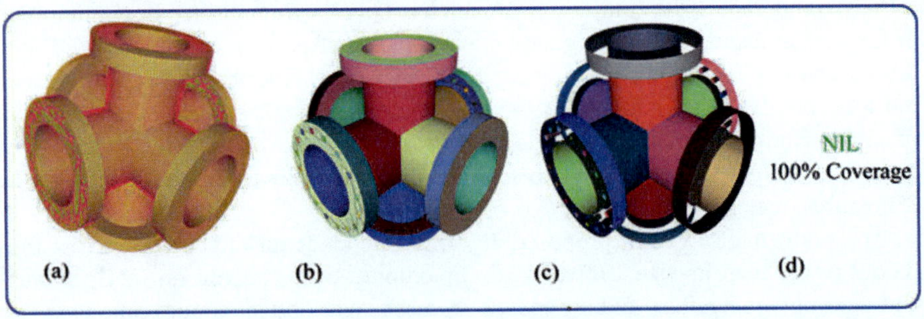

Fig. 21 Facet-based region growing (FBRG): Pipe model. (**a**) Input CAD Mesh model. (**b**) Result of FBRG. (**c**) Primitive extracted # planes are hidden. (**d**) Undetected regions

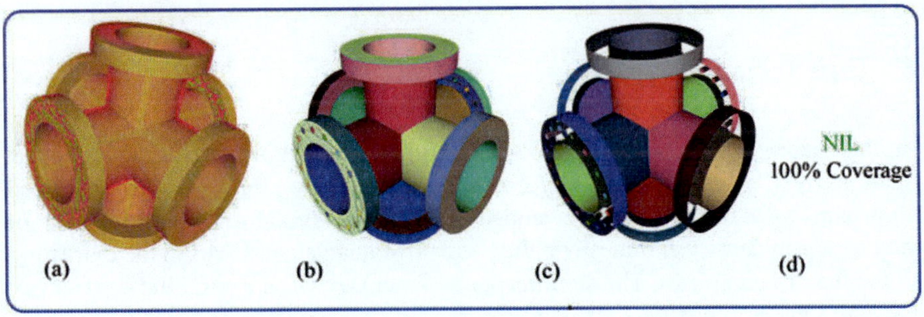

Fig. 22 Proposed hybrid mesh segmentation: Pipe model. (**a**) Input CAD Mesh model. (**b**) Result of FBRG. (**c**) Primitive extracted # planes are hidden. (**d**) Undetected regions

Compared to the current state-of-the-art methods, perhaps the nearest one among others is Le and Duan's method [9].

Time Complexity In this section, the computational complexity of the proposed algorithm is computed. A CAD Mesh Model consists of an ordered set of vertices $S = \{v_i\}_i \subset R^3$ and a set of faces $F = \{f_k = \Delta(v_{k1}, v_{k2}, v_{k3}, n_x, n_y, n_z)\}$ $S = \{V, F\}$

Let $v \epsilon V$ be a vertex of **M**. $T = \{t_1, t_2 \ldots t_k\}$ is the set of all triangles. For N_f and N_v denote the number of facets and the number of vertices, respectively of **M**. A_m and N_m are the area and the normal vector of f_k .

Let the number of facets in the model be F, and the number of vertices in the model be V. In the region growing stage, the complexity of curved facets segmentation is $O(V)$ using vertex-based clustering and $O(F)$ using facet-based clustering. Let S_V be the number of curved facets. Planar face segmentation is run on the remaining facets, the complexity of which is $O(F - S_V)$. Let N_T be the number of features of a specific type (e.g., cylinder) detected after the region growing. Then the complexity of the region growing for that feature type is $O(N_T)$. Let K be the total number of features after the region merging step. Let P be the number of undetected facets after the region growing step. The complexity of iterative reclamation algorithm for P undetected facets with a single feature is $O(P^2)$. With K features, the complexity is $O(K * P^2)$.

The performance of the proposed algorithm depends on the complexity of the model rather than its size. However, the maximum time is spent on mesh import and topology generation, and it depends on mesh size. The maximum time is spent on the curve and planer region segmentation whereas region merging takes the least time. From Tables 7 and 9, it is observed that the overall timing grows stiffly with varying in the number of facets.

6 Conclusion

In this paper, the performance of a hybrid segmentation algorithm has been quantitatively evaluated. The performance evaluated by comparing the proposed algorithm with the recently developed state-of-the-art algorithms, the effect of varying mesh density and mesh quality, and error analysis of HMS in the extraction of the feature parameter. The significance of Area Deviation Factor, the accuracy of FR, and time complexity is also discussed.

The experimentation indicates that HMS outperforms the existing state-of-the-art algorithms and achieves a promising performance. The quantitative results proved that HMS algorithm is efficient and competent, and found to be robust and consistent with coverage of more than 95%.

Future research work will be carried out in the direction of deep learning-based FR. One may develop boundary representation (B-rep) model by segmenting CMM using HMS.

Acknowledgments This research was supported by Centre for Computational Technologies (CCTech), Pune, India. Special thanks are given to Dr. Truc Le and Dr. Ye Duan [9] for helping us to quantify percentage coverage. The authors are grateful to the authors Attene et al. [4], RANSAC [5], and GlobFit et al. [6] who have made their code available to the public.

References

1. Rafibakhsh N, Campbell MI (2017) Hierarchical fuzzy primitive surface classification from tessellated solids for defining part-to-part removal directions. J Comput Inf Sci Eng 18:011006. https://doi.org/10.1115/1.4038144
2. Gao S, Zhao W, Lin H, Yang F, Chen X (2010) Feature suppression based CAD mesh model simplification. Comput Des 42:1178–1188. https://doi.org/10.1016/j.cad.2010.05.010
3. Wang J, Yu Z (2011) Surface feature based mesh segmentation. Comput Graph 35:661–667. https://doi.org/10.1016/j.cag.2011.03.016
4. Attene M, Falcidieno B, Spagnuolo M (2006) Hierarchical mesh segmentation based on fitting primitives. Vis Comput 22:181–193. https://doi.org/10.1007/s00371-006-0375-x
5. Schnabel R, Wahl R, Klein R (2007) Efficient RANSAC for point-cloud shape detection. Comput Graph Forum 26:214–226. https://doi.org/10.1111/j.1467-8659.2007.01016.x
6. Li Y, Wu X, Chrysathou Y, Sharf A, Cohen-Or D, Mitra NJ (2011) GlobFit: consistently fitting primitives by discovering global relations. In: ACM SIGGRAPH 2011 papers on - SIGGRAPH '11. ACM Press, New York, p 1
7. Yan D-M, Wang W, Liu Y, Yang Z (2012) Variational mesh segmentation via quadric surface fitting. Comput Des 44:1072–1082. https://doi.org/10.1016/j.cad.2012.04.005
8. Adhikary N, Gurumoorthy B (2016) A slice based approach to recognize and extract free-form volumetric features in a CAD mesh model. Comput Aided Des Appl 13:587–599. https://doi.org/10.1080/16864360.2016.1150703
9. Le T, Duan Y (2017) A primitive-based 3D segmentation algorithm for mechanical CAD models. Comput Aided Geom Des 52–53:231–246. https://doi.org/10.1016/j.cagd.2017.02.009
10. Shamir A (2004) A formulation of boundary mesh segmentation. In: Proceedings. 2nd international symposium on 3D data processing, visualization and transmission, 2004. 3DPVT 2004. IEEE, pp 82–89. https://doi.org/10.1109/TDPVT.2004.1335163
11. Attene M, Katz S, Mortara M, Patane G, Spagnuolo M, Tal A (2006) Mesh segmentation - a comparative study. In: IEEE international conference on shape modeling and applications 2006 (SMI'06). IEEE, p 7. https://doi.org/10.1109/SMI.2006.24
12. Agathos A, Pratikakis I, Perantonis S, Sapidis N, Azariadis P (2007) 3D mesh segmentation methodologies for CAD applications. Comput Aided Des Appl 4:827–841. https://doi.org/10.1080/16864360.2007.10738515
13. Shamir A (2008) A survey on mesh segmentation techniques. Comput Graph Forum 27:1539–1556. https://doi.org/10.1111/j.1467-8659.2007.01103.x
14. Chen X, Golovinskiy A, Funkhouser T (2009) A benchmark for 3D mesh segmentation. ACM Trans Graph 28:1. https://doi.org/10.1145/1531326.1531379
15. Theologou P, Pratikakis I, Theoharis T (2015) A comprehensive overview of methodologies and performance evaluation frameworks in 3D mesh segmentation. Comput Vis Image Underst 135:49–82. https://doi.org/10.1016/j.cviu.2014.12.008
16. Sunil VB, Pande SS (2008) Automatic recognition of features from freeform surface CAD models. Comput Des 40:502–517. https://doi.org/10.1016/j.cad.2008.01.006
17. Xiao D, Lin H, Xian C, Gao S (2011) CAD mesh model segmentation by clustering. Comput Graph 35:685–691. https://doi.org/10.1016/j.cag.2011.03.020
18. Jiao X, Bayyana NR (2008) Identification of and discontinuities for surface meshes in CAD. Comput Des 40:160–175. https://doi.org/10.1016/j.cad.2007.10.005

19. Di Angelo L, Di Stefano P, Morabito AE (2018) Secondary features segmentation from high-density tessellated surfaces. Int J Interact Des Manuf 12:801–809. https://doi.org/10.1007/s12008-017-0426-8
20. Razdan A, Bae M (2003) A hybrid approach to feature segmentation of triangle meshes. Comput Des 35:783–789. https://doi.org/10.1016/S0010-4485(02)00101-X
21. Xú S, Anwer N, Mehdi-Souzani C, Harik R, Qiao L (2016) STEP-NC based reverse engineering of in-process model of NC simulation. Int J Adv Manuf Technol 86:3267–3288. https://doi.org/10.1007/s00170-016-8434-6
22. Benkő P, Várady T (2004) Segmentation methods for smooth point regions of conventional engineering objects. Comput Des 36:511–523. https://doi.org/10.1016/S0010-4485(03)00159-3
23. Huang J, Menq C-H (2001) Automatic data segmentation for geometric feature extraction from unorganized 3-D coordinate points. IEEE Trans Robot Autom 17:268–279. https://doi.org/10.1109/70.938384
24. Csákány P, Wallace AM (2000) Computation of local differential parameters on irregular meshes. In: Cipolla R, Martin R (eds) The mathematics of surfaces IX. Springer, London, pp 19–33. https://doi.org/10.1007/978-1-4471-0495-7_2
25. Várady T, Facello MA, Terék Z (2007) Automatic extraction of surface structures in digital shape reconstruction. Comput Des 39:379–388. https://doi.org/10.1016/j.cad.2007.02.011
26. Peng YH, Gao CH, He BW (2008) Research on the relationship between triangle quality and discrete curvature. China Mech Eng 19:2459–2462, 2468
27. Peng Y, Chen Y, Huang B (2015) Region segmentation for STL triangular mesh of CAD object. Int J Sens Networks 19:62–68. https://doi.org/10.1504/ijsnet.2015.071383
28. Hase V, Bhalerao Y, Verma S, Vikhe G (2019) Blend recognition from CAD mesh models using pattern matching. AIP Conf Proc 2148:030029. https://doi.org/10.1063/1.5123951
29. Katz S, Tal A (2003) Hierarchical mesh decomposition using fuzzy clustering and cuts. ACM Trans Graph 22:954. https://doi.org/10.1145/882262.882369
30. Katz S, Leifman G, Tal A (2005) Mesh segmentation using feature point and core extraction. Vis Comput 21:649–658. https://doi.org/10.1007/s00371-005-0344-9
31. Mortara M, Patané G, Spagnuolo M, Falcidieno B, Rossignac J (2004) Blowing bubbles for multi-scale analysis and decomposition of triangle meshes. Algorithmica 38:227–248. https://doi.org/10.1007/s00453-003-1051-4
32. Muraleedharan LP, Kannan SS, Karve A, Muthuganapathy R (2018) Random cutting plane approach for identifying volumetric features in a CAD mesh model. Comput Graph 70:51–61. https://doi.org/10.1016/j.cag.2017.07.025
33. Shapira L, Shamir A, Cohen-Or D (2008) Consistent mesh partitioning and skeletonisation using the shape diameter function. Vis Comput 24:249–259. https://doi.org/10.1007/s00371-007-0197-5
34. Hase V, Bhalerao Y, Vikhe Patil G, Nagarkar M (2019) Intelligent threshold prediction for hybrid mesh segmentation through artificial neural network. In: Brijesh I, Deshpande P, Sharma S, Shiurkar U (eds) Computing in engineering and technology, advances in intelligent systems and computing, vol 1025. Springer, pp 889-899. https://doi.org/10.1007/978-981-32-9515-5_83
35. Kim HS, Choi HK, Lee KH (2009) Feature detection of triangular meshes based on tensor voting theory. Comput Des 41:47–58. https://doi.org/10.1016/j.cad.2008.12.003
36. Hase V, Bhalerao Y, Jadhav S, Nagarkar M: Automatic Interacting Feature Recognition from CAD Mesh Models based on Hybrid Mesh Segmentation. Int J Adv Manuf Technol (Communicated)
37. Hase V, Bhalerao Y, Verma S, Wakchaure V (2019) Automatic interacting hole suppression from CAD mesh models. In: Brijesh I, Deshpande P, Sharma S, Shiurkar U (eds) Computing in engineering and technology, advances in intelligent systems and computing, vol 1025. Springer, pp 855–865. https://doi.org/10.1007/978-981-32-9515-5_80
38. Kaick OVAN, Fish NOA, Kleiman Y, Asafi S, Cohen-or D (2014) Shape segmentation by approximate convexity analysis. ACM Trans Graph 34:1–11

39. Hase V, Bhalerao Y, Verma S, Vikhe G (2018) Intelligent systems for volumetric feature recognition from CAD mesh models. In: Haldorai A, Ramu A, Mohanram S, Onn C (eds) EAI international conference on big data innovation for sustainable cognitive computing. EAI/Springer innovations in communication and computing. Springer, pp 109–119. https://doi.org/10.1007/978-3-030-19562-5_11

Enhancing the Performance of Efficient Energy Using Cognitive Radio for Wireless Networks

D. Seema Dev Aksatha, R. Pugazendi, and D. Arul Pon Daniel

1 Introduction

The increased energy utilization is considered to be the key challenges of deploying the wireless networks. High data rate is utilized for wireless applications like multimedia service and interactive services which leads to excessive power usage. Energy expenditure can be minimized or utilized efficiently as much as possible. Different technologies and architectures have been utilized to meet out the challenges caused due to excessive power usage. Several initiative research steps have initiated for reducing power consumption of wireless networks.

Environmental concerns and economic point of view are the key criteria of efficient energy to intend wireless networks. Power utilization of a network increases with the density of access points in order to save the operational costs of the network operator. The increasing capacity of batteries for wireless terminals does not satisfy the user's expectations. The lifetime of the energy can be increased with energy saving schemes and devices should be introduced. In wireless communication, the ratio between total data speed and power used at the transmitter were the efficient energy [1]. Maximum utilization of energy does not mean that energy is employed effectively and efficiently, but we have to note that the budget limitations should not exceed the maximum.

D. Seema Dev Aksatha
Bharathiar University, Coimbatore, India

R. Pugazendi
Department of Computer Science, Government Arts College, Salem, India

D. Arul Pon Daniel (✉)
Department of Computer Applications, Loyola College, Mettala, Namakkal, India

A. Haldorai et al. (eds.), *2nd EAI International Conference on Big Data Innovation for Sustainable Cognitive Computing*, EAI/Springer Innovations in Communication and Computing, https://doi.org/10.1007/978-3-030-47560-4_11

The demand for connectivity, data rate, and QoS cannot be satisfied by increasing the transmit power of a base station (BS). The increase in the transmit power increases the signal strength and also the increases the interference received by the non-utilized base station (BS) during data broadcast. This leads to signal-to-interference-plus-noise-ratio (SINR) that had a great force of QoS. The efficient energy and better utilization can be attained with the help of cognitive radio (CR) technology. Cognitive radio itself an emerging technique in intelligently adapts of particular broadcast or reception parameters by sensing the environment. Cognitive radios can sense their environment and adapt it with resource constraints. CR finds the spectrum which are less interfered, and several channels are bonded together to achieve high broadcast speed. It saves the power during data broadcast and maintains the signal-to-noise ratio.

This paper mainly concentrates on the reflection of efficient energy that is fundamentally inclined by the trade-off between energy utilization and achievable quality of service (QoS). The aim of the paper is to meet out the required QoS while minimizing the required amount of energy. The objective of this paper is to present the metrics employed to measure and enhance the efficient energy and quality of service. The paper is designed in such a way that it is energy efficient when the wireless network has limited energy capacity and provides support for quality of service during multiple traffic in the wireless networks.

This organization of the paper has six sections. Sections 2 and 3 present the related work and proposed model of the system. Sections 4 and 5 present the performance measures of efficient energy and results simulated. Section 6 is the conclusion and presents the auxiliary research scope.

2 Related Work

In [2], the author also identified that for higher throughput targets, employing additional microsites is always beneficial. Further, discussed that for higher area throughput targets employing additional microsites is always beneficial, where the higher the throughput requirements, the higher the number of microbase stations. Higher user densities require more microsites in order to achieve better area power utilization figures was identified by the author.

In [3], Furthermore, the author had also identified that the three categories based on the application scenarios namely, energy-efficient architectures, energy-efficient resource management, and energy-efficient radio technologies, should be employed to improve the efficient energy of cellular networks.

In [4], the authors addressed energy efficient and low-power design within all layers of the wireless network protocol stack. In [5], the author outlined that recent efficient energy metrics and power utilization models should be employed to get better energy.

In [6], the authors observed that certain ratios of microcells and picocells per macro BS will result in sub-optimal of area-efficient energy. It also increases as the percentage area of macro BS overlaid by smaller cells and the density of micro/picocells increases.

In [7], the authors presented a comprehensive summary review of the state of the art concerning link and routing layer technologies developed for constrained wireless sensor network, IoT, and CPS applications.

In [8], the authors proposed an energy-aware heterogeneous cell deployment and operation framework that has theoretical results as well as practical guidelines on how mobile operators manage their BSs. The authors specifically focused on a problem pertaining to total energy utilization minimization while satisfying the requirement of area spectral efficiency (ASE) and decomposed it into deployment problem at peak time and operation problem at off-peak time.

In [9], the authors discuss an overview of efficient energy metrics in CRNs related to its design and operation. Metrics are categorized into the component, equipment, and network levels for easy analysis. Authors identified that the performance metrics of the network was also analyzed where the probability of false alarm, probability of detection, and probability of missed detection metrics were evaluated.

In [10], the authors presented different efficient energy techniques which are proposed to reduce the power utilization in cellular networks and took a detail literature survey on techniques that are employed to improve the efficient energy of cellular networks.

3 Proposed Model

Traditional communication system follows single system of packet delivery, and available bandwidth will be shared by the senders on the basis of the packet size. The aim of this paper is to identify the best nodes to transmit data from source without getting delay in reaching its destination in a wireless environment. The network consists of n nodes and CR nodes, and all the nodes use the same communication link to transmit data from source to destination.

Here, cognitive radio (CR) node plays a vital role and identifies the neighboring nodes to make data broadcast as a successful one. The nodes are monitored for data broadcast, and they can be categorized as loss monitor and drop monitor. If the nodes are identified as loss monitor for data communications in wireless network, it shows that direct connections are made from one node to another. If the node is found to be drop monitor, it shows how sub-connections are made within the nodes to make data broadcast.

In this case, last broadcast is found to be successful; then that node is employed for data transfer. CR node is employed to identify the neighboring nodes which make successful data broadcast and helps to improve the energy utilization.

Cognitive radio technology is actually employed to deal with the spectrum scarcity and spectrum usage concern. But, it has in-built properties which supports energy utilization. CR is a radio that can change its transmitter factors based on communication with the surroundings in which it controls in order to reason, plan, and choose future activities to convene various needs.

4 Performance Metrics

In this division, we seem to be the recital metrics which are employed to assess the efficient energy in the wireless networks. The metrics includes throughput, end-to-end delay, packet delivery ratio, and broadcast ratio.

4.1 Packet Delivery Ratio

Packet delivery ratio refers to the quantity ratio of transported data packet to the end. It illustrates the point of transported data to the end and calculated using the formula:

$$\text{PLR} = \frac{\sum_i \text{Packet delivery}}{\sum_i \text{Packet sent}} \times 100. \tag{1}$$

Packet delivery ratio is employed to survey notion through the framework. It denotes size among the packets in the network and explains how many packets will be sent from source to destination with certain speed; it is called as delivery ratio.

4.2 End-to-End Delay

It refers the standard time taken by the data packet to be broadcasted from source to destination across a network. It also includes the delay caused by route detection process and the queue in data packet broadcast [11]. Only the data packets that are effectively transported to the destinations are calculated and given by the formula:

$$\text{End-to-end delay} = \frac{\sum_i (\text{Arrive time} - \text{Send time})}{\sum_i \text{Number of connections}}. \tag{2}$$

4.3 Throughput Ratio

Throughput refers to the average rate of successful message delivery over a communication canal. The performance is usually measured in bits per second or in data packets per second or data packets per time slot. Performance is otherwise called as bit rate or bandwidth.

$$\text{Throughput} = \frac{\sum_i \text{Packet delivered}}{\text{Delay}}, \tag{3}$$

whereas the delay in time at distribution packets is delay $= \sum_i$ Packet arrival – Packet start time.

In order to find a QoS lane for sending real-time data to the gateway, end-to-end delay condition should be met.

4.4 Broadcast Ratio

Network performance is named as broadcast; it gives the maximum rate at which data can be broadcasted over a message lane or channel. It mainly depends on four factors, namely, data rate in bps; bandwidth, forced by transmitter and nature of broadcast medium, expressed in cycles per second, or Hz; noise, average noise level over channel; and error rate, percentage of time when bits are flipped.

4.5 Utilization of Energy

The asset performance and business success were based on the utilization.

5 Simulated Results

Here, the outcome is based on the performance of proposed framework. Making use of NS2, the overall performance of the efficient energy has been evaluated. In the experiment, 100 randomly placed nodes in a 1000 × 1000 m^2 area have been considered. The parameters employed in this work are shown in Table 1.

Each node has an initial energy of 5 J and well thought-out to be non-functional if its energy level arrives at 0. Each packet is time-stamped when it is allowed for the calculation of average delay per packet. Each packet has an energy field that has to be updated during the packet broadcast to calculate the average energy per packet. A packet drop probability is considered as 0.01, which makes the simulator more realistic.

In the proposed work, cell exposure is required to be 99% for each measured cell distance and every base station type that has been considered in order to provide high utilization of energy and minimized energy utilization.

Figure 1 describes the energy distribution of each and every node through runtime and the circulation flow utilization for packet broadcast.

Figure 2 describes the access node selection for secure data broadcast.

Table 1 Simulation constraint

Parameters	Values
Energy model	Energy model
Channel	Wireless channel
Propagation model	Two-ray ground model
Initial energy	100
Number of nodes	100
X value	1000
Y value	1000
Pause time	10.00
Packet size	512 mbps
Queue length	250
MAC type	802.11

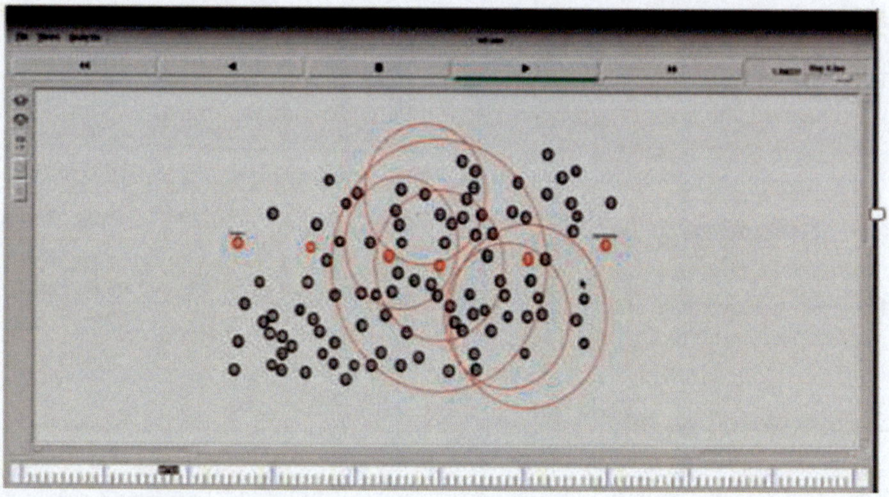

Fig. 1 Energy distribution

Figure 3 describes the model of a reproduction power equivalent of the regions in network nodes during data broadcast.

Figure 4 describes the power utilization during data broadcast from resource node to target node. Data is delivered from resource node to target node in a secure and fastest manner.

In this proposed work, data broadcasted from the source node to the destination node with reduce in the packet delivery ratio is shown in Fig. 5. It describes that the packet delivery ratio of the nodes in the network increases power utilization.

Under heavy traffic load, the packet delivery ratio varies with their traversing rate. The capability of delivery ratio will be from 20% to 40% for the data to reach its destination. The node collects and transmits the data that is received by the sensing nodes to the destination node using the cognitive radio. It is impossible to provide higher packet delivery ratio by the traditional method. The simulation

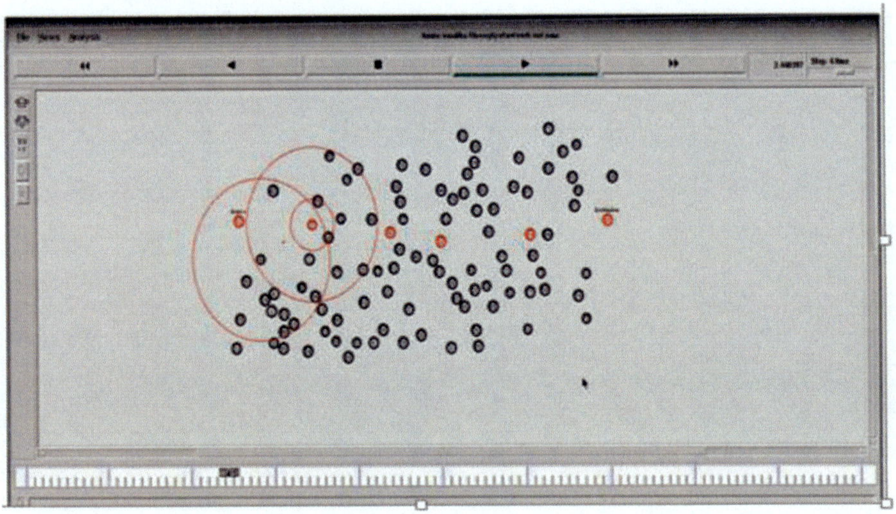

Fig. 2 Broadcast energy

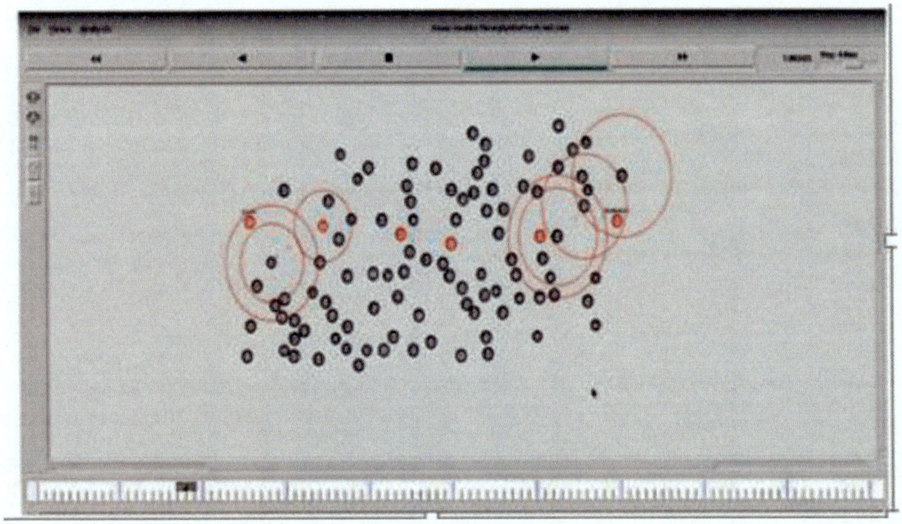

Fig. 3 Data broadcast

results show that using the cognitive radio technology, the packet delivery ratio has been decreased rather than traditional method because the cognitive radio node senses the data which has to receive the data and to be processed from the resource to target.

From Fig. 6, as the quantity of packets amplifies for data broadcast, end-to-end delay has been minimized.

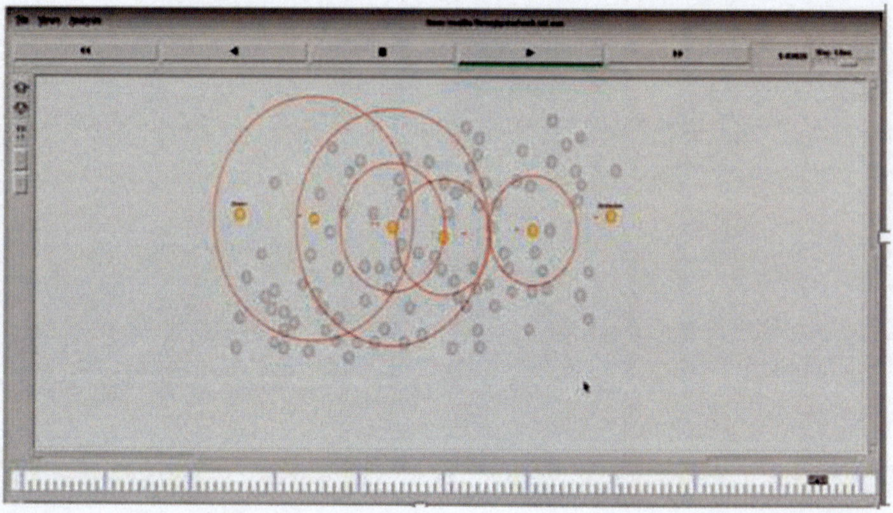

Fig. 4 Power utilization

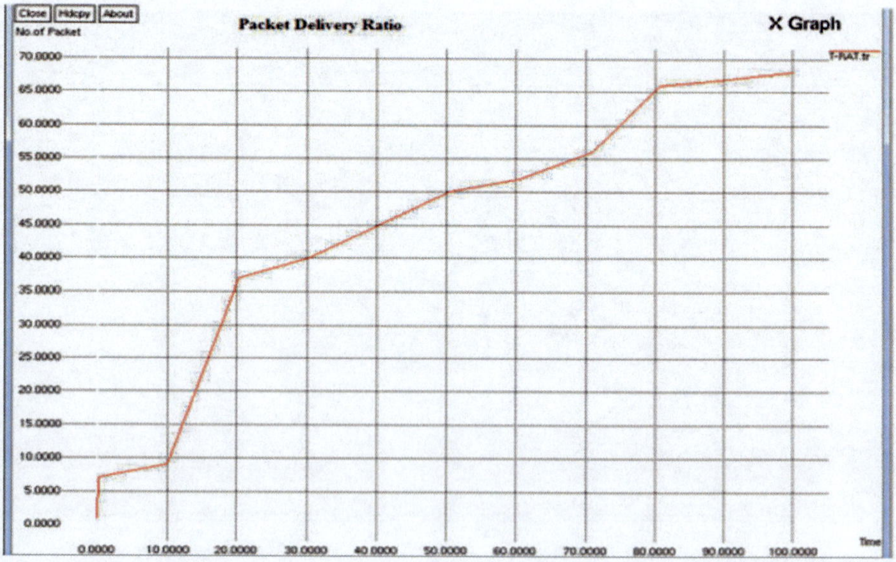

Fig. 5 Packet delivery ratio

The bandwidth utilization shows the maximum data rates of the Transceiver and Receiver of the node. During communication, the max data rate is affected by the noise in the system and depends on the noise of the propagation medium, the noise figure of the Receiver, the power level of the Transmitter, the broadcast loss, and the maximum tolerable error rate.

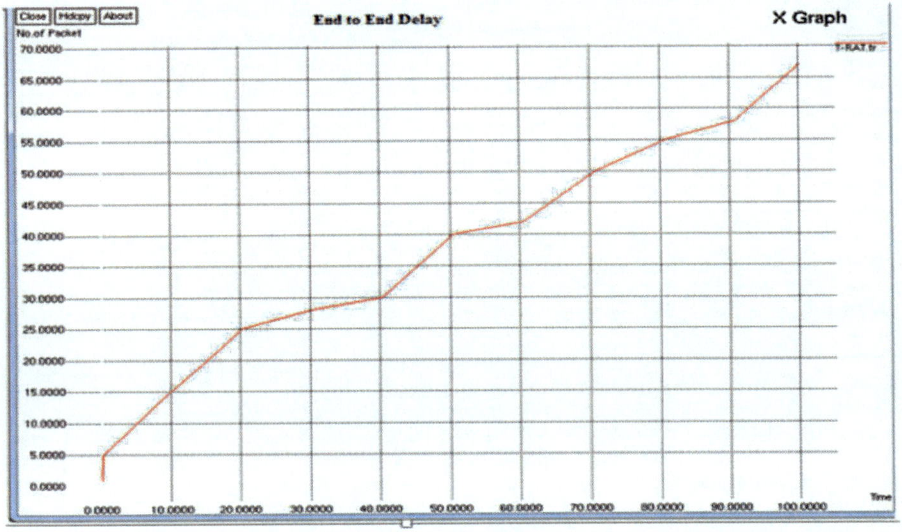

Fig. 6 End-to-end delay

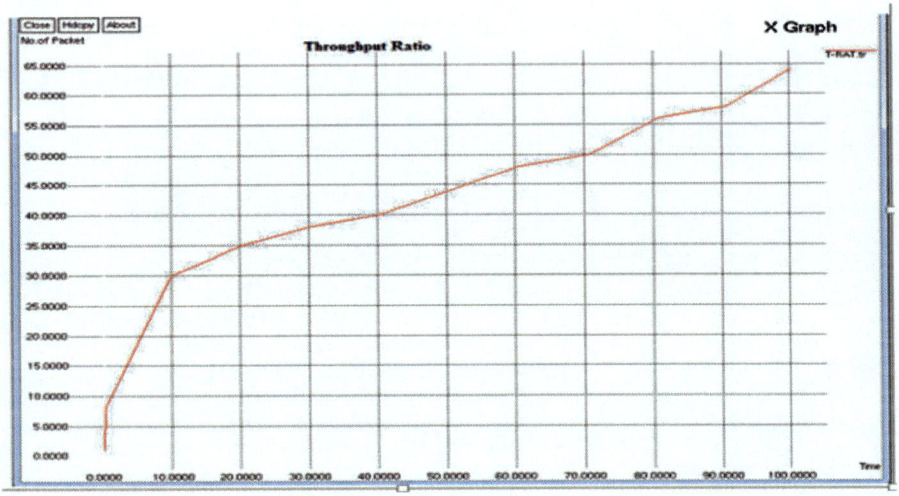

Fig. 7 Throughput ratio

In the proposed work, the overall data broadcast increases with maximized throughput as shown in Fig. 7.

Figure 8 describes the broadcast ratio of the node transmitting from source to destination in radio access technology.

The efficient energy and throughput are not only improved by means of efficient energy schemes, but also the performance of all users as shown in Fig. 9. The energy utilization reduced the interference and has less throughput loss.

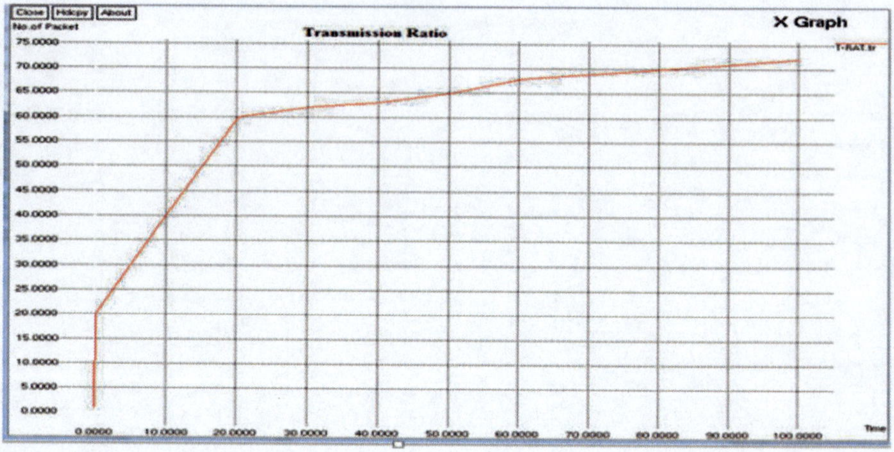

Fig. 8 Broadcast ratio

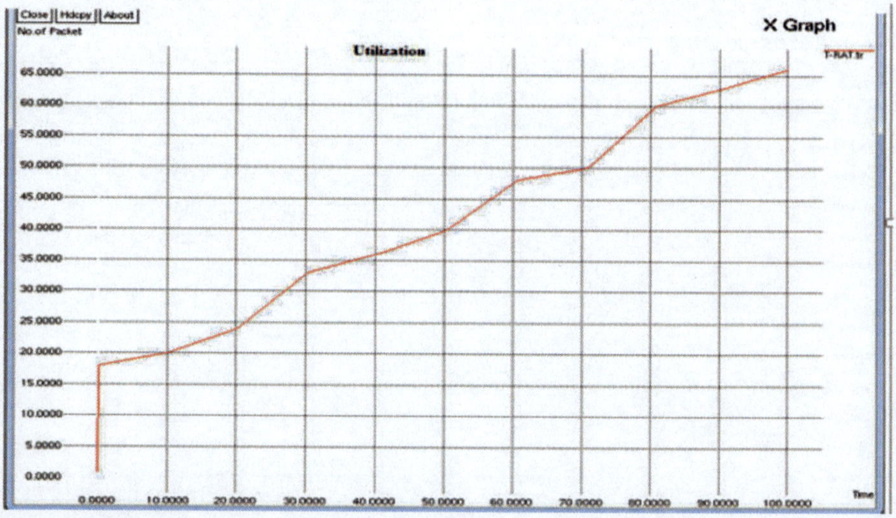

Fig. 9 Energy utilization

6 Conclusion

Efficient energy is considered the fast development in wireless networks because different stakeholders in wireless networking are more concerned in the design of the problem, green technology, cost, and final user satisfaction. The proposed method uses cognitive radio technology to achieve high signal-to-noise ratio with the similar broadcast power. It reduces the interference during data broadcast and allows selecting the node and frequency for data broadcast. The number of packets employed for data broadcast and different transfer modes are considered based on the size of the data to be broadcasted. The proposed work identifies the measures to improve the efficient energy by limiting the interference and delay, improving the throughput and data broadcast. With the help of selected QoS parameters, the power utilization, bandwidth scheduling, and energy utilization have been increased.

References

1. Sodhro AH, Chen L, Sekhari A, Ouzrout Y, Wu W (2018) Energy efficiency comparison between data rate control and transmission power control algorithms for wireless body sensor networks. Int J Distrib Sens Netw 14(1):1–18
2. Richter F, Fehske AJ, Marsch P, Fettweis GP. Traffic demand and energy efficiency in heterogeneous cellular mobile radio network
3. Xiaochen S, Sun E, Li M, Yu FR, Zhang Y (2013) A survey on energy efficiency in cellular networks. Commun Netw (Sci Res) 5:656–662
4. Jones CE, Sivalingam KM, Agrawal P, Chen JC (2001) A survey of energy efficient network protocols for wireless networks. Wirel Netw 7:343–358
5. Ibrahim AA, Kpochi KP, Smith EJ (2018) Energy consumption assessment of mobile cellular networks. Am J Eng Res (AJER) 7(3):96–101
6. Abdulkafi AA, Tiong SK, Chieng D, Ting A, Ghaleb AM, Koh J (2013) Modeling of energy efficiency in heterogeneous network. Res J Appl Sci Eng Technol 6:3193–3201
7. Boyle D, Kolcun R, Yeatman E (2017) Energy-efficient communication in wireless networks. Open Access Peer Reviewed Chapter, Mar 2017
8. Son K, Oh E, Krishnamachari B (2015) Energy-efficient design of heterogeneous cellular networks from deployment to operation. Comput Netw 78:95–106
9. Orumwense EF, Afullo TJ, Srivastava VM (2016) Energy efficiency metrics in cognitive radio networks: a hollistic overview. Int J Commun Netw Inform Security (IJCNIS) 8(2):75–85
10. Ahmed W, Junaid Arshad M (2017) Energy efficiency improvement through cognitive network cooperation. Int J Comput Sci Telecommun 8(3):6–12
11. Rashidha Begam K, Savitha Devi M (2014) Protected data transfer in wireless sensor network using promiscuous mode. Int J Comput Sci Mobile Comput 3(11):643–648

An Internet of Things Inspired Approach for Enhancing Reliability in Healthcare Monitoring

G. Yamini and Gopinath Ganapathy

1 Introduction

The major advancement in information and communication technologies led to development of all levels of innovative social environment in recent years and deploying into patients' daily lives has made Internet of things more effective technology rather than traditional methods. Especially, in healthcare sector, the need for real-time monitoring plays a decisive role in which patient's data gained from sensors provide unvarying monitoring of patient's health and in return enabling doctors to examine the health problems without worrying the underlying physical location of the patient's whereabouts; numerous research work have emphasized the importance of integrating Internet of things reliability factor into healthcare, where reliability metric plays an important role in intelligent healthcare. The only downside of providing a proper reliability in IoT and healthcare is data transfer cost overheads. In order to overcome this challenge, novel reliability metrics as well as methods should be deployed to enhance the reliability metric offered by Internet of things. The main focus of IoT-based healthcare applications should define its reliability factor and its method explicitly experimented and discussed in the paradigm of healthcare sector. The need for concentration of procurable data transmission methods and its selection of effective integration of reliability metric which can lessen the costs. An analytic technique is conferred supported with reliability (theoretical metric) and experimentally examined. In recent years, Internet of things technology has rapidly conquered the attention of researchers as well as engineers to the point of becoming one of the most effective technologies as described at the International Consumer Electronics. The current Internet, which

G. Yamini (✉) · G. Ganapathy
Department of Computer Science, Bharathidasan University, Trichy, India

A. Haldorai et al. (eds.), *2nd EAI International Conference on Big Data Innovation for Sustainable Cognitive Computing*, EAI/Springer Innovations in Communication and Computing, https://doi.org/10.1007/978-3-030-47560-4_12

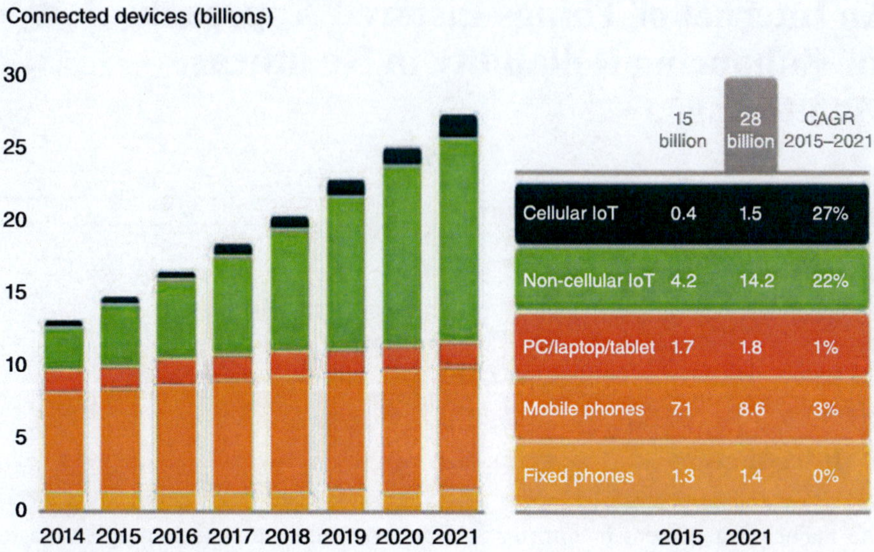

Fig. 1 Connected devices with IoT

simply consists of compute servers, desktops, laptops, smartphones and handheld, is bound to scale exponentially to have all kinds of devices and things.

Precisely speaking, the current Internet is the network of networked computers. But the future is the network of networked computers, devices and digitized entities. In other words, the Internet of devices, services, energy and things is to emerge and evolve rapidly to empower not only businesses but also people in their daily deals, decisions and deeds. When such kinds of connected and integrated devices, services, data sources, applications and things talk to one another, there will be massive amounts of interaction, transactional, commercial, operational and analytical data. Therefore, IT professors and professionals across the world are striving hard and stretching further to bring forth pioneering data analytic products, platforms, processes and patterns to meticulously capture, cleanse and crunch the growing data volumes using the batch as well as real-time processing methods to extricate actionable insights in time as shown in Fig. 1.

1.1 Demystifying High Reliability in Healthcare

In recent years, many methodologies have been developed to determine the reliability as well as performance in healthcare applications. Most of these methodologies concentrate on framing a standard group of metrics with the aim of employing them as "benchmarking" tools in healthcare. The Joint Commission in 2010 presented "The Specifications Manual for National Hospital Inpatient Quality Measures". It provided pre-existing guidelines for high reliability and quality care and combining them with the measurement frameworks. The main aim is to determine the adherence to clinical practice guidelines as a way of quantifying reliability and quality of care

The International Quality Improvement Program was created in Maryland, USA, which included 7 hospitals and the Maryland Hospital Association for this project. In the year 1994, this project became worldwide recognized and turned into Joint Initiative which includes more than 900 hospitals from the USA, 5 hospitals in England as well as 1 in Japan. In the year 2019, this initiative turned into largest data repository of quality indicators and serves the performance measurement and safety improvement for healthcare organizations worldwide. Another promising tool for assessing the quality enhancements is PATH known as Performance Assessment Tool for quality improvement in hospitals. The main aim of this tool is to assists hospitals to determine and assess their reliability as well as performance, question their results as well as act upon them, thereby enhancing the reliability as well as quality of care. However, PATH does not present any methodology for healthcare centre to create their own indicators but presents a path for the implementation, collection as well as analysis of a group of 18 "already-developed" indicators that determine reliability as well as performance in 6 various coordinated dimensions

2 Prominent Work in Healthcare

A compact home monitoring system device with low power consumption and with the entire functionality had been developed by the authors of [1] and is low in its cost and very easy to use. This utilizes the concept of distributed database storage, and it is said that only certain systems deal with the numerous devices. The patient data is examined by this system, and it is stored in the remote server. Certain limitations are found in the existing m-healthcare systems, and one of them is that the clinical data integration when needed by the professional is not provided by the closed systems that may arise from various sources. A new healthcare system that utilizes the concept of service-oriented architecture (SOA) called the SOAMOH presented in the paper [2], which integrates the clinical data by supporting HL7 and enables the people to attend the healthcare services through the wireless sensor network (WSN). But this system is never been implemented or tested at any aspect. The author in [3] had introduced a system that could measure the patient's heart rate with the software tools on the mobile platform. Moreover, an application known as the TuneWalk had been introduced that makes to patients to follow their cardiac rehabilitation exercise in their home. In the evaluation part, two of the authors had participated in the testing part. WBA measures the activity data and variability of the heart rate that had been recorded by the TuneWalk. The estimation made by the test subjects in TuneWalk was accurate at their walking speeds, but if the test subject runs at certain speed, the results become inaccurate. Due to this problem, the home-based CR program had been developed on the basis of walking and not suitable for running. An architecture based on the heart failure related to healthcare had been developed in the paper [4]. Several sensors are composed in this particular framework that could take various physical quantities and process them all, and the hub could be the smart sensors, which will transmit these data to the caretakers or the user.

The predictive and preventive device that could check the abnormality of the heart with the possible tachycardia had been proposed in [5] that utilize the method of advance prediction model in order to estimate the heartbeat rate of the sufficient patients. When the patient's heart rate exceeds beyond a certain limit, then notifications or alerts could be sent to the professionals or the caregivers so that appropriate actions could be taken. The authors of [6] introduced an extended version of traditional service-oriented architecture (SOA), and this framework integrates the personal health monitoring system data with an electronic healthcare network. This system could process complex event and forms a greatest advantage when compared to the traditional one. Wearable blood pressure sensors had been developed by [7] that monitor the health of the patient and sufficient treatment could be given at the patient locality for their hypertension.

AAL system had been built by the architecture of HERA which had been made in [8] that helps in detecting the early stages of certain cardiovascular diseases (Alzheimer's disease and/or other diseases) for the older or elderly person. This system is very low in their cost and could also enhance the quality of life. The developers of HERA system had done a complicated architecture and methodology with the sufficient evaluation process and are about to implement their plans. Hence, there is no sufficient proof for the system to work effectively. Certain research people had discussed this architecture in the hardware point of view. Implementation of a hand-off protocol could be done by the coordinates of the wireless body area sensor networks (WBASNs) and APs, when the RSS of the former falls below acceptable levels and it is presented in the paper [9].

The capacity of the system could be leveraged by implementing several radio channels that are promoted and employed, which permits multiple users monitoring from various rooms. Moreover, they also tried to build an effective and reliable health monitoring system based on WBASN that helps in monitoring the patients from their household. The data are collected through the Bluetooth that are said to be the offline data. Electrocardiograph (ECG) device provides the sufficient data, and the data is said to be offline since the real-time and online data monitoring is said to be highly difficult [10]. Body sensor network (BSN) utilizes a software network called SPINE (signal processing in-node environment) that is presented in [11]. This software requires data from various nodes and also enables emulation of a set of nodes forming a WBSN. Different healthcare systems are introduced in the literature [12]. Over the past decades certain equipment had been developed, and they had been modulated recently in certain research field, but still the capacity of monitoring the non-communicable diseases patients during their regular living activities is said to be tedious. Moreover, the patient is restricted to use the system within the room allocated in the specific area where the wireless body sensor network had been installed [13]. If the patient gets out of the area, then the device fails to record that particular data that results in the data loss. Additionally, the mobility could be limited by the usage of several hardware devices or PCs. Moreover, the improvement in these areas could enable the patient to view his/her medical data anywhere and anytime with limited hardware equipment [14].

3 An Effective High Reliability Organizational (HRO) Principles for Healthcare

Due to the catastrophic and potential risk sequences for the occurrence of failures and operation complexity in the field of healthcare, high reliability should be the attractive consequence in this particular field [15]. People interpret the high reliability as the concept of effective standardization in the field of healthcare as shown in Table 1. This concept runs beyond the idea of standardization, thus making it more reliable. According to an organization, high reliability is found to be the concept of persistent mindfulness as described. Resilience is highly cultivated that is cultivated by prioritizing safety relentlessly with the pressure over the performance, and this is mentioned as the persistent mindfulness among the organization [13]. In the field of healthcare, reliability could be maintained by making the system to think and making a failure-obsessed system. Certain industries should learn from certain platforms that follow their rule in order to maintain their person's safety (e.g. aviation). Therefore, continuous improvements should be met by the organizations, which are found to be the key for maintaining the reliability in healthcare. This roadmap is followed and being experimented with the Stanford University Hospital, ThedaCare and Stanford Children's. This manages and leads several ways by applying several principles and tools, which were operated in the field of healthcare manufacturing process.

This relies on dealing with everyday problem by relying the development of frontline staff, and then the frontline worker gets connected with their purpose of the organization [16].

Highly reliable playbook had been designed as a principle that becomes an outcome for the field of Indian healthcare organizations with the quality and the

Table 1 High reliability organization

Principle in guiding HRO	Relevant explanations
Failed preoccupations	Hold the weak signals and failures
Operational sensitivity	Accident identification without any error
Simplification disinclination	Situations that could not be avoided but taken lightly
Determination for the organization	Detect, change and terminate the occurrence of errors
Respect the relevant and the qualified experts	The advice could be taken from all level categories

Table 2 High reliability principles for healthcare

Guiding principles	Care takers behaviour	Examples
Failed preoccupations	Attitude	Physicians and other healthcare professionals marking the correct surgical site
Operational sensitivity	System-based value practices	Maintaining a good record of the team with the incoming and outgoing information and their present situation status to enhance the accuracy of the team
Simplification disinclination	Metacognitive skills	Patients admitted at the critical situation, and the fellow residents should know their roles and responsibilities
Determination for the organization	Emotional Intelligence and assertion	Nurses have the right to promote their advice to the physician regarding their allergies or any other physical information, which they might have known before
Respect the relevant and the qualified experts	Competency skills and leadership	Patient's healthcare should be monitored at the regular basis by the nurse who had been promoted to monitor their regular activities

cost problems in the clinic and in the hospitals. At last, the cost-effective with the safest system could be the result for every person. The question arises whether the healthcare system could be more reliable than the airline industry. In recent years, stakeholders, payers, providers and healthcare consumers have demanded better business outcomes and patient's care by achieving better reliable performances and organization status [17]. This industry should catch with other consumer-based industries and invest with several efforts and resources that could maintain a solid track to get operated with the reliable organization model to enhance the patient's care outcomes and business result performance. The future requirements had been solved with the mission-critical HRO model and physician-based practices followed in the clinic as shown in Table 2. The result of the sustainable business performance with the reliable performance outcomes could be consistently achieved by having a common a shared vision with the common destiny.

4 An Internet of Things Inspired Approach

Creating IoT-based healthcare applications offers the possibility to enhance the patient day today activities ranging from personal safety, reliable monitoring of the patient as well as effective use of resources [18]. Even though integrating IoT with healthcare provides several benefits and has gained major importance over a short period of time to point of releasing new IoT-based healthcare products, sensors which poses important drawbacks which appertain to the complexity of reliable designing such systems including heterogeneity as well as scalability of the rendered data. This development led to integration of ambient assisted living

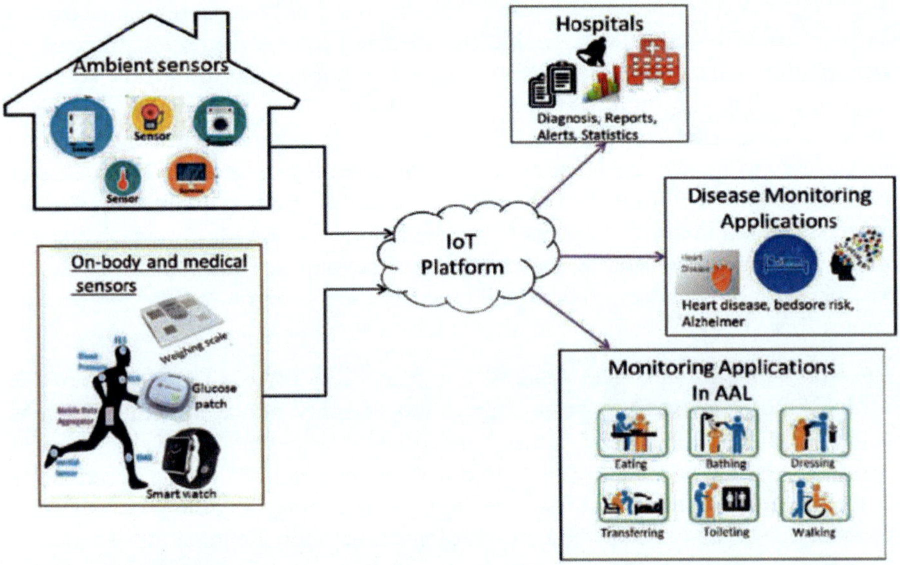

Fig. 2 Lacking of reliability in IoT-based healthcare activity recognition

(AAL), into healthcare in which AAL provides an effective IoT platform governed by artificial intelligence algorithms, thereby satisfying the reliability metric in monitoring patient's health in their place of living in a safe manner [19]. The AAL system includes activity monitoring of patients which is important for patient suffering from Alzheimer's disease, bedsore, diabetes, and osteoarthritis.

As shown in Fig. 2, an IoT-based healthcare system is implemented in the environment of monitored patient's wearable and ambient sensors in which each sensor consists of particular features as well as functionality as requested by any healthcare application connected to a dedicated network based on IoT communication platform, thereby providing real-time IoT data to the doctors for appropriate decision-making.

4.1 A Simple Method Using Metrics for Data Transfer for IoT-Based Healthcare Applications

However, creating an IoT platform lacks reliability metric because of high heterogeneity in the healthcare application. Several IoT applications use the data provider from various domain applications. The device scalability could be entirely supported by the replacing the sensor in the case of transparent system's components. This is also relevant that the information from the one sensor could be used for several other purposes by the not related systems. For example, the sensor that could detect the movement could be used at homes to monitor the regular activities and also could be used to determine the specified interest activities. In order to successfully maintain and make this evaluation possible, new sustainable software architecture should be built that makes the deployment easy with the existing ones. For instance, the physical device pooling is required in the multi-pathologies around the patients to reduce the amount of redundant deployed sensors. The message-oriented middleware (MOM) is the suitable publish/subscribe architecture for the above stated requirement. This could handle the distributed system which focuses on the information by sending and receiving of messages. The exchange of data is possible with its architecture that maintains an open access to all the sensors and applications.

Semantic representation: The semantic interoperability is the main principle of this architecture. In order to ensure the interoperability of the system, the MoM architecture had been improved with the semantic representation (SeMoM) relies on this particular model. The sensor and the metadata had been described in the semantic representation that enables in describing the domain concepts in observing the sensor with the same ontology. Due to this combination of semantic representation in the MoM architecture, the enhancement in the IoT system could be done with the smarter component that involves both the communication and the semantic features.

Domain requirements: The description of any application domain could be done by using Cognitive Semantic Sensor Network ontology (CoSSN) in the architecture. The CoSSN could be used in several healthcare concepts by using the on-body sensors with the determination of exact body conditions. However, if several redundant sensors are used, it could lead to the accumulation of several equipment that could bring certain financial concerns, and it disturbs the patient's comfort. Therefore, the IoT healthcare application could be bought with the MoM architecture that combines three concepts such as the semantic interoperability, information pooling and health requirements and the loosely coupling, which has the "independent, weakly coupled software components driven by semantic data".

The low coupling among the software components is promoted by the publish/subscribe mode of architecture since it is not mandatory for the sensor to know when, where and for what the data should be processed. Similarly, certain information could be obtained from various other means like the sleep disorder information obtained through the movement detectors or the pressure sensors and the system need not be aware of it.

As stated previously, the main purpose of this architecture is to promote the interoperability with the loosely coupled system and enhancing the scalability. Moreover, these components could also be named as the publishers, or it could be viewed as data providers or consumers or the subscribers. The broker is said to be the communication bus through which all the components communicate. The exchange of information is done in the form of token among the publisher and the receiver.

Publisher: Specific topics on the middleware are published or send by the publisher. In the healthcare as well as in the IoT domain, the sensor acts as the data providers in which enormous amount of data are collected. These sensors could be incorporated in the house as the ambient sensors else could be on body sensor that depends on either the wireless or the wired communication.

Subscriber: Here the specific topics are registered by the subscribers, and then every labelled message is consumed by each corresponding topic. The complex operations could be performed with the raw data consumed. The main subscriber in the healthcare domain is the monitoring applications. For example, the nurse will have a pad that severs as the monitoring application, which could alert by sending status regarding the patient's condition and is said to be the ultimate subscriber.

Broker: In MOM architecture, loosely coupled components means that the publisher and the subscribers do not communicate directly, and they communicate through the mediator called as the broker. Through this concept, the switching of provider could be possible with the data available without the knowledge of the providers or the consumers. As long as the information is still produced, the removal or the addition of the component is said to be transparent from the other equipment. This evolves the system capability and promotes the system with longer flexibility. The exchange of messages with the composition of several component of MOM architecture is shown in the figure given below. In this scenario, the broker gets connected with one publisher, and it sends a message with the specific property called the "topicX". Then the subscribers 1 and 2 get connected with the broker by subscribing the topic. Once they subscribed the broker, those subscribers started to receive the messages regarding the topic. The messages sent prior to the subscription (message1) will not be sent again.

4.2 An Effective Activity Recognition to Enhance Reliability in Healthcare

The proposed activity recognition (AR) employs three-axis accelerometer to detect the patient activity. The training traces are collected and labelled and then employed to create a model that can determine the activity label for fresh trace. These traces are created in the form of consecutive linear acceleration measurements of the patient's wearable sensor in x, y and z axes for various activities of the patient. After obtaining the labelled traces from the various patient activities, the trace contains successive readings obtained from the axes of the accelerometer sensor.

Consider t represents a matrix trace with columns as well as four rows. In order to represent the acceleration data from x, z and y directions, the first rows are employed, while the remaining last row represents sampling time which corresponds to the number of recording of the accelerometer present in the column.

The point value is asynchronous for different traces derived from the same activities. For instance, in single axis, a walking distance may arise from the positive acceleration, and at the same time, another walking distance may commence from the negative value. The proposed method processes the traces from every activity produced from the time series of probability distributions model, and the features are constructed from this model parameters.

Assuming the features $P1, \ldots, Pn$, a probabilistic model consisting of associated class (C) could be $F(C|P1, \ldots, Pn)$. The label for a specific group of features can be obtained for a class which possess maximum conditional probability:

$$\arg\max \ C = F\,(C|P1, \ldots, Pn)$$

According to the Bayes' theorem,

$$F\,(C|P1, \ldots, Pn) = F(C)F\,(P1, \ldots, Pn|C)\,/F\,(P1, \ldots, Pn)$$

Instead of obtaining the class with maximum $F(C|P1, \ldots, Pn)$, the greatest class $F(C)F(P1, \ldots, Pn|C)$ is selected.

Localization Method

The pedestrian dead reckoning (PDR) method works with the sampling frequency of 90 Hz with the three-axial accelerometer measurement for the indoor localization. According to this case study, the normal working position of accelerometer is vertical. Moreover, the working position that is taken previously is stored in the device memory for setting the calibration point automatically. Usually, the angle among the anchor and the unknown sensors are taken with the localization methods, and then the locations are estimated among the unknown sensors with certain geometry algorithms. Therefore, the main elements to be determined for the sensor localization are as follows: angle measurement, geometry measurement as well as the distance measurement.

Once the location of all the sensors involved is determined with the multidimensional scaling, the accurate estimation of the position is done with the error-prone and limited distance information. This is said to be the main advantage of using localization method in estimating the position of the unknown sensors. The coordinates are assumed with the distances that produce the dissimilarity of data, and this is the basic idea of this concept.

The position of each sensor could be determined by the multidimensional scaling method in an ad hoc network by collecting all the pairwise distances among the sensors. In a 2-D space, the true location for each set of sensor n is denoted by

$$T = \left[t_{ij}\right]n \times 2 \tag{1}$$

The distance among the sensor i and j depending on their positions is denoted as

$$d_{ij}(T) = \left(\sum_{z=1}^{2} \left(t_{iz} - t_{jz}\right)^2 \right)^{\frac{1}{2}} \tag{2}$$

If we define $H = TT'$, then

$$d_{ij}(T)^2 = \sum t_{ia}^2 + \sum t_{ia}^2 - 2\sum t_{ia}t_{ja} \tag{3}$$

$$= H_{ii} + H_{jj} – 2H_{ij}$$

The data is cantered at the coordinate matrix T without any loss of generality.

$$\sum_{i=1}^{n} H_{ij} = 0 \tag{4}$$

By summing Eq. (3) with i over j and both i and j, get

$$\frac{1}{n}\sum_{i=1}^{n} d_{ij}^2 = \frac{1}{n}\sum_{i=1}^{n} H_{ii} + H_{jj} \tag{5}$$

$$\frac{1}{n}\sum_{i=1}^{n} d_{ij}^2 = \frac{1}{n}\sum_{i=1}^{n} H_{ii} + H_{jj} \tag{6}$$

$$\frac{1}{n^2}\sum_{i=1}^{n}\sum_{j=1}^{n} d_{ij}^2 = \frac{2}{n}\sum_{i=1}^{n} H_{ii} \tag{7}$$

Now from Eq. (2),

$$H_{ij} = \frac{1}{2}\left[H_{ii} + H_{jj} - d_{ij}^2\right]$$

$$\frac{1}{2}\left[\frac{1}{n}\sum_{j} d_{ij}^2 + \frac{1}{n}\sum_{j} d_{ij}^2 - d_{ij}^2 - \frac{1}{n^2}\sum_{i}\sum_{j} d_{ij}^2\right] \tag{8}$$

First, from the known location, the device should be activated, and this could calibrate the system. For example, while the person is standing in the allotted place in the middle of the antechamber or while the person is sitting in the armchair. Several new sensors could be introduced into the running and the existing technology with the development of blooming equipment, which replaces or supersedes the old ones.

5 Evaluation

In order to evaluate the proposed activity recognition, we employed case-based activities derived from [20] which are repeated by several users' multiple times. The dataset consists of several actions, including simple movements. The trace includes information from all three axes of the accelerometer. The first three rows consists of acceleration data obtained from x, y as well as z directions, while last row represents the sampling time for its column. From this dataset of activity traces, the evaluation was performed, even though the traces were employed for testing traces and mainly for validating the results.

There were four activity classes which includes standing up, walking, walking downstairs as well as upstairs, for instance, the activity recognition for walking and standing up is shown in Fig. 3, similarly the walking upstairs and downstairs is determined.

As shown in Fig. 4, the proposed approach has been compared prominent and widely employed algorithms, namely, Decision Tree, Naive Bayes, K-Nearest Neighbour for four activity classes, respectively. As the results present the superiority of proposed method which is more significant as the number of classes increases.

The simulation result presents a reliable detection of activities with only few traces from each class.

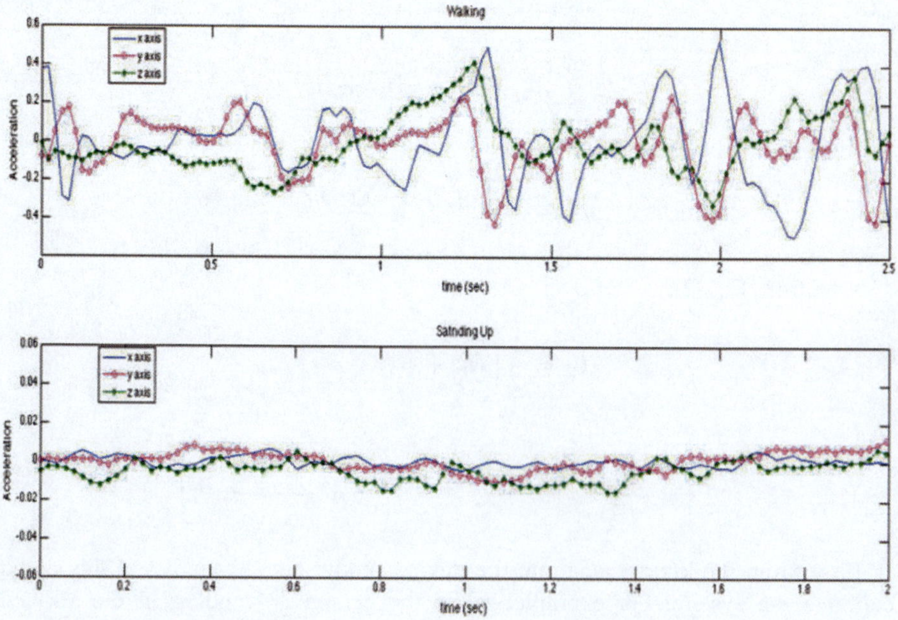

Fig. 3 Activity recognition classes

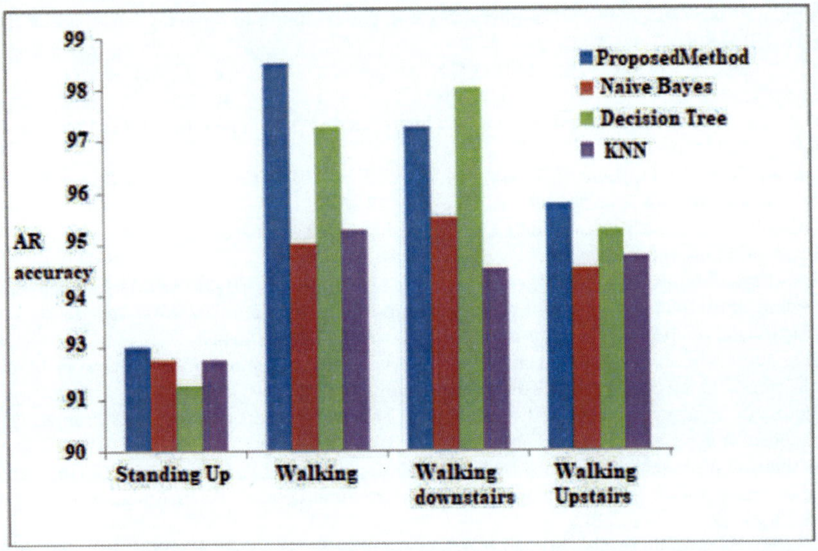

Fig. 4 Evaluation results

6 Conclusion and Future Work

With the high reliability, healthcare could enhance their characteristics by adapting their methodologies with other industries. By offering commitment with the resiliency, the patient safety culture is offered by eradicating the failures. This type of offer should be followed by every organization. Several organizations adapt this culture with the healthcare, and they promote reliability by offering high performance and promoting their patient's safety. But this is said to be a highly challenging task in order to sustain their continuity. The reliability of the IoT-based healthcare should reinforce the combination of highly distributed and miscellaneous data with its knowledge sources. There are many sensors which consist of huge amount of healthcare activities making IoT an inseparable component of effective healthcare. In this research paper, the need for high reliability for healthcare activity recognition is addressed, and it is evident from the results that the proposed approach achieved reliable performance. Activity recognition in mobile environments for healthcare monitoring will be conferred in future work.

References

1. Pasluosta F, Gassner H, Winkler J, Klucken J, Eskofier BM (2017) An emerging era in the management of Parkinson's disease: wearable technologies and the Internet of Things. IEEE J Biomed Health Inform 19(6):1873–1881

2. Sarkar S, Misra S (2016) From micro to nano: the evolution of wireless sensor-based health care. IEEE Pulse 7(1):21–25
3. Yin Y, Zeng Y, Chen X, Fan Y (2018) The Internet of Things in healthcare: an overview. J Ind Integr 1:3–13
4. Dimitrov DV (2017) Medical Internet of Things and big data in healthcare. Healthcare Inform Res 22(3):156–163
5. Islam MR, Kwak D, Kabir H, Hossain M, Kwak K-S (2016) The Internet of Things for health care: a comprehensive survey. IEEE Access 3:678–708
6. Schwab K (2018) The fourth industrial revolution: what it means and how to respond. World Economic Forum, Cologny
7. SmartEye SmartRep and RFID Technology—Westminster City Council—London (2017) [online]. http://www.smartparking.com/keep-up-to-date/case-studies/3-500-vehicle-detection-sensors-and-epermit-technology-in-the-city-of-westminster-london
8. Tokognon A, Gao B, Tian G, Yan Y (2017) Structural health monitoring framework based on Internet of Things: a survey. IEEE Internet Things J 4(3):619–635
9. Tan S, De D, Song W-Z, Yang J, Das SK (2017) Survey of security advances in smart grid: a data driven approach. IEEE Commun Surveys Tutorials 19(1):397–422
10. Wolgast G, Ehrenborg C, Israelsson A, Helander J, Johansson E, Manefjord H (2016) Wireless body area network for heart attack detection [education corner]. IEEE Antennas Propag Mag 58(5):84–92
11. Cretikos MA, Bellomo R, Hillman K, Chen J, Finfer S, Flabouris A (2008) Respiratory rate: the neglected vital sign. Med J Aust 188(11):657–659
12. Xu B, Xu LD, Cai H, Xie C, Hu J, Bu F (2014) Ubiquitous data accessing method in IoT-based information system for emergency medical services. IEEE Trans Ind Inform 10(2):1578–1586
13. Olaronke I, Oluwaseun O (2016) Big data in healthcare: prospects challenges and resolutions. In: Proceedings of the future technologies conference (FTC), Dec 2016, pp 1152–1157
14. Suresh A (2017) Heart disease prediction system using ANN, RBF and CBR. Int J Pure Appl Math (IJPAM) 117(21):199–216. ISSN: 1311-8080, E-ISSN: 1314-3395
15. Zhou J, Cao Z, Dong X, Vasilakos AV (2017) Security and privacy for cloud-based IoT: challenges. IEEE Commun Mag 55(1):26–33
16. Lee H, Ko H, Jeong C, Lee J (2017) Wearable photoplethysmographic sensor based on different LED light intensities. IEEE Sensors J 17(3):587–588
17. Shu Y, Li C, Wang Z, Mi W, Li Y, Ren T-L (2015) A pressure sensing system for heart rate monitoring with polymer-based pressure sensors and an anti-interference post processing circuit. Sensors 15(2):3224–3235
18. Wang D, Zhang D, Lu G (2016) An optimal pulse system design by multichannel sensors fusion. IEEE J Biomed Health Inform 20(2):450–459
19. Zuo W, Wang P, Zhang D (2016) Comparison of three different types of wrist pulse signals by their physical meanings and diagnosis performance. IEEE J Biomed Health Inform 20(1):119–127
20. Anguita D, Ghio A, Oneto L, Parra X, Reyes-Ortiz JL (2012) Human activity recognition on smartphones using a multiclass hardware-friendly support vector machine. In: International workshop of ambient assisted living (IWAAL 2012), Vitoria-Gasteiz, Spain, Dec 2012

Design of Deep Convolutional Neural Network for Efficient Classification of Malaria Parasite

M. Suriya, V. Chandran, and M. G. Sumithra

1 Introduction

Malaria is a form of blood-infected disease that is transmitted from female mosquitoes (*Anopheles*), which injects plasmodium parasitis to human bodies Microscopists often analyze the thickness and volume of a blood spray to diagnose disease. According to the World Health Survey, about 200 million malaria cases caused 429,000 deaths a year. Throughout 2016, malaria governments and foreign counterprojects spent US $2.7 billion throughout malaria prevention. In 2016, WHO countries from the African region contributed a total of 74% of investments. While investments have been stable since 2010, the malaria case has not decreased. The government planned to spend US $6.4 billion per year by 2020 to reduce the incidence of malaria.

The early diagnostic test and treatment was necessary in order to prevent malaria severity. Due to lack of knowledge and research of scientists, the risk factor for malaria treatment is not yet overcome. Febrile children received public sector services rather than private sector care. The statistics indicate that high numbers of febrile children are not properly cared for. There is the possibility of the malaria diseases with artificial intelligence systems during the screening process, in the recent advancement of profound learning theory.

Two important factors are parameter and hyperparameter in deep learning. The most important tuner for adjusting primary parameters, such as weights and distortions in the network, is the hyperparameter. They are different variables that determine the total structure of the network (i.e., number of hidden units) and show

M. Suriya (✉) · V. Chandran · M. G. Sumithra
KPR Institute of Engineering and Technology, Coimbatore, Tamil Nadu, India
e-mail: suriya.m@kpriet.ac.in; chandran.v@kpriet.ac.in

A. Haldorai et al. (eds.), *2nd EAI International Conference on Big Data Innovation for Sustainable Cognitive Computing*, EAI/Springer Innovations in Communication and Computing, https://doi.org/10.1007/978-3-030-47560-4_13

how the network is trained (learning rate). The key constraint is that the hyper-parameter is set before practice, i.e., before weights and bias are optimized. The hyperparameters automatically control parameters such as weight and distortion. Other hyperparameters include rate of learning, epochs, number of layers, number of hidden units, choices of functions of activation, percentage of decrease, weight of initialization, etc. The way these hyperparameters are tuned is to generalize the concept known as hyperparameter tuning.

In this paper Sect. 2 provides the information about the previous work related to proposed work. Sections 3 and 4 provide the theoretical background and proposed work of the paper, respectively. Sections 5 and 6 discuss the experimental results and evaluation metrics. The final section presents the conclusion and future scope of the work.

2 Literature Survey

May et al. [1] addressed automated detection by segmentation and identification of malaria diseases using photographs of the blood spray. The photos are turned into $L * a * b$ and the noise is filtered with the filter. By dilation and erosion, the author has eliminated the background features of a preprocessed image. As a segmentation technique, the Otsu method was used. The researcher clarified that because of the frequency variability, the pictures are misclassified. Although the results prove to be more accurate, sensitive, and specific, only the *P. vivax* parasites can be detected. The other parasites are not detected by the model.

The use of color enhancement techniques to identify malaria parasites was suggested by Abbas et al. [2]. The YCbCr color design is used to preserve the original shape and also to control the loss of features in the image.

The architecture status from the basis of deep learning architecture was suggested by Chollet [3]. The depth separable convolution is used in the Inception network to replace initial modules. For a large image classification data, the proposed network outpaced Incept V3 by approximately 350 million images. The separable convergence of depth may have various advantages over the starting network without reducing the number of parameters. The number of parameters is not reduced but is effectively treated.

In [4] He et al. proposed that it would not be possible to stack the layer to layer. The deeper network of layers also allows the models to automate network learning difficulties. The paper puts the residual value into the net. A short-circuit connection that connects the ingoing and outgoing networks to the remaining network is established through identity. The short relation maps the identity and the inputs to the output of stacked layers are summarized. The study consists of 152 layers, which are even less complex and 8 times more complex than the VGG 40 networks.

The cell counting problem with segmentation problem is discussed by Gopakumar et al. [5]. The two levels of segmentation problem strategy were followed, which means it increases the specificity and sensitivity of the model, not only in the exact part. It eliminates hand craftsmanship and offers an effective diagnosis level with minimal cost and efficient mechanism. The model shows the high Matthews correlation values.

Huang et al. in [6] explained about the traditional architecture called the Dense Web, which was designed to generate a more reliable and deeper network performance. Each network layer is connected to the other layer, and the network layer is linked in conventional networks with the network layer. The Dense Net performs highly competitive object recognition tasks with regard to computational costs over other traditional networks.

Lipton et al. [7] addressed strategies for optimizing the F1 score using the optimal threshold in binary and multi-class classification. It is not enough to describe only one performance metric to present a binary classification work but also to provide precision and metric retrieval that are informative in the real scenario.

3 Proposed Deep Convolutional Neural Network

The proposed project developed and trained an adequate convolutional neural network (DCNN) to classify blood cell images between the parasites and the unaffected ones. The conventional picture characteristics are extracted from a convolutional neural network where image functionality can be extracted in three categories: low, middle, and high. The database is downloaded from the Kaggle malaria cell disease datasets where it contains totally 26,188 images including 13,105 parasitized and 13,083 uninfected images. Figure 1 shows the block diagram of the proposed workflow in which data is divided according to the size of the dataset. The image is then pre-processed for measurement in 128 × 128. Image is fed into the neural network entry level, and the features are extracted and translated

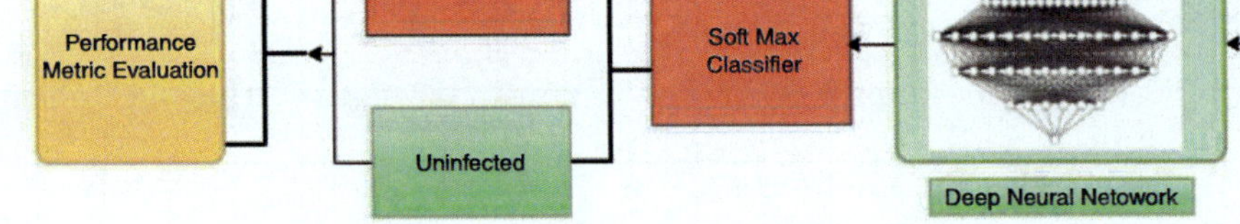

Fig. 1 Block diagram of the proposed DCNN

in deep neural network layers into softmax layer. In this case, the softmax layer is used to seek a prediction likelihood. Confusion matrix is used to determine measures of performance assessment.

4 DCNN Architecture Configuration

The deep convolutional neural network architecture can be modeled as a deep network or as a wider network in two ways. The proposed deep neural network architecture has 19 convolutionary layers to extract the image characteristics, 6 max pooling layers to minimize calculations and 3 standardization layers for batch standardization for the features that are regularly modified and four fully connected layers and one flattening layer that are required for modeling.

4.1 *Optimizer Function*

The optimizer feature eliminates the mistake by adjusting weights and distortions in the network using Adam and Adagrad measured by back propagation.

Adam Optimizer

The name imitates the adaptive time estimate when using the Adagrad and RMS prop optimizer function. The Adam optimizer offers better precision. It is a broader version of the stochastic downward gradient through which the network can update weights using the training data iteratively. With different parameters, Adam optimizer has a different learning speed. In the following you can find the algorithm for the Adam optimizer. To start the algorithm, the number of iterations was provided at first. Step 1 calculates the gradients. x and y are the moveable average measured equation that shifts exponentially. We now need to update the estimator. The bias correction is called x_hat and y_hat equations that are correction of the bias. The final step of the Algorithm for Adam Optimizer is to modify the network weight as z.

Algorithm for Adam Optimizer
for t in range(num_iterations):

1. $x\_hat = m/(1 - np.power(beta\_1, t)) + (1 - beta\_1) * w/(1 - np.power(beta\_1, t))$
2. $y\_hat = y/(1 - np.power(beta\_2, t))$
3. $z = z - step\_size * x\_hat/(np.sqrt(y\_hat) + epsilon)$

Adagrad Optimizer

With the help of past gradients, the Adagrad performs optimization, and the rate is modified for all steps and parameters. This therefore eliminates the manual adjustment of the learning rate. In practical terms, Eq. (1) has expressed the update of the parameter where θ is the current parameter. Equation (2) suggested the approximation of the gradient at time t. The Adagrad equation was given by Eq. (3).

$$\theta\,(t+1) = \theta(t) - \eta \cdot g(t)\sqrt{\varepsilon I + \text{diag}(Gt)} \tag{1}$$

$$g(t) = \frac{1}{\eta}\left\{\sum_{i=1}^{n} \nabla\theta L\Big(x \text{ power}(i), y \text{ power}(i), \theta(t)\right\} \tag{2}$$

$$Gt = \sum_{T-1}^{t} g(t) \text{ transpose}\,[g(t)] \tag{3}$$

where

θ = parameter need to be updated
η = initial learning rate
ε = small quantity the used to avoid division of zero
I = Identity matrix
$g(t)$ = gradient estimate
Gt = sum of the outer products of the gradients until time step (t)

5 Evaluation and Discussion

Table 1 demonstrates how the proposed method is being implemented by iterating through Epoch = 10, 20, and 25 template, with the parameter of hyper value set to check the efficiency of Adam and Adagrad optimizer. The batch size was 32, the learning rate is set to 0.001, and the loss function as the categorical cross entropy is allocated.

5.1 *For Adam Optimizer*

During epoch 10, validation and learning loss was 0.15, and validation and exercise accuracy were 0.95 and 0.96. At epoch 20, the model produced 0.13 losses for validation and 0.12 losses for learning. The loss of validation was 0.12 at epoch 25, and the loss of training at 0.09 and validation and learning precision, respectively,

Table 1 Performance measurement of DCNN at different run

Run no.	Epoch	Batch size	Loss function	Optimizer	Learning rate	Validation loss	Validation accuracy	Training loss	Training accuracy
1	10	32	Categorical cross entropy	Adam	0.001	0.15	0.95	0.15	0.96
2	20	32	Categorical cross entropy	Adam	0.001	0.13	0.96	0.12	0.96
3	25	32	Categorical cross entropy	Adam	0.001	0.12	0.97	0.09	0.97
4	10	32	Categorical cross entropy	Adagrad	0.001	0.14	0.96	0.13	0.96
5	20	32	Categorical cross entropy	Adagrad	0.001	0.17	0.94	0.08	0.97
6	25	32	Categorical cross entropy	Adagrad	0.001	0.26	0.95	0.03	0.99

were 0.95 and 0.96. A higher precision and a decrease in the loss compared to previous iterations culminated in an increasing epoch.

5.2 *For Adagrad Optimizer*

The validation and training loss of the model were 0.14 and 0.13 during the tenth period, while the accuracy of the validation and training was 0.96. At epoch 20 the design resulted in 0.17 amount of validation loss and a 0.08 of learning loss and 0.94 and 0.97, respectively. The validation loss at epoch 25 was 0.26, and the training loss was 0.03, and the estimation and learning quality were 0.95 and 0.99, respectively. The loss rate and reliability of Adagrad decreased at a minimum time of 10. With rising length, failure and accuracy have varied. The results showed that both deployment and configuration problems can be influenced by the Adagrad optimizer times.

The Adam and Adagrad optimizer are compared and are shown in Fig. 2 by changing the times and showing the hyper parameter values.

6 Conclusion

Malaria is one of the most deadly illnesses killing patients because they are unaware. Recent progress in deep education engineering helps to detect these diseases early. This article proposes a deep convolutionary neural network with an efficient tuning of fine hyperparameters to classify malaria disease efficiently. The suggested method is based on the basis of experimental findings that resulted a kappa coefficient of 93% and 95% and Matthews coefficient of correlation with a minimal loss of validity and increased precise value, respectively.

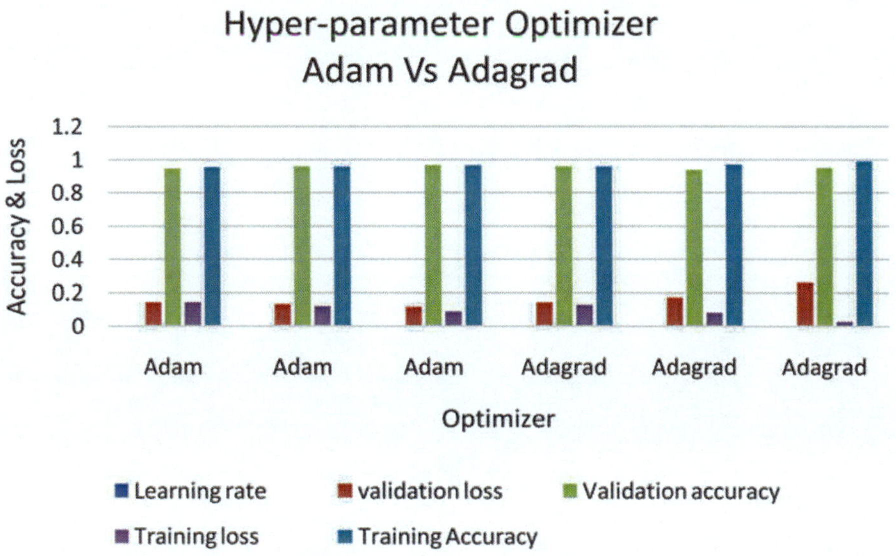

Fig. 2 Analysis of hyperparameter Adam vs Adagrad optimizer

References

1. May Z, Mohd Aziz SSA, Salamat R (2013) Automated quantification and classification of malaria parasites in thin blood smears. In: IEEE international conference on signal and image processing applications, Melaka, pp 369–373. https://doi.org/10.1109/ICSIPA.2013.6708035
2. Abbas N, Mohamad D (2013) Microscopic RGB color images enhancement for blood cells segmentation in YCBCR color space for k-means clustering. J Theor Appl Inf Technol 55(1):117–125
3. Chollet F (2016) Xception: deep learning with separable convolutions, pp 1–14. arXiv:1610.02357
4. He K, Zhang X, Ren S, Sun J (2016) Deep residual learning for image recognition. In: IEEE conference on computer vision and pattern recognition (CVPR), Las Vegas, NV. https://doi.org/10.1109/CVPR.2016.90
5. Gopakumar GP, Swetha M, Sai Siva G, Sai Subrahmanyam GRK (2018) Convolutional neural network-based malaria diagnosis from focus stack of blood smear images acquired using custom-built slide scanner. J Biophoton 11(3). https://doi.org/10.1002/jbio.201700003
6. Huang G, Liu Z, Weinberger KQ, Van der Maaten L (2016) Densely connected convolutional networks. arXiv:1608.06993
7. Lipton ZC, Elkan C, Naryanaswamy B (2014) Optimal thresholding of classifiers to maximize F1 measure. In: Machine learning and knowledge discovery in databases. Springer, Berlin. https://doi.org/10.1007/978-3-662-44851-9_15

Data Security and Sensitive Data Protection using Privacy by Design Technique

M. Suresh Babu, K. Bhavana Raj, and D. Asha Devi

1 Introduction

Information assurance by configuration is at last a methodology that guarantees you consider security and information insurance issues at the plan period of any framework, administration, item or procedure and after that all through the lifecycle.

As communicated by the GDPR, it expects to:

- Set up proper specialized and hierarchical estimates intended to actualize the information security standards and
- Integrate shields into your handling with the goal that you meet the GDPR's necessities and secure the individual rights

For the most part, this suggests you have to organize or 'plan in' data assurance into your taking care of activities and key approaches.

Information Protection by Design has far-reaching application. Models include:

- Developing new IT systems, organizations, things and techniques that incorporate taking care of individual data
- Developing hierarchical approaches, forms, strategic approaches as well as techniques that have security suggestions
- Physical structure
- Embarking on information sharing activities or

M. S. Babu (✉) · K. B. Raj
K.L. University off Campus, Hyderabad, India

D. A. Devi
SNIST, Hyderabad, India

A. Haldorai et al. (eds.), *2nd EAI International Conference on Big Data Innovation for Sustainable Cognitive Computing*, EAI/Springer Innovations in Communication and Computing, https://doi.org/10.1007/978-3-030-47560-4_14

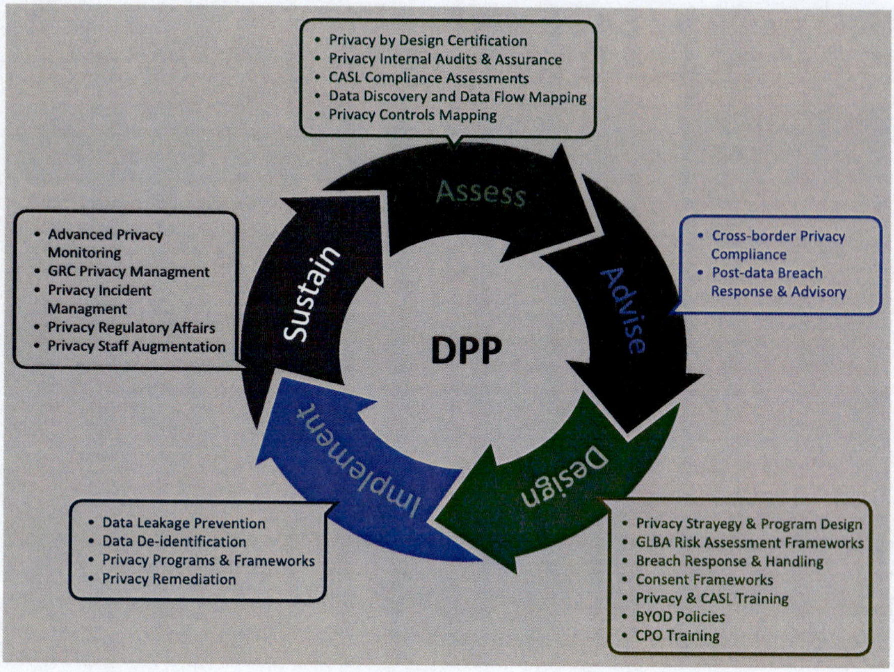

Fig. 1 Data protection and privacy services offered

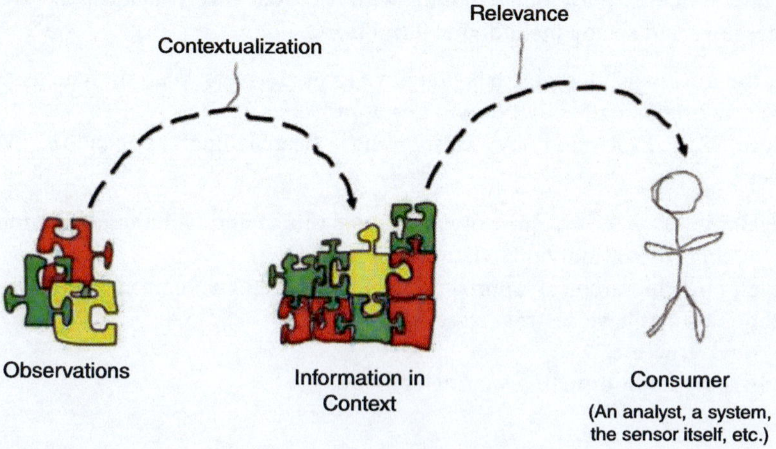

Fig. 2 Privacy by design context accumulating systems

- Using individual information for new purposes

The basic ideas of information insurance by configuration are not new. Under the name 'protection by plan' they have existed for a long time. Information assurance by configuration basically embeds the security by configuration approach into information insurance law [1].

Under the 1998 Act, the ICO upheld this methodology as it helped you to agree to your information assurance commitments. It is presently a lawful necessity. What is information insurance of course?

Information protection as per usual anticipates that you should ensure that you just methodology the data that is essential to achieve your specific reason. It connects to the crucial information security standards of information minimization and reason confinement. You need to process some close to home information to accomplish your purpose(s) [1]. Information security as a matter of course implies you have to indicate this information before the preparing begins, fittingly educate people and just procedure the information you require for your motivation. It does not expect you to embrace a 'default to off' arrangement. What you have to do rely upon the conditions of your preparing and the dangers presented to people. By the by, you should consider things like:

- Adopting a 'security first' approach with any default settings of frameworks and applications
- Ensuring you do not give a fanciful decision to people identifying with the information you will process
- Additional data aside of the individual are not ready
- Ensuring that individual data are not thus made unreservedly open to others with the exception of if the individual makes it so; and
- Providing individuals with satisfactory controls and choices to rehearse their rights. Who is in charge of agreeing to information assurance by plan and as a matter of course?

As per Article 25, the controller is responsible regarding consenting to information assurance by plan and as a matter of course [2]. Contingent upon your conditions, you may have various necessities for various zones inside your association. For instance:

- The senior administration, e.g. building up a culture of 'security mindfulness' and guaranteeing us to create strategies and methods in view of information insurance
- The programming engineers, framework modelers and application designers, – e.g. the individuals who plan frameworks, items and administrations should assess information security necessities and help us in agreeing to commitments and
- The strategic approaches, e.g. we ought to guarantee that we insert information security by plan in the entirety of our interior procedures and strategies

This may not make a difference to all organizations, obviously. In any case, information assurance by configuration is tied in with receiving an association-wide way to deal with information insurance, and 'preparing in' security contemplations into any handling action you embrace. It does not make a difference just on the off chance that you are the kind of association that has your very own product designers and frameworks planners [3].

In thinking about whether to force a punishment, the ICO will consider the specialized and hierarchical estimates you have set up in regard to information insurance by plan. Moreover, under the Data Protection Act 2018 (DPA 2018) we can issue an Enforcement Notice against you for any failings in regard to Article 25.

Shouldn't something be said about information processors?

In the event that you utilize another association to process individual information for your sake, at that point that association is an information processor under the GDPR [2].

Article 25 does not make reference to information processors explicitly. In any case, Article 28 indicates the contemplations you should take at whatever point you are choosing a processor. For instance, you should just utilize processors that give:

'Satisfactory affirmations to realize appropriate specific and progressive measures in such a way, that the dealing with will meet the necessities of this rule and assurance the security of the benefits of the data subject'.

This necessity covers the two information assurance by configuration in Article 25 just as your security commitments under Article 32. Your processor cannot really help you with your information insurance by structure commitments (not at all like with safety efforts), anyway you should just utilize processors that give adequate certifications to meet the GDPR's necessities.

Shouldn't something be said about different gatherings?

Information insurance by plan and as a matter of course can likewise affect organizations other than controllers and processors. Contingent upon your handling movement, different gatherings might be included, regardless of whether this is exactly where you buy an item or administration that you at that point use in your preparing. Models incorporate makers, item engineers, application designers and specialist co-ops.

Presentation broadens the ideas of information assurance by plan to different organizations, in spite of the fact that it does not put a prerequisite on them to go along—that remaining parts with you as the controller. It says: 'While making, arranging, picking and using applications, organizations and things that rely upon the treatment of individual data or technique singular data to fulfill their task, creators of the items, administrations and applications ought to be urged to consider the privilege to information security when creating and structuring such items, administrations and applications and, concerning the best in class, to ensure that controllers and processors can satisfy their information assurance commitments'. In this way, when thinking about what items and administrations you requirement for your handling, you should hope to pick those where the creators and engineers have considered information assurance. This can guarantee that your handling holds fast to the information security by structure necessities.

For an engineer or originator of items, administrations and applications, the GDPR puts no particular commitments on you about how you structure and assemble these items. In any case, you should take note of that controllers are required to consider information insurance by structure when choosing administrations and items for use in their information handling exercises—subsequently in the event that you plan these items in view of information assurance, you might be in a superior position.

The Big Difference with Big Data Big information is the following boondocks for development, rivalry and efficiency. The expression 'Huge Data' alludes to datasets whose size is past the capacity of run of the mill database programming devices to catch, store, oversee and examine But as mechanical advances improve our capacity to misuse Big Data, potential security concerns could mix an administrative backfire that would hose the information economy and smother development. These worries are reflected in, for instance, the discussion around the as of late proposed European enactment that incorporates a 'right to be overlooked' that is planned for helping people better oversee information insurance dangers online by expecting organizations to erase their information if there is no real reason for holding it.

Organizations are building up a more complete comprehension of their clients than any time in recent memory, as they better amass the information accessible to them. General well-being specialists, for instance, have a requirement for progressively point by point data so as to more readily educate arrangement choices identified with dealing with their undeniably restricted assets. The ability to collect bits of information from Big Data wills no doubt in the world be of massive money-related significance. Isolating bits of learning from Big Data has quickly transformed into an inside district for technologists around the globe. The articulation 'Colossal Data progresses' delineates another time of advances and models, planned to monetarily think a motivation from immense volumes of a wide grouping of data, by engaging fast catch, revelation and furthermore examination.

The present Big Data will give the crude material to tomorrow's advancements. Exploring this gigantic volume of data will expect us to consider information in new and inventive ways. More information, from heterogeneous sources, collects around a solitary individual – in spite of de-identification proof endeavours – endeavours to dependably ensure personality is undermined.

Envision an organizer that contains no references to the area you live in, the area where you work, your preferred coffeehouse and the make/model/year of your vehicle. Without individual identifiers, might it be able to be related with you? As an ever-increasing number of separately amiable realities are gathered, all things considered become unequivocally recognizing; to be sure, the correct arrangement of such information can approach your driver's permit number in its capacity to distinguish you. This does not, nonetheless, contend against utilizing systems to de-distinguish individual information. To be sure, de-recognizable proof procedures stay vital instruments in the security of protection. In any case, we should not overlook the way that Big Data can expand the danger of re-recognizable proof – and at times, coincidentally de-identification.

2 Sense-Making Systems

'Sense making' identifies with a developing class of innovation intended to enable organizations to understand their different observations. This perception space will frequently incorporate information they currently possess and control (e.g. organized ace information), just as information they cannot control (e.g. remotely created and less organized internet based life). Sensemaking frameworks will deal with amazingly enormous informational collections – possibly including tens to several billions of perceptions (exchanges) – being created from a consistently expanding assorted scope of information sources (e.g. from Twitter and OpenStreetMap to one's digital security logs). Clearly, these volumes are past the limit of human survey. Sensemaking frameworks will be utilized by organizations to settle on better choices, quicker.

From a sensemaking perspective, an association must be as shrewd as the total of its perceptions. These perceptions are gathered over the different endeavour frameworks, for example, client enlistment frameworks, money-related bookkeeping frameworks and finance frameworks. With each new exchange, an association gets the hang of something. When something is found out, an open door emerges to understand what this new bit of information implies, and to react suitably.

The powerlessness of an association to profit by the data it approaches or has produced in the past can bring about what has been alluded to as 'big business amnesia'. Studies, for instance, directed for a noteworthy retailer found that out of each 1000 representatives employed, two had been recently captured for taking from a similar store for which they had been rehired.

The test that organizations face in such manner is developing, in light of the fact that their perception space is developing as well – at an inconceivable rate. Today, these perceptions will in general be dispersed crosswise over various information sources, situated in physically better places, and composed in various structures. This dissemination of information makes it hard for an association to perceive the hugeness of related information focuses. Sense making tries to incorporate an association's different perception space – a developing goal if an association is to stay aggressive. Verifiably, progressed examination has been utilized, in addition to other things, to investigate huge informational indexes so as to discover designs that can help detach key factors to manufacture prescient models for basic leadership. Organizations utilize progressed examination with information mining to enhance their client connections; law authorization offices utilize progressed investigation to battle crime from fear mongering to tax avoidance to recognize burglary. Normally, these strategies have their cut-off points; for instance, information mining looking for new examples in counter-psychological warfare may yield little esteem.

Another class of logical capacity is rising that one may describe as 'broadly useful sense making'. These sensemaking strategies incorporate new exchanges (perceptions) with past exchanges – much similarly one takes a jigsaw astound piece and finds its buddies on the table – and utilize this setting collecting procedure to improve understanding about what's going on the present moment. Essentially, this procedure can happen quickly enough to allow the client to take care of whatever is going on while it is as yet occurring. Not at all like many existing scientific techniques that expect clients to pose inquiries of frameworks, these new frameworks work on an alternate guideline: the information finds the information and the importance finds the client. This is shown in the figure.

At the point when setting gathering frameworks are utilized with Big Data, three astounding wonders develop: False positives and false negatives both decline as setting diminishes equivocalness. This makes an interpretation of legitimately to higher quality business choices. Frameworks that are not working on setting collection will in general observe expanding false positives and false negatives as the size of the informational collection develops. Setting aggregation delivers the contrary impact as information sizes develop.

In setting aggregating frameworks blunders in the information (explicitly 'common inconstancy') are in certainty supportive. Conceivable varieties in a name, for example, Ann (additionally spelled Anne) might be entered by the information administrator and the exactness of setting aggregating frameworks can be improved because of collecting this fluctuation. One model that may be well-known to numerous individuals is that when looking through Google and Google reacts with 'Did you mean __?' This recommendation is not originating from an inner static word reference; rather, it has recollected everybody's error(s) previously. On the off chance that Google was not keeping this 'awful information' it would not be so savvy.

At last, maybe the most irrational shock as for setting amassing frameworks is that coordinating exchanges end up increasingly precise as well as quicker, even as the information store is getting greater. The most short-sighted approach to consider this is to think about why the last couple of bits of a riddle are about as simple as the initial not many when there is more 'information' before you than at any other time. This marvel is evidently new to examination and is well-suited to profoundly change what is conceivable in the Big Data time, particularly in the space of constant, sensemaking motors.

Be that as it may, in these new frameworks, the errand of guaranteeing information security ends up more diligently as more duplicates of data are made. Enormous information stores containing setting amassed data are increasingly helpful not exclusively to their main goal holders yet additionally to those with interests in abuse. That is, the more by and by recognizable data Big Data frameworks contain, the more noteworthy the potential hazard. This hazard emerges not just from potential abuse of the information by unapproved people, yet in addition from abuse of the framework itself. On the off chance that the examination framework is utilized for a reason that goes past its lawful mission, security might be in danger. Thus, organizations that need to exploit game-changing advances in investigation should remain back and consider the structure choices that can improve security and protection. By considering the protection suggestions at an opportune time, technologists have a superior shot of creating and preparing in security upgrading highlights and encouraging the arrangement and appropriation of these frameworks. Underneath, we layout the protection upgrading highlights of this new innovation, a 'Major Data investigative sensemaking' motor. This innovation has been intended to comprehend new perceptions as they occur, quick enough to take care of business while the exchange is as yet occurring. Since its systematic strategies, limit with respect to Big Data and its speed are down transforming from a security point of view, it has been planned starting from the earliest stage in light of security assurances. While the outcome may not be impeccable, it is plainly better than one structured without reference to protection. We trust it might motivate or direct others during the time spent making their very own cutting-edge examination.

3 Privacy by Design in the Age of Big Data

As advances advance, our experience and desires for protection additionally advance. Before, security was seen as an individual decent, as opposed to a societal one. In that capacity, protection was viewed as an issue of individual obligation. Wards far and wide received information security laws that reflected Fair Information Practices (FIPs) – widespread protection standards for the treatment of individual information. FIPs mirrored the essential ideas of information the executives. The main reason detail and use impediment required the explanations behind the gathering, use and exposure of by and by recognizable data should have been distinguished at or before the season of accumulation. Individual data ought not to be utilized or revealed for purposes other than those for which it was gathered, aside from with the assent of the individual or as approved by law [4]. The subsequent idea, client interest and straightforwardness, determined that people ought to be enabled to assume a participatory job in the lifecycle of their very own information and ought to be made mindful of the practices related with its utilization and revelation. In conclusion, FIPs featured the requirement for solid security to shield the privacy, trustworthiness and information accessibility as suitable to the affectability of the data. Reasonable information practices gave a fundamental beginning stage to sensible data the board rehearses. After some time, the assignment of securing individual data was seen essentially as an 'exercise in careful control' of contending business interests and protection necessities – a lose–lose outlook. These regulate approach stressed notice and decision as the essential technique for tending to individual information the executives. As innovations progressed, it may, the likelihood for people to genuinely apply command over their own data turned out to be increasingly troublesome. Numerous onlookers have since taken the view that FIPs were a fundamental however inadequate condition for securing protection. Likewise, the consideration of security controllers has since started to move from consistence with FIPs to proactively inserting protection into the plan of new advancements. A model may feature how current protection concerns identify with the powers of development, rivalry and the worldwide appropriation of data interchanges advancements. Security dangers to information about recognizable people may to a great extent be tended to with the best possible utilization of de-ID methods, joined with re-identification methodology. These systems limit the danger of unintended exposure and re-recognizable proof, while keeping up an abnormal state of information quality. In any case, mind-boggling and fast innovative change (e.g. rising investigation) may make security hurts as a side effect; for instance, progressively amazing examination may coincidentally cause it conceivable to distinguish people over huge informational collections. Preferably, at that point, protection should be inserted, of course, during the engineering, plan and development of the procedures. This was the focal inspiration for Privacy by Design which is planned for lessening dangers of security hurt from emerging in any case. PbD depends on seven [5] Foundational principles. It underlines regard for client security and the need to install protection as a default condition, yet saves a pledge to usefulness in a 'win–win', or positive-aggregate procedure. This methodology changes shopper protection issues from an unadulterated strategy or consistence issue into a business basic. Since getting security right has turned into a basic achievement factor to any association that manages individual data, adopting a strategy that is principled and innovation nonpartisan is currently more important than any other time in recent memory [6]. PbD is centred around procedures instead of a particular centre coordinating specialized results. This methodology mirrors the truth that it is troublesome by and to positively affect both buyer and client conduct sometime later. Or maybe, security is best proactively interlaced into business procedures and practices. To accomplish this, security standards ought to be presented early – during engineering arranging, framework structure and operational methodology. These standards, where conceivable, ought to be established into the code with defaults adjusting both security and business objectives. PbD supports that security be joined honestly with the arrangement and assignment, of advancement, yet also how a system is operationalized (e.g. work shapes, the administrators' structures, physical spaces and orchestrated establishment).

Today, PbD is generally perceived globally as the standard for creating security consistent data frameworks. As a structure for compelling security insurance, PbD's centre is progressively about urging organizations to both drive and exhibit their duty to protection than some severe specialized consistence definition. To put it plainly, in the time of Big Data, we unequivocally energize technologists occupied with the plan and arrangement of cutting-edge investigation to hold onto PbD as an approach to convey capable development.

Model: The creation of a Big Data Sensemaking System through PbD. In late 2008, Jeff Jonas set out on an aspiring voyage to make a sensemaking style framework. This exertion began with generally speaking engineering arranging and structure determinations. Over the primary year of this venture, while drafting and redrafting these plans, his group attempted to insert properties that would improve, as opposed to dissolve, the protection and common freedoms of information subjects. To design for security, his group gauged execution outcomes, default settings and which, assuming any, PbD highlights ought to be so difficult wired into the framework they actually cannot be impaired. Throughout the year, that traversed the primer and nitty-gritty structure, the group made a powerful suite of PbD highlights. Jeff's group accepts sensemaking framework designed for protection and common freedoms improving characteristics into this framework than any forerunner. On the off chance that others contrast, we welcome discussion and energetic challenge as more specialists respond to the call of PbD.

4 The Seven Commandments in Privacy by Design

1. Full attribution: PbD endorses that security be incorporated legitimately with the plan and task, of innovation, yet additionally how a framework is operationalized (e.g. work forms, the executives structures, physical spaces and arranged foundation). Each record contained in the database incorporates the metadata that focuses on the wellspring of the record—this pointer comprising of an information source and an exchange ID. Full attribution implies beneficiaries of knowledge from our motor can follow each contributing information indicate back its source. At the point when frameworks use combine/cleanse preparing it winds up hard to address prior mix-ups (when an alternate affirmation ought to have been made) as some unique information has been disposed of.
2. Data tethering: Adds, changes and erases happening in frameworks of record must be represented, progressively, in sub-seconds. Information cash in data sharing conditions is significant, particularly where information is utilized to make significant, hard to turn around choices that may influence individuals' opportunities or benefits. For instance, if harsh information is evacuated or rectified in an arrangement of record, such amendments ought to show up quickly over the data sharing biological system. In our sensemaking framework, each announced change brings about momentary remedy. Applying includes, changes and erases from information fastened frameworks of record cannot be killed.
3. Analytics on anonymized data: The capacity to perform progressed examination (counting some fluffy coordinating) over cryptographically changed information implies organizations can anonymize more information before data sharing. Each duplicate of information expands the danger of unintended divulgence. To diminish this hazard, information ought to be anonymized before exchange; upon receipt, the beneficiary will have no real option except to anonymize it very still (when set into a database). Also, on account of our full attribution prerequisite, re-distinguishing proof is by configuration, so as to guarantee responsibility, compromise and review. This element licenses information proprietors to share their data in an anonymized structure that in any case yields tangibly comparable outcomes when subject to cutting-edge investigation. Decrease of hazard without a material change in scientific outcomes makes for a convincing case to anonymize more information, not less. Security ensuring highlight will upgrade trust in data-sharing situations and will bring about positive win–win results.
4. Alter resistant audit logs: Every customer search should be marked in a change safe way – even the database executive should not have the choice to adjust the verification contained in this survey log. The request 'Who will watch the watchmen?' remains as pertinent today as when it was first displayed in Latin 2000 years earlier. Authenticated users can retrieve confidential data and access

records without a genuine business reason, e.g. an agent of a money-related system investigating his neighbour's record. Change safe logs make it possible to audit customer direct. Actualizing them may diminish infringement, since where workers know such reviews are conceivable, they might be more averse to surrender to allurement. Tamper-safe review log subsystem is necessary and compulsory segment of the sensemaking framework.

5. False-negative favouring methods: The capacity to all the more emphatically support false negatives is of basic significance in frameworks that could be utilized to influence somebody's respectful freedoms. In numerous business situations, it is smarter to miss a couple of things (false negatives) than coincidentally make guarantees that are not valid (false positives). False positives can nourish into choices that antagonistically influence individuals' lives – e.g. the police wind up thumping down an inappropriate entryway or a guiltless traveller is denied authorization to get onto a plane. Once in a while a solitary information point can prompt numerous ends. Frameworks that are not false-negative favouring may pick the most grounded end and negligence the remainder of the finishes. We have associated unprecedented effort to speak to such conditions by making remarkable figurings that help false negatives. This non-minor conduct is taking some work.
6. Self-correcting false positives: With each new information point displayed, earlier statements are rethought to guarantee they are as yet right, and if never again right, these prior affirmations can regularly be fixed – progressively. A bogus positive is an attestation (guarantee) that is made, however is not valid; e.g. consider somebody who cannot load onto a plane since the person shares a comparative name and date of birth as another person on a watch list. Where false positives are redressed by intermittent month to month reloading, incorrectly choices can persevere for as long as a month, despite the fact that the framework had adequate information indicates close by know previously. So as to avoid this, prior affirmations should be turned around continuously and at scale, as new information focuses present themselves. This happens to be the absolute most complex specialized part of our sensemaking framework. Envision having seen one billion records as of now and now one record arrives.
7. Information transfer accounting: Every optional exchange of information, regardless of whether to human eyeball or a tertiary framework, can be recorded to permit partners (e.g. information overseers or the shoppers themselves) to see how their information is streaming. So as to screen data streams, data move bookkeeping can be utilized to record both (a) who assessed each record and (b) where each record has been delivered off to. This log of outbound bookkeeping (out to eyeballs or out to frameworks) would work much like the U.S. credit revealing framework whereby at the base of the credit report is a log of who has pulled the record. This expands the straightforwardness into how frameworks are utilized. At some point, it could empower a shopper, at times, to demand a data review. When there is a progression of data spills (e.g. an insider risk), data move bookkeeping makes disclosure of who got to all records in the spilled arrangement an unimportant computational exertion. This can limit the extent of

an examination when searching for abusing individuals inside an association. Our data move bookkeeping ability is arranged at the tact of the framework chairmen. We empower selection by having planned our hidden sensemaking information structures to help this sort of utilization information effectively, which makes executing this component moderately straightforward. End Big Data can possibly produce tremendous incentive to society [6]. So as to guarantee that it does, chances to improve protection and common freedoms are best imagined at an opportune time. In this paper, we have investigated the development of Big Data sensemaking frameworks as a rising capacity with an exceptional capacity to coordinate recently enhanced information – and now and again, information about individuals and their day-by-day lives. The utilization of cutting-edge examination has made it conceivable to dissect huge informational indexes for developing examples. It is progressively evident, in any case, that these strategies alone will be inadequate to deal with the universe of Big Data – particularly given the requirement for organizations to have the option to react to dangers and openings continuously. Cutting-edge abilities like sense making offer a one of a kind way to deal with increasing important bits of knowledge from Big Data through setting collection. While these new improvements are exceptionally welcome, working in security upgrading components, by configuration, can limit the protection hurt, or even forestall the security hurt from emerging in any case. This will thusly cause more noteworthy trust and trust in the enterprises that utilize these new abilities. The dynamic pace of mechanical advancement expects us to secure protection in a proactive way so as to all the more likely shield protection inside our social orders.

5 Conclusion

This paper outlines how organizations can save time and money while improving data security and regulatory compliance and dramatically reduce the risk of a data breach or expensive penalties for noncompliance. What privacy by design means? Why it is so important? and how to implement it within any organization, using data protection and data access control technologies. Data sensemaking innovation was, from the beginning, built with security improving highlights. A portion of these highlights is so basic to precision that the group chose they ought to be obligatory – so profoundly heated in they cannot be killed. This paper shows how security and obligation can be progressed in this new time of Big Data examination. PbD's centre is increasingly about urging organizations to both drive and exhibit their duty to protection than some exacting specialized consistence definition.

References

1. Manyika J, et al (2011) Big data: the next frontier for innovation, competition, and productivity. McKinsey Global Institute. Online: http://www.mckinsey.com/Insights/MGI/Research/Technology_and_Innovation/Big_data_The_next_frontier_for_innovation
2. Commission Proposal for a Regulation of the European Parliament and of the Council on the Protection of Individuals with Regard to the Processing of Personal Data and on the Free Movement of Such Data (General Data Protection Regulation), COM (2012) 11 final (Jan. 25, 2012). Online: http://ec.europa.eu/justice/newsroom/data-protection/news/120125_en.htm
3. Tene O, Polonetsky J (2012) Privacy in the age of big data: a time for big decisions. Stanford Law Rev 64:63
4. Ohm P (2010) Broken promises of privacy: responding to the surprising failure of anonymization. UCLA Law Rev 57:1701
5. Jonas J, Sokol L (2009) Data finds data. Segaran T, Hammerbacher J. (eds.), Beautiful data: the stories behind elegant data solutions, O'Reilly Media Newton, MA 105
6. Jonas J (2011) Master Data Management (MDM) vs. Sensemaking. Online: http://jeffjonas.typepad.com/jeff_jonas/2011/11/master-data-management-mdm-vs-sensemaking.html

Effects of Customer Brand Engagement and Online Brand Experience: An Empirical Research

R. Mary Metilda and A. Grace Antony Rose

1 Introduction

The birth of internet has predominantly occupied the minds of the people and, in fact, a very prominent part of everybody's day-to-day life. In the twenty-first century, the retail sector is incessantly changing its concentration from brick and mortar to a click and mortar model. The customers who have born in the internet era are more connected to the online devices round-the-clock, and all their needs, expectations, aspirations, desires and wants are met through online experiences. It's again a very big challenge for the retailers in terms of stiff competition. The market value of retail industry has reached US $23,460 billion in the year 2017, and it is likely to record a CAGR of 5.3% during the prediction period 2018–2023, which, in turn, will have an impact on the growth of the global retail market to $31,880.8 billion by 2023. The increasing strength of online shopping has become a major economic driver and started penetrating the entire world through smart phones and Internet of Things (IoT). Internet retailing has become the fastest-growing segment in retail industry [1]. In spite of such a massive growth of online retailing, the relationship between the customer and the online retailer is not so easy, and it needs lot of efforts to build and maintain such relationships [2]. Nevertheless, online customer brand engagement (CBE) is the key factor to be concentrated by every online retail businesses to survive, sustain and succeed in the market for a long run. Unlike other studies, the prominent part of customer engagement on online brand experience is concentrated on this study [3–7]. Moreover, the earlier studies discovered the association amid customer engagement and customer

R. Mary Metilda (✉) · A. Grace Antony Rose
SREC Business School, Coimbatore, Tamil Nadu, India
e-mail: metilda@srec.ac.in

A. Haldorai et al. (eds.), *2nd EAI International Conference on Big Data Innovation for Sustainable Cognitive Computing*, EAI/Springer Innovations in Communication and Computing, https://doi.org/10.1007/978-3-030-47560-4_15

loyalty via online experiences in Toto [8]. The emphasis of this study is to embrace the three extents of brand engagement offered by Hollebeek and Chen [5], explicitly 'cognitive processing (COG), affection (AFF) and activation (ACT)', and compare them with online brand experience and learning performance factors including customer engagement behavior, brand loyalty and brand satisfaction. The present study will help to discover the effect of each dimension of customer brand engagement (CBE) in online experience, satisfaction and loyalty in online shopping websites/Apps wherein the prominence of conveying spectacular experience has been tremendously accepted [9–11].

2 Review of Literature

The study explores the drivers of customer engagement behaviors based on the U&G perspective in a social networking brands community setting. Initially, learning motivation was included in the framework as a forerunner as customer learning is created to understand the spectacle in networking brands community. Next, social interaction was given much preference especially the prominent role played by social media [12, 13]; hence, the study hypothesizes social interaction as collective learning in the suggested model. Further, as the study focused on customer engagement behavior and customer loyalty, brand satisfaction is the intermediary connecting between customer-learning motivation and collective learning to customer engagement behavior (CEB) and customer loyalty (CL) grounded on the U&G perspective [14].

In the previous decades, prominent research has explored the robust nature of customer's brand relationships [15, 16], and the Marketing Science Institute (MSI) discloses that customer brand engagement (CBE) has vigorous research urgency (MSI, 2014). Bowden [17] depicted that CBE means an impact of customers' understanding towards a particular brand. Mostly, it is understood that customer engagement with a particular brand is conceivably through the experience and that experience is crucial to customer loyalty. A brand retorts as long-term engagement amid both customer and retailer [18–21]. For this long-standing affiliation, the experience acts as input and the brand loyalty as the vital output. Former research has revealed that customer brand communications raise emotive state and cognitive and behavior responses, which establishes customer brand involvement, which in fact brings a smooth-edged brand valuation [22, 23]. Rapidly, in an online environment, there exists an inclination to uncover all the assurances for the interaction and a realistic period of brand experience to allow customers to get engaged with a particular brand [9, 24]. Today, business leaders are anxious about how the online setting impacts affiliation between brand satisfaction and customer loyalty and perhaps doubtful about how the customer engagement will be exerted through an online experience [6]. The emphasis of this study included three dimensions of brand engagement (BE) proposed by Hollebeek and Chen [5], namely, 'cognitive processing (COG), affection (AFF) and activation (ACT)', and measures them with brand experience (BE) exclusively in banking setting. Thus, the first objective of this study was to test the relation of CBE's dimensions, viz. 'cognitive processing (COG), affection (AFF) and activation (ACT)', with online brand experience (OBE) and the influence of OBE towards brand satisfaction (BS) and brand loyalty (BL).in banking sector. Moreover, satisfaction leads to brand loyalty (BL) and brand satisfaction. The second objective is to test the intervention of satisfaction between OBE and brand loyalty. And at final, the third objective is to measure the controlling effect of number of transactions made by customers on the relation of OBE and brand satisfaction (BS) and OBE and brand loyalty (BL). This has been very important to discover the outcome of each dimension of engagement in online experience, satisfaction and loyalty in banking industry, where the importance of conveying excellent experience has been exceptionally admitted [9–11].

3 Conceptual Framework

Hollebeek and Chen [5] have suggested three important extents of customer brand engagement (CBE), namely, 'cognitive processing (COG), affection (AFF) and activation (ACT)'. Customer brand engagement (CBE) is the rising archetype in the marketing literature by hovering the field of customer relationship marketing (CRM) [25]. CBE can be assumed as 'the conception of a deeper, more meaningful linking between the business and the consumer' [26].

3.1 H1: Online Brand Experience (OBE) Is Influenced by Cognitive Processing (COG)

Cognitive processing is delineated as the thought-provoking behavior of the consumer with higher level of interaction and engagement with the particular product or service of a particular brand. The term brand experience is originated from the multidisciplinary fields of rational knowledge, values and management. Brand experience is the customers' perception towards the brand, at times of contact, and it may be influenced by the image of the brand, promotional efforts and the dimensions of service quality received [27]. Due to the convergence of information and massively grown technology, online brand experience grabbed the advantages of such phenomenal growth of information technology to create a constructive experience with the online customers. Thus, the following hypothesis is framed to measure the level of customer engagement through cognitive processing and its impact on the online brand experience.

H1: Online brand experience (OBE) is influenced by cognitive processing.

3.2 H2: Online Brand Experience Is Influenced by Affection

Affection towards a brand is also referred as an emotional state of consumer behavior engagement. Any brand which stimulate a positive feeling, while the customer interacts in association with the brand. Hollebeek and Chen [5] state that affection is tested by using four scaled items. 'Affection' intensifies to 'user's degree of affirmative brand-related affect in a particular consumer/brand interface' [28]. Affection supports in crafting a distinctive and impressive experience in online purchase and enhances novel opportunities to the literature on online brand experiences and assignation. Furthermore, the results showed that customers are engaged with online brands through their perceptional elements which include cognitive, affection and activation and these perceptional elements impact their online practice and online experience, making the customers loyal.

Thus, the following hypothesis is framed to discover the effect of affection towards online brand experience (OBE).

H2: Online brand experience (OBE) is influenced by affection.

3.3 H3: Online Brand Experience (OBE) Is Influenced by Activation

Activation involves the customer's highest degree of interest, efforts and amount of period expended towards a specific brand [29]. In the study, the researchers have used four items to measure the influence of activation towards online brand experience adapted from Hollebeek and Chen [5]. Past research findings have proved that the self-expressive brands will have a negative impact towards activation [29]. The main reason behind such negative impact can be due to the online customers who involve themselves with socially self-expressive brands accept the wrong doing from a brand. In a study, it is examined that there is a positive relationship between activation and brand loyalty (BL), where brand loyalty (BL) has a direct influence on online brand experience [29]. Thus, the hypothesis is framed as follows.

H3: Online brand experience (OBE) is influenced by activation.

3.4 H4: Customer Engagement Behavior Is Influenced by Online Brand Experience

Customer engagement behavior entails a constructive bond between the short-term profits of a particular company's brand and long-term customer relationships [14]. Remarkable researchers [7, 30, 31] have defined customer engagement behavior

(CEB) as 'enduring and voluntary behaviors of online consumers derived from intrinsic drive of cognitive behavior stimulated by external environmental factors that are valued to be at high importance by the brand company beyond the business transaction' [14]. Latest technological trends are very useful to the consumers and the brand companies to explore the constructive relationship and remove the interactive barriers between them. The customers who are engaged well collaborate with the brand value co-creation and get satisfied with their needs, through better online brand experience [14]. Thus, the following hypothesis is evolved to measure the impact of online brand experience with customer engagement behavior.

H4: Customer engagement behavior is influenced by online brand experience.

3.5 *H5: Brand Loyalty Is Influenced by Online Brand Experience*

Brand loyalty is a significant result while involving in online customer engagement towards a particular brand/company [2]. Brand loyalty (BL) refers to 'the level of consumer affection towards a particular brand' [32]. To measure brand loyalty, four scales are used, adapted from Yoo and Donthu [33]. With respect to mobile users, Dwivedi [34] adopted energy, commitment and immersion dimensions of consumer brand experience (CBE) and proved that consumer brand experience (CBE) has an affirmative effect on brand loyalty (BL) [29]. This study has examined the affiliation between the level of brand experience and brand loyalty (BL) of online consumers. Leckie et al. [29] have proved that consumer interest, consumer involvement, self-expressive brands and consumer engagement behavior always have an affirmative outcome towards brand loyalty (BL). This study is influenced by such results and tried to examine the positive impact of brand loyalty towards online brand experience by framing the following hypothesis.

H5: Brand loyalty is influenced by online brand experience.

3.6 *H6: Brand Satisfaction Is Influenced by Online Brand Experience*

Satisfaction is an emotional reaction to acquire the state of affairs [35–37]. Satisfaction evolves from the brand experience [38–41]. The experience of any brand specially purchased through online includes sensory, responsive, rational and communicative spaces [22]. Brakus et al. [22] also mentioned that online brand experience (OBE) delivers value to the buyers and through this, consumer satisfaction towards a brand can be enhanced only if the online brand experience provides worth to the consumers. The various studies proved that online brand experience provides elite and amazing experiences and augments brand satisfaction in the setting of terminus branding [42], web grounded branding [39] and online brand setting [43, 44]. Thus, considering all the above statements, the following hypothesis is framed.

H6: Brand satisfaction is influenced by online brand experience.

Based on the hypothesis framed and examined, the following conceptual model is designed:

4 Methodology

A total of 100 questionnaires were distributed among the customers who shop online of which 93 useful questionnaires were collected. In the first part of the questionnaire, the respondents were asked on the frequency of purchase, age group and gender. In the second part of questionnaire, the online shoppers were asked about their attitudes and behavior towards online shopping. Partial least square (PLS) and structural equation modelling (SEM) approaches were used to examine simultaneously the structural components both measurement and structural model. The model shown in Fig. 1 was analysed using SmartPLS [45]. The convergent and discriminant validity are confirmed before assessment of the structural model. The next section explains the results of analysis.

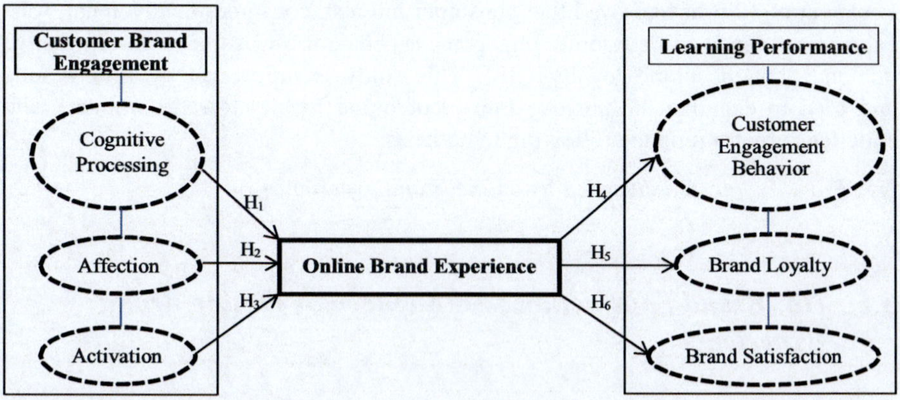

Fig. 1 Conceptual model. *CBE* customer brand engagement, *COG* cognitive processing, *AFF* affection, *ACT* activation, *OBE* online brand experience, *LP* learning performance, *CBE* customer engagement behavior, *BS* brand satisfaction, *BL* brand loyalty

5 Results and Discussion

Following Hair et al. [46] process, to assess the measurement models, we examine outer loadings, composite reliability, average variance extracted (AVE = convergent validity) and discriminant validity (Table 2). Our empirical results indicate the adequate reliability for all the measurements. The AVE values (convergent validity) are well above the minimum required level of 0.50, thus demonstrating convergent validity for all the constructs. The constructs were examined for collinearity which through the empirical results indicated that the indicators do not have problems with collinearity.

Another action taken in the assessment of the measurement model was the discriminant validity. Discriminant validity indicates the extent to which a given construct is different from other latent constructs [47, 48]. Diagonal elements are the square root of AVE between the constructs and their measures (Table 1). Off-diagonal elements are correlations between constructs. For discriminant validity, diagonal elements should be larger than off-diagonal elements in the same row and column. Table 2 shows the same. Thus, the results indicate that there is discriminant validity between all constructs based on the cross-loading criterion [46].

Once the construct measures have been confirmed as reliable and valid, the next step is to assess the structural model results which involve examining the model's predictive capabilities and the relationships between the constructs [46] (Fig. 2).

Table 1 Reliability results

Construct	Cronbach's alpha (CA)	rho_*A*	Composite reliability (CR)	Average variance extracted (AVE)
AC	0.815	0.834	0.877	0.641
AFF	0.921	0.935	0.938	0.716
BL	0.929	0.937	0.946	0.780
BS	0.868	0.868	0.911	0.719
CBE	0.890	0.896	0.912	0.564
COG	0.808	0.804	0.864	0.518
OBE	0.874	0.879	0.909	0.666

Table 2 Discriminant validity results

Construct	AC	AFF	BL	BS	CBE	COG	OBE
AC	0.801						
AFF	0.788	0.846					
BL	0.752	0.744	0.883				
BS	0.763	0.771	0.813	0.848			
CBE	0.758	0.717	0.763	0.722	0.751		
COG	0.627	0.602	0.502	0.534	0.614	0.720	
OBE	0.659	0.578	0.739	0.580	0.650	0.440	0.816

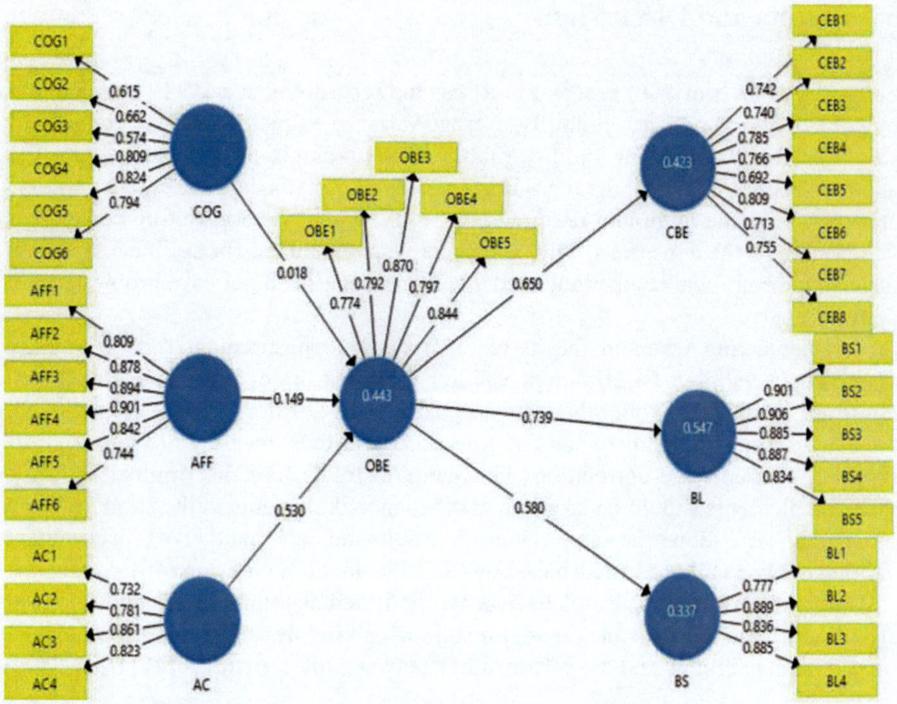

Fig. 2 Structural model

Table 3 Path coefficients

Hypothesis	Path	Path coefficients	Standard error	*t*-statistics	*p*-value	Decision
H1	AC OBE	0.530	0.121	4.383	0.000	Supported
H2	AFF OBE	0.149	0.120	1.246	0.213	Not supported
H3	COG OBE	0.018	0.102	0.166	0.868	Not supported
H4	OBE BL	0.739	0.056	13.099	0.000	Supported
H5	OBE BS	0.580	0.088	6.569	0.000	Supported
H6	OBE CBE	0.650	0.055	11.832	0.000	Supported

A bootstrap analysis was performed to assess the statistical significance of the path coefficients after computing the path estimates in the structural model. In using bootstrapping, the actual sample size is 93 and 500 resamples were performed. In addition, by applying the PLS-SEM algorithm, estimates were obtained for the structural model coefficients (the path coefficients), which represents the hypothesized relationships between the constructs. Table 3 presents the hypothesis testing.

The statistical results support the significant relationship between activation and online brand experience ($b = 0.530$, standard error = 0.121 and t-statistics = 4.383). There exists a strong relationship between online brand experience brand loyalty with a path coefficient of 0.739 and a low standard error of 0.056 and a t-statistic of 13.099. The relationship between online brand experience and brand satisfaction has a path coefficient of 0.580, standard error of 0.088 and a t-statistics of 6.569. It is also seen that there exists a strong relationship between online brand experience and customer brand engagement with a path coefficient of 0.650 with a low standard error of 0.055 and a high t-statistics of 11.832. It is observed from the results that two of the relationships are not supported by the statistical values, viz. affection and cognitive processing on online brand experience.

6 Limitations and Directions for Future Research

This research also has some limitations. The limitations will guide future researchers to explore the online brand experience and customer engagement of online shoppers. The first limitation was that the study did not limit to any specific group of products in the consumer goods. The result would have displayed a different pattern for different groups of products like apparel shopping, grocery shopping, home appliance shopping, etc. The future study could explore at specific products. The second limitation was that the research was focused on all age group online shoppers. In connection to this, online shopping experience among GenY and GenZ and the pattern of customer engagement behaviors would certainly vary, and hence the future research could examine and implement the research with GenY and GenZ group of customers who are the tech-savvy generations.

References

1. Mordor Intelligence report (2019). https://www.mordorintelligence.com/industryreports/retailindustry
2. Aksoy L, Van Doorn J, Lemon KN, Mittal V, Nass S, Pick D, Pirner P, Verhoef PC (2010) Customer engagement behavior: theoretical foundations and research directions. J Serv Res 13(3):253–266
3. Hollebeek L (2011) Exploring customer brand engagement: definition and themes. J Strat Mark 19(7):555–573
4. Hollebeek LD (2011) Demystifying customer brand engagement: exploring the loyalty nexus. J Mark Manag 27(7–8):785–807
5. Hollebeek LD, Chen T (2014) Exploring positively-versus negatively-valenced brand engagement: a conceptual model. J Prod Brand Manag 23(1):62–74
6. Nysveen H, Pedersen P (2014) Influences of co-creation on brand experience: the role of brand engagement. Int J Mark Res 56(6):807–832
7. Van Doorn J, Lemon KN, Mittal V et al (2010) Customer engagement behavior: theoretical foundations and research directions. J Serv Res 13(3):253–266
8. Khan I, Rahman Z, Fatma M (2016) The role of customer brand engagement and brand experience in online banking. Int J Bank Mark 34(7):1025–1041
9. Brun I, Rajaobelina L, Ricard L (2014) Online relationship quality: scale development and initial testing. Int J Bank Mark 32(1):5–27

10. Rajaobelina L, Brun I, Toufaily É (2013) A relational classification of online banking customers. Int J Bank Mark 31(3):187–205
11. Sunikka A, Bragge J, Kallio H (2011) The effectiveness of personalized marketing in online banking: a comparison between search and experience offerings. J Financ Serv Mark 16(3–4):183–194
12. Kaplan AM, Haenlein M (2010) Users of the world, unite! The challenges and opportunities of social media. Bus Horiz 53(1):59–68
13. Liang T-P, Ho Y-T, Li Y-W, Turban E (2011) What drives social commerce: the role of social support and relationship quality. Int J Electr Comm 16(2):69–90
14. Chiang C-T, Wei C-F, Parker KR, Davey B (2017) Exploring the drivers of customer engagement behaviours in social network brand communities: towards a customer learning model. J Mark Manag 33(17–18):1443–1464
15. Aaker J, Fournier S, Brasel SA (2004) When good brands do bad. J Consum Res 31(1):1–16
16. Fournier S (1998) Consumers and their brands: developing relationship theory in consumer research. J Consum Res 24(4):343–373
17. Bowden JLH (2009) The process of customer engagement: a conceptual framework. J Mark Theory Pract 17(1):63–74
18. Davis R, Buchanan-Oliver M, Brodie RJ (2000) Retail service branding in electronic-commerce environments. J Serv Res 3(2):178–186
19. Fournier S, Mick DG (1999) Rediscovering satisfaction. J Mark 63(4):5–23
20. Keller KL (1993) Conceptualizing, measuring, and managing customer-based brand equity. J Mark 57(1):1–22
21. Rao AR, Ruekert RW (1994) Brand alliances as signals of product quality. Sloan Manag Rev 36(1):87
22. Brakus JJ, Schmitt BH, Zarantonello L (2009) Brand experience: what is it? How is it measured? Does it affect loyalty? J Mark 73(3):52–68
23. O'Loughlin D, Szmigin I (2005) Customer perspectives on the role and importance of branding in Irish retail financial services. Int J Bank Mark 23(1):8–27
24. Moynagh M, Worsley R (2002) Tomorrow's consumer: the shifting balance of power. J Consum Behav 1(3):293–301
25. Vivek SD, Beatty SE, Morgan RM (2012) Customer engagement: exploring customer relationships beyond purchase. J Mark Theory Pract 20(2):122–146
26. Kumar V, Aksoy L, Donkers B et al (2010) Undervalued or overvalued customers: capturing total customer engagement value. J Serv Res 13(3):297–310
27. Alloza A (2008) Brand engagement and brand experience at BBVA, the transformation of a 150 years old company. Corp Reput Rev 11(4):371–379
28. Ashraf S, Iftikhar A, Yameen A, Younas S (2018) Empirical relationship of customer brand engagement with satisfaction and loyalty through online brand experience. IUP J Brand Manag 15(3):23–48
29. Leckie C, Nyadzayo MW, Johnson LW (2016) Antecedents of consumer brand engagement and brand loyalty. J Market Manag 32(5–6):558–578
30. Brodie RJ, Hollebeek LD, Jurić B, Ilić A (2011) Customer engagement: Conceptual domain, fundamental propositions, and implications for research. J Serv Res 14(3):252–271
31. Brodie RJ, Ilic A, Juric B, Hollebeek L (2013) Consumer engagement in a virtual brand community: an exploratory analysis. J Bus Res 66(1):105–114
32. Liu F, Li J, Mizerski D, Soh H (2012) Self-congruity, brand attitude, and brand loyalty: a study on luxury brands. Eur J Mark 46(7-8):922–937
33. Yoo B, Donthu N (2001) Developing and validating a multidimensional consumer-based brand equity scale. J Bus Res 52(1):1–14
34. Dwivedi A (2015) A higher-order model of consumer brand engagement and its impact on loyalty intentions. J Retail Consum Serv 24:100–109
35. Anderson JC, Narus JA (1990) A model of distributor firm and manufacturer firm working partnerships. J Mark 54(1):42–58

36. Bagozzi RP, Gopinath M, Nyer PU (1999) The role of emotions in marketing. J Acad Mark Sci 27(2):184
37. Bennett R, Härtel CEJ, McColl-Kennedy JR (2005) Experience as a moderator of involvement and satisfaction on brand loyalty in a business-to-business setting 02-314R. Ind Mark Manag 34(1):97–107
38. Anderson EW, Sullivan MW (1993) The antecedents and consequences of customer satisfaction for firms. Mark Sci 12(2):125–143
39. Ha HY, Perks H (2005) Effects of consumer perceptions of brand experience on the web: brand familiarity, satisfaction, and brand trust. J Consum Behav 4(6):438–452
40. Marinkovic V, Obradovic V (2015) Customers' emotional reactions in the banking industry. Int J Bank Mark 33(3):243–260
41. Zarantonello L, Schmitt BH (2010) Using the brand experience scale to profile consumers and predict consumer behavior. J Brand Manag 17(7):532–540
42. Barnes SJ, Mattsson J, Sørensen F (2014) Destination brand experience and visitor behavior: testing a scale in the tourism context. Ann Tour Res 48(Suppl C):121–139
43. Lee (Ally) S, Jeong M (2014) Enhancing online brand experiences: an application of congruity theory. Int J Hosp Manag 40(Suppl C):49–58
44. Morgan-Thomas A, Veloutsou C (2013) Beyond technology acceptance: brand relationships and online brand experience. J Bus Res 66(1):21–27
45. Ringle CM, Wende S, Will A (2005) SmartPLS 2.0
46. Hair JF, Hult GTM, Ringle C, Sarstedt M (2013) A primer on partial least squares structural equation modeling (PLS-SEM). SAGE Publications, Thousand Oaks, CA
47. Duarte P, Raposo M (2010) A PLS model to study brand preference: an application to the mobile phone market. In: Esposito Vinzi V, Chin WW, Henseler J, Wang H (eds) Handbook of partial least squares. Springer, Berlin, pp 449–485
48. Rezaei S, Ghodsi SS (2014) Does value matters in playing online game? An empirical study among massively multiplayer online role-playing games (MMORPGs). Comput Hum Behav 35:252–266

Bayesian Personalized Ranking-Based Rank Prediction Scheme (BPR-RPS)

J. Sengathir, M. Deva Priya, A. Christy Jeba Malar, G. Aishwaryalakshmi, and S. Priyadharshini

1 Introduction

Cloud computing is a style of computing where massively scalable IT-enabled capabilities are delivered 'As a Service' to external customers using Internet technologies [1]. From the recent past, the cloud service providers adore more and more prospects in the marketplace [2]. Cloud computing offers several beneficiary aspects for the users such as fast deployment of the services in the user's environment, providing access to services, easy pay for the usage of services leading to cost effectiveness and offering services in rapid provisioning and elasticity way [3]. The resources are shared through ubiquitous network access, and the cloud providers enable to access services in a resilient manner. They also provide mitigation mechanisms against network vulnerabilities. The cloud services include disaster recovery services, demand-based storage services, demand-based security control services and also demand-based rapid recomposition of services [4]. Though cloud

J. Sengathir
Department of Information Technology, CVR College of Engineering, Hyderabad, Telangana, India

M. Deva Priya (✉)
Department of Computer Science and Engineering, Sri Krishna College of Technology, Coimbatore, Tamil Nadu, India
e-mail: m.devapriya@skct.edu.in

A. Christy Jeba Malar · G. Aishwaryalakshmi · S. Priyadharshini
Department of Information Technology, Sri Krishna College of Technology, Coimbatore, Tamil Nadu, India
e-mail: a.christyjebamalar@skct.edu.in; aishwaryalakshmi.g@skct.edu.in; priyadharshini.s@skct.edu.in

A. Haldorai et al. (eds.), *2nd EAI International Conference on Big Data Innovation for Sustainable Cognitive Computing*, EAI/Springer Innovations in Communication and Computing, https://doi.org/10.1007/978-3-030-47560-4_16

offers many advantages, the main challenge in cloud computing is the risk involved in understanding the security aspects rightly and in case of any crisis, finding the right one responsible—either the cloud service provider or the user [5].

Apart from these issues, cloud business faces the challenge of addressing privacy issue raised in the new way of computing. Cloud computing transfers the application software and data of an organization to its data center [6]. However, the cloud data center does not guarantee 100% reliability in managing data and services [7]. Transferring an application of an organization to the cloud imposes new security issues in the cloud computing environment [8]. Some of the issues include the following:

(a) Accessing the data of an organization by an unauthorized user
(b) Attacks based on virtualization
(c) Attacks that can happen through web application, for example, Structured Query Language (SQL) injection attack and cross-site scripting attack
(d) Privacy and other control issues raised by the owner of the application software and data
(e) Identification-based issues in handling data
(f) Various issues related to data verification, data interfering, data integrity, message confidentiality, data loss or hacking
(g) Challenges related to authentication of an authorized user of the application and data
(h) Data or message spoofing [9]

Though cloud computing provides better service in sharing of resources, it suffers from high level of security risks [10].

2 Related Work

The application layer DDoS attack generally leads to resource starvation. The application layer DDoS generally affects the northbound Application Programming Interface (API) and the application. When compared to other types of DDoS attacks, the application layer DDoS attack will consume less amount of bandwidth. The application layer DDoS attacks are less detectable since they are clandestine in nature and resemble normal traffic. This attack will have an influence on the precise attribute of an application like Session Initiation Protocol (SIP), HyperText Transfer Protocol (HTTP) or Domain Name System (DNS) which makes it to have a parallel impact on other types of DDoS attacks. A novel mitigation mechanism using filtering tree is proposed for preventing HTTP- and Extensible Markup Language (XML)-based DDoS attacks on the cloud computing environment [11]. This filtering tree-based DDoS attack prevention scheme utilizes five levels of filters for reducing the impacts of the HTTP- and Extensible Markup Language (XML)-based DDoS attacks. In this filtering tree-based DDoS attack prevention approach, the suspicious packets are analyzed using a puzzle resolver for handling the issues that emerge due to the generated Simple Object Access Protocol (SOAP) header-based malicious data packets. Initially, this filtering tree scheme identifies the malicious message-initiated Internet Protocol (IP) addresses for sending the puzzles which are solved for determining the genuine client.

To protect from DDoS attack, there is a demand for complete access of the entire payload information. For a quicker preventive mechanism, the payload information has to be procured with a reduced amount of latency [12]. The key features present in a Software Defined Networking (SDN) architecture is useful for providing security at lower level of a network stack. In the seven layers of the network, the OpenFlow accepts only the network traffic information present in Layer 2 and Layer 3. The entire packet in general is not sent to the controller, but an exception is made if there is a nonavailability of buffers in a switch. The prevailing OpenFlow implementation is of no use to the applications that require access to the data payload. This is due to the fact that a thorough inspection of packets along with the data plane's aggressive polling will lead to a swift deterioration in performance of the network. FortNOX, an application layer DDoS defense mechanism is an extension of OpenFlow controller [13]. It is useful in imparting role-based authentication and also implementing security constraints. The purpose of role-based authentication is to identify the authorization level for every OpenFlow application that is also the rule producer. To maintain integrity, the authentication makes sure that the 'least privilege' principle is adhered. All requests based on OpenFlow insertion are passed through a detection engine that is used to detect a conflict on a rule. If a network flow is disabled, then the new OpenFlow rule used for enabling it is termed as the conflict, and it is removed by the conflict engine. The applications that are available in Layers 4 through 7 are found to be challenging with respect to SDN due to the difficulty in centralization and consolidation. The advantages of these applications are that they are malleable, less expensive and simple to manage.

To mitigate DDoS attack, several techniques like rate limiting, packet dropping, event filtering and time-out adjustment [3, 14–17] are proposed, out of which rate limiting is found to be the predominant one. When a DDoS attack occurs, the switches and the controllers are allowed to function normally due to rate limiting. But, rate limiting does not have the capability to protect other users. Another programmability-based dynamic rule updating mechanism called FRESCO [18] is propounded. It is optimally applicable to any kind of OpenFlow context and offers a programming architecture that derives the merits of click for monitoring data traffic.

3 Proposed System

The proposed framework is implemented by means of Rank_algorithm for performing service ranking. The training data for the proposed framework can be obtained from the QoS values (1) collected by monitoring the past response of other users and (2) obtained by monitoring cloud services. The proposed framework is a user-concerted mechanism which implements Bayesian Personalized Ranking (BPR) that improves the performance of the current system. The utilization details can be easily obtained in the cloud environment. Also client-side QoS performance of the called cloud services can be easily collected by monitoring the infrastructure. The cloud provider is responsible for collecting the client-side QoS values from various user applications easily. For simulating cloud environment, cloudlet toolkit is used. This toolkit enables modeling of cloud computing systems and application service environment.

3.1 *Ranking QoS Parameters of the Cloud Application*

In this section, ranking framework for cloud services is presented. Initially, the similarity of the active user with ad hoc data is computed. Thus, the similar users are recognized as follows. The most commonly used services are selected for which QoS rankings are obtained and compared. Based on the rankings for the same set of services or individual service, the K_rank correlation coefficient (K_RCC) is derived using Eq. (1).

The K_RCC for users '*i*' and '*j*' are computed by,

$$\text{Similarity}\,(i,\,j) = \frac{\text{CC}_\text{o} - \text{DC}_\text{o}}{S\,(S-1)\,/2} \tag{1}$$

where 'S' is the number of services, 'CC_o' is the number of concordant pairs between two lists and 'DC_o' is the number of discordant pairs. There are totally $S(S-1)/2$ pairs for 'S' cloud services.

An indicator function is defined as,

$$I(x) = \begin{cases} 1 \text{ if } x < 0 \\ 0 \text{ otherwise} \end{cases} \tag{2}$$

From the above equation, it is observed that the ranking similarity_index between two rankings for the services or two different clients lies in the interval of $[-1, 1]$, where '-1' is obtained when the order of user '*i*' is the exact reverse of user '*j*' and '1' is obtained when order of user '*i*' is equal to the order of user '*j*'.

3.2 *BPR-Based Optimal Personalized Ranking*

In BPR-RPS, the ranking based on optimal personalized ranking is estimated through maximum posterior probability. Generic learning scheme is employed using the concept of stochastic gradient descent. The steps involved in the implementation of this process are detailed as follows:

Step 1: Initialize the ranking threshold.
Step 2: Repeat until convergence.

Step 3: Derive the updated value of optimal personalized ranking using Eq. (3).

$$\theta = \theta + \alpha \left(\frac{e^{-x}}{1 + e^{-x}} \frac{\delta}{\delta \alpha} x_{ij}^{i} + \lambda \beta \right) \tag{3}$$

Step 4: Return updated optimal personalized ranking value.
Step 5: End procedure.

3.3 BPR-Based Ranking Prediction

The Rank_algorithm is designed to forecast the QoS ranking of optimal usage in cloud applications. The algorithm depicts the list of steps necessary to predict the ranking.

Algorithm 1: Rank_Algorithm

Input: An Employed Service set 'ES', a Full Service set 'IS', Preference Function 'ψ'

Output: A service ranking ρ′

```
BEGIN
F=ES
while F ≠ do
{
        t = arg max_{i ∈ F} Q_i
        ρ_c(t) = |E| − |F| + 1
        F=F-{t}
}
for each i ∈ IS do
{
        π (i) = Σ_{j ∈ IS} ψ (i, j);
}
n = |IS|
while IS ≠ do
{
        t = arg max_{i ∈ IS} π(i)
        ρ′(t) = n - |IS| + 1
        IS= IS − {t};
        for each 'i' ∈ IS do
        {
           π (i) = π (i) - ψ(i,t);
        }
}
while ES ≠ do
{
```

```
        e = arg min_{i ∈ E} ρ(i)
        index = min_{i ∈ E} ρ′(i)
        ρ′(e)= index;
        E = E −{e};
    }
    END
```

4 Results Analysis and Discussions for BPR-RPS

In experiment 1, the performance of BPR-RPS is studied by varying the number of tasks based on response time. The plots of response time under different ranking thresholds are depicted in Figs. 1, 2 and 3 respectively. From Fig. 1, it is evident that the response times of BPR-RPS, Quality Of Service in cloud using Reputation Potential Services (QOS-RPS) and Component Reputation Fault Segregation-based Swift Factor Technique (CRFS-SFT) decrease systematically with increase in the number of tasks [19]. This increase in the number of tasks incurs an additional time for processing the tasks, and BPR-RPS handles this context by enforcing a rapid task processing rate of 18% more than the traditional approaches like QOS-RPS and CRFS-SFT. It is transparent that BPR-RPS enhances the response time by 12–15% over QOS-RPS and 18–23% over CRFS-SFT. In addition, it is also clear that BPR-RPS improves the response time on an average by 15% in contrast to the benchmarked security approaches considered for study.

Figure 2 shows the response times of BPR-RPS, QOS-RPS and CRFS-SFT under the threshold of 0.4. This increase in threshold reduces the time taken for processing the tasks by ensuring a rapid task processing rate of 21% when compared to the considered traditional approaches like QOS-RPS and CRFS-SFT. It is transparent that BPR-RPS enhances the response time by 13–21% over QOS-RPS and 25–29% over CRFS-SFT. It is clear that BPR-RPS improves the response time on an average by 22% when compared to the benchmark security approaches considered for study.

Figure 3 presents the response times of BPR-RPS, QOS-RPS and CRFS-SFT under the threshold of 0.6. This increase in ranking threshold reduces the additional time for processing the tasks by increasing the task processing rate by 32% in contrast to the considered traditional approaches like QOS-RPS and CRFS-SFT. BPR-RPS improves the response time by 12–17% over QOS-RPS and 22–26% over CRFS-SFT. In addition, it is also clear that BPR-RPS improves the response time on an average by 23% in contrast to the benchmark security approaches.

In experiment 2, the performance of BPR-RPS, QOS-RPS and CRFS-SFT is analyzed by varying the tasks based on throughput. The plots of throughput on different ranking thresholds are portrayed in Figs. 4, 5 and 6, respectively.

From Fig. 4, it is evident that the throughputs of BPR-RPS, QOS-RPS and CRFS-SFT improve marginally with increase in the number of tasks at ranking threshold of 0.2. This increase in the number of tasks introduces overhead, but BPR-RPS improves the throughput by fastening the task processing capability by 14% in contrast to the considered traditional approaches like QOS-RPS and CRFS-SFT.

It improves the throughput by 16–20% over QOS-RPS and 23–27% over CRFS-SFT. BPR-RPS enhances the throughput on an average by 22% when compared to the baseline approaches.

Figure 5 shows the throughputs of BPR-RPS, QOS-RPS and CRFS-SFT under the ranking threshold of 0.4. BPR-RPS improves the throughput by 6–9% over QOS-RPS and 13–16% over CRFS-SFT. BPR-RPS enhances the throughput on an average by 17% in contrast to the benchmark security approaches.

Figure 6 presents the throughputs of BPR-RPS, QOS-RPS and CRFS-SFT under the threshold of 0.6, and this increase in ranking threshold improves the throughput by rapid data delivery rate. It is found that BPR-RPS improves the throughput by 15–19% over QOS-RPS and 23–26% over CRFS-SFT. BPR-RPS increases the throughput by 18% when compared to the benchmark security approaches.

In experiment 3, the performance of BPR-RPS is analyzed by varying the number of tasks based on cost function. The plots of cost function on different ranking thresholds are portrayed in Figs. 7, 8 and 9 respectively. From Fig. 7, it is evident that the cost functions of BPR-RPS, QOS-RPS and CRFS-SFT are potent in resolving the situation by reducing the cost function by 18% when compared to the considered traditional approaches like QOS-RPS and CRFS-SFT at ranking threshold of 0.2. It is transparent that BPR-RPS minimizes the cost function by 11–14% over QOS-RPS and 16–19% over CRFS-SFT.

Figure 8 shows the cost functions of BPR-RPS, QOS-RPS and CRFS-SFT under the ranking threshold of 0.4. BPR-RPS minimizes the cost function by 7–12% over

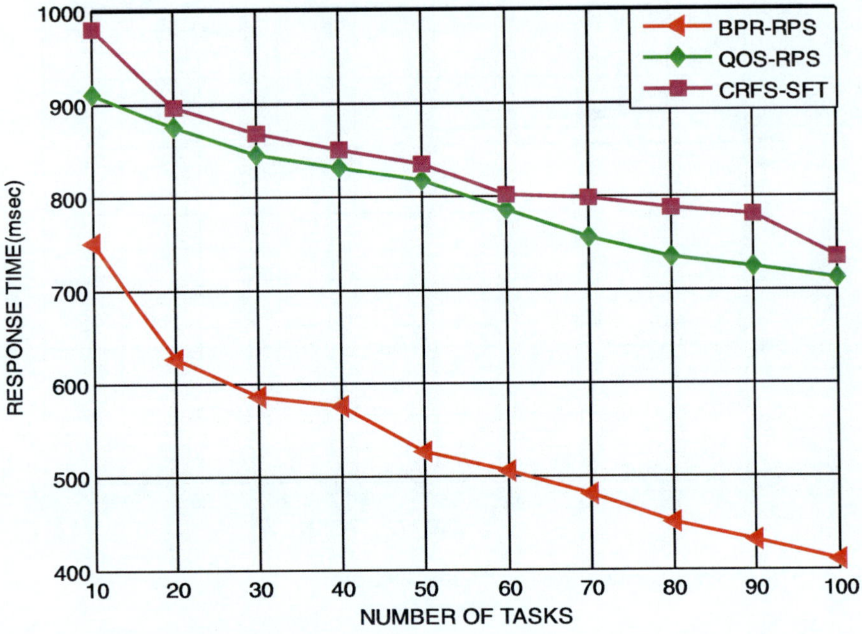

Fig. 1 BPR-RPS response time (ranking threshold = 0.2)

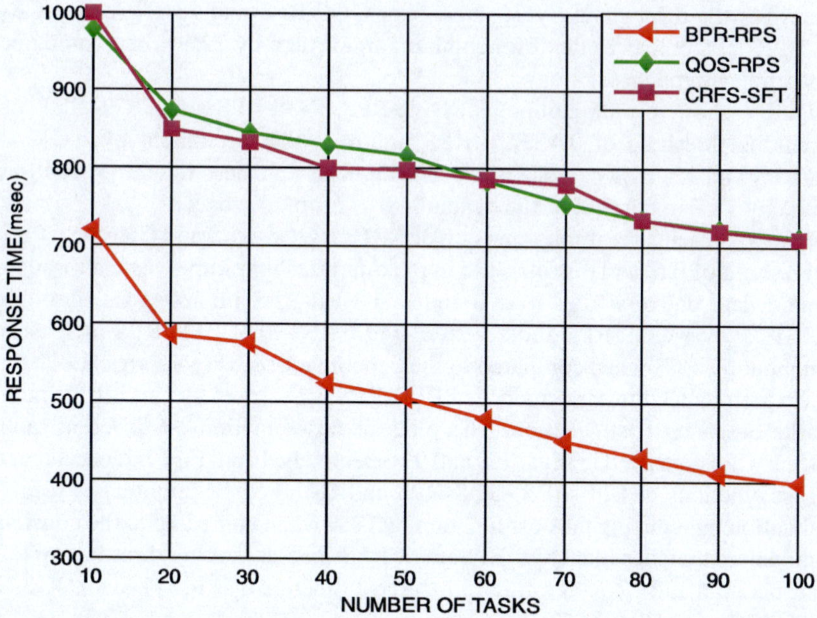

Fig. 2 BPR-RPS response time (ranking threshold = 0.4)

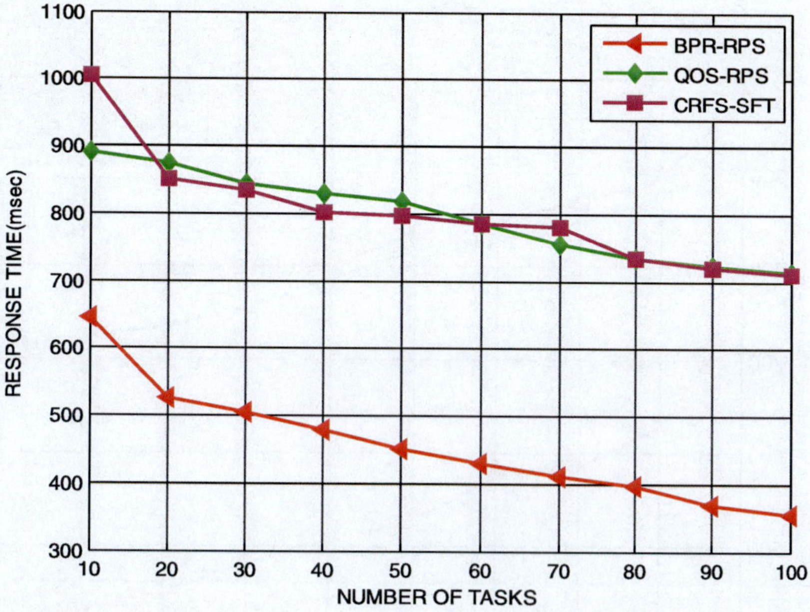

Fig. 3 BPR-RPS response time (ranking threshold = 0.6)

QOS-RPS and 16–21% over CRFS-SFT. BPR-RPS minimizes the cost function on an average by 14% when compared to the benchmark security approaches.

Figure 9 presents the cost functions of BPR-RPS, QOS-RPS and CRFS-SFT under the ranking threshold of 0.6. This increase in ranking threshold reduces the cost function by facilitating faster data delivery. It is found that BPR-RPS reduces the cost incurring function by 15–19% over QOS-RPS and 21–24% over CRFS-SFT. It is also clear that BPR-RPS minimizes the cost function by 18% in contrast to the benchmark security approaches.

Figure 10 shows the increase in throughputs of BPR-RPS, QOS-RPS and CRFS-SFT under different ranking thresholds. BPR-RPS enhances the rate of throughput by 6–8% over QOS-RPS and 9–13% over CRFS-SFT.

Further, Fig. 11 highlights the decrease in cost functions of BPR-RPS, QOS-RPS and CRFS-SFT under different ranking thresholds. BPR-RPS decreases the rate of the cost function by 5–9% over QOS-RPS and 12–16% over CRFS-SFT.

In addition, Fig. 12 presents the response times of BPR-RPS, QOS-RPS and CRFS-SFT under different ranking thresholds, and BPR-RPS is found to minimize the response time by 11–14% over QOS-RPS and 18–22% over CRFS-SFT.

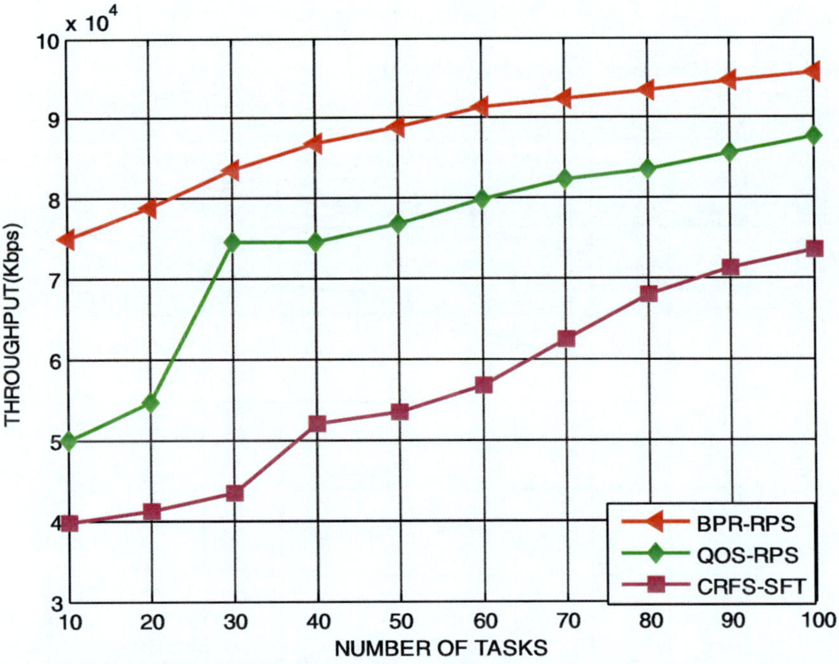

Fig. 4 BPR-RPS throughput (ranking threshold = 0.2)

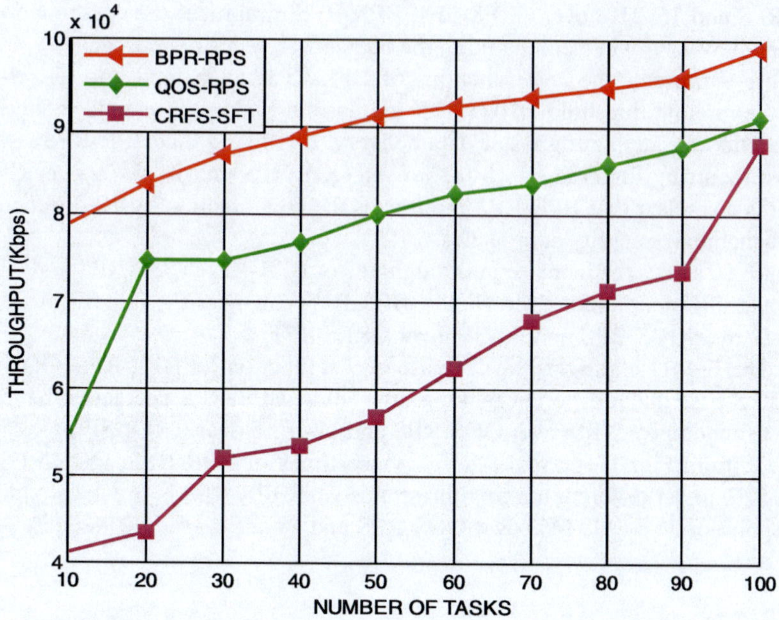

Fig. 5 BPR-RPS throughput (ranking threshold = 0.4)

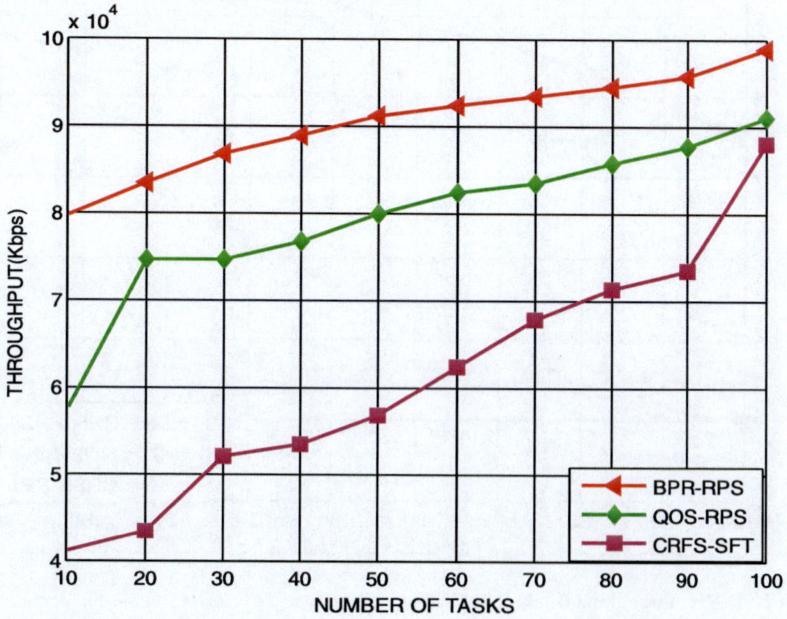

Fig. 6 BPR-RPS throughput (ranking threshold = 0.6)

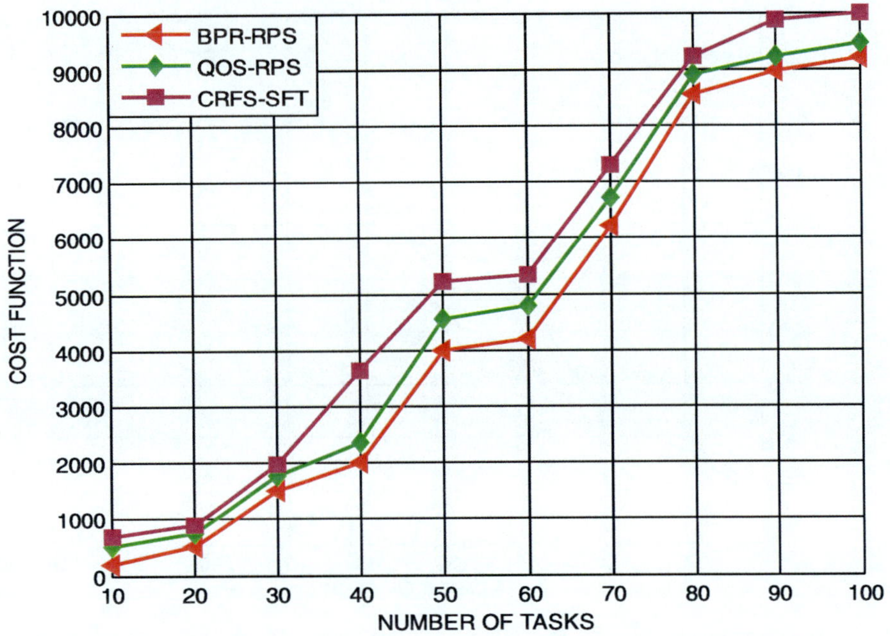

Fig. 7 BPR-RPS cost function (ranking threshold = 0.2)

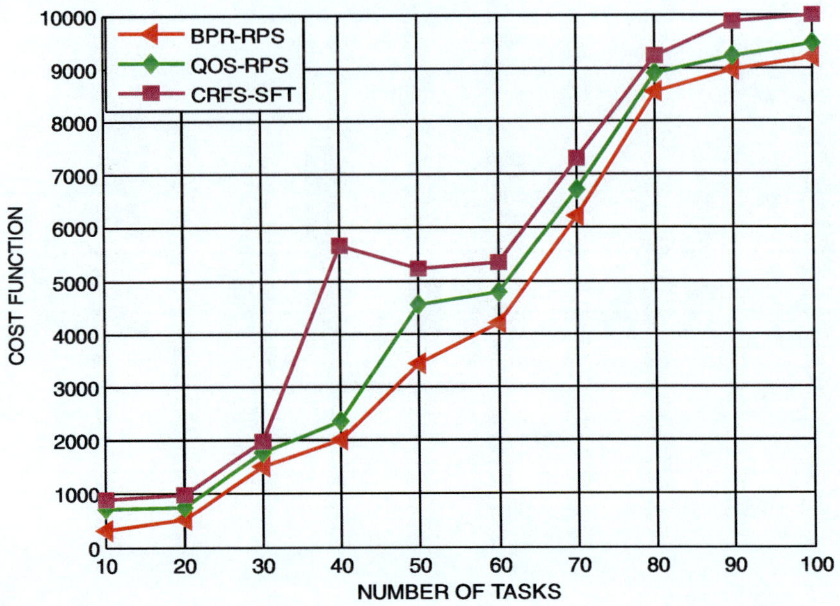

Fig. 8 BPR-RPS cost function (ranking threshold = 0.4)

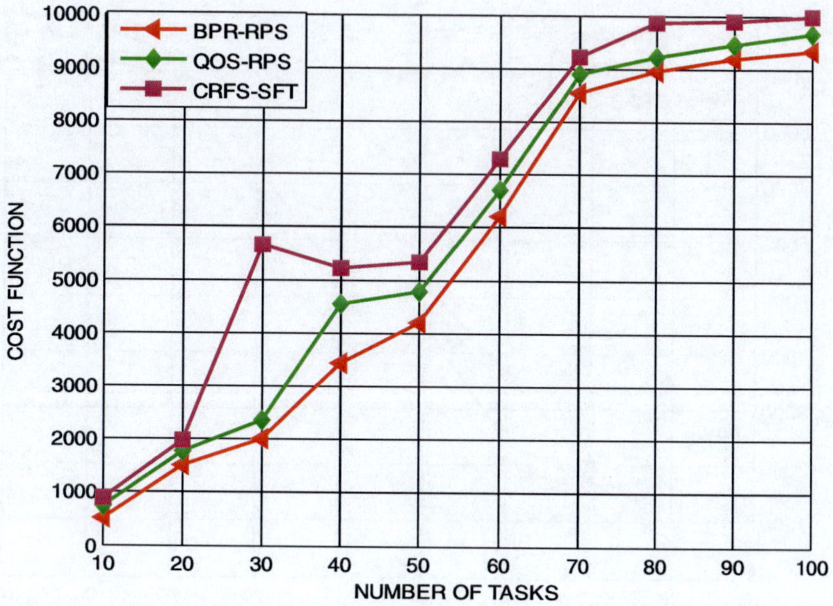

Fig. 9 BPR-RPS cost function (ranking threshold = 0.6)

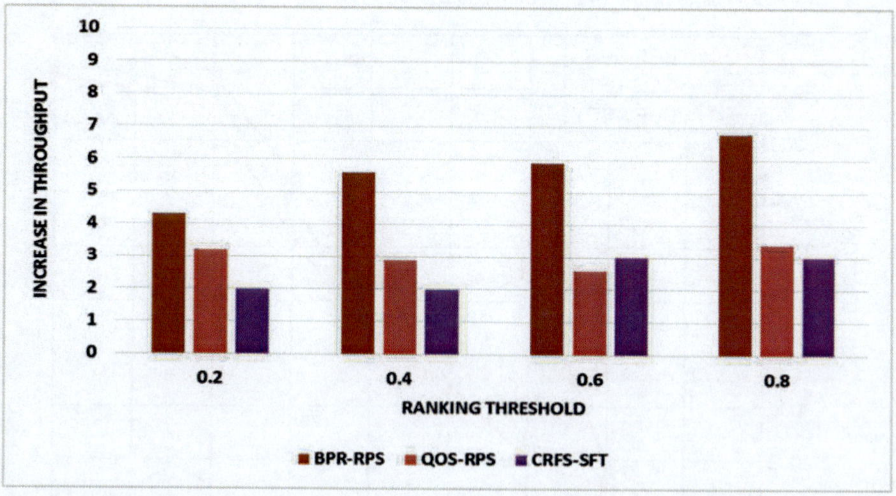

Fig. 10 BPR-RPS increase in throughput

5 Conclusion

The proposed Bayesian Personalized Ranking-based Rank Prediction Scheme (BPR-RPS) is a reliable rank prediction approach that estimates QoS properties on the client side. This proposed Rank Prediction framework is determined to

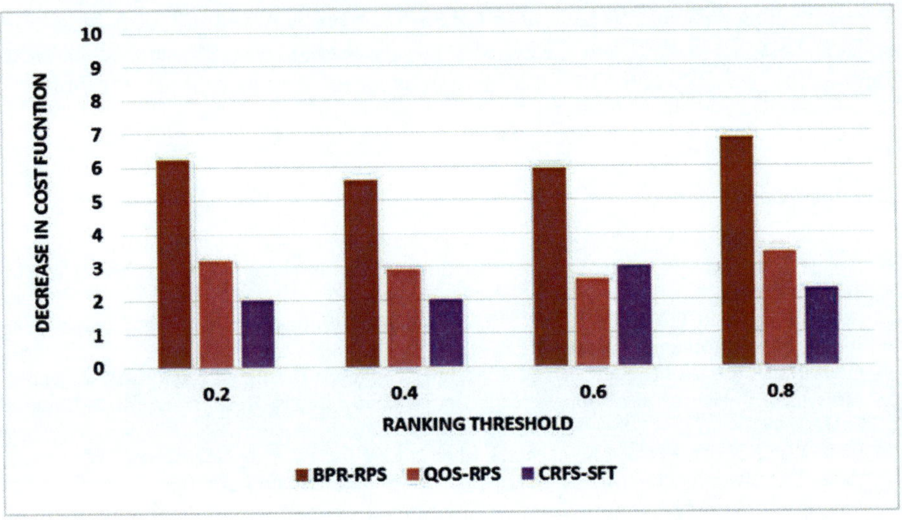

Fig. 11 BPR-RPS decrease in cost function

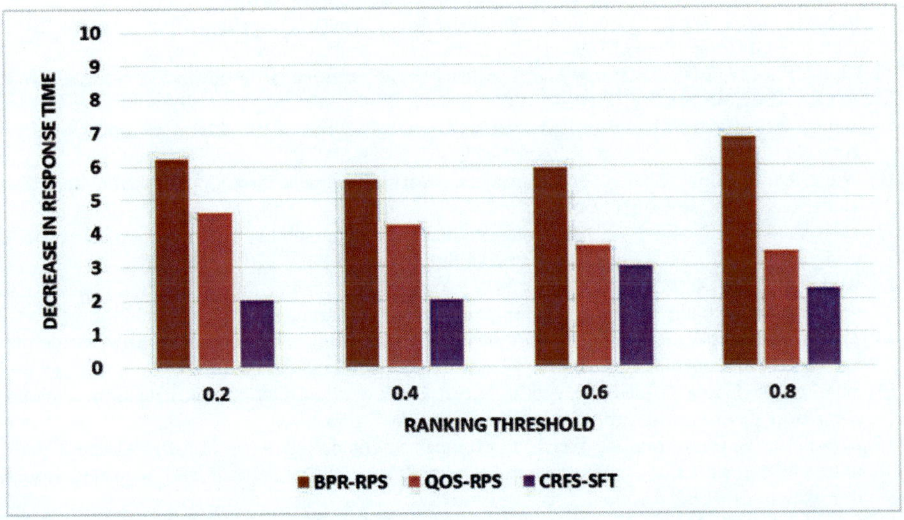

Fig. 12 BPR-RPS decrease in response time

facilitate ranking for services when specific cloud services are requested by the client. The simulation results portray the significance of BPR-RPS based on query time, computation overhead, storage overhead, cost function and response time. The performance of CRFS-SFT and BPR-RPS shows the improved response time and reduced cost function as it is phenomenal by 13% in contrast to QOS-RPS and CRFS-SFT. Further, the proposed BPR-RPS enhances the rate of throughput by 6–

8% over QOS-RPS and 9–13% over CRFS-SFT. Furthermore, the cost functions of BPR-RPS, QOS-RPS and CRFS-SFT are decreased by 5–9% and 12–16% in contrast to QOS-RPS and CRFS-SFT respectively for varying ranking thresholds.

References

1. Sehgal NK, Bhatt PC (2018) Cloud computing and information security. Cloud Comput 1(2):93–113
2. Sehgal NK, Bhatt PC, Acken JM (2019) Foundations of cloud computing and information security. Cloud Comput Security 1(1):13–48
3. Gonzales D, Kaplan JM, Saltzman E, Winkelman Z, Woods D (2015) Cloud-trust—a security assessment model for infrastructure as a service (IaaS) clouds. IEEE Trans Cloud Comput 5(3):523–536
4. Li X, Wang Q, Lan X, Chen X, Zhang N, Chen D (2019) Enhancing cloud-based IoT security through trustworthy cloud service: an integration of security and reputation approach. IEEE Access 7:9368–9383
5. Majumdar S, Madi T, Wang Y, Jarraya Y, Pourzandi M, Wang L, Debbabi M (2017) User-level runtime security auditing for the cloud. IEEE Trans Inform Foren Security 13(5):1185–1199
6. Yang Y, Liu R, Chen Y, Li T, Tang Y (2018) Normal cloud model-based algorithm for multi-attribute trusted cloud service selection. IEEE Access 6:37644–37652
7. Varadharajan V, Tupakula U (2016) Securing services in networked cloud infrastructures. IEEE Trans Cloud Comput 6(4):1149–1163
8. Choi C, Choi J (2019) Ontology-based security context reasoning for power IoT-cloud security service. IEEE Access 7:110510–110517
9. Sehgal NK, Bhatt PC, Acken JM (2019) Analytics in the cloud. Cloud Comput Security 2(1):217–233
10. Wu Y, Lyu Y, Shi Y (2019) Cloud storage security assessment through equilibrium analysis. Tsinghua Sci Technol 24(6):738–749
11. VivinSandar S, Shenai S (2012) Economic denial of sustainability (EDoS) in cloud services using HTTP and XML based DDoS attacks. Int J Comput Appl 41(20):11–16
12. Wang X, Chen X, Wang Y, Ge L (2019) An efficient scheme for SDN state consistency verification in cloud computing environment. Concurr Comput Pract Exp 1(1):34–46
13. Tiwari V, Parekh R, Patel V (2014) A survey on vulnerabilities of openflow network and its impact on SDN/openflow controller. World Acad J Eng Sci 01(01):1005
14. Bhushan K, Gupta B (2018) Hypothesis test for low-rate DDoS attack detection in cloud computing environment. Proc Comput Sci 132(1):947–955
15. Benkhelifa E, Bani Hani A, Welsh T, Mthunzi S, Ghedira Guegan C (2019) Virtual environments testing as a cloud service: a methodology for protecting and securing virtual infrastructures. IEEE Access 7:108660–108676
16. Emeakaroha VC, Fatema K, van der Werff L, Healy P, Lynn T, Morrison JP (2016) A trust label system for communicating trust in cloud services. IEEE Trans Serv Comput 10(5):689–700
17. Jiang Q, Ma J, Wei F (2016) On the security of a privacy-aware authentication scheme for distributed mobile cloud computing services. IEEE Syst J 12(2):2039–2042
18. Shin SW, Porras P, Yegneswara V, Fong M, Gu G, Tyson M (2013) Fresco: modular composable security services for software-defined networks. In: 20th Annual network and distributed system security symposium, NDSS
19. Arunarani A, Perkinian DM (2018) Intelligent techniques for providing effective security to cloud databases. Int J Intell Inf Technol 14(1):1–16

Optimal Placement Techniques of Mesh Router Nodes in Wireless Mesh Networks

S. Seetha, Sharmila Anand John Francis, and E. Grace Mary Kanaga

1 Introduction

A wireless mesh network (WMN) is a mesh network formed via the connection of wireless access points deployed at each network user's locality. Figure 1 illustrates the architecture of WMN consisting of mesh gateways, routers, and clients. Mesh routers are used to support the network access of mesh clients and to forward traffic to and from the gateways. WMN requires few mesh routers connecting to the wired network and enables the integration of WMNs with other wireless networks such as the Internet, cellular, wireless sensor, Wi-Fi, and WiMAX. Those specialized mesh routers are called gateways. These are dominant devices not having restrictions of energy, computing power, and memory and are generally scattered in a static. Mesh clients are nothing but wireless devices such as cell phones, laptops, and other wireless devices. The benefits of WMNs offer plentiful applications, for instance, enterprise networks, transportation systems, health and medical systems, security surveillance systems, public safety, rescue and recovery operation, and so on.

The placement of mesh routers is one of the most challenging problems [1, 2]. Traditionally the positions of mesh routers are predetermined [3]. This leads to the service provider every time need to determine where to place the gateway devices to be connected as part of the Internet, which results in diverse gateway positioning reasons different mesh backbone and which in turn affects the network throughput [4].

S. Seetha (✉) · E. Grace Mary Kanaga
Karunya Institute of Technology and Sciences, Coimbatore, India
e-mail: grace@karunya.edu

S. Anand John Francis
King Khalid University, Abha, Saudi Arabia

A. Haldorai et al. (eds.), *2nd EAI International Conference on Big Data Innovation for Sustainable Cognitive Computing*, EAI/Springer Innovations in Communication and Computing, https://doi.org/10.1007/978-3-030-47560-4_17

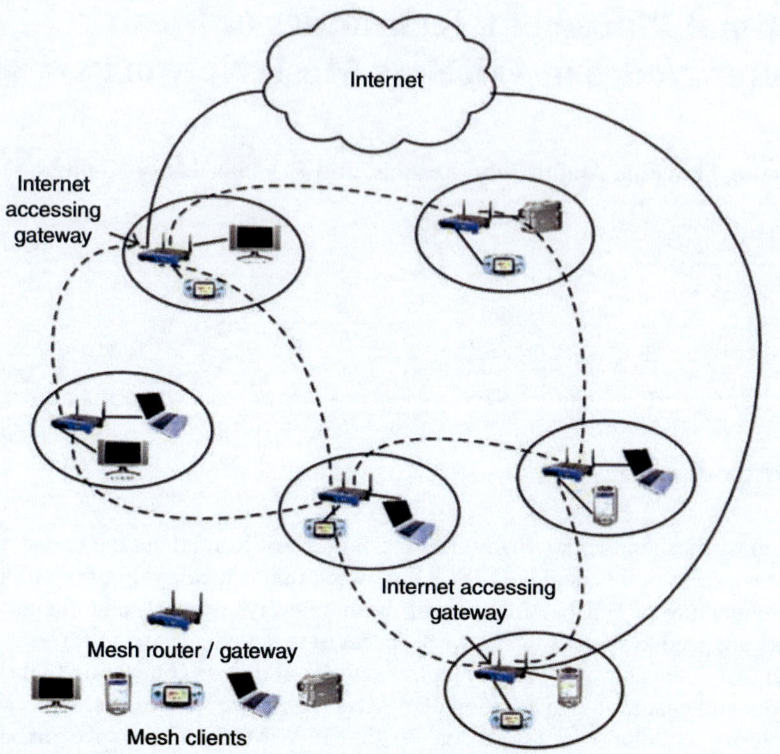

Fig. 1 The architecture of wireless mesh networks [1]

In WMN, all the traffic movements are either to or from a gateway, as conflicting to ad hoc networks where all the traffics are within any pairs of nodes. Since the mesh gateways aggregate the traffic flows, if they are misdirected, they may become the hurdle for the data communication. Therefore the placement of mesh gateways heavily affects the outcome of a network, like a throughput, connectivity, and coverage of a WMN [3]. Poor placement of mesh routers can not only result in significant interference but also create unfavorable hotspots that could withstand unpredicted heavy traffic overload. The mesh router placement system is often a tradeoff between cost and traffic demand [5, 6].

Hence, prudently placing and connecting the gateways to the Internet is important for the effective process of a WMN. A practical mesh router placement scheme should meet the following fundamental objectives as depicted in Fig. 2 in such a way that quality of service (QoS) demands of service clients are guaranteed.

The rest of the article is organized as follows: in Sect. 2, we briefly discuss the existing works for the problem of placement of mesh routers in WMN. In Sect. 3, the performance analysis of existing approaches has been done along with some of the factors that influence the design of WMN are described. The article concludes with Sect. 4.

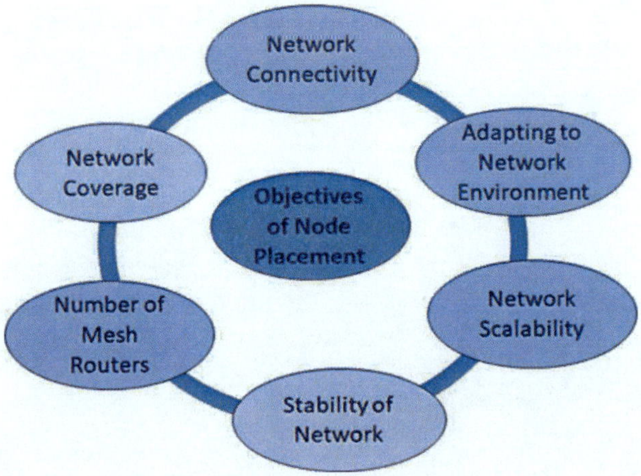

Fig. 2 The fundamental objectives of practical mesh router placement

2 Existing Approaches to the Placement of Mesh Router Nodes in WMN

Very limited works have been found in the literature about the mesh router placement problem in a wireless mesh network. The existing works are classified based on the various approaches to solve the node placement problem, viz., heuristic, meta-heuristic, and multi-objective approaches, which are shown in Fig. 3.

2.1 Heuristic and Meta-heuristic Approaches

The heuristic algorithm aims to discover an optimal solution as possible for all occurrences of the problem in a feasible time, whereas the meta-heuristic is an advanced practice, aimed to find a partial search algorithm that may offer a satisfactorily best solution to an optimization problem. Some of the heuristic and meta-heuristic algorithms that have been implemented to solve node placement problems are described in the following subsections.

Greedy Method

Greedy algorithms always make the choice that appears to be good at that instant by hoping that it will guide to an optimal solution. In the grid-based deployment approach [4], at first, the placement region is partitioned into a $a \times b$ grid with the condition of only placing the gateways in the cross points on the grid. Then it tries every likely grouping of the k-gateway assignment and evaluating them by computing the maximum throughput can be achieved by this grouping. The major

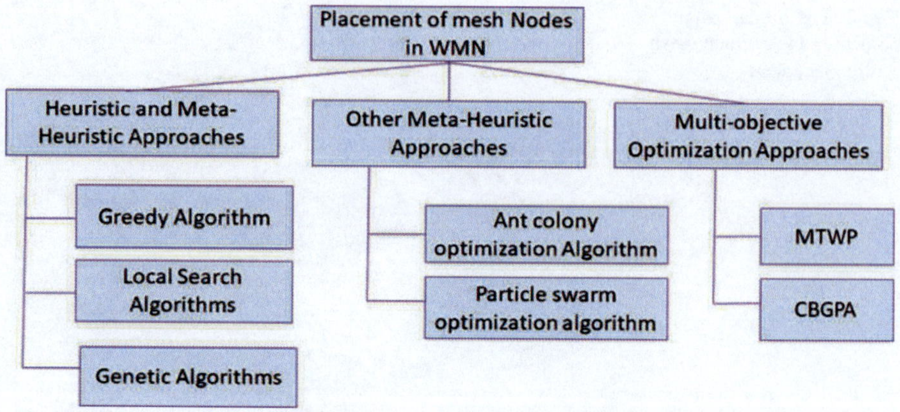

Fig. 3 The taxonomies of mesh router placement approach in WMN

advantage of this approach is that it performs well than random deployment and fixed deployment approach, but the demerit is that it may increase the computational cost when the grid size is larger.

Local Search Method

Local search algorithms change from one solution to another solution in the search space by trying limited changes, prior to an optimal solution is reached. The approach [7] using the hill-climbing algorithm aimed at the excellent assignment of mesh routers in WMNs with a dual purpose. The one is to maximize network connectivity, and the other is to ensure user coverage. The local search algorithm discovers diverse local movements and gradually increases the quality of the mesh router placement. To attain closely optimal assignment of mesh routers, this approach considers several movement types such as Uniform, Normal, Exponential, and Weibull which can be used to plan WMNs of great network connectivity. The merits of this approach are that it improves the quality of router node placement by ensuring high network connectivity and user coverage. However, for the optimal assignment of mesh router nodes in WMN, there is a need to consider more diversity positions for mesh client nodes on the grid region.

Genetic Algorithms

A genetic algorithm (GA) is a technique for resolving optimization problems based on natural selection. The approach using GA [8] works in a situation where we are given an amount of mesh client devices that are distributed in a grid region and given the amount of mesh router devices that need to be placed in the cells of the grid region. Using GAs, the solution starts with information about the present position of routers, clients, and connection between the routers and clients which are then evaluated, selected, crossed, and mutated to reproduce new individuals of the best quality. The fitness of individuals is then computed concerning network connectivity and user coverage for improving the throughput of WMNs. Also, this approach considers diverse client node positions such as Normal, Uniform, Weibull and Exponential. The advantage of this technique is that it is very effective at calculating the assignment of mesh router devices and establishes connectivity of almost all mesh router devices. The demerit is that genetic algorithms are computationally expensive algorithms as it requires a huge amount of iterations to attain high-quality solutions.

2.2 *Other Meta-heuristic Approaches*

This section deals with the other two meta-heuristic approaches [8], namely, ant colony and particle swarm optimization algorithms for solving mesh router node problems.

Ant Colony and Particle Swarm Optimization Algorithms

The ant colony is an algorithm for discovering the best paths that are founded on the behavior of ants to fetch food [9]. On the other hand, the particle swarm optimization is analogous to the genetic algorithm in that instead of focusing on a single individual, a population of individuals is considered.

Based on the ant colony optimization algorithm [10], the location of gateways is randomly generated, and the algorithm then estimates the probability and pheromone values of ants. This process is repeated for the current gateway to the following client. After each repetition, the pheromone values are modified by all the ants that have arrived at the receiver positively and got the best solution.

The algorithm using particle swarm optimization estimates the fitness cost of each system and modifies them gradually with the foremost system to rapidly discover the optimal solution. Compared to ant colony optimization, the particle swarm optimization algorithm yields the best result concerning minimum cost and consuming time.

2.3 *Multi-objective Optimization Approaches*

Multi-objective optimization is the field of multiple criteria decision-making that carry out mathematical optimization problems that involve multi-objective function to be solved concurrently. Two approaches in solving such mesh node placement problems are described below.

The first approach deals with the optimal placement of a gateway using a design metric called multi-hop traffic-flow weight (MTW) [1]. The MTW calculation considers many issues that influence the throughput of WMNs, such as the amount of mesh routers and mesh clients, the amount of gateways, traffic requirement of clients, the position of gateways, and potential interference between nodes. Based on MTW, a repetitive algorithm is used to determine the optimal location of the gateway. Every time a gateway is chosen to jointly locate with the mesh router that has the uppermost MTW. This technique increases the throughput of WMNs through the appropriate placement of gateways.

The clustering-based gateway placement algorithm (CGPA) [11] determines the strategic location for mesh gateway based on cluster radius which is constrained by the greatest amount allowable hops (H) among cluster nodes and the located mesh gateway. In this technique, the dual objectives of placement cost and congestion of gateways are concurrently reduced at the same time assuring complete coverage to mesh clients.

3 Performance Comparison

This portion discusses the performance comparison of existing approaches in the placement of mesh router nodes in WMN concerning different parameters, viz., optimization algorithms, design metrics, objectives, merits, and demerits, which are shown in Table 1.

From the literature survey, it was noted that local search methods discover top solutions to the problem for small-scale to medium-sized networks, while genetic algorithms can discover best results for large-scale networks, even though its execution time is bigger. The review also discovered that genetic algorithms achieved improved results than hill-climbing algorithm and it is most suitable for enhancing the dimension of the giant module at the same time ensuring end-users' network coverage.

Nevertheless, local search methods are best suited for resolving the mesh router placement problem in WMNs for small- to medium-sized networks under movement constraints. From another point of view, concerning user coverage, the local search methods attained the best outcome than genetic algorithms.

The multi-objective optimization approaches are useful in preventing local optima [12] issues and have drawn substantial focus in network optimization where it considers multiple objectives such as minimum interference level along with minimum deployment cost and better throughput. On the other side, multiple objective optimizations are an NP-hard optimization problem that cannot be solved in polynomial time [11], and sometimes it is difficult because of the several conflicting objectives such as minimizing the number of mesh routers and at the same time ensuring the stability and coverage of a network.

Table 1 Comparison of existing approaches

Name of the protocol	Optimization algorithm	Performance metrics	Objectives	Merits	Issues to be addressed
Grid-based deployment protocol	Greedy link scheduling algorithm	Inference transmission ratio	To place mesh gateway in the WMN such that overall throughput is improved	Achieves better than random deployment and fixed deployment methods	To consider reduction of execution cost when the grid size is larger
Protocol using local search methods	Hill-climbing algorithm	Fitness function	To discover an optimal and strong topology for WMN to provision Internet communication facilities	Improves the quality of router node placement concerning network connectivity and user coverage	Need to consider more diversity of mesh client locations on the grid region
Genetic algorithms for efficient placement of router nodes	Genetic algorithm	Intermediate population size	To support flexible and resilient wireless Internet provision to mobile data, voice, and video	Very effective at computing assignment of mesh router nodes and create connectivity of nearly all mesh router devices	Genetic algorithms are computationally costly algorithms as it require huge amount of repetitions to achieve high-quality solutions
ACO and PSO approach to gateway placement	Ant colony and particle swarm optimization algorithms	Throughput of the mesh clients	To improve the throughput and to design cost-effective WMN	Particle swarm algorithm attained optimal results than ant colony optimization	To consider optimization of gateway placement along with throughput expansion
MTWP (multi-hop traffic-flow weight protocol)	TDMA traffic scheduling algorithm	MTW (multi-hop traffic-flow weight)	To design a gateway assignment algorithm to considerably increase throughput of WMNs	Provides a framework of considerably increasing throughput of WMNs over appropriate placement of gateways	To consider the cross-optimization between gateway placement and throughput of WMNs
Clustering-based approach to optimal placement of gateways	Clustering-based gateway placement algorithm (CGPA)	Average number of nodes per cluster (NC_{clus})	To design WMN topology by concurrently optimizing objectives of placement cost and congestion of gateways	Provides modular and restricted delay planning solutions with cost reduction assurance	To consider the impact of the amount of radio interfaces on CBGPA effectiveness

3.1 Design Metrics

The following metrics play major role in determining the performance of networks during optimal assignment of mesh router nodes in WMN.

Distance

The length of the route from a traffic area to the Internet gateway (IGW) can be estimated by the Euclidean distance, which is calculated using Eq. (1) as follows:

$$d = \sqrt{(x - x0)^2 + (y - y0)^2} \tag{1}$$

where x and y are coordinates at which IGW is located.

Multi-hop Traffic-Flow Weight (MTW)

The MTW metric considers several aspects, viz., the amount of mesh routers, the amount of mesh clients, the amount of gateways, traffic requirements from clients, localities of gateway, and potential interference between gateways. Equation (2) defines MTW as follows:

$$\begin{aligned} \mathrm{MTW}(i) = (G_r + 1) \times T_r(i) \\ + G_r \times (\text{traffic requirements on all one-hop neighbors of MR}) \\ + (G_r - 1) \times (\text{traffic requirements on all two-hop neighbors of MR}) \\ + (G_r - 2) \times (\text{traffic requirements on all three-hop neighbors of MR}) \\ + \cdots \end{aligned} \tag{2}$$

where G_r is the gateway radius and $T_r(i)$ is the traffic requirements on each mesh router (MR) i.

Node Cluster ($\mathbf{NC_{clus}}$)

The average number of nodes per cluster is estimated as defined in Eq. (3) as follows:

$$\mathrm{NC}_{\mathrm{clus}} \le R \prod_{i=1}^{H-1} (R - 1) \le R^H \tag{3}$$

where R is the count of radio interfaces and H is the count of hops among the mesh nodes.

Throughput Calculation

The throughput of the ith mesh client when gateways (N_g) are positioned, denoted as $TH(i, N_g)$, is calculated as follows in Eq. (4):

$$TH\left(i, N_g\right) = \min_{i=1,\ldots,N_c} \left\{TH_{W1}\left(i, N_g\right), TH_{W2}(i)\right\} \tag{4}$$

where $TH_{W1}(i, N_g)$ is denoted as the throughput of the ith mesh client in backbone transportations and $TH_{W2}(i)$ is the throughput of ith mesh client in local transportations.

4 Conclusion

The optimal placement of mesh router nodes is gaining more attention among the research community as there is a continuous requirement for creating a robust and inexpensive WMN. In this article, the major classifications of mesh router placement methods, their architecture, objectives, design metrics, strengths, and weaknesses are discussed. There are many challenges and research opportunities in the field of placement of mesh router nodes in WMN as per Table 1. The network performance can be achieved at a greater level if networks are deployed with fully optimal solutions that address all the fundamental objectives along with QoS requirements of end-users.

References

1. Zhou P, Wang X, Manoj BS, Rao R (2010) On optimizing gateway placement for throughput in wireless mesh networks. EURASIP J Wirel Commun Netw 2010:368423
2. Seetha S, Francis SAJ, Kanaga EGM, Daniel E, Durga S (2019) A framework for multi-constraint multicast routing in wireless mesh networks. In 2019 Fifth international conference on advanced computing & communication systems (ICACCS), IEEE, Mar 2019, pp 445–451
3. Hui SY, Yeung KH, Wong KY (2008) Optimal placement of mesh points in wireless mesh networks. In: International conference on research in networking, May 2008. Springer, Berlin, pp 848–855
4. Li F, Wang Y, Li XY, Nusairat A, Wu Y (2008) Gateway placement for throughput optimization in wireless mesh networks. Mobile Netw Appl 13(1–2):198–211
5. Wang J, Xie B, Cai K, Agrawal DP (2007) Efficient mesh router placement in wireless mesh networks. In: IEEE international conference on mobile adhoc and sensor systems, Oct 2007, pp 1–9
6. Rezaei M, Sarram MA, Derhami V, Sarvestani HM (2011) Novel placement mesh router approach for wireless mesh network. In: Proceedings of the international conference on wireless networks (ICWN). The Steering committee of the world congress in computer science, computer engineering and applied computing (WorldComp), p 1
7. Xhafa F, Sánchez C, Barolli L (2012) Local search methods for efficient router nodes placement in wireless mesh networks. J Intell Manuf 23(4):1293–1303

8. Fatos X, Sánchez C, Barolli L (2010) Genetic algorithms for efficient placement of router nodes in wireless mesh networks. In: 2010 24th IEEE international conference on advanced information networking and applications, Apr 2010. IEEE, pp 465–472
9. Macura WK. Ant colony algorithm. From *MathWorld*—a Wolfram web resource, created by Eric W. Weisstein. http://mathworld.wolfram.com/AntColonyAlgorithm.html
10. Le DN, Nguyen NG, Le ND, Dinh NH, Le VT (2012) ACO and PSO algorithms applied to gateway placement optimization in wireless mesh networks. In: International proceedings of computer science and information technology, Jan 2012, vol 57, p 8
11. Benyamina D, Hafid A, Gendreau M (2009) Optimal placement of gateways in multi-hop wireless mesh networks: a clustering-based approach. In: IEEE 34th conference on local computer networks, Oct 2009. IEEE, pp 625–632
12. LINGO 17.0 user's manual on mathematical modeling. https://www.lindo.com/doc/online_help/lingo17_0/local_optima_vs__global_optima.htm

Predicting Replication Time in Cassandra

Abhin S. Lingamaneni, Adarsh Mishra, and Dinkar Sitaram

1 Introduction

It becomes important to store and process enormous amounts of data and query the data reliably and efficiently. Apache Cassandra [9] is a distributed, decentralised, scalable database which bases its distribution design on Amazon's Dynamo [5] and its data model design on Google's Bigtable [3]. Cassandra has a peer-to-peer architecture, i.e., all the nodes on the cluster have similar functionalities and there is no fixed orchestrator or master node. This also ensures that there is no single point of failure. Cassandra also supports elastic scalability, which means that the system can seamlessly upscale and downscale number of nodes without disrupting any running service and processes.

Data integrity is extremely important in a storage system. Most Big Data storage systems ensure availability by replicating the data [11]. However, in a heavily loaded cluster, the replication time may lag when compared to the client requests by some amount of time. There are numerous papers which have benchmarked various NoSQL databases [1, 7]. But most of them have just compared how the read and write latency varies with different workloads and the number of instances. In this paper we have presented our study on how replication time behaves in an overloaded cluster.

A. S. Lingamaneni (✉) · A. Mishra · D. Sitaram
PES University, Bengaluru, Karnataka, India

A. Haldorai et al. (eds.), *2nd EAI International Conference on Big Data Innovation for Sustainable Cognitive Computing*, EAI/Springer Innovations in Communication and Computing, https://doi.org/10.1007/978-3-030-47560-4_18

1.1 *Our Contribution*

In MongoDB replication lag plays an important role in data integrity. A secondary node falls behind on replication any time it cannot keep up with the rate at which the primary is writing data. This is known as replication lag. Replication lag makes it more likely that results of any read operations distributed across secondaries will be inconsistent. Replication time is important as it impacts data integrity. Just as in the impact of replication lag in MongoDB even in Cassandra the replication time can have a significant impact on data integrity. We are attempting to model the replication time in a Cassandra cluster as a function of system parameters, internal Cassandra parameters, and type of read/write workload. In this paper we present the behaviour of replication time under different read/write workloads and have tried to come up with a machine learning model to predict the replication time of Cassandra. Important parameters have been identified which impact the replication time. As far as we are aware this is the first paper to attempt to do this.

2 Related Work

It has been shown replication in NoSQL databases like Cassandra and MongoDB plays an important role to achieve an accurate evaluation of the performance of these data stores [6]. Abramova et al. [2] evaluates the scalability and scalability limitations of Cassandra under different YCSB workloads. Cooper et al. [4] provides benchmarks for different cloud systems. An alternate approach has been proposed in [10] helps in understanding advanced features like function shipping filters and speedup techniques from client to server. Kishore et al. [8] describes a QoS aware architecture for Cassandra to improve the resource utilisation. Though some papers speak about the importance of replication time, none of them attempts to predict it.

3 Background: Cassandra Architecture

Cassandra is a widely used database which can be used when there is a necessity for high availability and scalability on the fly without compromising the performance. Writes are extremely fast in Cassandra. Whenever a write operation is performed in Cassandra first it is immediately written to the commit log. This ensures durability of the data as even if the system crashes the commit log would have the necessary information to redo the operation. This ensures fault tolerance in case of a failure. After the data is written to the commitlog the write is essentially written to the memtable. Memtable is stored in the RAM and all the update operations are appended in memtable. The memtable has a limit to the data it can hold. Once memtable is full, the contents of memtable are flushed to SStable which is a file on the disk. Every memtable flush creates a new SStable. At a particular time more than one SStable can exist. Different SStables might contain both old value and new value of the same cell, or an old value for a cell later deleted. That is fine, as Cassandra uses timestamps on each value or deletion to figure out which is the most recent value the one which is updated last is stored and the other is discarded. In case of multiple SStables Cassandra has to read multiple SStable to compose the results. This becomes very costly operation and to avoid this compaction is needed. Compaction reads several SStables and outputs one SStable containing the merged, most recent information.

Cassandra is built in such a way that it can ensure very fast writes. The writes operation are all sequential which means there is no disk seek or read operation. However, the trade-off is that there is a background compaction operation that combines all the data and reorganises them which reduces the read latency. If writes were not append operations the write clients have to pay read and seek costs upfront.

On a read operation the database reads both memtable and SStables to ensure that no data is left unread. Whenever a query is fired by the client it first goes to the coordinator node which directs it to the required nodes as shown in Fig. 1. Any node in the cluster can act as coordinator node and there are different policies for choosing the coordinator node. The coordinator node hashes the key to examine the hashed range to determine the required node where the query has to be directed. Cassandra also allows the end users to tune consistency of read and write queries. In case of strong consistency all the replicas are updated or read before sending acknowledged message back to the client. In case of eventual consistency as soon as the number of replicas specified by the consistency level is updated or read acknowledgement message is sent back to the client and the other replicas are eventually updated. By controlling read and write consistencies, a trade-off is established between consistency and availability.

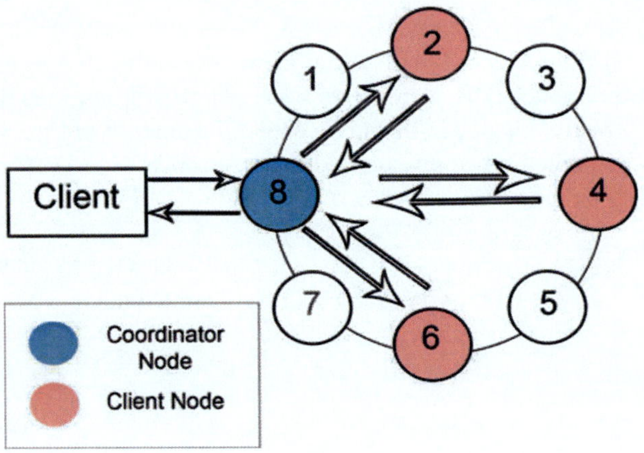

Fig. 1 Cassandra query path

Cassandra follows a Stage Event Driven Architecture [12]. A stage is a basic unit of work, and a single operation may internally state-transition from one stage to the next. Each stage is a background task and has a thread pool associated with it. A stage consists of an incoming event queue, an event handler, and an associated threadpool.

4 Experiment Setup

4.1 Workloads

In our experiment evaluation we have used YCSB (Yahoo! Cloud Service Benchmark) [4] as our client which generates queries against the database. YCSB mainly consists of two parts, first a client which acts as a workload generator and second a set of standard workloads to be executed by the generator. Client provides metrics such as number of target operations per second, which is useful to load and simulate stress on the server. We have used the Zipfian distribution, which assigns more popularity to the records that are the head of distribution while most records will be unpopular (the tail). This distribution reflects the real world scenario where most client requests are for those objects that are popular and trending. Our main goal is to analyse and predict replication time for standard YCSB workloads. Some standard workloads which we use in our experiments are shown in Table 1.

Workload A is write heavy and Workload B is read intensive. Workload C depicts the scenario when an update request comes in a completely read heavy environment.

4.2 Parameters

Table 2 shows the Threadpool and system parameters collected in each and every node to build the model. The parameters were recorded at an interval of 1 s. The read latency and write latency in that node depict the amount of time for which data is being read or updated on a particular node per second.

Table 1 Workloads configuration for YCSB

Workloads	Read(%)	Write(%)
Workload A	50	50
Workload B	95	5
Workload C	100	0

Table 2 Important system parameters and threadpool values collected

Parameters	Description
MemtableFlushWrite	Number of jobs writing memtable contents to disk
MutationStage	Number of local writes
Native-Transport-Requests	Number of requests to the server
PendingRangeCalculator	Number of jobs to calculate pending ranges
ReadRepairStage	Number of update of replicas of a key
ReadStage	Number of local reads
RequestResponseStage	Number of responses from other nodes
ValidationExecutor	Number of schema validation jobs
Number of records	Number of records added/changed
Type of request	Read/Write/Scan
CPU utilisation	Utilisation of all cores of the CPU
IO-Read rate	Total bytes read every 0.5 s
IO-Write Rate	Total bytes written every 0.5 s
Read latency	Read latency on that particular node
Write latency	Write latency on that particular node
Memory utilisation	Memory utilisation of each node

4.3 Replication Time

A three node cluster is set up each consisting of a 4 cores Intel Xeon CPU and 16GB of RAM. The client which runs two YCSB process's also has the same configuration. 70 million records were inserted in our cluster with a replication factor of 3. Each field is 100 bytes and a record has 10 fields. In each node of the Cassandra cluster there are scripts running which are responsible for collecting system parameters and system values which would be used as attributes in our machine learning model. The important collected parameters and metrics are described in Table 2.

Two client YCSB processes are running on server 1 and Cassandra process runs on server's 2, 3, and 4. For one of the client process we set the consistency level equal to replication factor and it sends an update requests every second. This is done to ensure that this client does not put much load on the Cassandra cluster and does not affect the replication time. The coordinator node is responsible for forwarding the update request to all other nodes simultaneously. The client driver was configured to use DCAwareRoundRobinPolicy (the default policy). In this policy coordinator node is selected in a round robin fashion which ensures load balancing in the cluster. The coordinator node then waits for the response from all the other nodes. Once the coordinator node gets update successful response from all the other nodes an OK message is sent to the client and the time for the request to be processed is recorded. We are treating this time as replication time. We made the assumption that the time required to send the update request to coordinator node and gets the response back from the coordinator node is negligible as they

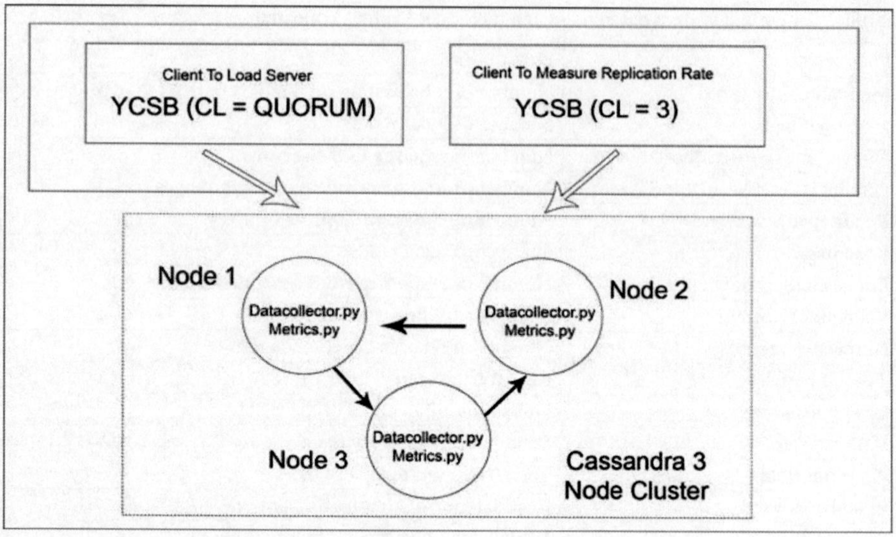

Fig. 2 Experiment Setup

are set up in the same rack. This is a proxy method to measure replication time as there is no direct way to measure replication time in Cassandra. The second YCSB client adds load to the cluster and performs read and update operation as defined by custom workloads. To overload the servers the YCSB client bombards the severs with 10,000 ops/s with different workloads. Figure 2 shows the cluster setup. The replication, read and write latencies are collected in microseconds.

5 Experiment Results and Accuracy

5.1 Analysis of Replication Time

After conducting the runs it was clear that replication time varied more compared to read and write latencies of the client for all the workloads. Figures 3, 4, and 5 give us a visual idea of how the replication time is more volatile compared to the read and write latencies of the workloads.

Tables 3, 4, and 5 give a statistical description of replication, read and write latencies of workload A, workload B, and workload C, respectively. It can be seen that the standard deviation of replication time is significantly more than read and write latencies for all the workloads. From Table 3 we can observe that the standard deviation of replication time is 8 times more than standard deviation of write latency and 4 times more than read latency. The mean values of all the three operations are comparatively similar. This might be due to the fact that the number of points with

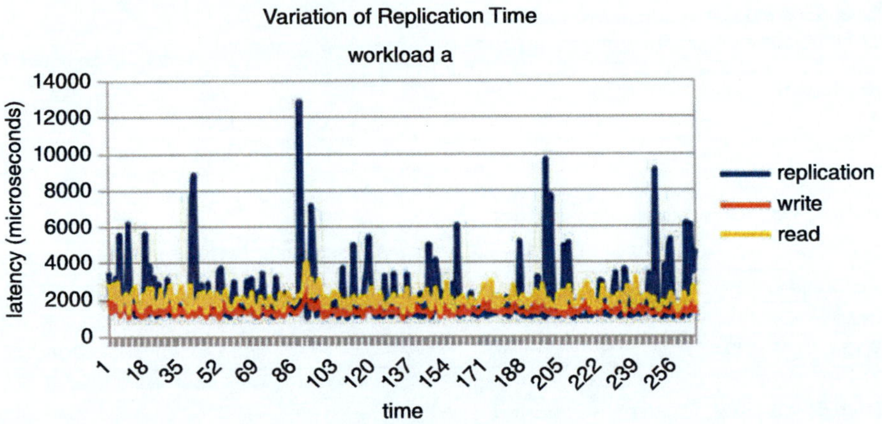

Fig. 3 Variation of replication for workload A

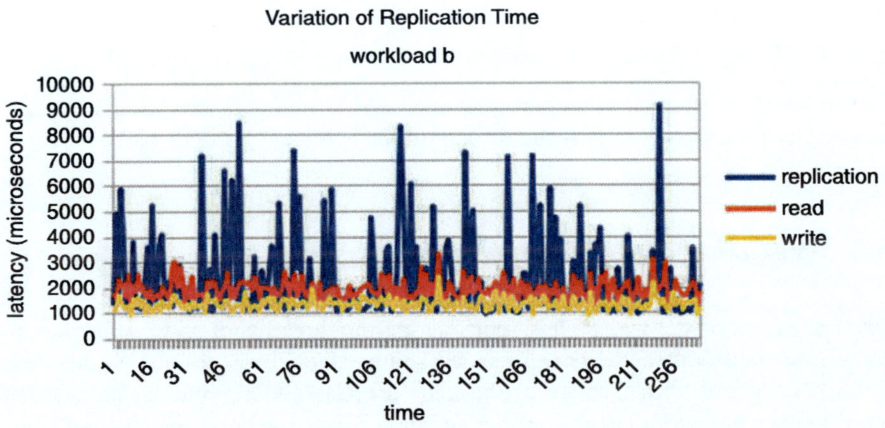

Fig. 4 Variation of replication for workload B

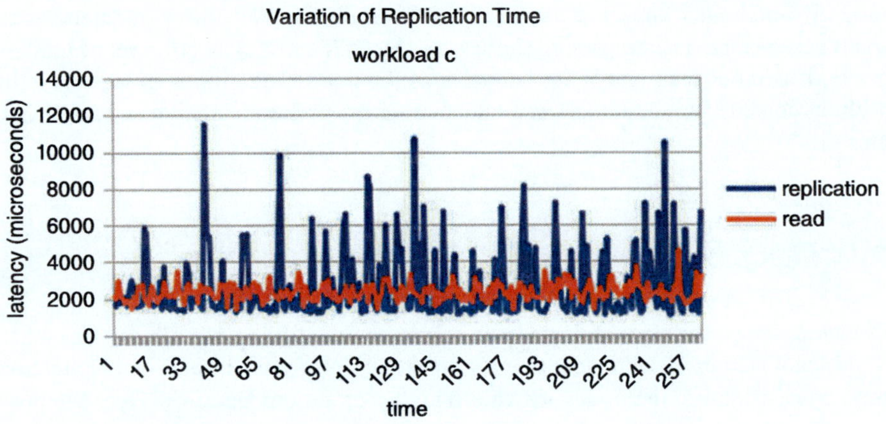

Fig. 5 Variation of replication for workload C

Table 3 Workloads A descriptive statistics

Type	Average	Standard deviation	>5000	>10,000	Minimum	Maximum
Replication	1763	1730	277	37	600	29, 495
Read	1916	391	8	3	1140	5221
Write	1348	204	7	2	894	3235

Table 4 Workloads B descriptive statistics

Type	Average	Standard deviation	>5000	>10,000	Minimum	Maximum
Replication	1911	3217	365	11	562	181, 223
Read	1903	418	4	2	1153	10, 249
Write	1347	256	2	1	830	7018

Table 5 Workloads C descriptive statistics

Type	Average	Standard deviation	>5000	>10,000	Minimum	Maximum
Replication	2047	2122	357	11	716	71, 921
Read	2274	477	6	0	1281	9171

high latency for replication is more than read and write operations. Similar results can be observed from Tables 4 and 5 for workloads B and C, respectively.

5.2 *Prediction of Replication Time*

The regression model was built by considering the thread pool and system parameters. Each node's data points were added and our model was trained on this data. Read latency and write latency of the node are Cassandra's internal parameters. Read latency and write latency are the amount of time the node spends on reading and updating, respectively, per second. The read and write latencies of the node depend on other system parameters like Read and Write IO [8], thus two sets of models were built. One set of model is built considering read and write latencies as input parameters. The graphs are shown in Figs. 6, 7, and 8. The other set of models are built without read and write latencies of the node. The graphs of models built without considering read and write latencies of the nodes are shown in Figs. 9, 10, and 11.

5.3 *Significant Parameters*

Decision Trees are powerful when it comes to identifying significant variables. Significant variables were identified which have profound impact on replication time by considering the nodes which are higher up on the Decision Tree. Memory

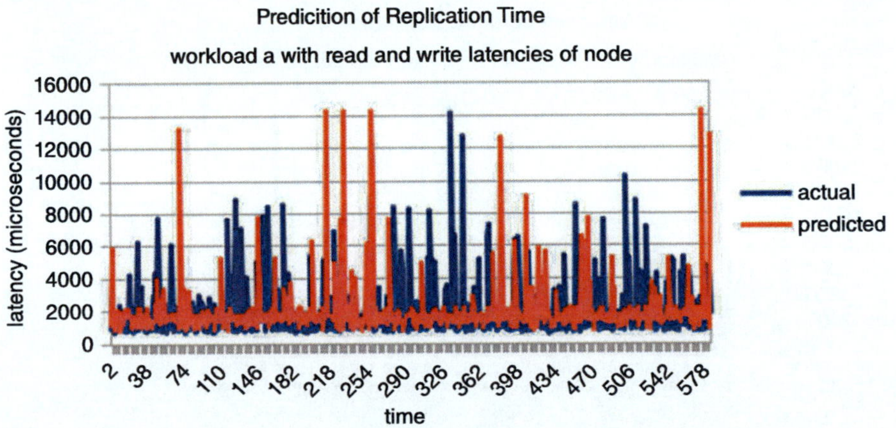

Fig. 6 Prediction of replication time considering read and write latencies for workload A

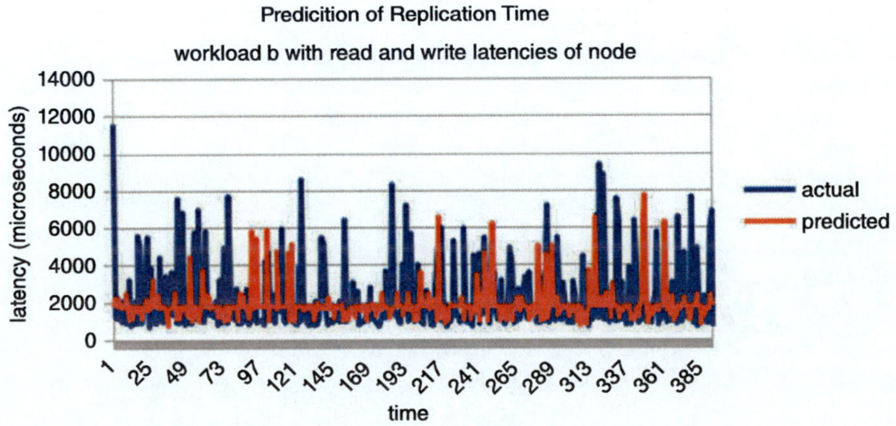

Fig. 7 Prediction of replication time considering read and write latencies for workload B

utilisation occupied was the root node for update intensive workload in both the scenarios. For workload B and workload C most of the significant parameters are related to read background tasks (threadpool) and Read-IO system parameter as the workloads are read intensive. In all of the workloads under both the scenarios NativeTransportRequest occupies one of the topmost nodes in the decision tree. As coordinator node's role is to send and accept the requests to the required replica and from other nodes the number of threads allocated (NativeTransportRequestActive) and the queue length (NativeTransportRequestPending) affects the behaviour of replication time.

Tables 6 and 7 depict the significant parameters for different workloads in both the scenarios. For workload A we can see from Tables 6 and 7 that memory

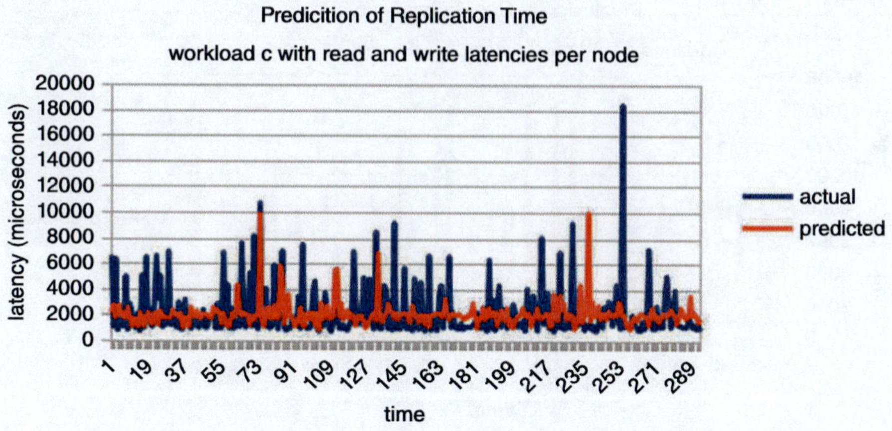

Fig. 8 Prediction of replication time considering read and write latencies for workload C

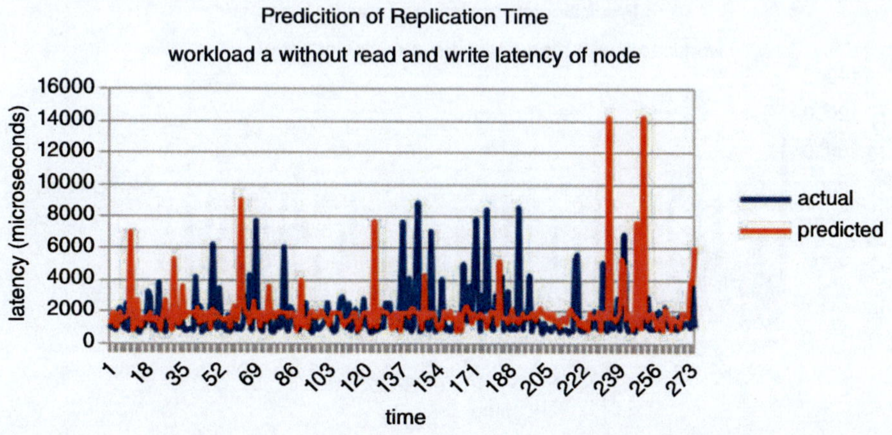

Fig. 9 Prediction of replication time without considering read and write latencies for workload A

utilisation occupies the root of the tree. For workloads B and C read related system and threadpool parameters have an impact on replication time.

Table 6 Important parameters affecting replication time considering read and write latencies in descending order of importance

Workload A	Workload B	Workload C
Memory utilisation	Write latency	NativeTransportRequestsActive
Read latency	CPU utilisation	IO-Read-Rate
ReadRepairStagePending	IO-Read-Rate	RequestResponseStagePending
IO-Write-Rate	NativeTransportRequestsActive	ReadLatency
NativeTransportRequestsActive	Read latency	ReadRepairStagePending

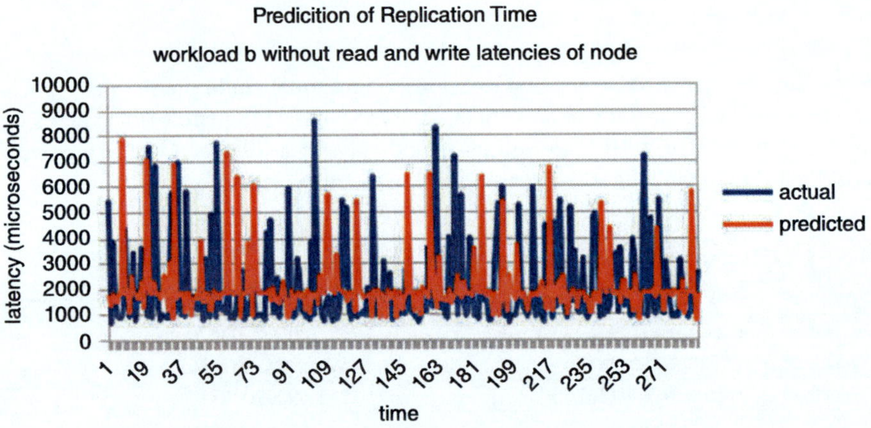

Fig. 10 Prediction of replication time without considering read and write latencies for workload B

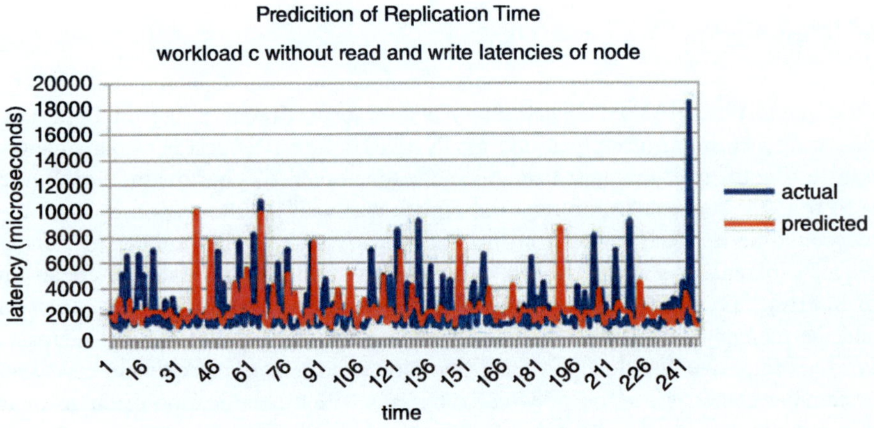

Fig. 11 Prediction of replication time without considering read and write latencies for workload C

Table 7 Important parameters affecting replication time without considering read and write latencies in descending order of importance

Workload A	Workload B	Workload C
Memory utilisation	CPU utilisation	NativeTransportRequestsActive
ReadRepairStagePending	MutationStagePending	IO-Read-Rate
IO-Write-Rate	CompactionExecutorPending	RequestResponseStagePending
NativeTransportRequestsActive	ReadStageActive	ReadRepairStagePending
IO-Read-Rate	NativeTransportRequestPending	

5.4 Accuracy

Mean Squared Error measures how accurately our model could predict the replication time. From Table 8 it can be seen that the MSE (Mean Square Error) values for workload A and workload C are similar in both the scenarios. The MSE value under workload B is much higher when compared to the other workloads.

Table 8 MSE of different workloads

Workload	MSE (inclusive of read and write latency)	MSE (excluding read and write latency)
Workload A	2225.568401995179	2022.0060700067065
Workload B	6908.778752678336	9576.575850679259
Workload C	2115.3687899736296	2211.690070017744

6 Conclusion

From the results above we observe that NativeTransportRequest plays an important role in the behaviour of replication time. NativeTransportRequest is responsible for transferring the query request from one node to another. As coordinator node's role is to send the requests to the required replica and wait for the responses from the respective nodes. The replication time becomes worse as the number of remote requests increase. As workload A is write intensive and due to Cassandra's decision of inserting the records in the memtable. So when an update to replica comes in and the memory utilisation in the node is more the replication time can increase. For read intensive workloads B and C more of Read related background tasks and system parameters are activated which interferes with the replication mechanism as it is mostly concerned with updating the record. In these workloads more threads are allocated for read related tasks which can have an inversely proportional effect on updation of required records during replication, hence increasing the replication time.

7 Future Work

A monitoring tool can be made which proactively monitors the active number of threads allocated to significant parameters. Depending on the important parameters and how they affect replication time number of threads for the background tasks can be varied which can result in lower replication time. As remote requests have an inverse effect on the replication time a policy can be developed to choose the node where data is available as the coordinator node rather than choosing the coordinator node in a round robin fashion every time and then making remote requests. Additionally table metrics like compaction rate, compaction ratio, and SStable flush rate can be collected and analysed to see how it affects the replication time. Also sequence Machine Learning models like Recurrent Neural Networks and Hidden Markov Models can be used to build the model.

References

1. Abramova V, Bernardino J (2013) NoSQL databases: MongoDB vs Cassandra. In: Proceedings of the international C* conference on computer science and software engineering. ACM, New York, pp 14–22
2. Abramova V, Bernardino J, Furtado P (2014) Testing cloud benchmark scalability with Cassandra. In: 2014 IEEE world congress on services. IEEE, Piscataway, pp 434–441
3. Chang F, Dean J, Ghemawat S, Hsieh WC, Wallach DA, Burrows M, Chandra T, Fikes A, Gruber RE (2008) Bigtable: a distributed storage system for structured data. ACM Trans Comput Syst 26(2):4
4. Cooper BF, Silberstein A, Tam E, Ramakrishnan R, Sears R (2010) Benchmarking cloud serving systems with YCSB. In: Proceedings of the 1st ACM symposium on cloud computing. ACM, New York, pp 143–154
5. DeCandia G, Hastorun D, Jampani M, Kakulapati G, Lakshman A, Pilchin A, Sivasubramanian S, Vosshall P, Vogels W (2007) Dynamo: amazon's highly available key-value store. In: ACM SIGOPS operating systems review, vol 41. ACM, New York, pp 205–220
6. Haughian G, Osman R, Knottenbelt WJ (2016) Benchmarking replication in Cassandra and MongoDB NoSQL datastores. In: International conference on database and expert systems applications. Springer, Cham, pp 152–166
7. Hendawi A, Gupta J, Jiayi L, Teredesai A, Naveen R, Mohak S, Ali M (2018) Distributed NoSQL data stores: performance analysis and a case study. In: 2018 IEEE international conference on big data (big data). IEEE, Piscataway, pp 1937–1944
8. Kishore Y, Datta NV, Subramaniam K, Sitaram D (2016) QoS aware resource management for Apache Cassandra. In: 2016 IEEE 23rd international conference on high performance computing workshops (HiPCW). IEEE, Piscataway, pp 3–10
9. Lakshman A, Malik P (2010) Cassandra: a decentralized structured storage system. ACM SIGOPS Oper Syst Rev 44(2):35–40
10. Patil S, Polte M, Ren K, Tantisiriroj W, Xiao L, López J, Gibson G, Fuchs A, Rinaldi B (2011) YCSB++: benchmarking and performance debugging advanced features in scalable table stores. In: Proceedings of the 2nd ACM symposium on cloud computing. ACM, New York, p 9
11. Shvachko K, Kuang H, Radia S, Chansler R et al (2010) The Hadoop distributed file system. In: MSST, vol 10, pp 1–10
12. Welsh M, Culler D, Brewer E (2001) SEDA: an architecture for well-conditioned, scalable internet services. ACM SIGOPS Oper Syst Rev 35:230–243. ACM, New York

A 600 mV + 12 dBm IIP3 CMOS LNA with G_m Smoothening Auxiliary Path for 2.4 GHz Wireless Applications

D. Sharath Babu Rao and V. Sumalatha

1 Introduction

The Institute of Electrical and Electronics Engineers (IEEE) created the first WLAN standard in 1997, with 2 MBPS support. With the evolution of wireless communication technology, home and business, owners preferring an array of choices, which conform to the 802.11a, 802.11b/g/n and/or 802.11ac wireless standards collectively known as Wi-Fi technologies. The IEEE 802.11ac/g/n/ac/ad standard gives abutment to 2.5/5 GHz short range wireless communication in WSNs [2, 3]. RF design has become increasingly important due to rapidly growing wireless markets. The main goal of a designer is to obtain a design that meets the specifications in quick time. Designers are faced with shorter design cycles and with increasingly difficult specifications. From the past decade, the title role of RF front end in transceiver design throughout and is widely influencing the amount of data that can be drawn by the transceivers.

The low-noise amplifier (LNA) [1] is the essential discriminating part in the simple front-end of a radio frequency (RF) receivers. LNAs are found in radio communications systems, medical instruments and electronic equipment. The primary role of LNA (receiver front end) is to provide voltage gain, impedance matching with tolerable noise figure and linearity. As a part of 802.11 wireless RF receiver, a low-noise amplifier (LNA) is used to amplify signals of weak strength, i.e. of order −30 to −120 dBm [4], usually from receiving antenna where signals are barely recognizable and should be amplified without adding any noise, without any information loss. The LNA is accountable for offering sufficient amplification to

D. Sharath Babu Rao · V. Sumalatha (✉)
Electronics and Communication Engineering Department, JNTUA, Anantapuramu, India
e-mail: vsumalatha.ece@jntua.ac.in

A. Haldorai et al. (eds.), *2nd EAI International Conference on Big Data Innovation for Sustainable Cognitive Computing*, EAI/Springer Innovations in Communication and Computing, https://doi.org/10.1007/978-3-030-47560-4_19

noise susceptible signals while limiting the amount of introduced electronic noise and distortion. As a result, the characteristics of the LNA set the higher limit at the performance of the overall communication system. Low latency, high data rate and speed considerations have to converge for motivated potential design.

Feedback low-noise amplifiers [6] are notorious for gain enhancement applications [5] offering improvement in the A_{IIP3} by $(1 + g_1\omega L_s)^{3/2}$. However, due to second-order interaction to the third-order nonlinearity component, IMD3 decreases the A_{IIP3} performance. In Ref. [6] good gain of 19.2 dB is achieved, but the A_{IIP3} is −20.1 at 1.2 V, which indicates strong interference. An IM_3 cancellation technique [7] generates IM_2 and injects it in out of phase to IM_3 to suppress IM_3. The proposed technique makes use of the second-order nonlinearity of the same devices in the main path for effective cancellation with excess noise figure of 4.1 dB, with 1.5 V supply voltage. This paper describes an IM_3 cancellation technique for differential transconductors with simple circuit implementation and negligible extra power consumption that is well-suited for fully integrated receiver. Even this technique is going to offer A_{IIP3} of only 17.5 to improve its linearity and to remove third-order nonlinearity, a harmonic [8] rejection technique is proposed using RC feedback at the gain stage. The third harmonic component of the drain node of the common-gate transistor is fed back to the source node of the common-gate transistor to restrict the generation of the third harmonic component at the output of the LNA, but this technique requires an additional circuit, i.e. cross-coupled feedback loop in addition to the LNA stage which consumes more power, i.e. 23 mW. This method is going to offer gain of 19 dB with only A_{IIP3} of only 4. With optimum gate biasing [9], a bias circuit is used to generate the gate voltage for zero third-order nonlinearity of the FET transconductance. One of the major drawbacks of this technique is that a significant IIP3 improvement occurs in a very narrow V_{gs} range (about 10–20 mV) around the optimum voltage. Gain enhancement techniques with folded cascade configuration [10] can be operated at low voltage, i.e. 600 mV offers a good gain of 15.4 dB with A_{IIP3} of 4. This method g_m boosting method to increase the gain with moderate weak IIP3 performance. With linearity improvement as objective, robust derivative superposition method [11] is proposed which reduces third-order nonlinearities effectively, but the second-order nonlinearity to the third-order nonlinearity is never reduced effectively. This technique offers gain of 11.3, A_{IIP3} of 18.4 at 2.4 V of power sully and consumes a huge power of 17.7 mW.

In modified derivative superposition methods [12], the degrading effect of the IIP3 due to feedback in the circuit in the derivative superposition (DS) method was reduced. The contribution to the IMD3 due to second-order nonlinearity is reduced by adding the composite third-order nonlinearity with the second-order nonlinearity contribution, which maximizes the IIP3 performance. The limitation in this technique is the operation of the transistor in auxiliary path in moderate inversion region contributing enough individual third-order nonlinearity to the composite third-order nonlinearity in the proposed architecture.

In modified derivative superposition method [12], linearity is important design consideration along with noise, gain and impedance matching. To reduce inter-modulated harmonics, the MOS transistor and the degenerative inductors are split

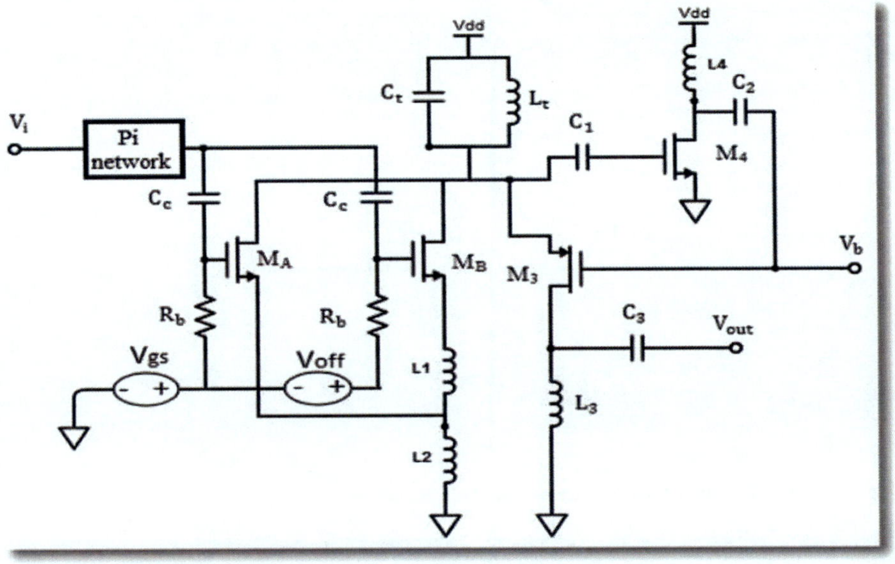

Fig. 1 LNA with MDS method

into two stages (M_A, M_B) and (L_1, L_2) as shown in Fig. 1. Based on the VI (DC characteristics) characteristics of the MOS device, the operating point can be selected in three regions, i.e. weak inversion, moderate inversion and strong inversion. The transistor M_A is biased in moderate inversion region, and the transistor M_B is operated in strong inversion region. The weak inversion region is not preferred, due to increase in gate induced noise, the drain current reduces and noise figure increases, as it is inversely proportional to drain current. The second-order and third-order nonlinearities as a function of gate to source voltage are plotted in Fig. 2.

The dominant nonlinearity introduced by the transconductance stage of common source stage of LNA is reduced by noise cancelling by adding currents in main and auxiliary paths at output nodes, which increases the power dissipation. Due to the source, degeneration feedback second-order nonlinearity component (g_{m2}) is added and contributes the third-order nonlinearity. In this, the degrading effect of the IIP3 due to feedback in the circuit in the derivative superposition (DS) method was reduced (Fig. 3).

In MDS design methodology if the transistor is biased at zero crossing of the third-order nonlinearity component of the transconductance (g_{m3}), then third-order nonlinearity will be reduced. The contribution to the IMD3 due to second-order nonlinearity is reduced by adding the composite third-order nonlinearity with the second-order nonlinearity contribution in 180° out of phase, which maximizes the IIP3 performance.

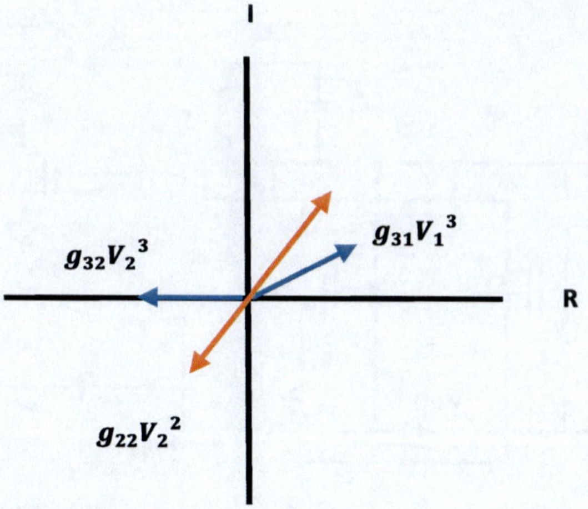

Fig. 2 Vector diagram of IMD components of DS technique

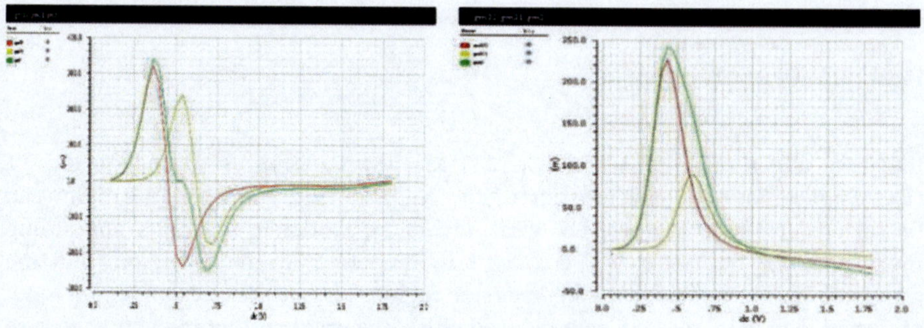

Fig. 3 Third- and second-order nonlinearities in MDS method

The MOS transistor drain current is given by

$$I_d = g_{m1} V_{gs} + g_{m2} V_{gs}^2 + g_{m3} V_{gs}^3 \tag{1}$$

The second-order nonlinear coefficient (g_{m2}) and third-order nonlinear coefficient (g_{m3}) are obtained as a function of gate to source voltage (V_{gs}) is shown as

$$g_{m1} = \frac{\partial I_d}{\partial V_{gs}}, \quad g_{m2} = \frac{1}{2}\frac{\partial^2 I_d}{\partial V_{gs}^2}, \quad g_{m3} = \frac{1}{6}\frac{\partial^3 I_d}{\partial V_{gs}^3} \tag{2}$$

The second-order and third-order nonlinearities of the MOS transistors M_A and M_B are given as g_{m21}, g_{m22} and g_{m31}, g_{m32}. The total output current from the two transistors, i.e. i_{out}, is represented as

$$i_{out}(v_i) = g_{m31}{v_1}^3 + g_{m12}v_2 + g_{m22}{v_2}^2 + g_{m32}v_1{v_2}^3 \quad (3)$$

where

$$g_{m2} = g_{m21} + g_{m22}, \quad g_{m3} = g_{m31} + g_{m32} \quad (4)$$

The operating biasing conditions are chosen such that if the positive peak of the g_{m31} lies with the negative peak of the g_{m32}, then overall g_{m3} is reduced, as shown in Fig. 2. The composite resultant magnitude of g_{m31} and g_{m32} is g_{m3}, and that resultant should be out of phase with the g_{m2} (=$g_{m21} + g_{m22}$). Thus with the source degeneration inductor L_1, g_{m2} contribution to g_{m3} nonlinearity reduced by adjusting of their phase by 180°, as shown in Fig. 2. Thus, IIP3 is strengthened by minimizing the g_{m3}.

In receivers third-order intermodulation product is the most troublesome, as they can fall within the required frequency band and interfere with the wanted signal. Third-order intercept point is the measure of intermodulation distortion products and indicates how well a receiver performs in the presence of strong interferers. Two-tone test is commonly used to measure the third-order intercept point. As shown in Eq. (5), the third-order intermodulation products increase in proportion to cube of input power. The second-order intermodulation products can be removed by the use of differential architecture.

The IIP3 of the LNA is given as

$$\mathrm{IIP3} = \sqrt{\frac{4}{3}\left|\frac{g_{m1}}{g_{m3}}\right|} \quad (5)$$

The IIP3 represented as

$$\mathrm{IIP}_3(2\omega_2 - \omega_1) = \frac{1}{6\,\mathrm{Re}(Z_{in}(s))}\left|\frac{4{g_{m12}}^2\omega^2\left[L_2C_{gsT} + L_1C_{gs2}\right]}{|\varepsilon|}\right| \quad (6)$$

$$\varepsilon = g_{m31}\left(1 + j\omega L_1 g_{m12}\right)\left[1 + (\omega L_1 g_{m12})^2\right]\left[1 + \frac{L_1C_{gs2}}{L_2C_{gsT} + L_1C_{gs2}}\right] + g_{m32} - \frac{2{g_{m22}}^2}{3g_{m12}}\left(\frac{j2\omega g_{m12}[L_1 + L_2]}{1 + j2\omega g_{m12}[L_1 + L_2]}\right) \quad (7)$$

1.1 *Limitations of the MDS Technique*

The transistors M_{A} and M_{B} used in MDS technique are operated in strong inversion and moderate inversion region, possessing enough transconductance to contribute second-order and third-order nonlinearities (g_{22}, g_{21}, g_{31}, g_{32}); no one of these nonlinearities can't be neglected for analysis purpose. The range of applied gate bias voltage V_{gs} over which the third-order nonlinearities can be cancelled is very less, i.e. from 520 to 540 mV, beyond which the third-order nonlinearities are prone to high IIP3.

2 Proposed Method to Increase the Linearity with Differential MOS Loads and Auxiliary Common Source Stage

In the proposed method, in common-source stage, three transistors M_1, M_2 and M_3 are biased such that one transistor M_3 is operated in strong inversion region, and the remaining two transistor M_2 and M_1 are operated in weak inversion region with differential configuration to eliminate nonlinearity. As one transistor is enough to provide gain, M_3 is operated in strong inversion. Transistors M_1 and M_2 are biased in differential configuration with PMOS loads, which are effective to reduce the introduced second-order nonlinearity contribution to third-order contribution, as well as the third-order nonlinearity. The residual nonlinearity components are added out of phase by 180° as shown in Fig. 2, which improves IIP3 performance can be modelled with the following equation.

$$i_{\mathrm{out}}\left(v_i\right) = \left(g_{\mathrm{m31}}\right) v_1{}^3 + g_{\mathrm{m13}} v_2 + g_{\mathrm{m23}} v_2{}^2 + g_{\mathrm{m33}} v_2{}^3 \tag{8}$$

where

$$g_{\mathrm{m2}} = g_{\mathrm{m21}} + g_{\mathrm{m22}} + g_{\mathrm{m23}}, \quad g_{\mathrm{m3}} = g_{\mathrm{m31}} + g_{\mathrm{m32}} + g_{\mathrm{m33}} \tag{9}$$

In this proposed method, induced third-order nonlinearity due to second-order interaction is cancelled by using complementary PMOS loads in the auxiliary path of LNA. The factor ε for the proposed method in Fig. 4 can be written as shown in Eq. (12). Precise choice of L_1 and L_2 contributes the reduced value of ε and G_{m} independent IIP3 improved performance.

The IIP3 represented as

$$\mathrm{IIP}_3\left(2\omega_2 - \omega_1\right) = \frac{1}{6\,\mathrm{Re}\left(Z_{\mathrm{in}}(s)\right)} \left| \frac{4 g_{\mathrm{m12}}{}^2 \omega^2 \left[L_2 C_{\mathrm{gsT}} + L_1 C_{\mathrm{gs2}}\right]}{\omega\,|\varepsilon|} \right| \tag{10}$$

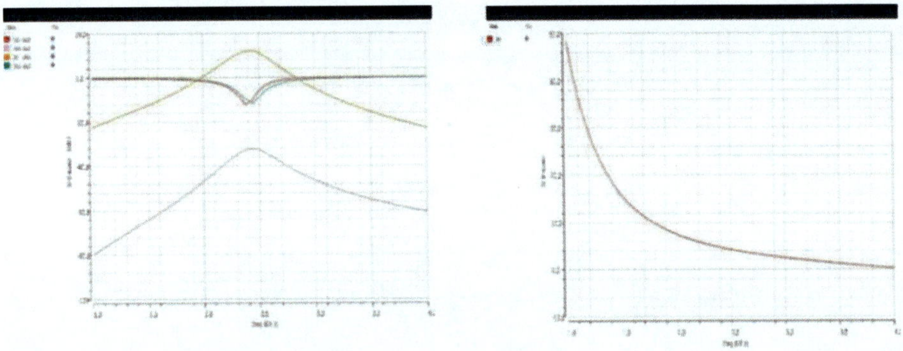

Fig. 4 *S* Parameter and noise figure

where $Z_{in}(s)$ is the input impedance,

$$Z_{in}(s) = s\left[\left(\frac{C_{gs2}}{C_{gsT}}\right)L_1 + L_2\right] + \frac{1}{sC_{gsT}} + \frac{G_m L_2 + g_{m13}\,(L_1 + L_2)}{C_{gsT}}$$

$$\begin{aligned} G_m &= \text{Trans-conductance of auxiliary path} \\ &= (g_{mn})\,\{r_{OP}/\,(1 + g_{mn} r_{ON})\, J\omega L_2\}\left(\tfrac{1}{1+g_{mn} J\omega L_2}\right) \end{aligned} \tag{11}$$

$$C_{gsT} = C_{gs1} + C_{gs2} + C_{gs3}$$

$$\begin{aligned} \varepsilon = {} & G_{m3}\left(1 + j\omega L_1 g_{m13}\right)\left[1 + (\omega L_1 g_{m13})^2\right]\left[1 + \frac{L_1 C_{gs2}}{L_2 C_{gsT} + L_1 C_{gs2}}\right] + g_{m33} \\ & - \frac{2{g_{m23}}^2}{3g_{m13}}\left(\frac{j2\omega g_{m13}\,[L_1 + L_2]}{1 + j2\omega g_{m13}\,[L_1 + L_2]}\right) \end{aligned} \tag{12}$$

The value of IIP3 in equation [6] is inversely proportional to input impedance $Z_{in}(s)$; further input impedance is directly proportional to the gain of the auxiliary path transistor g_{11}. The choice of the auxiliary path, which smoothens the transconductance of the auxiliary path, improved the IIP3 performance as shown in the equation [12].

The drain currents of the auxiliary paths are represented as below:

$$I_{dp} = g_{m1p} V_{gsp} + g_{m2p} V_{gsp}^2 + g_{m3} p V_{gsp}^3$$

$$I_{dn} = g_{m1p} V_{gsn} + g_{m2p} V_{gsn}^2 + g_{m3} p V_{gsn}^3$$

$$g_{m1p} = \frac{\partial I_{dp}}{\partial V_{gsp}}, \quad g_{m2p} = \frac{1}{2}\frac{\partial^2 I_{dp}}{\partial V_{gsp}^2}, \quad g_{m3p} = \frac{1}{6}\frac{\partial^3 I_{dp}}{\partial V_{gsp}^3}$$

$$g_{m1n} = \frac{\partial I_{dn}}{\partial V_{gsn}}, \quad g_{m2p} = \frac{1}{2}\frac{\partial^2 I_{dn}}{\partial V_{gsn}^2}, \quad g_{m3p} = \frac{1}{6}\frac{\partial^3 I_{dn}}{\partial V_{gsn}^3}$$

The output of the can be taken as

$$V_{out} = V_{outp} - V_{outn}$$

$$V_{outp} = G_1 V_{IN} + G_2 V_{IN}^2 + G_3 V_{IN}^3$$

$$V_{outn} = G_1 V_{IN} - G_2 V_{IN}^2 + G_3 V_{IN}^3$$

$$V_{out} = V_{outp} - V_{outn} = 2 G_2 V_{IN}^2$$

From the above equation, it is clear that the auxiliary path used in proposed method able to reduce the external induced third order nonlinearity, in turn increasing the IIP3.

2.1 *Input Matching*

The small signal model of the proposed technique is considered for high-frequency analysis, and the input impedance is given as

$$Z_{in}(s) = s\left[\left(\frac{C_{gs1} + C_{gs2}}{C_{gsT}}\right) L_1 + L_2\right] + \frac{1}{sC_{gsT}} + \frac{g_{m11} L_2 + g_{m12}\,(L_1 + L_2)}{C_{gsT}} \tag{13}$$

$$C_{gsT} = C_{gs1} + C_{gs2} + C_{gs3}$$

The Π-network (consisting of Xp, Xs1, Xs2) shown in above Fig. 4 is chosen for matching the impedances. Z_{in} becomes real at resonance, and it should be equal to 50 Ω. In proposed technique, the input impedance of the circuit with pi matching network is observed as 50.116–*j*5.32 (Figs. 5 and 6).

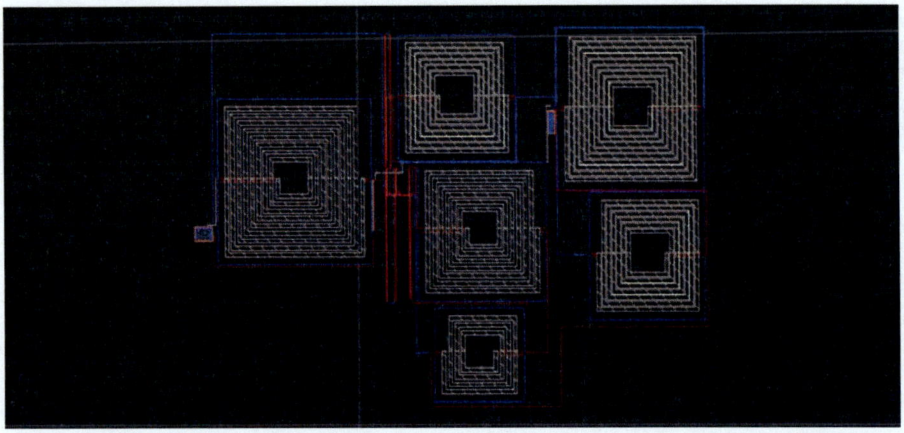

Fig. 5 Layout of the proposed LNA

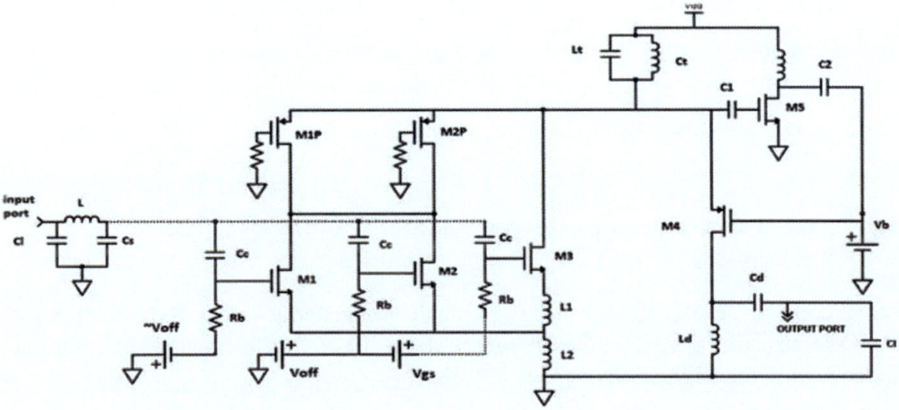

Fig. 6 Schematic of proposed method

2.2 *Voltage Gain*

The transconductance G_M of the folded cascode LNA is given by the sum of common-source and common-gate stagess (Fig. 7)

$$G_M = G_{M,CS} + G_{M,CG}$$

$$G_{M,CS} = \frac{g_{mN1} + g_{mP1}}{1 + L_{S1} j\omega (g_{mN1} + g_{mP1})} + \frac{g_{mN2} + g_{mP2}}{1 + L_{S1} j\omega (g_{mN2} + g_{mP2})} + \frac{g_{mN3}}{1 + L_{S2} j\omega g_{mN3}}$$

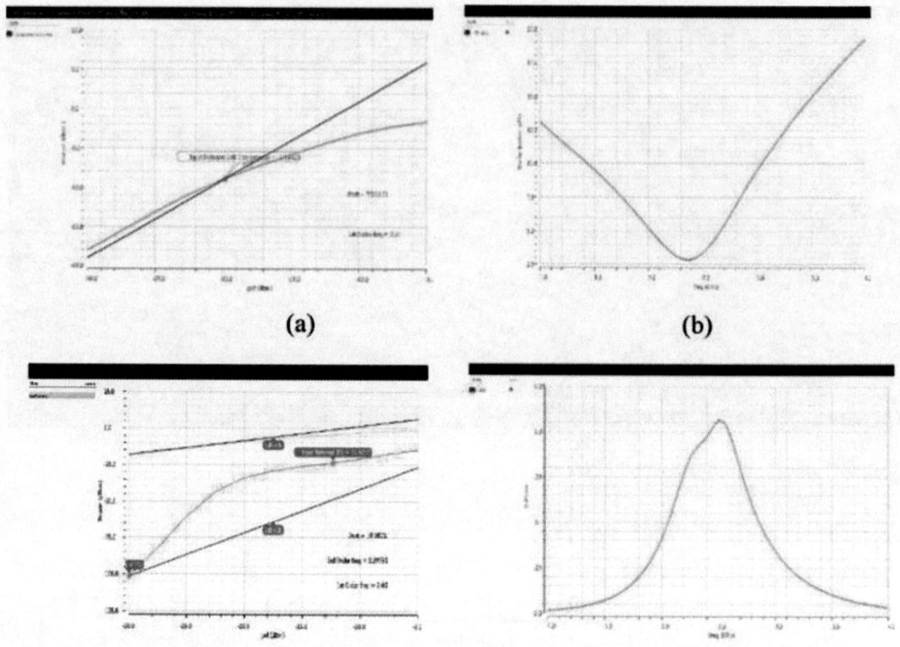

(a) (b)

(c) (d)

Fig. 7 (**a**) 1-dB compression point. (**b**) Noise figure. (**c**) IIP3. (**d**) Stability factor bf

where $g_{mN1}, g_{mN2}, g_{mN3}, g_{mP1}, g_{mP2}$ are the transconductances of the NMOS and PMOS transistors used in common-source stage of LNA. Effective gain of common-gate stage in the LNA is $G_{M,\,CG} = (A + 1)g_{m,\,CG}$

$$A_V = \frac{V_{out}}{V_{in}} = G_M R_D = 24\ \text{mA/V} * 500 = 12$$

2.3 *S-Parameters*

With power constrained design procedure, the obtained *S*-parameters are

S_{11} (power/voltage reflected back to port1, not transmitted) $= \frac{b_1}{a_1}$ at $(a_2 = 0) = -15$ dB

S_{21} (power/voltage transmitted to port2 from port1) $= \frac{b_2}{a_1}$ at $(a_2 = 0) = 12$ dB

S_{12} (power/voltage transmitted to port1 from port2) $= \frac{b_2}{a_2}$ at $(a_1 = 0) = -32$ dB

S_{22} (power/voltage reflected back to port2, not transmitted) $= \frac{b_2}{a_2}$ at $(a_1 = 0) = -11$ dB

2.4 Stability

Stability is an important characteristic of an amplifier design. The stability of a circuit is characterized by Stern stability factor given in

$$K = \frac{1 - |S_{11}|^2 - |S_{22}|^2 + |\Delta|^2}{2 \mid S_{11} \| S_{22} \mid} = 5.8, \quad \Delta = S_{11}S_{22} - S_{12}S_{21}$$

Another stability measure is described by the B_{if} factor given by

$$B_{if} = 1 + |S_{11}|^2 - |S_{22}|^2 + |\Delta|^2 = 0.935$$

A circuit is unconditionally stable if $K > 1$ and $B_{if} > 0$.

2.5 Noise Figure of LNA

$$F_{min} = 1 + 2.4\frac{\gamma}{\alpha}\left[\frac{\omega}{\omega_T}\right] + \underbrace{\frac{\gamma}{g_m R_S}}_{\text{Common source}} + \underbrace{\frac{R_S}{R_1}\left[1 + \frac{1}{g_m R_S}\right]}_{\text{Common gate with amplifier feedback}}$$

$$= 2.84\,\text{dB}$$

2.6 1-dB Compression Point

The definition of P1dB is 1 dB below expected output power by giving input signal as shown in equation:

$$= 10\log\left|\alpha_1 + \frac{3\alpha_3 A^2}{4}\right| = 10\log|\alpha_1| - 1\ \text{dB} = -21.12\ \text{dBm}$$

3 Summary

A Narrow Band (NB) FC-LNA with re-modified MDS technique is designed with cadence GPDK 180 nm in this article and its relevant s-parameters, 1 dB compression point, IPN curves are analysed. As foremost block of Wi-Fi, receiver front end, the main requirement is high linearity and better noise figure. The dominated source of nonlinearity in FC-LNA is CS stage is replaced by the re-modified derivative superposition (MDS) technique. This method boosted the IIP3 for the folded cascode LNA. It provides the good noise figure and high gain at 2.4 GHz frequency.

In this paper, an architecture of folded cascode LNA with re-modified MDS technique is designed in cadence 0.18 μm GPDK standard CMOS platform. The results are summarized in Table 1. The proposed LNA single with gain (S_{21}) of 16 dB operates at ultra-low voltage supply headroom of 600 mV with good stability deemed as a favourable requirement for low power wireless applications.

Table 1 Performance summary with the reported literature

Gain boosting, linearization techniques comparison with proposed method: summary							
	Cascode IEEE (2017) [6]	Folded cascode IEEE (2013) [10]	MDS IEEE (2005), Elsevier (2016) [12]	Harmonic termination IEEE (2014) [8]	IM2 injection IEEE (2008) [7]	Optimum biasing IEEE (2004) [9]	This work
	Gain boosting		Linearity enhancement				
IIP3	−20.1	4.09	4.19	1	14.3	−3.2	12.5
Gain	19.3	15.4	14.6	20	17.5	14.6	12
NF (dB)	3.2	1.74	2.9	2.87	4.1	1.8	2.84
Power (mW)	2.4	5.16	3.8	22.8	15.6	5.4	1.63
Supply voltage (V)	1.2	0.6	600 m	1.2	1.5	2.7	600 m
Frequency (GHz)	2.4/5.2	0.9	2.44	4.6	0.9	0.880	2.4
Process	130 nm	130 nm	180 nm	180 nm	180 nm	0.25 µm	180 nm
1-dB compression point	−29.6		4.19	−14	−4.5		−21

References

1. RF Microelectronics (second Edition), by Behzad Razavi, Prentice Hall Communications Engineering and Emerging Technologies Series from Ted Rappaport.
2. D Sharath Babu Rao, Sumalatha V (2016) The journey of Ethernet towards 1 TBPS and various fields of applications. Ind J Sci Technol 9(11). https://doi.org/10.17485/ijst/2016/v9i11/89293. ISSN (Print): 0974-6846; ISSN (Online): 0974-5645
3. D Sharath Babu Rao, Sumalatha V (2016) The elements of the Ethernet: proliferation of data rates through the bridge layer & MAC protocol stack. In: 2016 International conference on inventive computation technologies (ICICT), Aug 2016
4. T. H. Lee, the Design of CMOS Radio-Frequency Integrated Circuits. New York: Cambridge Univ. Press
5. Sattar S, Zulkifli TZA (2017) A 2.4/5.2-GHz concurrent dual-band CMOS low noise amplifier. IEEE Access 5:21148–21156. ISSN: 2169-3536
6. Zhang H, Sinencio ES (2011) Linearization techniques for CMOS low noise amplifiers: a tutorial. IEEE Trans Microw Theory Techn 58(1):22–36
7. Lou S, Luong HC (2008) A linearization technique for RF receiver front-end using second-order-intermodulation injection. IEEE J Solid State Circuits 43(11):2404–2412
8. Yoon J, Park C (2014) A CMOS LNA using a harmonic rejection technique to enhance its linearity. IEEE Microw Wirel Comp Lett 24(9):605–607
9. Aparin V, Brown G, Larson LE (2004) Linearization of CMOS LNAs via optimum gate biasing. In: Proceedings of the IEEE international circuits and system symposium, Vancouver, BC, Canada, May 2004, vol 4, pp 748–751
10. Kim YM, Han H, Kim TW (2013) A 0.6-V +4 dBm IIP3 LC folded cascode CMOS LNA with Gm linearization. IEEE Trans Circ Syst II Exp Briefs 60(3):122–126
11. Geddada HM, Park JW, Silva-Martinez J (2009) Robust derivative superposition method for linearizing broadband LNAs. IEE Electron Lett 45(9):435–436
12. Kumaravel S, Kukde A, Venkataramani B, Raja R (2016) A high linearity and high gain folded cascode LNA for narrowband receiver applications. Microelectron J (Elsevier) 54:101–108

A Survey on Non-small Cell Lung Cancer Prediction Using Machine Learning Methods

S. Shanthi

1 Introduction

Lung cancer is the primary cause of death and it is a challenging one to detect early since most of the symptoms appear only in its advanced stages and thus this type of cancer is the highest in terms of mortality rate compared to other types of cancer. There are plenty of deaths owing to lung cancer compared to other forms of cancer like prostate, colon or breast cancers. A considerable amount of evidence that indicates an early recognition of lung cancer can bring down the mortality rate to a very significant extent. The recent estimates that have been made on the basis of the latest statistics by the World Health Organization have shown about 7.6 million deaths all over the world every year owing to this cancer. Also, the mortality from cancer is likely to increase and go up to 17 million by the year 2030 worldwide [1].

Cancer is a disease that has an anomalous growth of the cell with the capacity to attack and further spread to other different parts of human body. It is a disease that is life threatening and its accurate prediction is crucial for a proper clinical analysis and treatment recommendation. Currently, its correct detection continues to remain a challenge even for the clinicians with plenty of experience. It is challenging to make an early diagnosis which is reliable and this is done based on the diagnosis of the symptoms of cancer by physicians [2].

Several techniques can be identified to diagnose the lung cancer like chest radiograph (X-ray), computed tomography (CT), magnetic resonance imaging (MRI scan), and sputum cytology. But they are quite expensive and take a lot of time for completion. This means most techniques have been used for detection of lung cancer in its progressive phases wherein the chances of existence for the patient are quite

S. Shanthi (✉)
Sri Eshwar College of Engineering, Coimbatore, India

A. Haldorai et al. (eds.), *2nd EAI International Conference on Big Data Innovation for Sustainable Cognitive Computing*, EAI/Springer Innovations in Communication and Computing, https://doi.org/10.1007/978-3-030-47560-4_20

low. Thus, there has been a need to employ another new technology for diagnosing lung cancer earlier. A tool of good quality to improve the manual analysis of cancer is the technique of image processing. There are several medical researchers using an analysis of the sputum cells to identify lung cancer at an initial phase and most of the recent research depends on its quantitative information like shape, size or ratio of the affected cells.

Speaking classically, the decisions of treatment in patients of lung cancer are based on the tumour and its histology. There are two major subtypes of histology of lung cancer pointed out by the World Health Organization (WHO). They are (1) small-cell lung cancer (SCLC) and (2) non-small cell lung cancer (NSCLC). The SCLC accounts for about 25% of lung cancer and the NSCL accounts for the remaining 75%. The NSCLC is defined at its molecular level itself by means of continuous driver mutations that indicate changes to the DNA and its sequence [3]. This is even more common compared to the SCLC and it normally grows and spreads slowly. The SCLC has been related to smoking and will grow faster forming large tumours spreading all over the body. This normally starts in the chest centre on the bronchi. The death rate of lung cancer is normally connected to the total cigarettes smoked.

Cessation of smoking, modification of diet and prevention of chemo are the main activities of prevention [4]. Screening indicates secondary prevention and the method of identifying possible patients of lung cancer has been based on its systematic investigation of risk factors and symptoms. The non-clinical symptoms along with the risk factors are certain generic indicators of these types of cancer. There is a major role played by the environment in cancer among humans. There are several carcinogens present in the air, food and water. Now and then, we have experience to carcinogens in the environment that is unavoidable resulting in cancer among humans. The human cancer is very complex and is quite challenging for the cancers that have a long latency that is associated with its exposure to the environmental carcinogens that are ubiquitous.

Adenocarcinoma (ADC) along with the squamous cell carcinoma (SCC) has been identified as the commonest among the histological subtypes among various lung cancers. Both are types of cancers developing in epithelial cells (carcinoma) that belong to non-small cell cancer of the lung. The ADC of the lung will develop in glands secreting products within the bloodstream or any other cavity found in the body—the glands secreting mucus. Generally, the lung ADC arises in the peripheral or outer layers of the lung and the lung SCC in the flat surfaces that cover the cells. The squamous will allow a transmembrane movement such as diffusion or filtration and its exchange of air among the lung alveoli. The squamous tends to serve as the boundary for protecting different organs. Most of the squamous cell cancer types tend to arise in the central area of the chest on the bronchi [5].

The DNA methylation is a diagnostic technology that is emerging for the purpose of measuring any epigenetic changes made to the DNA that has been characterised by adding a methyl group within the DNA regions that are known by means of having the CpG islands. Speaking traditionally, the gene expression is used as a new prognostic biomarker to the lung carcinoma and is expressed differentially among the genes between the subtypes of lung cancer which is found. But there is a suggestion that the methylation of the DNA signatures to cancer has to be taken into consideration as a diagnostic biomarker for this disease. There is a distinct DNA methylation of signatures existing between the ADC and the SCC, and further between its tumour tissue and normal surrounding tissue. As DNA methylation has a significant role to play in regulating gene expression, there may be a value added to investigate both types of data.

This has resulted in interest to the development of approaches of computation like the machine learning (ML) which is a subfield to artificial intelligence (AI), for improving medical management such that the outcomes of patients and their workflow are enhanced. The ML is both an analysis and an interpretation used by the machine algorithms permitting classification, its prediction, and finally the segmentation of all information that provides insight that is not available readily to human cognition. Even though the ML algorithms had been developed more than 20 years back, there has been a recent surge in its applications which have shown tremendous potential for this being used in cancer care [6]. With an increase in the use of the AI this chapter offers a review of the applications of the ML which has been developed in order to develop or treat the NSCLC and all the challenges that are faced in its clinical adaptation.

Radiomics has been identified as a novel technique employing a quantitative image feature of high throughput for both diagnosis and prognosis. The radiomics takes the images to be data and also achieves data mining for predicting clinical phenotype and gene data. For a certain application, radiomic approach will proceed using two phases: the first will be training or a phase of feature selection after which there is a second phase of either testing or application. The phase of training will typically proceed in the following way. The features are extracted from its large corpus belonging to the training data in which an object of interest like a tumour is described and the computer algorithm is able to extract all its quantitative features in an automatic manner [7].

Finally, there is a feature selection that is applied aiming to choose a subset that is smaller which are the features capturing efficiently all imaging characteristics found in their biological phenomena. For instance, in the classification of nodules into either benign or malignant, there may be certain features that have been picked either individually or as a combination performing its best on the training data task. In testing stage, there are radiomics that have been applied to the image of a certain patient and this is the process which is very similar to that of the training phase and the chosen features are thus identified by its algorithm and are also extracted and used for classifying patients. But both the training step and the testing step of a classification algorithm are to be defined to translate the values of radiomics into various classifications.

The researchers make use of various CT images that are quantitative for characterising its gene expression data and for predicting the rate of survival of the NSCLC [8]. In recent times, there is a radiomic signature which is a very significant biomarker of classification. But even now there is no other method which is quantitative for the non-invasive lung distinction as to whether it is ADC or SCC.

Until now, there has been hardly any research done on using radiomic signatures for the prediction of both lung ADC and lung SCC. So, for the purpose of this work, the radiomic signature based on CT had been constructed to be used as a factor of diagnostics to discriminate the lung ADC from the lung SCC. Here, feature selection making use of the MRMR and the CFS for the diagnosis of lung cancer has been made.

2 Literature Survey

Hawkins et al. [9] had applied radiomics for choosing the 3-D features that were from the CT images of the lung in order to provide some predictive data. By focusing on the various cases of adenocarcinoma, the subtype of NSCLC tumours from a dataset that is large, the classifiers that are built for predicting the time of survival can be shown. This is compared with the approaches to feature selection. The accuracy that was the best at the time of predicting its survival was about 77.5% that made use of a decision tree which was in a cross-validation of leave-one-out that had been obtained once five different features were chosen for a fold from a total of 219. To diagnose a condition of lung cancer may be a challenging task among medical researchers. In order to be able to overcome such challenging tasks, there are several researchers using the techniques of data mining applied for predicting several diseases.

Sathiya Priya [10] had made an investigation and had made a comparison of various types of classification for classifying and further predicting cases of lung cancer. For the purpose of this work of research the naïve Bayes, the J48, the K-nearest neighbour (KNN), and the support vector algorithm (SVM) are used. By means of using the algorithms in classification, it will produce other results for the datasets of lung cancer. The resulting quality is measured and is dependent on instances that are both correct and incorrect that have been classified correctly using the techniques of classification.

Pradeep and Naveen [11] had made an analysis of the electronic health records (EHRs) along with the rate of survivability that was predicted as lung cancer. The machine learning techniques (MLT) were used for the prediction of the rate of survivability in order to ensure the provision of chemotherapy for people affected by cancer. The MLT has been accepted by doctors and this works well for the diagnosis and prediction of cancer. Another SVM ensemble with the classification trees (C4.5) was used for the evaluation of patterns which are the risk factors used for the investigation of lung cancer. Lung cancer datasets with new data of patients were used to estimate the presentation of the C4.5, the NBs, and the SVMs brought about by the North Central Cancer Treatment Group (NCCTG). This comparison was on the basis of its accuracy, the receiver operating characteristic (ROC), and the area under the curve (AUC) where the results proved the C4.5 to perform better in the prediction of lung cancer along with an increase in the training dataset.

The process of microscopic examination of the biopsy has certain drawbacks. When the diagnosis is made, the medical practitioners tend to continue using visual observation in order to ensure that the result analysis is subjective to take a very long time.

Adi et al. [12] had developed a microscopic analysis made on the biopsy along with the techniques of digital imaging. The identification process of cancer cells that were made in the sample of biopsy was made through various stages of extraction of features by using a GLCM along with a classification that was made by employing the naïve Bayes' algorithm. The classification biopsy results had shown an accuracy of about 88.57% which was a combination of various parameters that are contrasting and homogenous. The techniques of digital image processing may be implemented during a microscopic examination of a biopsy.

The oligometastatic NSCLC was a condition that was heterogeneous having some factors named as risk stratification. The quantitative imaging feature (QIF) based on the positron-emission tomography (PET) and the GLCM energy was linked to the outcome of the non-metastatic NSCLC. There was a hypothesis that the GLCM energy can improve the models comprising a standard of clinical prognostic factors (CPF) for stratifying the oligometastatic patients that were based on the overall survival (OS) made by *Jensen et al.* [13]. The energy has been an important predictor of the OS ($P = 0.028$) which was in addition to the chosen CPFs. C-indexes used for the CPF-only and the CPF + energy models had been 0.720 and 0.739, respectively.

Kirienko et al. [14] have been aiming at identifying image-based radiomic signatures that are capable of predicting the disease-free survival (DFS) found among NSCLC patients who are undergoing surgery. A cohort of a total of 295 patients were chosen. The clinical parameters (age, sex, type of histology, grade of tumour, and stage) had been recorded for the patients. The work had an endpoint that was the DFS and both the CT and the fluorodeoxyglucose PET-based images created from its PET/CT scanner had been analysed. There were textural features that had been computed by making use of the LifeX package. There was also a statistical analysis that had been performed by making use of the R platform. These datasets had been separated into two different cohorts by a random selection for performing both training and validation made to the statistical models. The predictors had been fed into the multivariate Cox which was proportional to the model of hazard regression along with the ROC curve and its corresponding AUC that had been computed for every model that was built.

Zhang et al. [15] had included about 180 patients who were diagnosed with NSCLC along with their pre-therapy scans. By employing a radiomic method, a total of 485 features reflecting both heterogeneity and a prototype of the tumours had been extracted. Once this was done, the radiomic features had been employed for the prediction of the epidermal growth factor receptor (EGFR) and its mutation status using the least absolute shrinkage and the selection operator (LASSO) which was based on the multivariable logistic regression. The result of this was that the radiomic features were found to have an ability of prognostics in the prediction of the mutation status of the EGFR. Additionally, it made use of a nomogram with a curve of calibration for testing the model and its performance.

Choi et al. [16] had further examined another set of a total of 72 pulmonary nodules (PNs) (among which 31 were benign and 41 were malignant) taken from the Lung Image Database Consortium image collection (LIDC-IDRI). About 103 of the CT radiomic features had been extracted from every PN. Even before the model building takes place, there were some very distinctive features that had been identified by using the method of hierarchical clustering. After this, a model of prediction was constructed by employing the SVM classifier which was coupled with the LASSO. Another tenfold cross-validation (CV) had been repetitive about ten times (10×10-fold CV) for evaluation of the accuracy of the model of SVM-LASSO. Lastly, the model that was the best that was from the 10×10-fold CV had been evaluated by the 20×5- and 50×2-fold CVs.

Paul et al. [17] had made a presentation of advancement of deep learning along with convolutional neural networks (CNNs), where there are deep features that are recognised for analysing the CT of the lung for the purpose of prediction of prognosis and its diagnosis. With the number of images available being limited in the field of medicine, the concept of transfer learning is quite helpful. By using the subsets of the participants taken from the National Lung Screening Trial (NLST), it employed an approach to transfer learning for differentiating the nodules of lung cancer versus the positive controls. The feature combination from the pre-trained and the CNNs that were trained based on the NLST data along with the classical radiomics had been used for building the classifiers. The accuracy that was the best (76.79%) had been obtained by making use of the feature combinations. There was also an AUC-ROC of about 0.87 that had been obtained by using the CNN that was trained on an augmented data cohort of the NLST.

Shakeel et al. [18] had introduced another real and enhanced technique of neural and its soft computing for minimising challenges found. In the initial stages, the biomedical data of the lung had been collected using the ELVIRA Biomedical Data Set Repository. Any noise present in the data set has been eliminated by applying bin smoothing to the process of normalisation. There was a minimum repetition with Wolf heuristic features that had been chosen subsequently for minimising the complexity and dimensionality of these features. The chosen lung features had been analysed by using a discrete AdaBoost which was an optimised ensemble learning method among the generalised neural networks that was analysed successfully. The system's efficiency had been further evaluated by employing the MATLAB experimental setup as regards the rate of error, rate of prediction, *F*-measure, recall, and precision.

3 Methodology

3.1 *Grey-Level Co-occurrence Matrix (GLCM) Feature Extraction Method*

The GLCM feature extraction is a matrix describing the frequency of occurrence for two different pixels that have a certain amount of intensity at a distance d with an angular orientation of θ for an image. This GLCM feature extraction has been carried out using a four-angular direction where each one has 45° intervals: 0°, 45°, 90°, and 135°. The feature extraction employs a new texture analysis that is conducted considering the characteristics of the greyscale of objects to differentiate them from others. All these characteristics extracted will include entropy, energy, correlation, homogeneity, and contrast.

Entropy denotes a statistical measure of its randomness used for characterising the input image and its texture as in (3):

$$\text{Entropy} = -\sum\sum p\,(i, j) \log p\,(i, j) \tag{1}$$

wherein p denotes the actual number of the GLCM matrices.

Contrast features will be used for calculating the actual degree of difference of that of the greyness found in the image. The higher the difference between greyness, the more the contrast will be. At the same time, the difference of greyness that is less significant among two different pixels will have lower contrast. The contrast has been defined as per (2):

$$\text{Contrast} = \sum_i\sum_j (i - j)^2 p\,(i, j) \tag{2}$$

wherein $p(i, j)$ denotes the GLCM matrix.

Correlation will bring out the method in which a reference pixel gets correlated to its neighbour on a certain image is correlated to the neighbour over that of an image. Correlation has been defined as (3)

$$\text{Correlation} = \sum_i\sum_j \frac{ijP_\mathrm{d}\,(i, j) - \mu_x\mu_y}{\sigma_x\sigma_y} \tag{3}$$

wherein μ_x, μ_y and σ_x, σ_y are the mean and the standard deviation of a probability matrix of the GLCM with the row-wise x and the column-wise y.

Energy value is the greyness degree of distribution and is written as (4)

$$\text{Energy} = \sum_i\sum_j p^2\,(i, j) \tag{4}$$

Homogeneity will be the feature that computes the actual degree of the greyness homogeneity within the image. The value of homogeneity was higher in the images of a similar degree of greyness. It is defined as per (5) the following:

$$\text{Homogeneity} = \sum_i \sum_j \frac{p\,(i,\, j)}{1 + |\, i - j\,|} \tag{5}$$

3.2 Maximum Relevance and Minimum Redundancy (MRMR) Feature Selection

Aside from the ranking of the ability of discrimination for every feature by the multi-ROC, there was also a ranking of features in accordance with their relevance to that of the target classes using the MRMR approach that has been applied effectively to various problems in classification. The dependency which is $I(x, y)$ among two random variables that are x and y is (6)

$$I\,(x,\, y) = \iint p\,(x,\, y) \log \frac{p\,(x,\, y)}{p(x)p(y)} dxdy \tag{6}$$

wherein $p(x)$ and $p(y)$ denote the probability of the density functions of both x and y, respectively. $p(x, y)$ will be its joint probability density function. So the relevance of feature x and its target class $c = (c_1, c_2, \ldots, c_k)$ is signified as $D = I(x, c)$, which is the maximum that is applied to the rank features for the purpose of classification. In order to bring down redundancy owing to correlations among all ranked features using the strategy of maximum relevance, average mutual information $R = \frac{1}{m}\sum_{j=1}^{m} I\left(x_i, x_j\right)$ that is between the candidate feature x and the ranked features $x_j(i = 1, \ldots, m)$ will have to be minimised (MRMR). These methods of incremental search are used to rank features by means of problems of optimisation that are based on the principle of MRMR (7):

$$\max_{x \in C} \left(I\,(x,\, c) \frac{1}{m} \sum_{j=1}^{m} I\left(x,\, x_j\right) \right) \tag{7}$$

wherein C denotes the set of candidate features that are unranked, and $\{x_1, \ldots, x_m\}(m \geq 1)$ denotes the set of such ranked features.

3.3 Correlation-Based Feature Selection (CFS)

The CFS is a filter algorithm that is used for simple ranking of the feature subsets and discovering the merits and the features of the subsets in accordance with the heuristic evaluation function and its correlation. The primary purpose of the CFS was to identify subsets with features that have been highly correlated along with a class that is uncorrelated with one another. The remaining features are duly ignored and the redundant features will have to be excluded. Accepting of features is dependent on the prediction of classes in the areas of the instance space that has not been predicted by other features. The evaluation function of the feature subset of the CFS has been shown as per (8) the following:

$$\text{Merit}_s = \frac{kr_{\text{cf}}}{\sqrt{k + (k + 1)\, r_{\text{ff}}}} \tag{8}$$

wherein the merits denote the heuristic "Merits" belonging to a feature subset S that consists of the k features, r_{cf} denotes its mean feature-class correlation ($f \in s$), and the r_{ff} denotes its average feature-to-feature intercorrelation. The equation is nothing but Pearson's coefficient of correlation that has its variables standardised. The numerator is the one that gives an indication of the predictive manner of the class of group features and the denominator denotes the level of redundancy. This heuristic will handle all the features that are irrelevant since they are poor predictors in this class. The redundant attributes will be discriminated against this since they are highly correlated with the other features.

3.4 Decision Tree Classifier

The decision trees have been used for delineating the process of decision-making. This denotes a classifier that is embodied using a flow chart such as the construction of a tree which is utilised extensively for the purpose of embodying the association models and owing to their graspable nature, holding in mind their human reasoning. These are further used for categorising the various instances by means of sorting them. Each node will duly specify the examination of an instance and every deviation will correspond to a probable benefit of the attribute. The decision tree will build the models of classification or regression as a tree structure. This will further divide the dataset into subsets that are tinier and tinier. The consequence will have two other extra divisions and a leaf node may embody either the association or the decision. The decision node that is the topmost in the tree corresponds to its best predictor in the origin node.

The decision trees are extensively analysed in machine learning. Irrespective of the gains such as its skill to clarify the procedure of decision and its low computation cost, the decision trees produce a good aftermath in the analogy that is alongside its supplementary contraption that discovers the algorithm. The decision trees are extremely well suited for the problems that are found together with their pursuing features:

- The examples have been signified using the feature value pair. The examples are the attribute—temperature, and value is—hot or cold.

- A target function will have a discrete output variable.
- There may be some disjunction explanations that are needed.
- There may be some errors in training data.
- The training data may be able to cover the missing attribute values.

The ID3, the C4.5, and the classification and regression trees (CART) are the common algorithms of decision trees found in data mining used for various splitting criteria that split the nodes in each level that forms a homogeneous (consisting of objects that belong to a similar category) node. For instance, its data instance will be classified and described by a tuple (age = 23, gender = female, intensity of symptoms = medium, goal = ?), where "?" signifies an unknown value for the goal for instance. A gender attribute will be irrelevant to a certain task of classification. The tree tests the symptom value and its intensity and in case the answer is medium, the instance will be pushed down by means of the corresponding branch and will reach the age node. After this, the tree will test the age value and in case the answer is 23, it is pushed again using the corresponding branch. Finally, the instance will reach its leaf node and will be classified as yes.

The naïve Bayes' algorithm is a probabilistic classifier which computes a new set of probabilities by counting both frequency and combination of values for a dataset. These probabilities are derived by means of a frequency calculation of every feature in a training dataset. This was used to train a new classifier algorithm by the identified values that calculate future.

3.5 *Naïve Bayes' Classifier*

This algorithm makes use of the Bayes' theorem and assumes the attributes that are to be independent with the value of its class variable. This type of assumption of conditional independence will hold good in the applications of the real world and so the characterisation as the naïve Bayes will have it to perform well and will rapidly learn different problems in classification. Such "naivety" permits the algorithm to be able to construct the classifications among large datasets that do not resort to any complicated parameter of the schemes of estimation.

If X denotes a data sample that has the class label which is unknown and H is a hypothesis, a data sample X will be part of a certain class C. The Bayes' theorem was used for the calculation of the posterior probability $P(C|X)$, which is from $P(C)$, $P(X)$, and $P(X|C)$ in (9):

$$P\,(C|X) = \frac{P\,(X|C) \cdot P(C)}{P(X)} \tag{9}$$

wherein

$P(C|X)$ denotes the posterior probability for the target class.
$P(C)$ is the prior probability of a class.

$P(X|C)$ denotes the chances of the probability of prediction for a class.
$P(X)$ denotes the prior probability for a class predictor.

An NB classifier will work as below:

- If D is the training dataset that is connected with the class labels, every tuple will be the n-dimensional element vector, $X = (x1, x2, x3, \ldots, xn)$.
- The m classes $C1$, $C2$, $C3$, ..., Cm are available. If it wants to be able to classify the unknown tuple X, a classifier predicts X belonging to a class having a high posterior probability that is conditioned on the D to a class Ci only if $P(Ci|X) > P(Cj|X)$. For $1 \leq j \leq m$, and the $i \neq j$, all of the above posterior probabilities are calculated with the Bayes' theorem.

4 Conclusion

Lung cancer is a painful and a deadliest disease in the world and understanding the use of different machine learning algorithms in detecting the disease will save millions of lives. Lung diseases are disorders affecting the lungs which are the breathing organs. This condition is very common all over the world and more in India. This work aims at detection and classification of such lung diseases using effective methods of feature extraction and their classifiers. A GLCM is a tabulation by which there are several combinations in the values of pixel brightness (the grey levels) occurring in the image. CNNs along with deep learning find deep features of an image for predicting lung cancer. This chapter explains various techniques of machine learning methods based on efficiency in classification in order to detect lung cancer in various classes.

References

1. Adetiba E, Olugbara OO (2015) Lung cancer prediction using neural network ensemble with histogram of oriented gradient genomic features. Sci World J 2015:786013
2. Najafabadipour M, Tuñas JM, Rodríguez-González A, Menasalvas E (2018) Lung cancer concept annotation from Spanish clinical narratives. In: International conference on data integration in the life sciences, Nov 2018. Springer, Cham, pp 153–163
3. Taher F, Sammouda R (2011) Lung cancer detection by using artificial neural network and fuzzy clustering methods. In: 2011 IEEE GCC conference and exhibition (GCC), Feb 2011. IEEE, pp 295–298
4. Krishnaiah V, Narsimha DG, Chandra DNS (2013) Diagnosis of lung cancer prediction system using data mining classification techniques. Int J Comput Sci Inform Technol 4(1):39–45
5. Pineda AL, Ogoe HA, Balasubramanian JB, Escareño CR, Visweswaran S, Herman JG, Gopalakrishnan V (2016) On predicting lung cancer subtypes using 'omic' data from tumor and tumor-adjacent histologically-normal tissue. BMC Cancer 16(1):184
6. Rabbani M, Kanevsky J, Kafi K, Chandelier F, Giles FJ (2018) Role of artificial intelligence in the care of patients with nonsmall cell lung cancer. Eur J Clin Investig 48(4):e12901

7. Kadir T, Gleeson F (2018) Lung cancer prediction using machine learning and advanced imaging techniques. Transl Lung Cancer Res 7(3):304–312
8. Zhu X, Dong D, Chen Z, Fang M, Zhang L, Song J et al (2018) Radiomic signature as a diagnostic factor for histologic subtype classification of non-small cell lung cancer. Eur Radiol 28(7):2772–2778
9. Hawkins SH, Korecki JN, Balagurunathan Y, Gu Y, Kumar V, Basu S et al (2014) Predicting outcomes of nonsmall cell lung cancer using CT image features. IEEE Access 2:1418–1426
10. Sathiya Priya E (2017) A study on classification algorithms and performance analysis of data mining using cancer data to predict lung cancer disease. Int J New Technol Res 3(11):88–93
11. Pradeep KR, Naveen NC (2018) Lung cancer survivability prediction based on performance using classification techniques of support vector machines, C4.5 and Naive Bayes algorithms for healthcare analytics. Proc Comput Sci 132:412–420
12. Adi K, Widodo CE, Widodo AP, Gernowo R, Pamungkas A, Syifa RA (2017) Naïve Bayes algorithm for lung cancer diagnosis using image processing techniques. Adv Sci Lett 23(3):2296–2298
13. Jensen GL, Yost CM, Mackin DS, Fried DV, Zhou S, Court LE, Gomez DR (2018) Prognostic value of combining a quantitative image feature from positron emission tomography with clinical factors in oligometastatic non-small cell lung cancer. Radiother Oncol 126(2):362–367
14. Kirienko M, Cozzi L, Antunovic L, Lozza L, Fogliata A, Voulaz E et al (2018) Prediction of disease-free survival by the PET/CT radiomic signature in non-small cell lung cancer patients undergoing surgery. Eur J Nucl Med Mol Imaging 45(2):207–217
15. Zhang L, Chen B, Liu X, Song J, Fang M, Hu C et al (2018) Quantitative biomarkers for prediction of epidermal growth factor receptor mutation in non-small cell lung cancer. Transl Oncol 11(1):94–101
16. Choi W, Oh JH, Riyahi S, Liu CJ, Jiang F, Chen W et al (2018) Radiomics analysis of pulmonary nodules in low-dose CT for early detection of lung cancer. Med Phys 45(4):1537–1549
17. Paul R, Hawkins SH, Schabath MB, Gillies RJ, Hall LO, Goldgof DB (2018) Predicting malignant nodules by fusing deep features with classical radiomics features. J Med Imag 5(1):011021
18. Shakeel PM, Tolba A, Al-Makhadmeh Z, Jaber MM (2019) Automatic detection of lung cancer from biomedical data set using discrete AdaBoost optimized ensemble learning generalized neural networks. Neural Comput Appl 32:1–14

A Study of Internet-Based Cognitive Platforms for Psychological and Behavior Problems

Aashish A. Gadgil, Pratijnya S. Ajawan, and Veena V. Desai

1 Introduction

The World Health Organization (WHO) has estimated that in a single year almost 34 million people suffer from depressive disorder and go untreated. The data mostly comprises of people in the land of America and Europe. There are many hindrances to treatment, some of which include geographic distance, psychological state insurance, the prohibitory price of treatment, and perceived stigma [1]. This outlines the importance of Internet-based cognitive behavioral therapy (iCBT) in medicine that involves mental disorders. iCBT's effectiveness is being explored thoroughly in treatment of patients with depression, generalized anxiety disorder (GAD), anxiety disorder, obsessive-compulsive disorder (OCD), post-traumatic stress disorder (PTSD), adjustment disorder, emotional disorder, chronic pain, and phobias [2]. This article presents a review of the use of Internet-based platforms used in treatment of patients with psychological and behavioral health [3]. Iterapi was developed at the Department of Behavioral Sciences and Learning at Linköping University, Sweden. It has been employed in several randomized controlled trials and patient treatments. The role of Internet-based therapy programs for mental disorders is growing. Through study it has been found that programs using human support yield higher outcomes than those with no such support. Therapeutic alliance may also be a significant component during this support. Currently, the importance of therapeutic alliance in case of guided Internet-delivered cognitive behavioral therapy programs (iCBT) needs to be investigated.

A. A. Gadgil (✉) · P. S. Ajawan · V. V. Desai
Department of E&C Engineering, KLS Gogte Institute of Technology, Belagavi, Karnataka, India
e-mail: aagadgil@git.edu; psajawan@git.edu; veenadesai@git.edu

A. Haldorai et al. (eds.), *2nd EAI International Conference on Big Data Innovation for Sustainable Cognitive Computing*, EAI/Springer Innovations in Communication and Computing, https://doi.org/10.1007/978-3-030-47560-4_21

1.1 Healthcare World from Reactive to Proactive Care

Healthcare has been a reactive approach to patient care. Individuals go for annual physicals or checkups, but they search out a healthcare skilled only when they have a particular symptom or issue that must be addressed. A lot of patients and practitioners alike are realizing that an additional proactive approach is important. More importantly early detection is very much necessary in case of brain health issue before the symptoms become more pronounced in patients so as to avoid distress to family members so as to explore all potential interventions. For a health practitioner conducting periodic—e.g., annual—physicals, schedules offer you a simple and automatic technique of getting necessary cognitive knowledge throughout the year, permitting you to closely monitor any potential problems or worrying trends.

1.2 Form More Meaningful, Longer-Term Relationships with Your Patients

In treatment of people with short- or longer-term neurologic disorders or conditions, the patient care cycle is often finite. In alternative words, a patient's will are available in to ascertain you many times over a selected period—say, three visits over a span of 8 weeks. The relationship of the patient with a medical practitioner is restricted except for some additional follow-up visits or checkups within the months and years posttreatment. Though they do come back to your clinic years later, there's usually a significant knowledge gap—you will typically have no means that to judge however their cognition has trended over that amount of time.

Wherever machine-controlled assessment schedules come into play, it permits the medical practitioner to monitor a patient's brain health even when they leave the clinic.

For example, if you're a neurofeedback practician, medical specialist, or psychologist—anyone serving to a personal life through or manage their mental health—setting up an assessment schedule for patients when treatment might enable you to proactively get sooner than potential relapses. If you're a neurologist, concussion specialist, or occupational therapist—anyone operating to assist people pass though traumatic or nonheritable brain injuries—you will still gather knowledge posttreatment, and with no further effort, to confirm the patient is maintaining the acceptable level of cognitive operate months, or maybe years, once injury.

1.3 Mobile Networks

Considerable modification has been seen in the mobile networks. They will change further with the advent of 5G technology and IoT.

The operators face most of the impact of the various changes in network complexity and technologies. The operator success will depend on factors such as less latency, handiness of the network, and surprising high-throughput rates.

2 Internet-Based Cognitive Platforms

2.1 Cambridge Brain Sciences (CBS)

Easily Create and Administer Custom Assessments

CBS Health community has released a new feature. Using a CBS Health account, one can instantly produce and administer custom assessments. This provides a person with the pliability to assess patients quickly by creating numerous combinations of the Cambridge Brain Sciences tasks.

Why It Matters

Many clinics use CBS Health to conduct investigation on the patients visiting for the particular treatment. In-depth understanding of patient's cognitive strengths and weakness is analyzed, before making treatment or medical plans. When the patient visits the clinic for the follow-up, faster custom assessments are done that concentrate on the cognitive domains.

The patient is provided with options to keep the track of him. If the tasks completed by the patients are wrong or missing, the doctor tracks and rectifies the treatment. This helps constant monitoring the patients through task scores.

Working

The procedure starts by adding new assessment and following the steps in correct pattern. The patient must be selected and "+ produce New Assessment" must be selected.

Use of CBS Research by Brain Health Trials to Fight Dementia

Challenges

- Thousands of participants are involved in the study. Hence manually carrying out the test with pen and paper is not feasible.
- Correct conclusions regarding brain health are necessary.

- The test must be conducted regularly to analyze the brain heath condition, which is difficult to monitor without a specific tool.

Solutions

- CBS Research.
- Engaging tests administered at scale via email that guarantee high-quality information is gathered from thousands of participants and collective into one file for simple analysis.
- Collecting the data from email so as to ensure large amount of data is collected, which will further be used to evaluate appropriate results.
- Refereeing to various standard papers and study different criteria for brain health condition monitoring.

Results

- To scale back risk of dementia with a program 100% delivered over the web.
- In case of older adults. The tests can reveal visions into lifestyle factors.
- Potential for results that cause widespread reduced risk of Alzheimer's and other dementia.

Real Outcomes: Finding the Lifestyle Changes That Reduce Cognitive Deficits

Lifestyle changes include monitoring the treatment which includes diet and exercise, which helps fight dementia.

The tests were good for the long pursuit of brain health. They will be taken in participants' homes, in a very short quantity of your time, creating it simple to achieve associate degree huge quantity of knowledge with virtually no effort on the part of the investigators. As a result of the tests, square measure proved reliable over time, and puzzles square measure every which way generated for every take a look at to attenuate follow effects, the investigators will be certain that changes in brain health square measure a right away results of their interventions.

The test analysis provided outstanding technical experience associate degreed support as an external partner within the huge IT development effort.

When the MYB trial is complete, the knowledge associate degreed strategies gained can give real hope for an aging population to continue thriving in returning years.

Research Makes It Easy to Run Trials: Large or Small

The analysis causative to associate degree economical trial will study the participants. However, analysis scales to any size—from little pilot trials to cooperative international studies.

2.2 AVA Cognitive Services Platform

AVA offers prophetic services throughout care and operations.

Nokia tackles the network density using three ways:

- Analysis, virtualization, and self-learning: Analysis is done on huge amounts of data. To recognize problems affecting network performance, machine learning is used that automatically corrects problems related to configuration [2].
- Fast and flexible delivery for services: AVA [2] is cloud-based. New use cases are deployed within days or in some case even in a matter of hours.
- Automation is used to process and filter data at a rapid rate to smoothly run services at any time.

Features and Benefits

Analysys Mason has rated Nokia AVA as the leader in Telco Analytics Ecosystem. Cognitive hubs are used to hold customer meetings to determine new solutions.

To maintain the ecosystem, the following components are necessary:

Cognitive framework: Use cases that are based on AVA are used to create a fast and easy environment for developers [2].

Use case library: Use cases are shared among collaborators to find solutions and fit to the need of customers [2].

Telco Expertise: The value that Nokia brings to the table with the help of partners [2, 3].

New revenue streams: Commercial modules can be used for pricing [2].

Unlocking Value from Telco Data Through Expanded Application of Augmented Intelligence

Nokia expands its analytics services providing to unlock worth from public utility information, facultative improved network issue resolution, and fewer born needs operators. Millions of crowdsourcing measurements are provided using services of analytics [4]. Machine learning is used to provide fast and rapid analysis.

Use cases used by Nokia are:

More advanced and agile and strategies square measure needed to deal with the necessities of complex networks [2]. The same approach is used to modify granular capability.

Operators square measure evolves toward AI-driven operations resulting in a 30–40% improvement in resolution.

Nokia came up with optimization and said: “Our analytics services facilitate to deal with the quality of today’s networks. We have a tendency to [2], at Nokia give insights to enhance network convenience and quality. We will augment human intelligence to enhance potency and cut back the value of operations. Additionally, we will give deeper insights to enhance quality of expertise supported subscriber, device and application usage patterns."

Nokia's next-generation devices have scientific intelligence infused in them [5]. Bell Labs, the celebrated analysis and development engine of Nokia, is associate degree skilled in increased intelligence.

2.3 *IBM Watson*

For cognitive computing to withstand as a competitive and main revenue obtaining are all the firms must adapt analytics in there forefront ventures [6]. CMOs, CIOs, CTOs, and CDOs operate with each other to narrow the gaps between various groups and interconnect the data, working tools, and shared insights. This practice boosts the decision-making, forces worker expertise, and obtains a complete implementation of the product which caters to execution of services on the data available in business practices.

Cognitive technology helps to strengthen and extend social safety nets by addressing a number of the key challenges that generally impede provision and delivery, like knowledge inaccessibility, complexness, and therefore the rate of caseworker churn. With this technology, insight will be extracted from existing knowledge to not solely develop customized services plans but additionally to assist understand vulnerability from a macroscopic read. Cognitive technology is like having the simplest caseworkers on each case—translating into higher protection for at-risk teams.

The development of AI is gaining momentum in the recent years. But when it is said that AI is cutting edge next-generation technology, it also has cons associated with it. Not all the application domains have adapted to the AI technology. IBM Watson is involved in one such application domain, i.e., the healthcare.

Cognitive computing is used by IBM Watson for analyzing large chunk of information [7]. The data is not in the required format always when it comes to healthcare. There is image data to be tested for evidence, which has to be analyzed and obtain valid conclusion from the analysis.

Watson healthcare is a project that adapts cognitive computing and machine learning. The machine surfaces various ideas by analyzing the data be it personal/medical/practical/pharmaceutical, etc.

Below listed are some of Watson the real-life application provided.

Medicine

New York's Memorial Sloan engineer Cancer Center in collaboration [8] results in powerful resources for the patient care. The analyzed dataset provides the doctor with the most preferred choices, scientific action plans, experiments, and valid results, thus reducing time period of the treatment.

At first, breast and lung cancers were treated. Watson from its own expertise learns, up the accuracy and creating additional assured treatment selections, this machine will supply treatment for pancreas, prostate, uterine, liver, ovarian, colon, bladder, kidney, cervical, etc.

Medtronic

There are millions of individuals with diabetes. In line with the information provided by experts, these numbers will grow by 500 throughout consequent 20 years [9]. So as to alter this example and supply training for individuals with diabetes, Medtronic in collaboration with IBM created a cognitive app. This app caters to easy diabetes management.

Collaboration with Apple

Our sleep habits are one of the parameters to decide the performance of the individual. To be more productive concentrating on our sleep habits is a must. Improper sleep affects the individual's health, energy, and efficiency. However, it is going to be tough for someone to work out individuals sleep quality. This problem has to be resolved, and to address a solution to this problem [10], an app named SleepHealth was developed. It is a product of the collaboration with Apple. The app is formed on ResearchKit, using open supply framework. The data from the gyro, pulse monitor, and accelerometer is recorded. The data recorded also includes sleep, movements, shifting position, etc. This product study will enhance high-quality sleep practices, improve health, and even predict some medical conditions.

The best outcomes of the app measure as described below [5]:

1. Access to the most important health & fitness community
2. Food ingesting pursuit
3. Management of nutrition
4. Food visual recognition
5. Recommendations for heath monitoring based on weather, surroundings, and data

2.4 *Azure Cognitive Services*

It aids in development of Internet sites and bots with brainy algorithms to ascertain, listen, utter, perceive, and understand your user desires through natural ways of communication.

The advent of the cloud and sensible technologies is revealing new eventualities that were merely unacceptable hitherto [6]. Sensible sensors and connected Internet of Things (IoT) devices currently permit Northern American countries to capture new knowledge from industrial equipment: from factories to farms and from sensible cities to homes. And whether or not it's an automobile or maybe an icebox, new devices square measure more and more cloud connected by default.

Services

Some of the services that it offers square measure mentioned below. The tip goal is to grasp a small amount additional regarding what these cognitive services APIs will do [7].

- Applications that already use Azure Cognitive Services
- Project template creation
- Using a Face API detect faces in images
- Determining what Azure Services can offer
- Text-to-speech conversion using the Bing Speech API

Neural Text to Speech

The conversion application is accurate to a larger extent; there is minute difference between the recorded and the translated voice [10]. The smart system enhances the voice over text conversion experience.

Oral Communication Transcription

This facility combines oral communication transcript with the SDK to support meeting eventualities; thus participants will equally interact within the discussion and apprehend World Health Organization same what once and quickly follow informed next steps. Azure Cognitive Services offer you that ability [11].

Azure APIs

Azure Cognitive Services may be a set of Azure APIs that build it simple to boost your applications in five areas: vision, speech, language, data, and search; thus you'll make the most of the probabilities that AI needs to supply [3, 5].

- Vision

  APIs beneath the vision flags permit your apps to grasp pictures and video content. They permit you to retrieve data regarding faces, feelings, and different visual contents. You'll stabilize videos and acknowledge celebrities. You'll scan text in pictures and generate thumbnails from videos and pictures. There square measure four APIs contained within the vision domain.

- Speech

  Adding one in all the Speech APIs permits your application to listen to and speak to your users. The APIs will filter noise and establish speakers. Supported by the recognized intent, they'll drive any actions in your application. The speech domain contains three APIs.
- Language

  APIs that are associated with the language domain permit your application to method tongue and find out how to acknowledge what users wish. You'll add matter and linguistic analysis to your application, still as natural language understanding. The subsequent five APIs will be found within the language domain.
- Knowledge

  When we observe data APIs, we have a tendency to square measure talking regarding APIs that permit you to faucet into wealthy data. This might be data from the net or from domain or it's going to be your own knowledge. Victimization these APIs, you may be able to explore the various nuances of data. There are four APIs contained within the data API domain.
- Search

  Search APIs provide you with the power to form your applications additional intelligent with the ability of Bing. Victimization these APIs, you'll use one decision to access knowledge from billions of web content, images, videos, and news articles.

Azure Cognitive Services may be a nice service giving for anyone World Health Organization desires to boost and expand their applications, conveyance the advantages of AI and Machine Learning to your organization. If you'd prefer to learn additional regarding this service, or any Azure service, click the link below—we'd like to facilitate.

2.5 *Cosmos*

Cosmos AI permits the development of customer-satisfied apps that help you to tap into the ability of easy-to-use APIs for computer, speech, video, face and emotion recognition, vision, understanding, language, text, and data exploration [9]. Our cognitive knowledge pipeline platform onboards client knowledge during a straightforward and ascendable manner and creates advanced processing workloads that are fault-tolerant, repeatable, and extremely obtainable; thus brands will build knowledge integration and simply remodel and integrate huge processing and machine learning. Transcend initial knowledge onboarding to phase, activate, and live knowledge while not the leak and frustration of managing multiple partners. Create, integrate, and manage multiple knowledge workloads in real time, interactive, and batch. Syndicate knowledge workloads across multiple pipelines to supply a sturdy knowledge governance framework to manage compliance with rules, restrictions, and policies associated with knowledge usage.

Benefits

- Unify and own all omni-channel, cross-device client knowledge with low latency pipelines.
- 80+ knowledge instrumentality APIs offer screw integrations along with your client knowledge intelligence.
- Flexible knowledge uptake accepts knowledge loaded from cloud storage or streamed to modify period of time analysis.
- Data offer chain quickly and simply orchestrates cloud knowledge collections from any supply to activate in any destination.
- Data protection and accessibility eliminate knowledge operations burden by providing automatic knowledge replication for disaster recovery and high accessibility.
- Built-in knowledge pipeline analytics enables you to instantly discover however every of your channels square measure playing, in conjunction with recommendation.

3 Discussion

Research in iCBT aims to extend work in several aspects. More precise control groups (face-to-face CBT, active placebo controls) enable testing for differential precision. With rapid advances in mobile technology, there is great opportunity for swift evolution in which iCBT programs are delivered [7, 8]. Primarily with the rapid large number of smartphone apps, it's likely that they will become the dominant mode of accessing such kind of programs compared to computer and web-based methods [5]. Therapeutic association in iCBT is soaring and may even be stronger than in face to face therapy. Conversely, relationship among therapeutic alliance in iCBT and its outcome is still understudied [9, 10]. Here, we have considered on the increasing potential of iCBT in the management of different psychiatric disorders and its effectiveness in patient care. Therapeutic-guided iCBT entails checking via emails with the therapist or weekly online sessions [6]. Alternatively patients can use the self-guided iCBT with the help of an Internet program like "e-Ouch" or "MoodHacker" [5]. They can also use web pages such as "Living Life ToThe Full" that deliver cognitive behavioral therapy (CBT). The therapy provided includes maladaptive thoughts, anxiety control, sleeping disorders, and exercise, diet, and relaxation techniques [11]. The methodologies used in most of the studies use a self-help-guided program that is administered using an Internet-based treatment platform wherein the participants and therapists will interact in a secure way. As part of their treatment, the participants perform many activities such as answering self-report questionnaires on treatment progress, watching videos, reading specific treatment modules doing interactive homework, and listening to audio files [6].

4 Conclusion

Although there is research to provide overview on iCBT and other psychological and health disorders that are treated through an Internet-based cognitive platform, there are no papers that a comprehensive overview of the actual online/Internet-based cognitive platforms can be used for such purposes. Our paper provides such an overview of these platforms.

The beneficial outcomes on the patients treated using ICBT, as well as cost-effectiveness, ease of access, improved screening, and assessment as well as reduced stigma due to online anonymity, there are now many solutions for delivering health services through the Internet via web platforms and mobile applications. Some of the examples include Deprexis, Minddistrict, SilverCloud, ICT4DEPRESSION, Online-Therapy.com, THIS WAY UP, Moodbuster, and MindSpot.

References

1. Webb CA et al (2017) Internet-based cognitive behavioral therapy for depression: current progress & future directions. Harv Rev Psychiatry 25(3):114–122. https://doi.org/10.1097/HRP.0000000000000139
2. Kumar V, Sattar Y, Bseiso A et al (2017) The effectiveness of internet-based cognitive behavioral therapy in treatment of psychiatric disorders. Cureus 9(8):e1626. https://doi.org/10.7759/cureus.1626
3. Vlaescu G et al (2016) Features and functionality of the Iterapi platform for internet-based psychological treatment. Internet Interv 6:107–114
4. Papafragou A, Hulbert J, Trueswell J (2008) Does language guide event perception? Evidence from eye movements. Cognition 108:155–184
5. Birney AJ, Gunn R, Russell JK, Ary DV (2016) MoodHacker mobile web app with email for adults to self-manage mild-to-moderate depression: randomized controlled trial. JMIR 4:8. https://doi.org/10.2196/mhealth.4231
6. Andersson G (2014) The internet and CBT: a clinical guide. CRC Press, Boca Raton, FL
7. Hedman E, El Alaoui S, Lindefors N, Andersson E, Rück C, Ghaderi A, Kaldo V, Lekander M, Andersson G, Ljótsson B (2014) Clinical effectiveness and cost-effectiveness of Internet- vs. group-based cognitive behavior therapy for social anxiety disorder: 4-year follow-up of a randomized trial. Behav Res Ther 59:20–29. https://doi.org/10.1016/j.brat.2014.05.010
8. Muñoz RF (2010) Using evidence-based internet interventions to reduce health disparities worldwide. J Med Internet Res 12(5):e60
9. Griffiths KM, Christensen H (2007) Internet-based mental health programs: a powerful tool in the rural medical kit. Aust J Rural Health 15(2):81–87
10. Houston TK, Cooper LA, Vu HT, Kahn J, Toser J, Ford DE (2001) Screening the public for depression through the internet. Psychiatr Serv 52(3):362–367
11. Gega L, Marks I, Mataix-Cols D (2004) Computer-aided CBT self-help for anxiety and depressive disorders: experience of a London clinic and future directions. J Clin Psychol 60(2):147–157

Virtual Machine Migration and Rack Consolidation for Energy Management in Cloud Data Centers

I. G. Hemanandhini, R. Pavithra, and P. Sugantha Priyadharshini

1 Introduction

Cloud computing is a disruptive technology which allows clients to access the computing resources of third parties in pay for what you utilize model. Users are allowed to use the distributed and heterogeneous resources despite of their underlying hardware and are paid only for what they use. Thus cloud became one of the most popular computing models. There are several technologies that make cloud computing possible. Virtualization is one among them. Virtualization is a deployment infrastructure services for physical assets and coverts all hardware resources into software resources. Using virtualization we can consolidate servers into few pieces of hardware and save money on power and cooling. Cloud computing resources are stored in data centers. Since cloud became easy to use and user friendly, cloud solutions are provided by even small enterprises for their clients, and thus the demand for the resources increases. This laid to the foundation of a huge number of data centers that contribute to the worldwide energy consumption and as a result to the environmental drawbacks like carbon emission and global warming. As mentioned in McKinsey statement on "Revolutionizing Data Center Energy Efficiency": The energy consumed by a classic data center is equal to 25,000 household's uses. The overall energy for data centers in 2010 amounts to over $11 billion; in addition to it, the energy cost in a classic data center increases every 5 years. About 50% of the power usage in the data center is by the servers. Also a study on data centers showed that servers which have no running resources are still kept active with no use. They are called "comatose" servers. The frequency of the

I. G. Hemanandhini (✉) · R. Pavithra · P. Sugantha Priyadharshini
Sri Ramakrishna Engineering College, Coimbatore, Tamil Nadu, India
e-mail: hemanandhini@srec.ac.in; pavithra.r@srec.ac.in; sugantha.ravee@srec.ac.in

A. Haldorai et al. (eds.), *2nd EAI International Conference on Big Data Innovation for Sustainable Cognitive Computing*, EAI/Springer Innovations in Communication and Computing, https://doi.org/10.1007/978-3-030-47560-4_22

comatose servers varies from one data center to another like 150 is used in 1800 in one survey and 354 is used in 3500 in another. Kenneth Brill states "unless you have a rigorous program of removing obsolete servers at the end of their lifecycle is very likely that between 15% and 30% of the equipment running in your data center is comatose. It consumes electricity without doing any computing." So, one of the techniques to reduce the set of inactive physical machine is to power off them by shifting the virtual machines from the inactive physical machine to another active physical machine. Our contributions in this paper are scheduling the VMs to find best servers and migrating those VMs to those servers and then consolidating the servers into reduced number of active racks.

Rest of our work is ordered as follows: Sect. 2 confers the allied works. Section 3 alludes to implementation, and Sect. 4 discuss about results. Section 5 refers overall summary and further research.

2 Related Works

The allied work deals with the existing works related to the different techniques available for energy conservation in data centers.

Consolidation of VMs is a technique that is always used with live migration of VMs. Consolidation is used for reducing the power consumed by the servers by migrating some of the VMs from underutilized or overutilized physical servers to other non-underutilized servers. VM Consolidation could be categorized as static consolidation as well as dynamic consolidation. Static consolidation does not involve migration where dynamic consolidation always used migration technique along with consolidation. Ashraf et al. [1] used the consolidation technique for energy efficiency. Here [1] initially the VMs are placed in the servers using bin packing algorithm, and utilization of CPU is predicted. When utilization exceeds the threshold, the VMs are shifted and consolidated using metaheuristic online algorithm, namely, Ant Colony System (ACS).

As CPU works in less voltage, energy can be accumulated, so Patel and Bheda [2] used the combined approach of DVFS algorithm and live migration mechanism. Live migration is a technique that is often used for data center power consumption problem. It allows VMs to be shifted to another physical server from an overused physical server. It increases the resource usage of the physical servers while conserving energy. VM machine migration process involves four steps. The primary step is to choose the overutilized or underutilized physical machine, then, one or above VMs are chosen for migration, and then choose the appropriate physical machine where chosen VMs will be accommodated also the last stage is to shift the VMs. Deciding the proper host is a complicated job under migration method, as erroneous host selection (physical machine) can amplify the number of migrations, energy consumption along with resource wastage. Here DVFS is applied by monitoring the CPU utilization.

Chauhan and Gupta [3] used scheduling algorithm for the cloud data center with DVFS technique. The proposed system can satisfy minimum requirement of resource for a job and can avoid over use of resource.

Wangy et al. [4] suggested a system which aims in minimizing system cost by thermal-aware scheduling that minimizes power consumption and temperature of the servers. Here the thermal-aware scheduling algorithm (TASA) schedules the "active" tasks on "inactive" compute servers that are hosted on racks in the data centers and tries to minimize the temperature of the compute servers. The temperature of a server is calculated by the temperature sensors which get input from the environment, the compute server, and the online job temperature.

Beloglazov et al. [5] proposed a framework in which the resources are assigned to the physical servers using Modified Best Fit Decreasing (MBFD) algorithm. The PMs are assigned with resources in such a manner that the power needed to run that VM is minimal. Here the resources are arranged in nonincreasing manner of the CPU utilization; also they are allocated to a physical machine which has minimal increase in energy consumption. The virtual machines are migrated based on double threshold strategy.

Though physical machine consolidation can help to reduce overall power utilization of data centers, Ilage et al. [6] convey that aggressive PM consolidation on server can create reverse effects in energy utilization and consistency of the system and tries to address the issue through thermal-aware physical machine consolidation methods. They proposed an energy and thermal-aware scheduling algorithm (ETAS) which instantly consolidates the active servers without any local hotspots that result in bad energy consumption in the data center.

Ismaeel et al. proposed a consolidation technique [7] which proactively consolidates the servers using clustering methodology. The VMs are clustered based on their historical workload and user behavior and made available for migration and placement.

Also in [8] Nasim proposed a robust optimization technique that overcomes the uncertainties in resource demands and overbooking, improbability in the migration-related overhead, improbability in the resource demands for VMs, and proposed a robust scheduling and optimization technique for VM placement.

These works focus the energy utilization of the data centers and attempt to trim down the quantity of powered on physical servers in several ways.

3 Proposed System

The proposed system reduces the active physical servers by migrating the VMs from overutilized or underutilized servers to other non-underutilized servers. Also to decrease the number of migrations and to improve the migration strategy, the virtual machines are first scheduled using particle swarm optimization (PSO) algorithm. After scheduling VMs, they are migrated to an appropriate physical server. The energy aware scheduling aims at minimizing the power consumed by the PMs for processing the VMs. Also, scheduling improves migration process. The number of migrations required for server consolidation is minimized by scheduling. After scheduling and migration, the servers running are consolidated into racks using Hybrid Server and Rack Consolidation (HSRC) algorithm.

Initially the VMs are placed into the physical servers using a novel bin packing algorithm. Here VMs are placed initially using Modified Best Fit Decreasing (MBFD) algorithm. The bin packing problem comes under combinatorial NP-hard problem. Here, the objects with varying volumes should be packed into a fixed set of bins with capacity M. This gives a feasible solution in most cases but may not be the optimal solution. For example, the first fit algorithm gives a fast but often nonoptimal solution, concerning the placement of every item in first bin in which it fits. It entails ($n \log n$) time, here n denotes set of elements which are to be filled. Hence the proposed system uses a modified bin packing algorithm whose time complexity is ($n * m$) here n denotes set of objects along with m denotes the set of bins used. Here VM migration is compared with bin packing problem because the virtual machines are considered as objects and the physical servers are considered as the bins into which the VMs must be placed. There are many other heuristic algorithms in practice that are used for VM placements, but the reason for using Power Aware BFD in the proposed system is that it allocates the VMs based on the power need for them to be processed by each PMs. In this way it is one step ahead of the other heuristic algorithms. This way is an advantage of the proposed system since the initial VM placement itself is energy effective.

The algorithm for Modified Best Fit Decreasing is given below:

1. **Input** : server list, Virtual Machine list
2. **Output** : allotment of Virtual Machines
3. SortVMsInDecreasingorder()
4. for each VM in VM list do
5. minpower = MAX
6. AllocatedServer = NULL
7. foreach server in server list do
8. if server is empty then
9. append server to emptyserverQueue
10. else
11. if server is an underutilizedserver then
12. append server to underutilizedserverQueue
13. else
14. if server has enough resource then
15. power = estimatedPower by equ.(1)
16. AllocatedServer = server
17. minPower = power
18. if allocatedserver = NULL then
19. foreach server in UderutilizedserverQueue
20. do lines 14-17
21. if allocatedserver = NULL then

```
22. foreach server in EmptyServerQueue
23. do lines 14-17
24. if allocatedserver != NULL then
25. allocate VMs to allocatedServer
26. else
27. error : NO server has enough resources
28. return allocation of VMs.
```

$$\text{power} = k * \text{Pmax} + (1\text{-}k)\ \text{Pmax} * U_i \tag{1}$$

After the initial placement, the virtual machines are scheduled for exact mapping of virtual machine and physical server pairs. Most of the existing works up to the best of our knowledge didn't consider scheduling of virtual machines while migrating. Here we try to optimize the migration process and decrease the migration count required during consolidation by scheduling the VMs. The VMs are scheduled using particle swarm optimization algorithm. PSO are of different types and are used in more computing areas. Here conventional PSO that is used in power system reliability and security has been used. It [9] is an approach used to investigate the search space of a given problem to discover the settings or factors essential to increase a particular object. In PSO algorithm several solutions will be maintained concurrently in the search space. Throughout every iteration of the algorithm, every candidate result is assessed by the goal function being optimized, finding the fitness of that result. Every candidate result can be considered as a particle "flying" during the fitness background deciding the high or low of the goal function. The two fundamental disciplines of PSO [9] are social science and computer science. While accumulating, PSO utilize the swarm intelligence idea, which is the characteristic of a system, thereby the cooperative behaviors of unrefined agents that are interrelated nearby with the situation and generate logical overall useful patterns. The PSO algorithm has three steps [6]:

1. Assess the fitness of every particle.
2. Renew individual and overall best fitness's in addition with position.
3. Renew velocity along with position of every particle using PSO.

PSO [7] replicates the characteristics of flock of birds. Assume a subsequent circumstance: flock of birds arbitrarily starts finding foodstuff in some region. There is single portion of foodstuff in region found. PSO is well-read and furthermore utilized to resolve optimization issues. Under PSO, "bird" refers individual result, and it is named as a "particle." The entire particle contains fitness values assessed by fitness function to be enhanced and holds velocities which deviate particles. Every particle flocks throughout the problem region by subsequent favorable particles. PSO is assigned among a set of arbitrary solutions (particles) and finds optima via renewing upcoming generations. In individual iteration, every particle gets restructured as a result of subsequent diploid "best" values. Here obtained foremost value is the finest solution which has attained up to now. In addition the fitness value gets accumulated and now the obtained value known as pbest. Further "best" value is obtained by following the particle swarm optimizer is attained till by some particle in the community. The obtained best value determines global best and known as gbest. Once a particle is identified in community, the obtained best value is local best and known as lbest.

Once pair of best values discovered, particle renews its velocity in addition to positions with following equations:

$$\text{v}[\,] = \text{v}[\,] + \text{c1} * \text{rand}() * (\text{pbest}[\,] - \text{present}[\,]) + \text{c2} * \text{rand}() * (\text{gbest}[\,] - \text{present}[\,]) \quad (1)$$

$$\text{present}[\,] = \text{present}[\,] + \text{v}[\,] \quad (2)$$

v[] denotes particle velocity, present[] denotes current particle. The terms namely pbest[] with gbest[] are defined already. rand() denotes a random figure among (0,1). c1, c2 denotes learning features. Typically c1 = c2 = 2.

The fitness function can be calculated using below formula:

$$\text{fitness} = \text{Math.pow}((2.8125 - \text{x} + x * \text{Math.pow}(\text{y}, 4)), 2) + \text{Math.pow}((2.25 - \text{x} + \text{x} * \text{Math.pow}(\text{y}, 2)), 2) + \text{Math.pow}((1.5 - \text{x} + \text{x} * \text{y}), 2) \quad (3)$$

The PSO algorithm is given below:

1. **Input** :VMsList, ServerList
2. **Output** : VM – PM pairs
3. Initialize all parameters
4. Initialize particles positions randomly and velocity of each particle randomly
5. repeat For i =1 to iterations
6. Compute fitness using fitness evaluation function using equation (3)
7. if fitness of a schedule is greater than previous, update it to new best solution
8. if f(Xi) > f(pbesti) , then Pbesti = Xi
9. end
10. Now compute gbest that is globally best from all the pbest schedules
11. if f (pbesti) > f (gbesti), then gbesti = pbesti
12. end
13. Velocity updation using equation (1)
14. Position updation using equation (2)
15. end for loop until stopping criteria is met

After migrating the VMs to appropriate servers, the servers are consolidated into minimum number of racks using Hybrid Server and Rack Consolidation (HSRC).

Hybrid Server and Rack Consolidation (HSRC) algorithm is utilized for rack consolidation. Firstly, all the racks are sorted in nonincreasing fashion of rack utilization. Subsequently, arrange the under loaded servers of every rack in reducing fashion of server used. Thus, the under loaded servers present in every rack results maximum operation are placed on head of the arranged list. Then, the virtual machine is allocated to former server starting in arranged list which will have the sufficient resources for the VM to be accommodated. Thus, the virtual machines are positioned on servers in racks commencing head of the list by utilizing first-fit algorithm so in turn operation of every rack can be enhanced as much as promising. After rack consolidation the number of active racks is reduced. The idle servers and racks are put in sleep mode.

HSRC algorithm is given below:

1. **Input :** RackList, VMsList, ServerList
2. **Output :** Allocation of Virtual Machines
3. SortVMsInDecreasingOrder()
4. for each Virtual Machines in the VM list do
5. SrvList = All Non – Underutilized Servers
6. AllocatedServer = Find Appropriate Server by MBFD
7. if AllocatedServer = NULL then
8. SortedServers = Sort underutilized servers in decreasing order of utilization of their racks and servers respectively
9. AllocatedServer = find server by First Fit
10. if AllocatedServer != NULL then
11. Assign Virtual Machine to server
12. else
13. Error : No server has enough resources
14. **Return** Allocation of VMs

After rack consolidation the number of active racks will be minimized, and all the idle racks are consolidated to turn off cooling system, network routers to save more power.

4 Results and Discussions

The experimental tests were conducted among existing and proposed methodology with varying number of tasks. The comparison is made between the existing methodology and the proposed rack consolidation technique. The performance metrics considered for comparing the existing and proposed approaches are power consumption and resource utilization. The comparison and evaluation made are discussed in the following section.

The Initial VM placement is important because an improper VM placement could degrade the performance. Initial VM allocation aims at allocating more virtual machines on less physical machines to allow maximum utilization of the running physical machine without degradation on performance.

From Fig. 1 it is observed that the proposed methodology can complete the task execution with more resource utilization than the existing method. In this graph, x-axis plots values of PMs and y-axis plots number of VMs.

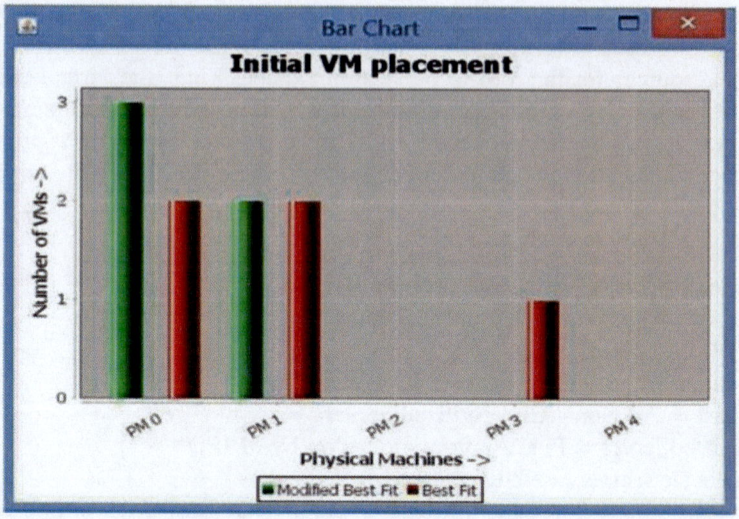

Fig. 1 Active number of servers

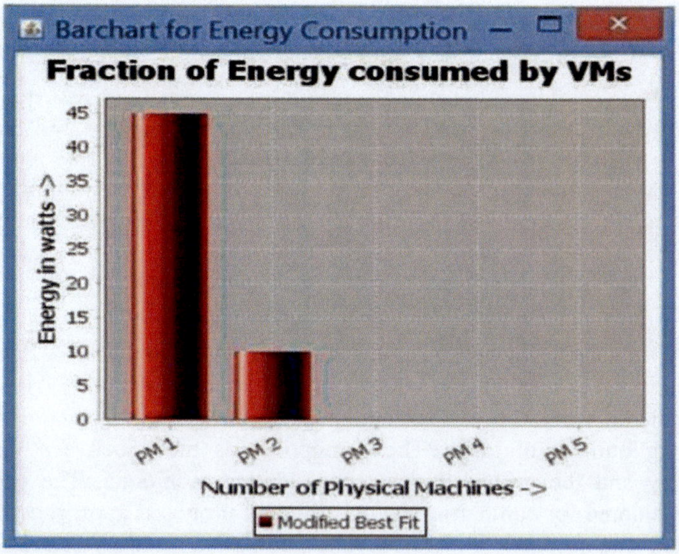

Fig. 2 Power consumed by VMs after initial placement

Figure 2 shows the fraction of energy consumed by the virtual machines after allocated to a physical server using Modified Best Fit Decreasing algorithm. In Fig. 2, the *x*-axis denotes physical machines, and the *y*-axis indicates energy consumed by the corresponding physical machines. PM1 and PM2 are active, and

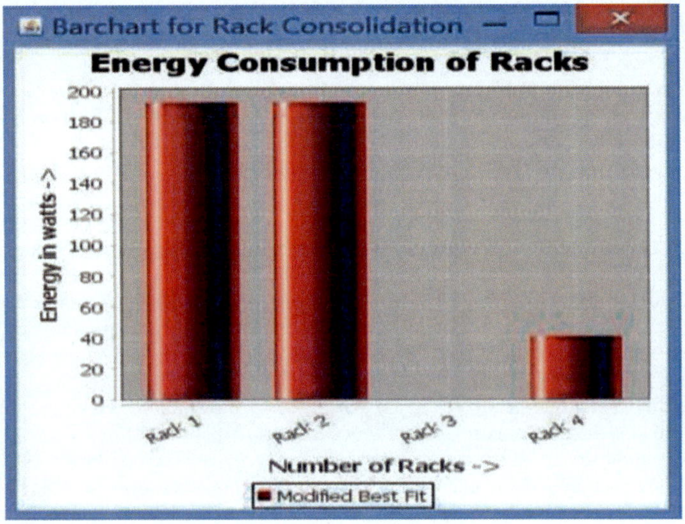

Fig. 3 Energy consumed by the racks

the corresponding power consumed by them is calculated by Eq. (1) which is given under MBDF algorithm.

Figure 3 illustrates the amount of energy utilized by the active racks. Lesser amount of active racks, smaller the quantity of energy consumed. Finally the idle racks are switched off.

In Fig. 3, x-axis denotes racks, and y-axis denotes power consumed by corresponding racks in the data center. The power consumed by the racks is calculated by the fitness function.

5 Conclusions and Future Work

The overall work utilizes technique intended for reducing the energy used by the physical machines in data center by scheduling and migrating the VMs from one or more idle physical servers to other active non-underutilized servers based on the threshold values and also consolidating servers into non-underutilized racks.

Our future work will be implemented using CloudSim and compares the results obtained with the literature instances and also utilizes the thermal model in data center for dropping the power consumed in datacenters that belongs to cloud.

References

1. Farahnakian F, Ashraf A, Pahikkala T, Liljeberg P, Plosila J, Porres I, Tenhunen H (2015) Using ant colony system to consolidate VMs for green cloud computing. IEEE Trans Serv Comput 8(2):187–198
2. Patel VJ, Bheda HA (2014) Reducing energy consumption with Dvfs for real-time services in cloud computing. IOSR J Comput Eng 16(3):53–57, Ver. II. e-ISSN: 2278-0661, p-ISSN: 2278-8727
3. Chauhan P, Gupta M (2014) Energy aware cloud computing using dynamic voltage frequency scaling. IJCST 5(4):195–199
4. Wangy L, von Laszewskiy G, Dayaly J, Hey X, Younge AJ, Furlaniz TR. Towards thermal aware workload scheduling in a data center. Service Oriented Cyber infrastructure Lab, Rochester Institute of Technology, Center for Computational Research, State University of New York at Buffalo, Buffalo, NY.
5. Kim KH, Buyya R, Beloglazov A (2013) Power-aware provisioning of virtual machines for real-time cloud services. Concurr Comput Pract Exp 23(13):1491–1505
6. Ilage S, Ramamohanarao K, Buyya R (2019) ETAS: energy and thermal-aware dynamic virtual machine consolidation in cloud data center with proactive hotspot mitigation. Cloud Computing and Distributed Systems (CLOUDS) Laboratory, School of Computing and Information Systems, University of Melbourne, Melbourne, pp 1–15
7. Ismaeel S, Karim R, Miri A (2018) Proactive dynamic virtual-machine consolidation for energy conservation in cloud data centres. J Cloud Comput Adv Syst Appl 7:10
8. Nasim R, Zola E, Kassler AJ (2018) Robust optimization for energy-efficient virtual machine consolidation in modern datacenters. Clust Comput 21:1681–1709
9. Ahmad RW, Gani A, Hamid SHA, Shiraz M, Yousafzai A, Xia F (2015) A survey on virtual machine migration and server consolidation frameworks for cloud data centers. J Netw Comput Appl 52:11–25

Adaptive Uplink Scheduler for WiMAX Networks

M. Deva Priya, A. Christy Jeba Malar, N. Kiruthiga, R. Anitha, and G. Sandhya

1 Introduction

IEEE 802.16, the Worldwide Interoperability for Microwave Access (WiMAX) standard provides Broadband Wireless Access (BWA) in Metropolitan Area Networks (MANs). It supports high bandwidth applications by providing wireless communications with QoS guarantees.

It provides "last mile" connectivity in MAN where other methods fail or are not cost-effective. It acts as a replacement to satellite Internet services in remote areas. It supports high mobility and provides a communication link between the Mobile Stations (MSs) and the Base Stations (BSs).

A MS gets authenticated and connects to the Access Service Network Gateway (ASN-GW), and the requests are inter and intra-class scheduled. The slots are logically and physically assigned. Logically, the demand in terms of number of slots needed is computed based on the service class. Physically, the subchannels and time intervals that are appropriate for each user are computed. Many authors have proposed scheduling schemes for IEEE 802.16 standards with the aim of enhancing throughput and guaranteeing fairness and QoS.

M. Deva Priya (✉) · N. Kiruthiga · R. Anitha · G. Sandhya
Department of Computer Science and Engineering, Sri Krishna College of Technology, Coimbatore, Tamil Nadu, India
e-mail: m.devapriya@skct.edu.in; n.kiruthiga@skct.edu.in; r.anitha@skct.edu.in; sandhya.g@skct.edu.in

A. Christy Jeba Malar
Department of Information Technology, Sri Krishna College of Technology, Coimbatore, Tamil Nadu, India
e-mail: a.christyjebamalar@skct.edu.in

A. Haldorai et al. (eds.), *2nd EAI International Conference on Big Data Innovation for Sustainable Cognitive Computing*, EAI/Springer Innovations in Communication and Computing, https://doi.org/10.1007/978-3-030-47560-4_23

The BS assigns access slots to the contending MSs. The time slot is dynamically assigned to the MS. The BS controls the QoS parameters by balancing time slot assignments for different applications. A scheduler is said to be efficient if it is stable and bandwidth efficient.

In this paper, the existing Deficit Weighted Round Robin (DWRR) scheduling scheme is enhanced. Performance is observed in terms of Packet Delivery Ratio (PDR), Packet Loss Ratio (PLR), throughput, jitter and delay for different traffic. Improved DWRR (IDWRR) is capable of handling increased number of requests by assigning a queue's outstanding DeficitCounter to the upcoming queues.

2 Scheduling Mechanisms in IEEE 802.16 Networks

The scheduling and Call Admission Control (CAC) schemes are open for research in the IEEE 802.16 standard. Ensuring QoS for each service class is the main challenge while designing a scheduler. The incoming traffic is assigned to a scheduler based on its type. Based on the QoS requirements and instantaneous channel conditions, the scheduler assigns resources. Orthogonal Frequency-Division Multiple Access (OFDMA) aids in ensuring efficiency and fairness [1].

The scheduling schemes are categorized into the following:

- *Channel-unaware scheduler*: As the channels are assumed to be error-free, the channel state parameters including channel error, energy level and loss rate are not involved while making decisions. They are classified into intra-class and interclass schedulers [2].
- *Channel-aware schedulers*: This scheduling suits faultless channel conditions that are free from losses and power constraints. The nature of wireless medium and mobility demands a challenging environment. For Downlink (DL) scheduling, the scheduler at the BS considers the Carrier-to-Interference-and-Noise Ratio (CINR) at the MS. For UL scheduling, the CINR of former transmissions are taken into account.

2.1 *QoS Service Classes*

To ensure QoS, efficient management of traffic is essential. IEEE 802.16 prioritizes traffic and categorizes services into five different classes. The classes include Unsolicited Grant Service (UGS), real-time Polling Service (rtPS), extended real-time Polling Service (ertPS), non-real-time Polling Service (nrtPS) and Best Effort (BE) service [2] intended for specific applications [3].

- *Unsolicited Grant Service (UGS)*: UGS traffic supports real-time, periodic, fixed size packets like Voice over IP (VoIP) and is suitable for Constant Bit Rate (CBR) services.

- *Real-time Polling Service (rtPS)*: It supports real-time, periodic, variable-sized packets including Variable Bit Rate (VBR) services like Moving Picture Experts Group (MPEG) videos. The rtPS service flows are delay sensitive.
- *Extended real-time Polling Service (ertPS)*: It is an amalgamation of UGS and rtPS service classes. It involves adaptable data rates and is suitable for VoIP traffic with silence suppression.
- *Non-real-time Polling Service (nrtPS)*: It is appropriate for non-real-time VBR services that are bandwidth intensive and tolerate moderate delay like File Transfer Protocol (FTP). The nrtPS, unlike rtPS, is delay insensitive and requires minimum amount of bandwidth and supports variable data rates.
- *Best Effort (BE) service*: It supports BE traffic like Hypertext Transfer Protocol (HTTP), Simple Mail Transfer Protocol (SMTP), etc. They do not demand any service guarantee.

3 Literature Survey

There are several scheduling algorithms proposed for WiMAX networks. Some schedulers such as Weighted Fair Queuing (WFQ) [4], Weighted Round Robin (WRR) [5], packet-by-packet Generalized Processor Sharing (GPS) scheme [6], Deficit Weighted Round Robin (DWRR) [7], and Earliest Deadline First (EDF) [8], and their variants are adopted and evaluated for intra-class scheduling for both delay sensitive rtPS and throughput guaranteed nrtPS applications [9].

DWRR and WFQ queuing schemes maintain fairness [10]. WFQ gives better performance when compared to Priority Queuing (PQ) for real-time application such as video streaming. It ensures good QoS for multimedia applications [11] and maintains low latency for applications with high temporal constraints. It guarantees minimum average delay for UGS applications.

Chandur et al. [12] have studied the performance of Proportional Fair (PF) scheduler and have introduced a new algorithm named EXPonential rule for Queue Length and Waiting time (EXPQW), an upgradation of exponential rule which assigns weights to subscribers based on the waiting time and the queue length. Comparable throughput is obtained in moderately loaded situations.

Singla and Kamboj [13] have compared WFQ with priority scheduling (PS) based on packet delay and traffic parameters. Simulation results show that WFQ involves less end-to-end delay in contrast to PS.

Puranik et al. [14] have provided a comprehensive analysis on homogeneous and hybrid scheduling algorithms. WRR scheduling algorithms work well with different classes of packets allocated to different queues with different quantum of bandwidth. WRR outperforms RR and WFQ in terms of throughput.

Teixeira and Guardieiro [15] have designed scheduling algorithms that are directly applied to bandwidth request queues to support real-time applications.

Sharma and Panjeta [16] have compared the performance of first-in first-out (FIFO), PQ, WFQ, round robin (RR), Deficit Round Robin (DRR) and Modified

Deficit Round Robin (MDRR) scheduling algorithms. They have concluded that the choice of scheduling algorithms plays a vital role in improving QoS.

Ahmed and AlMuhallabi [17] have evaluated and compared the performance of Maximum Throughput (MT), First Maximum Expansion (FME) and Round Robin (RR) scheduling algorithms. From the results, it is obvious that throughput and fairness cannot be improved simultaneously.

Dighriri et al. [18] have compared the performance of PQ, FIFO, and WFQ for data traffic in 5G networks. Performance degradation occurs in WFQ due to multiple flows in a single queue.

Shareef et al. [19] have presented a Class-based QoS Scheduling (CBS) to ensure QoS guarantee in downlink stream communication. It supports diverse types of traffic and offers considerable throughput.

Yadav et al. [20] have propounded a hybrid scheduler based on WFQ and RR which has a splitter for FTP traffic. Proper load distribution is provided by two-stage priority scheduling. It is suitable for heterogeneous traffic applications, but the dynamic nature of wireless networks demands throughput optimization.

Lal and Kumar [21] have propounded a Heuristic Earliest Deadline First (H-EDF) scheduling scheme to ensure fairness. The performance of ICN-WiMAX is assessed using H-EDF scheme.

Mahesh et al. [22] have proposed an Adaptive Uplink Scheduling (AUS) by improving the MAC layer using Multi-Objective Genetic Algorithm (MOGA). The scheduler is based on both the physical and application layers. This adaptive modulation technique based on user application supports QoS provisioning at the application layer.

Khanna and Kumar [23] have designed a hybrid scheduling algorithm of First Come First Serve (FCFS) and RR for the Long-Term Evolution (LTE)-based networks for handling heterogeneous traffic services. Signal-to-Noise Ratio (SNR) and the Bit Error Rate (BER) are measured. Though the proposed algorithm is hybrid in nature, it is not capable of dealing with dynamic loads and congestion.

Ahmed et al. [24] have designed a QoS framework by providing a two-level scheduling algorithm to handle video traffic. They have dealt with ensuring QoS by evading starvation and improving scalability.

4 Deficit Weighted Round Robin

The existing Deficit Weighted Round Robin (DWRR) is a modified version of Round Robin (RR) scheduling. The non-empty queues are visited, and the packets with varying sizes are considered without any knowledge of their mean sizes. It is a variant of DRR in its capacity to allocate weights to flows [7]. DWRR involves the following quantities.

- *Weight*: Proportion of bandwidth assigned to a queue
- *DeficitCounter*: Queue's transmission capacity in bytes
- *Quantum*: Weight given to a queue stated in bytes

It does not support fragmentation of packets and deals with only whole packets. Larger packets cannot be serviced in a round. If the DeficitCounter is insufficient, then the packet has to be serviced only in multiple rounds. The maximum packet size is deduced from the DeficitCounter, and the outstanding slice of the packet is serviced in the next round. The queue is left unserved, and its credit is increased by the quantum. The DeficitCounter for the next round is computed based on the increased credit. The packets in a queue are scheduled, and the value of the DeficitCounter is dynamically computed. This continues until the size of the packet exceeds the value of the DeficitCounter or the queue turns out to be empty.

If a queue is empty, its DeficitCounter is made zero, and the queue is deactivated. The empty queues are taken from the ActiveList, and the buckets' token rates are correspondingly modified. This ensures proper utilization of bandwidth. Token accumulation is based on the amount of bandwidth assigned to a queue, which in turn depends on the number of queues waiting to be served.

In contrast to Weighted RR (WRR), DWRR does not demand the knowledge of the sizes of the incoming packets. As the queues contain packets of different sizes based on the type of traffic, DWRR supports weighted uniform distribution of resource to the flows. A misbehaving service class in a queue does not affect the performance of other service classes on the same output port. It is simple and inexpensive and does not demand the upkeep of service class state.

5 Improved Deficit Weighted Round Robin

In the existing DWRR algorithm, insufficient quantum leads to dropping of the packet at the front of the queue in the current round. The queue is left unserved, and the packet is forced to wait until next round.

There are chances for the queue to contain packets with size less than the DeficitCounter. If they are served, the overall delay will be reduced to a greater extent. The proposed Improved DWRR (IDWRR) algorithm overcomes this challenge by considering those packets by sorting the queues based on the size of packets. By doing so, packets with sizes less than the DeficitCounter are transmitted. Initially, the DeficitCounters of the queues are set to zero, and the quantum is taken as the product of queue weight and bandwidth. The arriving packet is put into the appropriate queue based on the type of traffic. If the corresponding queue is not in the ActiveList, it is activated and initialized, and the packet is enqueued. Enqueue and dequeue algorithms of IDWRR are given in Fig. 1.

If the ActiveList contains queues, then the one at the top is selected. The value of the quantum is added to the DeficitCounter of the queue. When the queue contains packets, the size of the packet at the front is determined and checked with the DeficitCounter. If it is less than the DeficitCounter, the packet is transmitted, and the DeficitCounter is reduced by the packet size. Else, the packets are sorted based on their sizes. The remaining DeficitCounter is moved to the ensuing queue with

```
Algorithm IDWRR
initialize(Q)/* Initialization */
for every queue
    Set DeficitCounter=0
    Compute Quantum = weight * Bandwidth
end /*for*/
end /* initialize() */

enqueue(Q, P) /* Insert packet 'P' into Queue 'Q' */
Select the queue of the service flow pertaining to the incoming packet
if (!InActiveList(Q)) then
        activate(Q)
        initialize(Q)
end /* if*/
Add P to Q
end    /* enqueue()*/

dequeue(Q) /* Removing a packet from the Queue Q*/
while (!isEmpty(ActiveList)) then
        Select Q from the ActiveList
        DeficitCounter[Q]+= Quantum[Q]
        while(!isEmpty(Q))
                if (DeficitCounter[Q] ≤ Size(P))  then
                    Arrange packets in the increasing order of Size(P)
                else
                    Transmit P
                    DeficitCounter[Q]-= Size(P)
                End  /*if*/
        End /*while*/
        if (isEmpty(Q)) then
                Transfer the outstanding DeficitCounter[Q] to the
                ensuing Queue in the ActiveList
                deactivate(Q)
        else
                activate(Q)
        End  /*if*/
End    /* while*/
End   /*dequeue*/
```

Fig. 1 IDWRR algorithm

packets waiting to be served. An empty queue is deactivated, and the non-empty ones are added to the ActiveList.

In the existing DWRR algorithm, the DeficitCounter of a queue is made zero once it becomes empty. Instead, in IDWRR, in case a queue becomes empty, the DeficitCounter is transferred to the ensuing queue. This increases the DeficitCounter of the current queue that is served, which in turn improves the number of packets served in a round.

6 Results and Discussion

The system was simulated using ns2. The algorithms were analyzed for all the service classes in terms of QoS parameters like Packet Delivery Ratio (PDR), Packet Loss Ratio (PLR), throughput, jitter, and delay. The frame duration is taken as 0.02 ms. Table 1 shows the simulation parameters.

From the results, it is obvious that IDWRR outperforms the existing algorithm (Figs. 2–5). IDWRR is suitable for certain service classes offered by WiMAX. The performance of the scheduling algorithms is analyzed for different types of traffic.

This section shows how BE, UGS, rtPS, and nrtPS traffic adapt to DWRR and IDWRR algorithms. As the available DeficitCounter in each queue is transferred to the next queue, the proposed scheduler services more packets and hence yields better throughput and PDR involving less delay, jitter and PLR. IDWRR is suitable for BE and nrtPS services as they do not have acceptable delay constraints. The following graphs show the performance of DWRR and IDWRR for increasing packet sizes. It is seen that PDR and throughput decrease with increase in packet size, while PLR, delay, and jitter show an increase with packet sizes.

IDWRR offers 1.1 times better PDR and 6.9 times better throughput, 8.7 times less delay, 4.6 times less jitter and 5.1 less PLR for rtPS traffic (Fig. 2). IDWRR

Table 1 Simulation parameters

Parameter	Value
MAC protocol	IEEE 802.16e
Routing protocol	DSDV
Modulation scheme	OFDM_QPSK
Queue length	50
Queue type	Drop Tail/WFQ
Bandwidth	50 Mbps
Packet size	
UGS & nrtPS	1024
BE & rtPS	512
Transmission range	250–400 m
Number of mobile stations	100
Speed	1–40 ms^{-1}
Simulation time	80 s

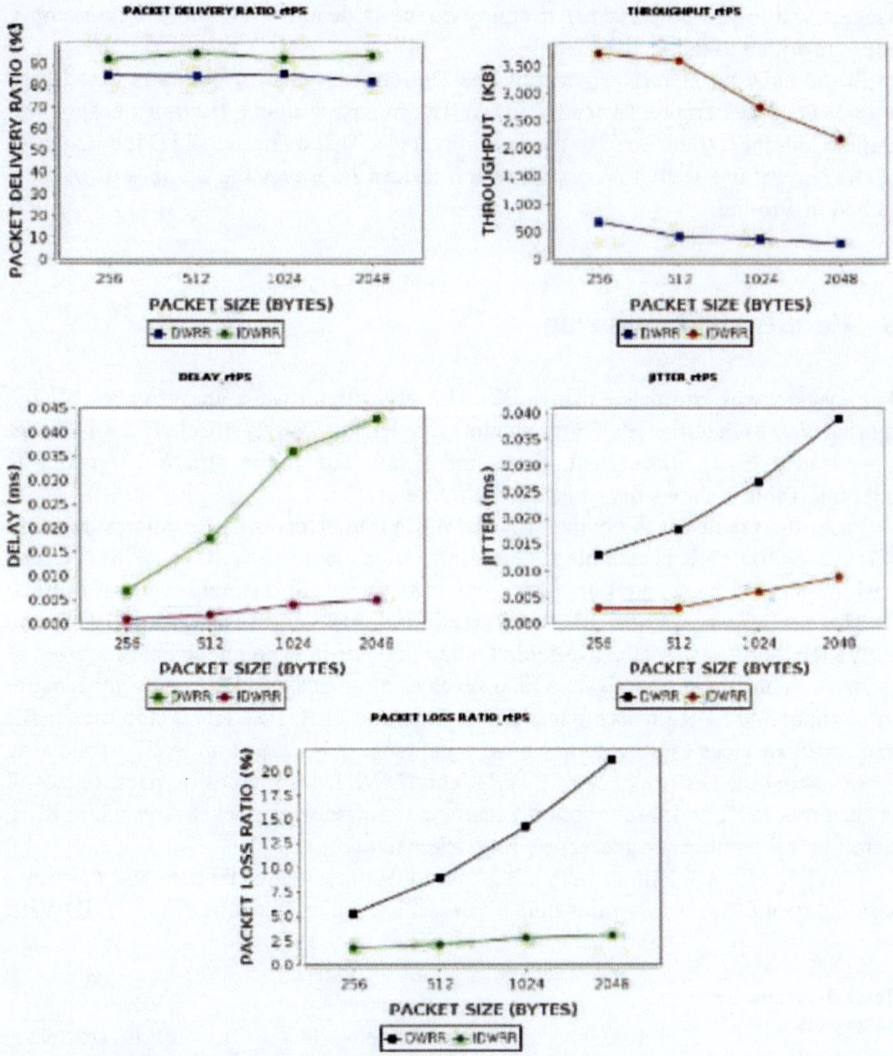

Fig. 2 Performance of DWRR and IDWRR for rtPS traffic

offers 1.1 times better PDR and 5.3 times better throughput, 2.8 times less delay, 3.5 times less jitter and 3.9 less PLR for nrtPS traffic (Fig. 3).

IDWRR offers 1.1 times better PDR and 2.3 times better throughput, 12.2 times less delay, 11.4 times less jitter and 6.2 less PLR for UGS traffic (Fig. 4).

IDWRR offers 1.1 times better PDR and 5.4 times better throughput, 4.1 times less delay, 3.2 times less jitter and 3.9 less PLR for BE traffic (Fig. 5).

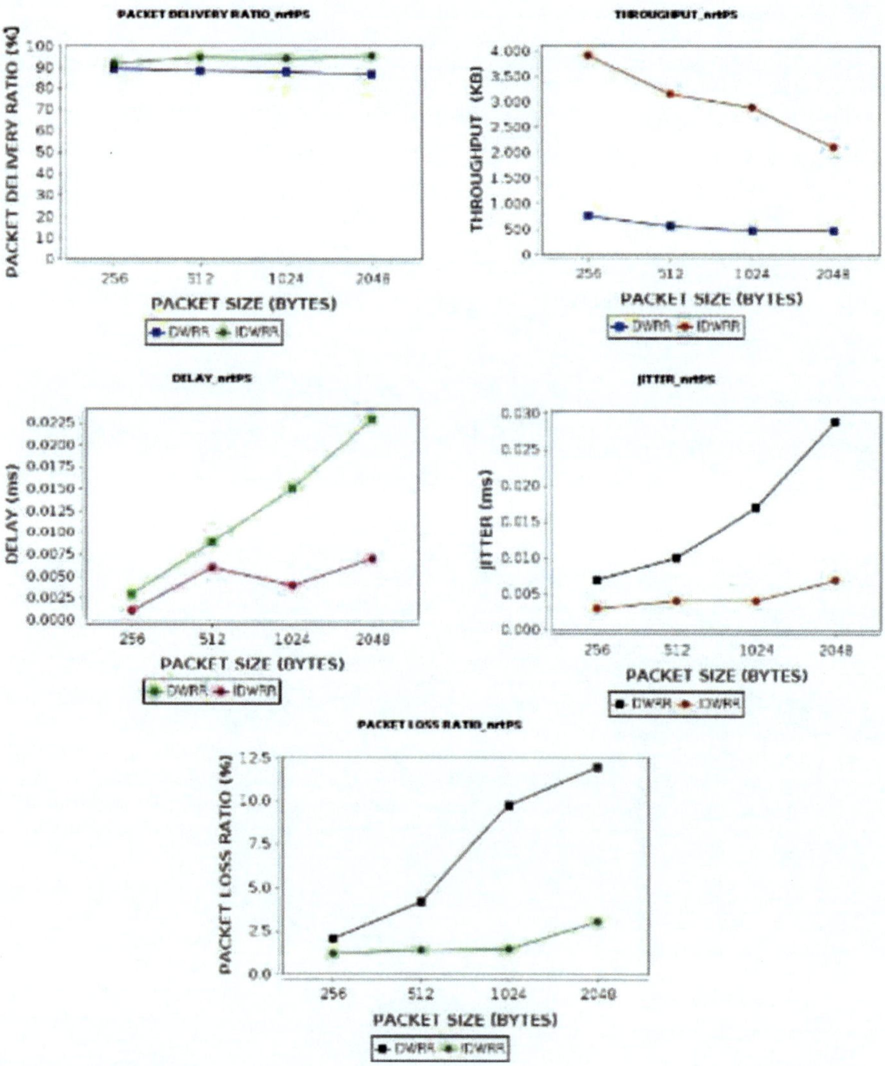

Fig. 3 Performance of DWRR and IDWRR for nrtPS traffic

7 Conclusion

In DWRR, a queue is left unserved if the size of the packet at the front of a queue exceeds the available quantum. A smaller packet may be present in the queue waiting to be serviced. Improved DWRR (IDWRR) searches for packets with sizes less than the DeficitCounter, sorts the queue, and serves smaller packets in the current round. Further, once the queue becomes empty, the DeficitCounter is moved

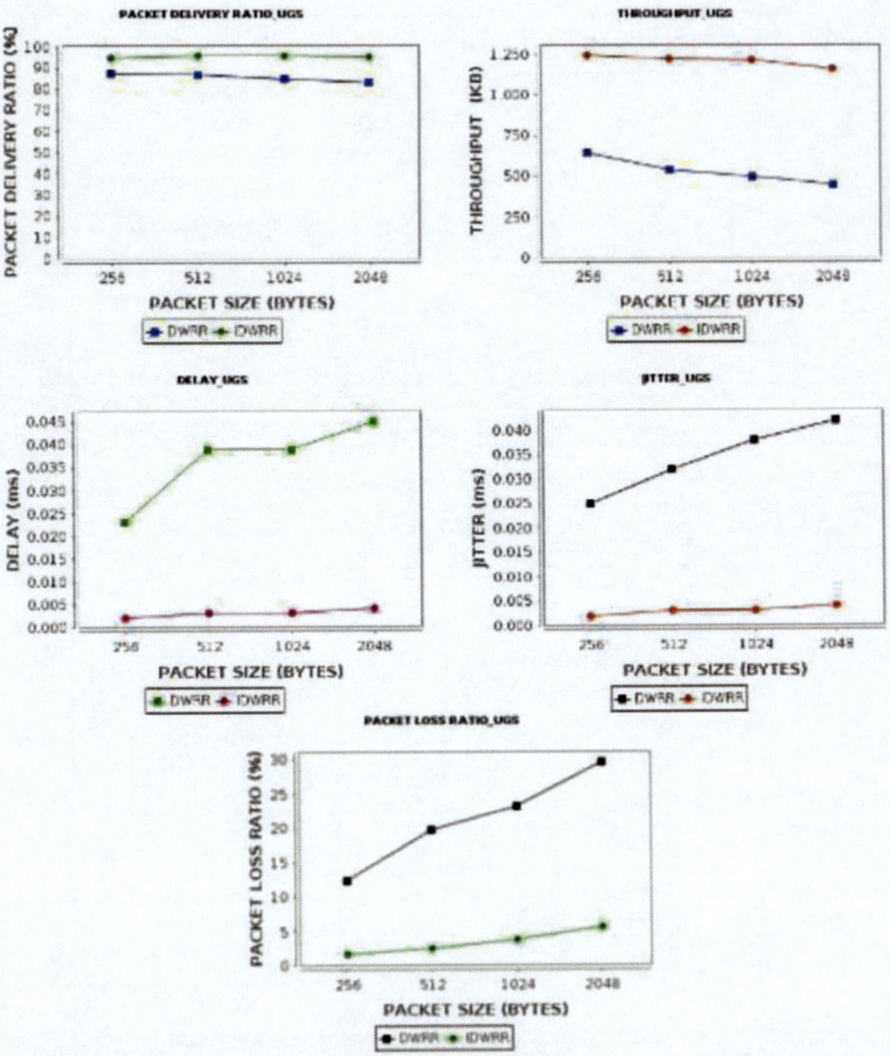

Fig. 4 Performance of DWRR and IDWRR for UGS traffic

to the next active queue rather than making it zero. This helps in servicing more number of packets in a round, thus offering better Packet Delivery Ratio (PDR) and throughput. The proposed IDWRR scheduling algorithm involves less delay, jitter, and Packet Loss Ratio (PLR) and is suitable for BE and nrtPS services.

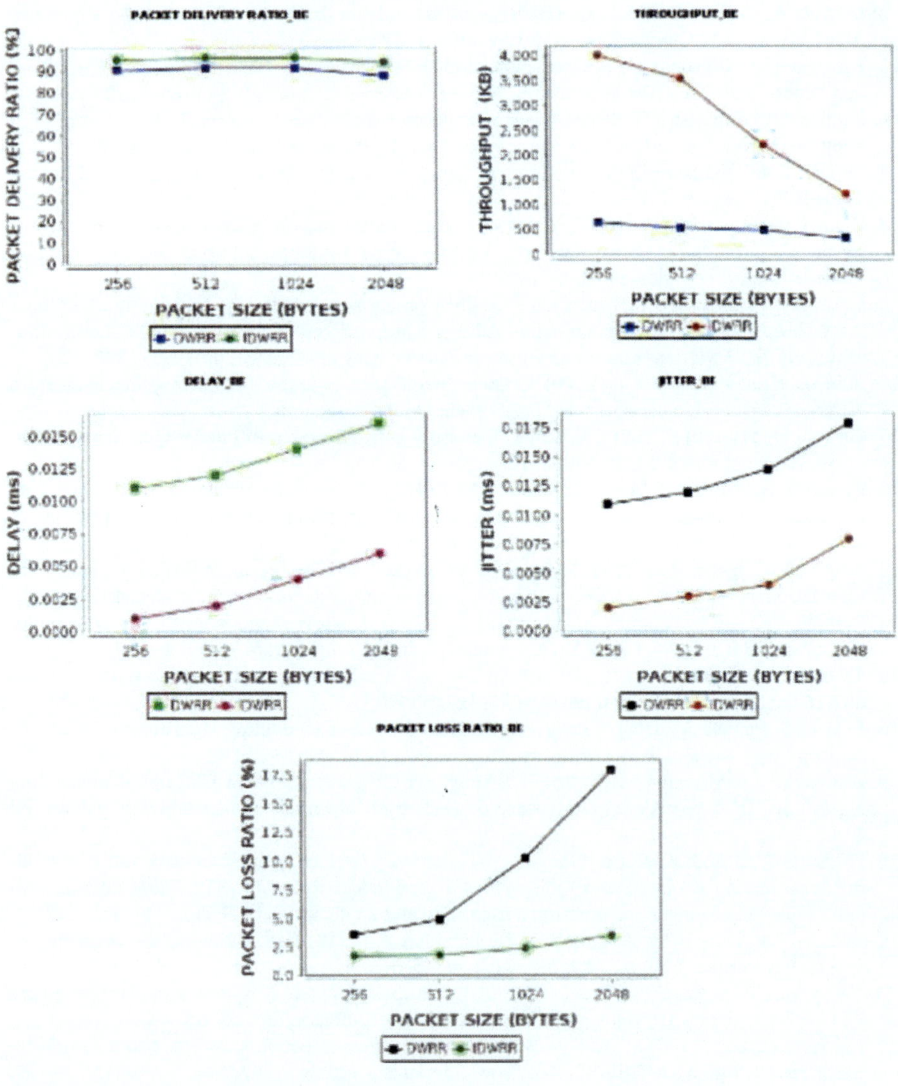

Fig. 5 Performance of DWRR and IDWRR for BE traffic

References

1. Bo B, Wei C, Zhigang C, Khaled BL (2010) Uplink cross-layer scheduling with differential QoS requirements in OFDMA systems. EURASIP J Wirel Commun Netw 2010:1–10
2. So-In C, Jain R, Tamimi AK (2009) Scheduling in IEEE 802.16e mobile WiMAX networks: key issues and a survey. IEEE J Sel Areas Commun 27(2):156–171
3. Cicconetti C, Erta A, Lenzini L, Mingozzi E (2007) Performance evaluation of the IEEE 802.16 MAC for QoS support. IEEE Trans Mobile Comput 6(1):26–38

4. Demers A, Keshav S, Shenker S (1989) Analysis and simulation of a fair queueing algorithm. ACM SIGCOMM Comput Commun Rev 19(4):1–12
5. Katevenis M, Sidiropoulos S, Courcoubetis C (1991) Weighted round-robin cell multiplexing in a general-purpose ATM switch chip. IEEE J Sel Areas Commun 9(8):1265–1279
6. Parekh AK, Gallager RG (1993) A generalized processor sharing approach to flow control in integrated services networks: the single-node case. IEEE/ACM Trans Netw 1(3):344–357
7. Shreedhar M, Varghese G (1996) Efficient fair queuing using deficit round-robin. IEEE/ACM Trans Netw 4(3):375–385
8. Ruangchaijatupon N, Wang L, Ji Y (2006) A study on the performance of scheduling schemes for broadband wireless access networks. In: International symposium on communications and information technologies, pp 1008–1012
9. Lakkakorpi J, Sayenko A, Moilanen J (2008) Comparison of different scheduling algorithms for WiMAX base station: deficit round-robin vs. proportional fair vs. weighted deficit round-robin. In: IEEE wireless communications and networking conference, pp 1991–1996
10. Shin J, Kim J, Kuo CCJ (2000) Content-based packet video forwarding mechanism in differentiated service networks. In: IEEE packet video workshop
11. Guesmi H, Maaloul S, Tourki R (2011) Design of scheduling algorithm for QoS management on WiMAX networks. J Comput Sci Eng 1(2):43–50
12. Chandur P, Karthik RM, Sivalingam KM (2012) Performance evaluation of scheduling algorithms for mobile WiMAX networks. In: IEEE international conference on pervasive computing and communications workshops, pp 764–769
13. Singla S, Kamboj MS (2012) Analysis of packet scheduling in WiMAX network. In: International conference on recent advances and future trends in information technology
14. Puranik SR, Vijayalakshmi M, Kulkarni L (2013) A survey and analysis on scheduling algorithms in IEEE 802.16e (WiMAX) standard. Int J Comput Appl 79(12):1–10
15. Teixeira MA, Guardieiro PR (2013) Adaptive packet scheduling for the uplink traffic in IEEE 802.16e networks. Int J Commun Syst 26(8):1038–1053
16. Sharma S, Panjeta S (2016) A review on quality of services scheduling algorithms in WiMAX. Int Res J Eng Technol 3(5):1–4
17. Ahmed RE, AlMuhallabi HM (2016) Throughput-fairness tradeoff in LTE uplink scheduling algorithms. In: International conference on industrial informatics and computer systems, pp 1–4
18. Dighriri M, Alfoudi ASD, Lee GM, Baker T, Pereira R (2017) Comparison data traffic scheduling techniques for classifying QoS over 5G mobile networks. In: 31st IEEE international conference on advanced information networking and applications workshops, pp 492–497
19. Shareef ZA, Hussin M, Abdullah A, Muhammed A (2018) Class-based QoS scheduling of WiMAX networks. J High Speed Netw 24(4):345–362
20. Yadav AL, Vyavahare PD, Bansod PP (2018) Proposed WiMAX hybrid scheduler with split FTP traffic and its performance evaluation. Int J Wirel Microw Technol 6:1–14
21. Lal KN, Kumar A (2018) ICN-WiMAX: an application of network coding based centrality-measures caching over IEEE 802.16. Proc Comput Sci 125:241–247
22. Mahesh DS, Chandramouli H, Sanjay RC (2019) Adaptive uplink scheduling model for WiMAX network using evolutionary computing model. Indonesian J Electr Eng Comput Sci 14(3):1345–1355
23. Khanna R, Kumar N (2019) Quality of service aware traffic scheduling algorithm for heterogeneous wireless networks. Doctoral dissertation
24. Ahmed Z, Hamma S, Nasir Z (2019) An optimal bandwidth allocation algorithm for improving QoS in WiMAX. Multim Tools Appl 78(18):25937–25976

Mask R-CNN for Instance Segmentation of Water Bodies from Satellite Image

S. Dhyakesh, A. Ashwini, S. Supraja, C. M. R. Aasikaa, M. Nithesh, J. Akshaya, and S. Vivitha

1 Introduction

Instance segmentation is the technique of clubbing pixels of various instances of an object in an image which could be boiled down into two subtasks, namely, object detection and image segmentation. Image segmentation is one of the difficult problems of image processing, and it involves clubbing pixels of the desired object. In the idea of gaining more accuracy, traditional image segmentation algorithms such as level set segmentation and threshold segmentation turned out to be unsatisfactory. With the aim of getting better results, few clustering techniques in machine learning such as k-means clustering [1], Markov random fields [2], and SVMs [3] were used. The main problem with the abovementioned methods was that they used manual selection of ROIs rather than automated selection. Object detection could be made as automated selection of ROI instead of the traditional manual method. Image segmentation and automated ROI selection using object detection are combined to form Mask R-CNN [4]. In order to identify and locate water bodies, we have used object detection and classification. To club the pixels containing water bodies, we have used image segmentation. Here, we have used fully convolutional network for this process. For localization and classification, we have used Faster R-CNN [5] which is a successor of R-CNN [6] and Fast R-CNN [7]. In order to extract features from the image, we have used Resnet (Residual Network) [8] and Feature Pyramid Network [9]. We have tweaked Faster R-CNN by replacing ROI pooling with ROI align which resembles the Mask R-CNN model proposed by Kaiming. Instead of

S. Dhyakesh (✉) · A. Ashwini · S. Supraja · C. M. R. Aasikaa · M. Nithesh · J. Akshaya
S. Vivitha
Sri Krishna College of Engineering and Technology, Coimbatore, India
e-mail: 15eucs026@skcet.ac.in; 16euit021@skcet.ac.in; 16eucs003@skcet.ac.in

A. Haldorai et al. (eds.), *2nd EAI International Conference on Big Data Innovation for Sustainable Cognitive Computing*, EAI/Springer Innovations in Communication and Computing, https://doi.org/10.1007/978-3-030-47560-4_24

using fully connected network, FCN [10] uses convolutional network to get the segmentation done. The overall aim of this is to couple Faster R-CNN and fully convolutional neural network (segmentation network) to provide a better accuracy in the problem of instance segmentation. Traditional CNN [11] serves as a base for computer vision techniques.

2 Background

2.1 *Mask R-CNN*

Mask R-CNN is built on the well-known method for object detection, Faster R-CNN. Mask R-CNN adds another branch for the task of segmentation. The total three branches are as follows:

1. Classification
2. Bounding box regression
3. Segmentation

Mask R-CNN also enhances the ROI pooling step in Faster R-CNN and proposes a ROI align layer instead. Thus, Mask R-CNN outputs a mask over the detected object using the pixels identified as the object's class.

2.2 *Instance Segmentation*

The process of detecting outlines of the object at the pixel level is called as instance segmentation. It is one among the hardest vision tasks, when compared to others. Consider the following tasks in extracting water bodies from a given satellite image: Classification, identify the water bodies (objects). Semantic segmentation, identify each and every pixels of the water body. Object detection, identify the number of the water bodies in the image along with their locations and also account for the objects that overlap. Instance segmentation means identifying the objects along with their locations and the pixels.

2.3 *Core Idea of Mask R-CNN*

Mask R-CNN is a dual step process. The image is scanned, and the output proposals are produced at the end of the first step. Classification of the proposals and generation of the bounding boxes and masks are done in the second.

3 The Proposed System

At a high level, the proposed system consists of the following four modules:

Level 1—Backbone Structure
Level 2—Region Proposal Network (RPN) Structure
Level 3—ROI Classifier and Bounding Box Regressor
Level 4—Segmentation Masks

3.1 Backbone: Level 1

The CNN extracts features from the input image. The beginning layers extract lower-level traits, and the successive layers extract the high-level traits.

Level 1 converts the input satellite image from $1024 \times 1024 \times 3$ (band—RGB) to a tensor of dimensions $32 \times 32 \times 2048$. The upcoming stage takes this tensor as input. The Feature Pyramid Network (FPN) augments traits to identify objects of various sizes.

The standard ConvNet is improvised by augmenting FPN. The initial layers could get higher level traits which are given to them by FPN. There is a tensor at every level of the second pyramid which is chosen dynamically to use based on the objects' attributes.

3.2 Region Proposal Network (RPN): Level 2

The Level 2 network is a lightweight neural network (RPN) that glances over the image in a sliding window fashion and identifies the regions that hold objects. The RPN scans over areas (called anchors) in the image. They are like boxes spread over the image. There'll be approximately 200,000 anchors of various dimensions, and they overlap the image to conquer maximum of it. The RPN scans all the anchors quickly. The RPN's convolutional nature handles the sliding window, thereby allowing it to glance each and every area in parallel. The RPN does not directly glance over the image. The tensor output from Level 1 is where the RPN scans, thereby allowing the RPN to use the extracted traits again optimally. With these improvements, the Faster R-CNN paper states that the Level 2 network (RPN) takes around 10 ms to run. On the other hand, we use bigger images, and thus more anchors are generated, and so the process may be a little slower when compared to Faster R-CNN in general. The RPN generates two outputs for each anchor: foreground and background. The foreground class states that there is a possibility that the box holds an object.

Bounding Box Refinement

In order to fit the anchor box perfectly on the object contained in the image, there's a change in the dimension of the anchor box which is done by calculating delta (percentage change in x coordinate, y coordinate, height, width) by RPN.

The object containing anchors are picked using RPN predictions. Their locations and sizes are refined. If several anchors overlap, the anchor with the maximum FG value is retained, and the others are discarded (non-max suppression). Then, the ultimate ROIs are passed to the upcoming level.

3.3 *ROI Classifier and Bounding Box Regressor: Level 3*

RPN produces ROIs on which this stage works. It produces two outputs for every ROI, like the RPN. This represents the class of the object in the ROI. This network is quite deep, having the capability to classify areas to specific classes, whereas RPN has only (FG/BG). The ROI is discarded when the background class is generated.

Bounding Box Refinement

Its task is to fine-tune the size as well as the location of the bounding boxes to hold the object. The ROI pooling classifiers don't work with inputs of varying dimensions. They require inputs of unvarying dimensions. The ROI pooling is introduced to tackle the different dimensions of ROI boxes produced by the bounding box refinement. Truncating the tensor and standardizing its dimension are done in ROI pooling. It is the same as the principle of cropping a part of an image and then resizing it. Thus, we prefer the method of ROI align, in which the feature map is sampled at different points and is applied with bilinear interpolation. During implementation, the TensorFlow function for cropping and resizing is used for simple nature, and it satisfies for most cases.

3.4 *Segmentation Masks: Level 4*

The Mask R-CNN utilizes an add-on to the Faster R-CNN, called as the mask network. The mask branch consists of a convolutional network. The masks are produced based on the positive areas picked by the ROI classifier. In our approach, the masks generated are of 28×28 pixels. They contain more details than binary masks by using float and called as soft mask. During inferring, the final masks are given by scaling up the predicted masks to the dimension of ROI bounding box. But, during training, to compute the loss, we opted to scale down the ground-truth masks to 28×28 (Fig. 1).

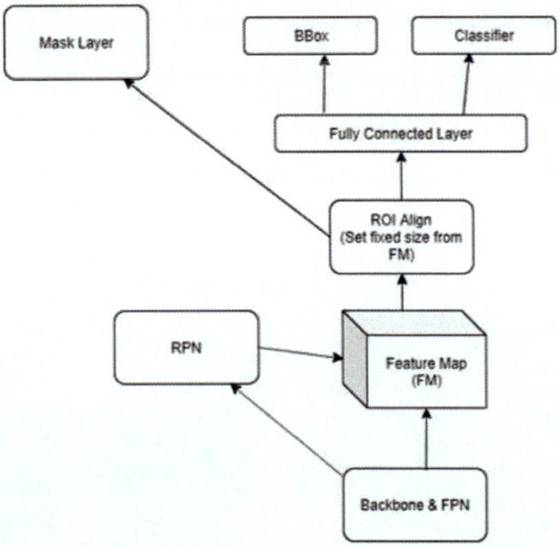

Fig. 1 Mask R-CNN model

Fig. 2 Sample image from the dataset

4 Experimental Results

The input images are obtained from National Remote Sensing Centre, ISRO, Hyderabad, India (Fig. 2).

The given image is converted into 1024 × 1024 × 3 size using the bilinear interpolation and padding technique. It is then passed through Level 1 to obtain the tensor. This tensor is fed to the RPN to get the ROI. The non-max suppression technique is used to eliminate the overlapping anchors. Then, the ROIs are fed into the ROI classifier. Finally, the masks are generated using a convolution layer (Fig. 3).

Fig. 3 Loss after 16 epochs

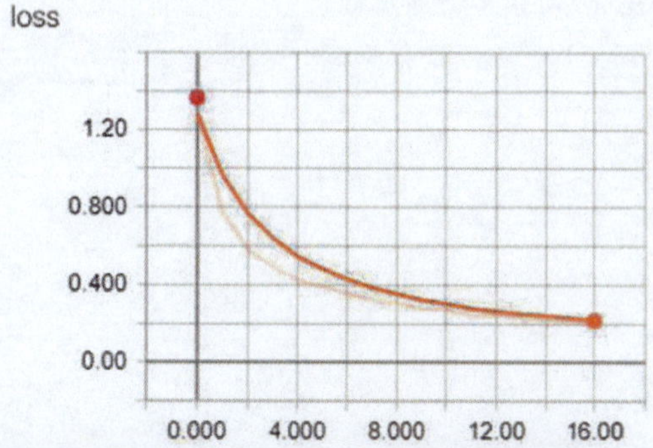

Fig. 4 Masked output

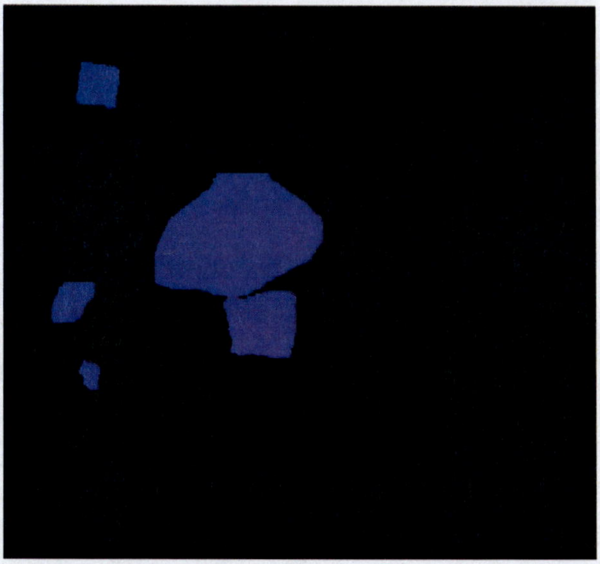

The visualization tool used here is TensorBoard, Fig. 3. A small set of 100 cropped images were used to train the system, and 50 images were used to evaluate the system. After 30 epochs the loss was found to be 0.2010 (x-axis, epochs; y-axis, loss). Figure 4 shows the segmented output image.

5 Conclusion

The current project attempted to implement Mask R-CNN in the field of remote sensing. Object detection requires a humungous amount of images for accurate results. Hence, it must be used only when important objects are needed to be detected. Better results can be given by global features and shape-based method. Thus, it was concluded that further advancement in the algorithms is needed to increase the precision. In object detection, it is desired that the exact object is detected. This is possible by instance segmentation, which has been achieved by implementing Mask R-CNN. This allows for more fine-grained information about the extent of the object within the box, unlike in Faster R-CNN where the end results are in terms on bounding boxes enclosing objects. Pixel level segmentation can be achieved by using Mask R-CNN on remote sensing images, which extends Faster R-CNN by augmenting a branch for identifying an object mask in parallel with the existing network. Once these masks were generated, Mask R-CNN combined them with the classifications and bounding boxes from Faster R-CNN and generated precise segmentations. This can have numerous applications in remote sensing such as delineation of water bodies for flood analysis, extraction of road networks for navigation, estimation of total cultivated area, etc. The generated masks could also be used to calculate the area of any desired instance. Since it is instance segmentation, we could also use this to find the number of instances in the image.

References

1. Samundeeswari ES, Saranya PK, Manavalan R (2016) Segmentation of breast ultrasound image using regularized K-means (ReKM) clustering. In: International conference on wireless communications, signal processing and networking. IEEE, pp 1379–1383
2. Li L, Lin J, Li D, Wang T (2007) Segmentation of medical ultrasound image based on Markov random field. In: International conference on bioinformatics and biomedical engineering. IEEE, pp. 968–971
3. Nguyen TD, Sang HK, Kim NC (2007) Surface extraction using SVM-based texture classification for 3D fetal ultrasound imaging. In: International conference on communications and electronics. IEEE, pp 285–290
4. He K, Gkioxari G, Dollár P, Girshick R (2017) Mask R-CNN. In Proceedings of the IEEE international conference on computer vision, pp 2961–2969
5. Ren S, He K, Girshick R, Sun J (2015) Faster R-CNN: towards real-time object detection with region proposal networks. In: International conference on neural information processing systems, vol 39. MIT Press, Cambridge, pp 91–99
6. Girshick R (2015) Fast R-CNN. In: Proceedings of the IEEE international conference on computer vision, pp 1440–1448
7. Lin TY, Dollar P, Girshick R, He K, Hariharan B, Belongie S (2016) Feature pyramid networks for object detection, pp 936–944
8. Hariharan B, Arbeláez P, Girshick R, Malik J (2014) Simultaneous detection and segmentation. In: Fleet D, Pajdla T, Schiele B, Tuytelaars T (eds) ECCV 2014, LNCS, vol 8695. Springer, Cham, pp 297–312
9. Long J, Shelhamer E, Darrell T (2015) Fully convolutional networks for semantic segmentation. In: Computer vision and pattern recognition, vol 79. IEEE, pp 3431–3440
10. He K, Zhang X, Ren S, Sun J (2016) Deep residual learning for image recognition. In: CVPR, pp 770–778
11. Liu F, Lin G, Shen C (2015) CRF learning with CNN features for image segmentation. Elsevier Science Inc., Amsterdam

Dynamic Spectrum Access for Spectrum Handoff in Cognitive Radio Networks Using Optimization Techniques

M. Kalpana Devi and K. Umamaheswari

1 Introduction

In the past, the spectrum allocation was based on the service required by the secondary user. Due to the static assignment of spectrum, there occurs a problem of inadequacy toward the spectrum allocation. So in this aspect, the cognitive radio network (CRN) has emerged a new trend in wireless communication to isolate the white spaces (spectrum hole) which are not accessible by the licensed user. The known white space is allocated for the secondary user to transmit the data smoothly without any interruption [1]. CRN works on the progress of cognitive radio cycle, namely [2, 3]:

- *Spectrum sensing*: Senses the vacant space in the spectrum.
- *Spectrum sharing*: The licensed channel is shared by the SU.
- *Spectrum decision*: The sensed channel is selected for transmission by the SU.
- *Spectrum handoff*: During the arrival of the PU, the SU has to identify the free channel and vacate.

The dynamic spectrum access (DSA) is used for reusing the spectrum in opportunistic manner by the secondary user, and its efficiency gets improved without the interference of the primary user [4]. In this aspect, to sense the channels, various methods of topologies and usage amount of spectrum utilizations by the secondary users are done effectively with the help of DSA [5]. The utilization of spectrum for

M. Kalpana Devi (✉)
Department of Information Technology, Sri Ramakrishna Institute of Technology, Coimbatore, India

K. Umamaheswari
Department of Information Technology, PSG College of Technology, Coimbatore, India

A. Haldorai et al. (eds.), *2nd EAI International Conference on Big Data Innovation for Sustainable Cognitive Computing*, EAI/Springer Innovations in Communication and Computing, https://doi.org/10.1007/978-3-030-47560-4_25

the secondary user is allocated based on the concept of dynamic spectrum access using swarm-based algorithm [6].

2 Literature Review

2.1 *Cognitive Radio Network*

CRN is an intelligent network used to detect the channels which are currently active or inactive in the spectrum. This network is used to sense the white space available in the spectrum and allocates that identified space for the unlicensed user (i.e., secondary user) [7]. One method used to solve the problem of scarcity in cognitive radio network is spectrum handoff. It is applied during the arrival of primary user for accessing its licensed channel. When it occurs, the handoff operation is required for the secondary user to swift from corresponding channel to the next white space channel [8].

Components of Cognitive Radio Networks

The two leading components of cognitive radio network are the authorized user and unauthorized user. The authorized user is denoted as the primary user who is fully qualified for accessing the channel, and the unauthorized user is called the secondary user or cognitive user who can access only during the absences of the primary user. The CR network is divided into three categories:

Interleave networks: Once the white or spectrum hole is identified, the SU will start activating its transmission till the arrival of the PU.

Underlay network: The PU and SU will work simultaneously in the same spectrum slot with different frequency bandwidths.

Overlay network: The PU and SU will work concurrently with interference. The PU will send the message before transmitting the data, and similarly the SU also performs the same before transmission.

The primary user and secondary user in cognitive radio network are illustrated in Fig. 1.

2.2 *Dynamic Spectrum Access*

The assignment of frequency bandwidth in dynamic spectrum access is built on dynamic process instead of static. It is a main source to solve the hypothetic problem that occurs during the handoff by the secondary user. This dynamic spectrum access

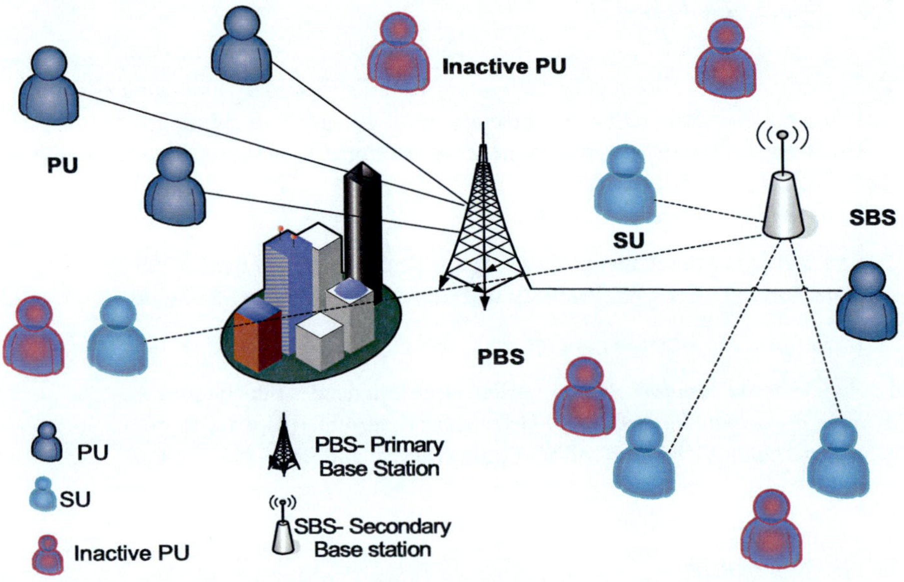

Fig. 1 Architecture concept of cognitive radio network

is used to share the channel for both the PU and SU without any interference [9]. It senses the readiness of white space in spectrum and allocates the free channel for the SU without interrupting the service during the appearance of the PU [10].

2.3 *Spectrum Sensing*

Sensing the white space in cognitive radio network is accomplished by spectrum sensing with the support of various techniques such as narrow band sensing and wide band sensing. The issues and challenge faced by the spectrum sensing are as follows [11]:

(a) *Channel uncertainty*: The presence of PU is wrongly interpreted due to the delay, energy detection, and performance of the channel utilization.
(b) *Interference temperature measurement*: The transmission data, time, and the bandwidth level is known by the SU, but the arrival of PU is not known.
(c) *Error prone channel*: Since the SU is moving on to the accessible free channel when PU comes, there may occur erroneous channel detection.
(d) *Mobility*: Often changing the channel by SU is a big issue.

2.4 *Spectrum Handoff*

The spectrum handoff is mainly used to perform the secondary user work efficiently without any interference of the primary user. In order to moderate the loss of information and throughput and to increase the signal to noise interference (SINR), the spectrum handoff is used [12].

Two types of spectrum handoff:

1. *Soft handoff*: Interchangeability with the same frequency bandwidth
2. *Hard handoff*: Interchangeability with different frequency bandwidths.

Techniques in spectrum handoff:

1. *Pure reactive handoff*: Getting prior request to access the channel.
2. *Pure proactive handoff*: The PU arrival is determined and acted upon.
3. *Hybrid handoff*: Mixture of both pure reactive and proactive handoff.

3 Methodology

3.1 *Proposed Algorithm*

Improved PSO Algorithm

The local minima and premature convergence are the two main drawbacks that arise in spectrum particle swarm optimization (SpecPSO). To overcome the problem of SpecPSO, this proposed algorithm was suggested. The two parameters were used in order to enhance the efficiency of throughput and data transmission. One parameter is inertia weight and the other is learning factor. The chaotic optimization is used for initializing the parameter [13].

Chaotic Optimization

The large range of population is generated from this chaotic optimization. Using this, the global best and the local best value is identified with perfectly [14].

The following steps are architectural flow of the proposed algorithm which is shown in Fig. 2:

- The spectrum hole is identified by spectrum sensing.
- The primary user and the secondary user are initialized.
- Spectrum mobility is done when the PU is in need of the channel used by the SU.
- The channel identification takes place with the help of iPSO efficiently.
- For smooth transmission, the next channel is traced quickly for the SU to work.

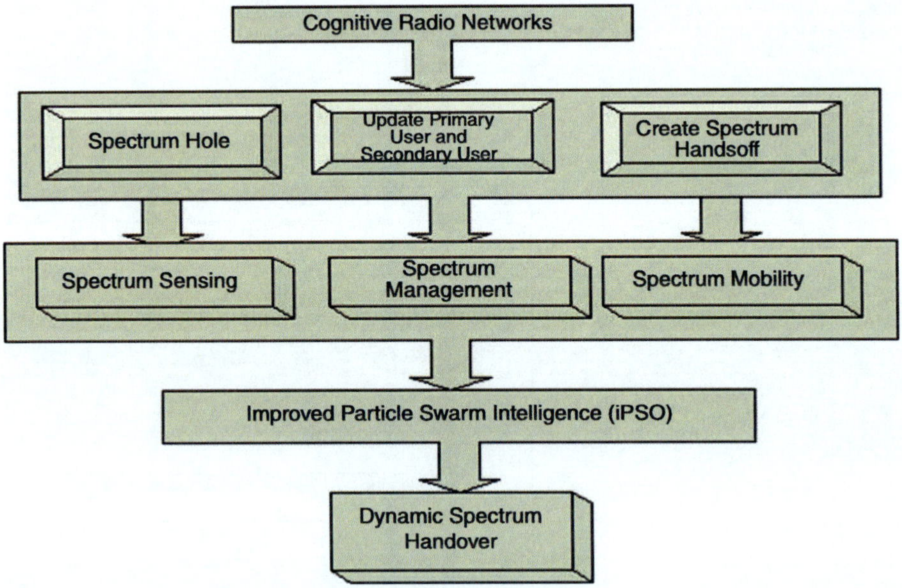

Fig. 2 The proposed diagram of improved PSO

Figure 3 demonstrates the implementation of swarm-based optimization by using the parameters which is calculated as follows:

Step 1: Consider the number of population, particles, and parameters are to be initialized.

Step 2: The present iteration with the previous iteration is compared and calculated for considering the value of pbest. The objective function value which is low is treated as pbest as in Eq. (1):

$$\text{pbest}_{d+1} = \begin{cases} \text{pbest}_d & \text{if } f_{d+1} \geq f_d \\ x_{d+1} & \text{if } f_{d+1} \leq f_d \end{cases} \tag{1}$$

Here k is treated for iteration value and f is function of the objective value.

Step 3: The gbest value is calculated by comparing with pbest value, and the minimum rate is gbest as in Eq. (4):

$$\text{gbest}_{d+1} = \begin{cases} \text{gbest}_d & \text{if } f_{d+1} \geq f_d \\ \text{pbest}_{d+1} & \text{if } f_{d+1} \leq f_d \end{cases} \tag{2}$$

Step 4: The next iteration should be modified with the velocity updation, after the calculation of the pbest and gbest by using Eqs. (3)–(5):

$$v_{d+1} = v_d + w * v_d + c_1 * r_1 (\text{pbest} - x_d) + c_2 * r_2 (\text{gbest} - x_d) \tag{3}$$

Fig. 3 Implementation of improved PSO (iPSO)

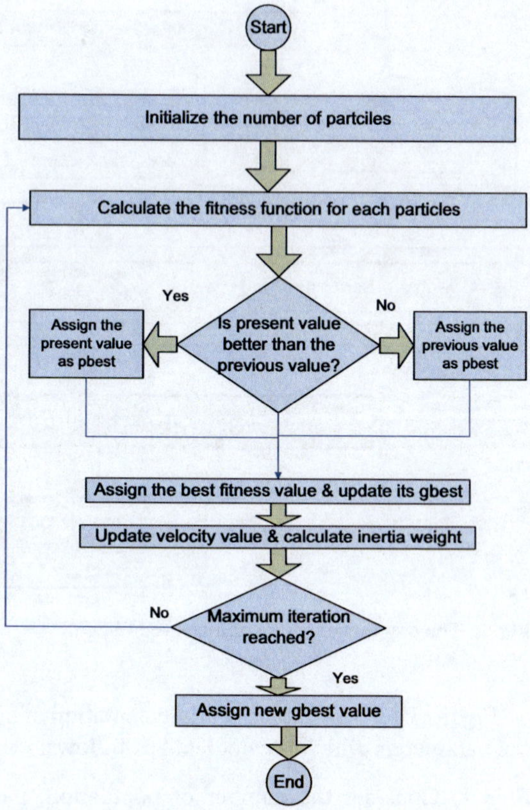

$$x_{d+1} = x_d + v_{d+1} \quad (4)$$

$$\omega = \omega_{\text{maximum}} - \frac{\omega_{\text{maximum}} - \omega_{\text{minimum}}}{i_{\text{maximum}}} * i \quad (5)$$

where i represents iteration and ω is inertia weight which is used to identify the best global best value.

Step 5: If i values with maximum no of iteration is performed then, move on to step 6 else step 2.

Step 6: Optimal solution for each and every individual population is generated to find the new global best.

4 Simulation and Result

The simulation work is done through MATLAB R2014a with the configuration of 4 GB RAM. The initial population size is 50; the assumed user is 20, and channel size considered for execution is 200. The frequency bandwidth of each and every channel is 30 kHz. The data transfer rate is 256 kbits.

The results are compared with genetic algorithm, specPSO, and iPSO. The best optimal solution is found with minimal iteration. The data transfer rate is high when compared with other evolutionary algorithms such as GA and specPSO. The total simulation time is faster in identifying the white space, in order to perform mobility during the PU's arrival. This representation is shown in Figs. 4 and 5, respectively.

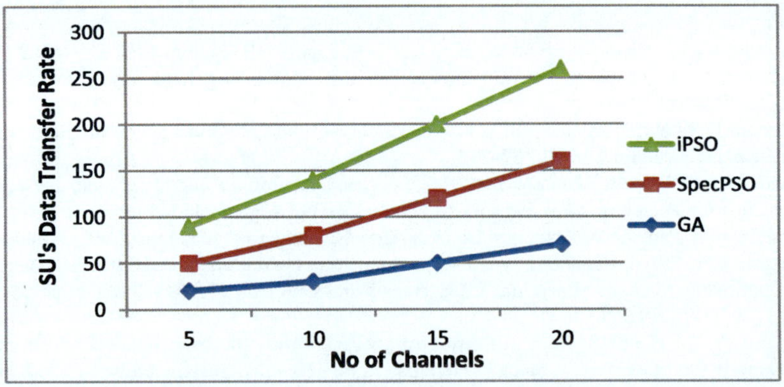

Fig. 4 Data transmission rate of secondary user after the handoff

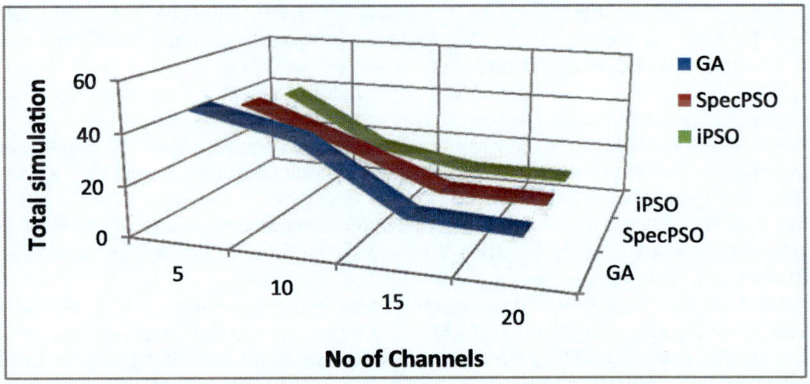

Fig. 5 Total simulation time of GA, SpecPSO, iPSO

5 Conclusion

This proposed work focuses on the improvement of channel utilization for SU in order to progress its activity without any interruption. It also identifies the unused channel where the PU is not accessing it for transmission. It provides a high data transfer with limited iteration. Thus iPSO algorithm concentrates on good network communication for transmitting the data and also performs handoff during the arrival of the PU in an efficient manner. The future work focuses on handoff optimization in cognitive radio network using BPSO and iPSO techniques.

References

1. Ali A, Abbas L, Shafiq M, Bashir AK, Afzal MK, Liaqat HB, Siddiqi MH, Kwak KS (2018) Hybrid fuzzy logic scheme for efficient channel utilization in cognitive radio networks. IEEE Access 7:24463–24476
2. Koroupi F, Talebi S, Salehinejad H (2012) Cognitive radio networks spectrum allocation: an ACS perspective. Sci Iran 19:767–773
3. Kumar K, Prakash A, Tripathi R (2016) Spectrum handoff in cognitive radio networks: a classification and comprehensive survey. J Netw Comput Appl 61:161–188
4. Bhardwaj P, Panwar A, Ozdemir O, Masazade E, Kasperovich I, Drozd AL, Mohan CK, Varshney PK (2016) Enhanced dynamic spectrum access in multiband cognitive radio networks via optimized resource allocation. IEEE Trans Wirel Commun 15:8093–8106. https://doi.org/10.1109/TWC.2016.2612627
5. Khalid W, Yu H (2018) Sum utilization of spectrum with spectrum handoff and imperfect sensing in interweave multi-channel cognitive radio networks. Sustainability 10:1764. https://doi.org/10.3390/su10061764. www.mdpi.com/journal/sustainability
6. Kalpana Devi M, Umamaheswari K (2019) Intelligent process of spectrum handoff for dynamic spectrum access in cognitive radio network using swarm intelligence. Int J computers and applications, Taylor & Francis, pp. 1–9. https://doi.org/10.1080/1206212X.2019.1704483
7. Feng C, Wang W, Jiang X (2012) Cognitive learning-based spectrum handoff for cognitive radio network. Int J Comput Commun Eng 1:1–4
8. Lala NA, Balkhi AA, Mir GM (2017) Spectrum handoff in cognitive radio networks: a survey. Orient J Comput Sci Technol 10:765–772
9. Liu X, Zhang W (2011) A novel dynamic spectrum access strategy applied to cognitive radio network. In: 7th International conference on wireless communications networking and mobile computing (WiCOM), pp 1–5
10. Awoyemi BS, Maharaj BTJ, Alfa AS (2016) Solving resource allocation problems in cognitive radio networks: a survey. EURASIP J Wirel Commun Netw 2016:176–183. https://doi.org/10.1186/s13638-016-0673-6
11. Yucek T, Arslan H (2009) A survey of spectrum sensing algorithms for cognitive radio applications. IEEE Commun Surv Tutorials 11:116–130. https://doi.org/10.1109/SURV.2009.090109
12. Yawada PS, Dong MT (2019) Intelligent process of spectrum handoff/mobility in cognitive radio networks. J Electr Comput Eng 2019:1–12. https://doi.org/10.1155/2019/7692630
13. Jiang Y, Hu T, Huang CC, Wu X (2007) An improved particle swarm optimization algorithm. Appl Math Comput 193:231–239. https://doi.org/10.1016/j.amc.2007.03.047
14. Prema Kumar N, Mercy Rosalina K (2015) IPSO algorithm for maximization of system loadability, voltage stability and loss minimisation by optimal DG placement. Int J Innov Res Electr Electron Instrum Control Eng 3:73–77. https://doi.org/10.17148/IJIREEICE.2015.31115

VLSI Implementation of Color Demosaicing Algorithm for Real-Time Image Applications

S. Allwin Devaraj, Stalin Jacob, Jenifer Darling Rosita, and M. M. Vijay

1 Introduction

An image is digitized to convert it to a form which can be stored in a computer's memory or in some form of storage media such as a hard disk or CD-ROM [1–5]. This conversion procedure is regularly executed by using a scanner, or with the aid of a video camera linked to a body connected to a frame grabber board in a computer. Once the image has been digitized, it is frequently operated upon with the aid of numerous image system operations [6–8].

The drawback of that procedure is that the effect of mosaicking artifacts reduces image sharpness simultaneously and a baseline nonmosaiced original image would provide a highly inaccurate prediction of actual image quality [1, 2, 9, 10, 12].

The drawback of that procedure is complexity due to usage of large amount of add and shift operations. The frequency response is lower in conventional approaches, mainly at high frequency. Real-world picture simulation additionally shows that the proposed technique produces worst PSNR and best performance photograph [9].

S. Allwin Devaraj (✉)
Department of ECE, Francis Xavier Engineering College, Tirunelveli, Tamil Nadu, India

S. Jacob
Engineering Department, Botho University, Gaborone, Botswana

J. Darling Rosita
Electrical and Electronics Engineering Department, New Era College, Gaborone, Botswana

M. M. Vijay
Department of ECE, V V College of Engineering, Tisaiyanvilai, Tamil Nadu, India

A. Haldorai et al. (eds.), *2nd EAI International Conference on Big Data Innovation for Sustainable Cognitive Computing*, EAI/Springer Innovations in Communication and Computing, https://doi.org/10.1007/978-3-030-47560-4_26

The disadvantage of that procedure is the high computational complexity because of large mathematical expressions. It characterizes the resultant image artifacts due to inadequate use of both correlations. It has enhanced two modern-day demosaicking methods, ECI and AP methods, which do not totally make use of the spatial or spectral correlation and these greater techniques now not only successfully suppress demosaicking artifacts in side areas [10].

In textured regions, where edges have a tendency to be shorter and in specific directions, the nearby education set is not a consultant of the neighborhood statistics and the algorithm introduces extra errors. The images window and statue have large texture regions with poorly described edges. So vision is not as sensitive in textured regions, because its performance deteriorates [12–14].

2 Demosaicing Algorithm

2.1 *Block Diagram*

The proposed method consists of input image, preprocessing unit, pixel array selector, defective pixel array, threshold array generator, white balance unit, contrast compensation, color optimization, and output generator (see Fig. 1).

The overall performance of this sketch used to be expanded with the aid of a pipeline timetable. Although the local gain that is obtained by the edge-direction weighting information, it can efficiently improve the quality of the interpolated images. (see Figs. 2 and 3).

The module 3 represents FPGA kit to be implemented to VLSI design (see Fig. 4), which consists of excessive decision CCD camera, RS-232 cable, and FPGA unit. Here, CCD digital camera is working as an image acquisition unit. RS-232 cable connects the FPGA device and camera. VLSI implements digital photo processing in real-time image applications.

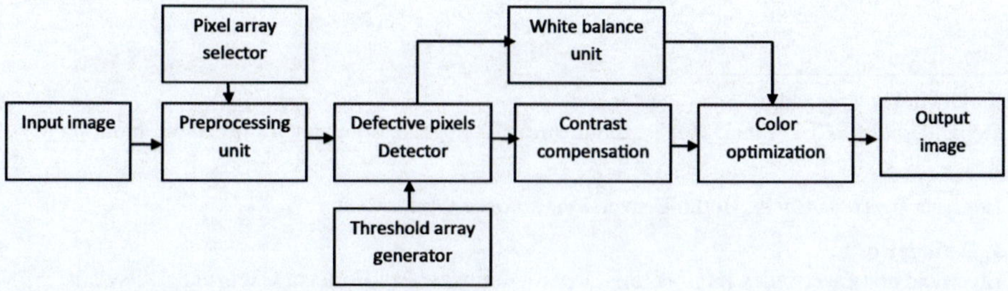

Fig. 1 Block diagram of demosaicing algorithm

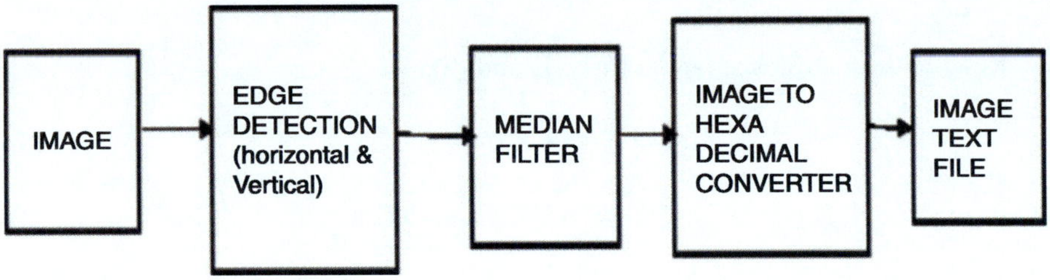

Fig. 2 Module 1

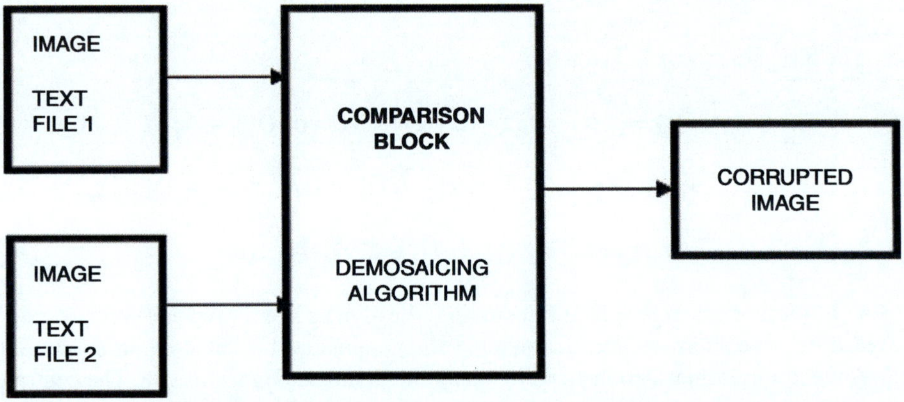

Fig. 3 General block diagram for module 2 to identify corrupted image

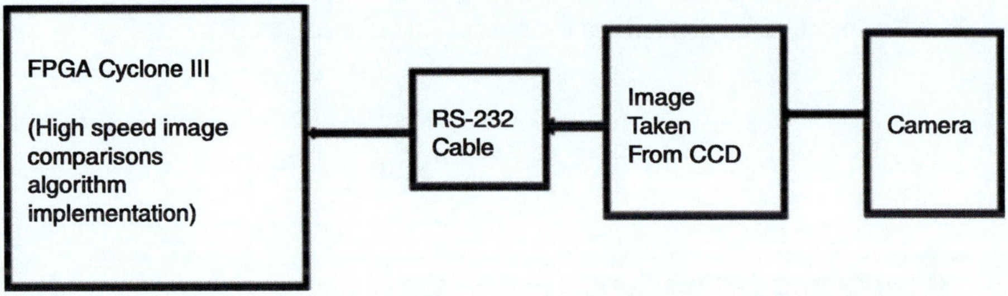

Fig. 4 Module 3

2.2 *Demosaicing Algorithm*

$$X(t) = F(t) * X(t-1) + U(t) \tag{1}$$

$$Y(t) = ADH(t) * X(t) + V(t) \tag{2}$$

Rewrite the state space model Eqs. (1) and (2):

$$Z(t) = F(t) * Z(t-1) + e(t) \tag{3}$$

$$Y(t) = ADZ(t) + V(t) \tag{4}$$

The covariance matrix is

$$\pi(t) = F(t)\pi(t-1) + Ce(t) \tag{5}$$

The KF gain matrix is given by

$$K(t) = \pi(t) * AD * [Cy(t) + AD\pi(t)DT]{-1} \tag{6}$$

$$Z(t) = F(t)Z(t-1) + K(t)[y(t) - ADF(t)Z(t-1)] \tag{7}$$

$$\pi(t) = \mathrm{Cov}(Z(t)) = [1 - K(t)AD]\pi(t) \tag{8}$$

It is partner instance of demonstrating the dynamic multiframe demosaicing algorithm. The increase and decrease in images inside the left column show the low-resolution frames demosaiced by using the demosaicing technique. The central snapshots show the ensuing color snapshots during the algorithmic dynamic shift-and-add process. A number of resolutions have been rehabilitated at this time. The SSIM index is calculated on various windows of an image (Fig. 5).

PSNR is most easily defined via the mean squared error (MSE):

$$\mathrm{PSNR} = 10\log_{10}\frac{L^2}{\mathrm{MSE}}$$

3 Results and Discussion

First the noisy image is given as an input in MATLAB and then this image can be converted into two hexadecimal values; these hexadecimal values appear in two text files in Xilinx.

3.1 *True Input Image*

Figure 6, that identify the image is corrected or not. The loading photo is pattern image 1 and the checking-out snapshots are sample photograph 2, and pattern image 3 (i.e.) is demosaiced image.

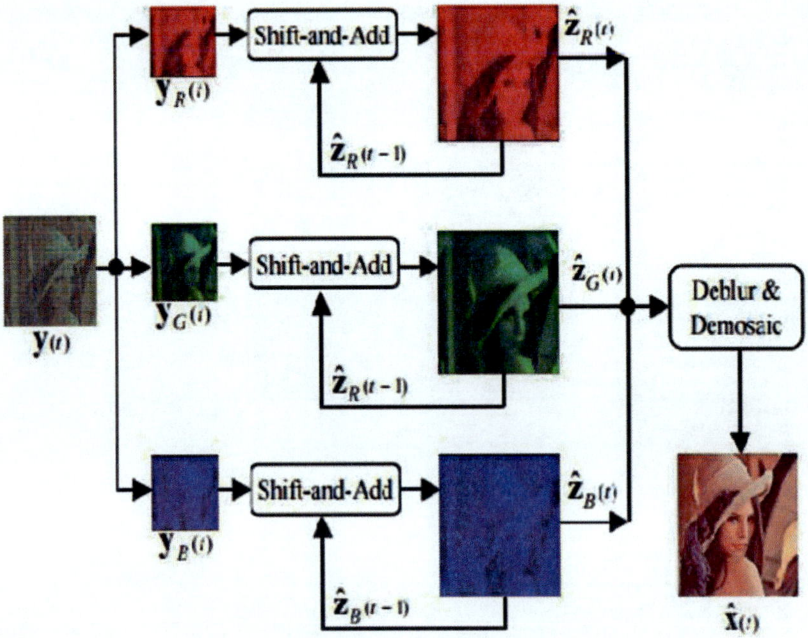

Fig. 5 Fast dynamic multiframe demosaicing (FDMD) process

Fig. 6 True input image

3.2 Demosaicked Image

Figure 7 indicates that sample 1 photograph (original photograph not affected) is resized to 256×256 pixel values.

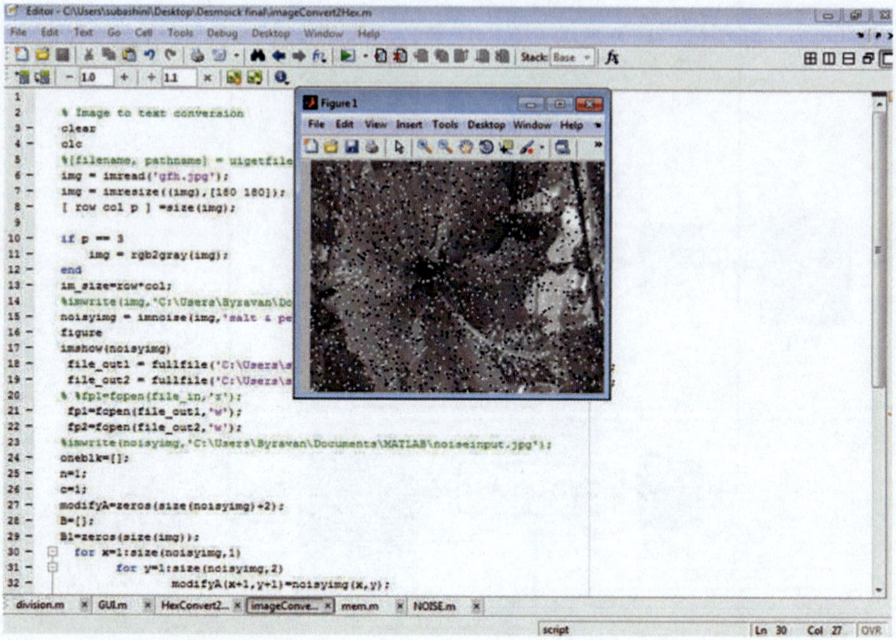

Fig. 7 Demosaicked image

3.3 *Image Converted into Hexadecimal Value*

Figure 8 shows that the hexadecimal value is given to Xilinx software in text file format. In coding the hexadecimal value is created as a text file document.

3.4 *Xilinx Output for Edge Detection (Fig. 9)*

3.5 *Output Text File*

In Xilinx software generate output as text document file. First find the text output valid for generating reconstructed image (Fig. 10).

3.6 *Reconstructed Image*

Figure 11 shows that the reconstructed image is produced at MATLAB. The image enhancement value can be determined by comparing properties of input and output images.

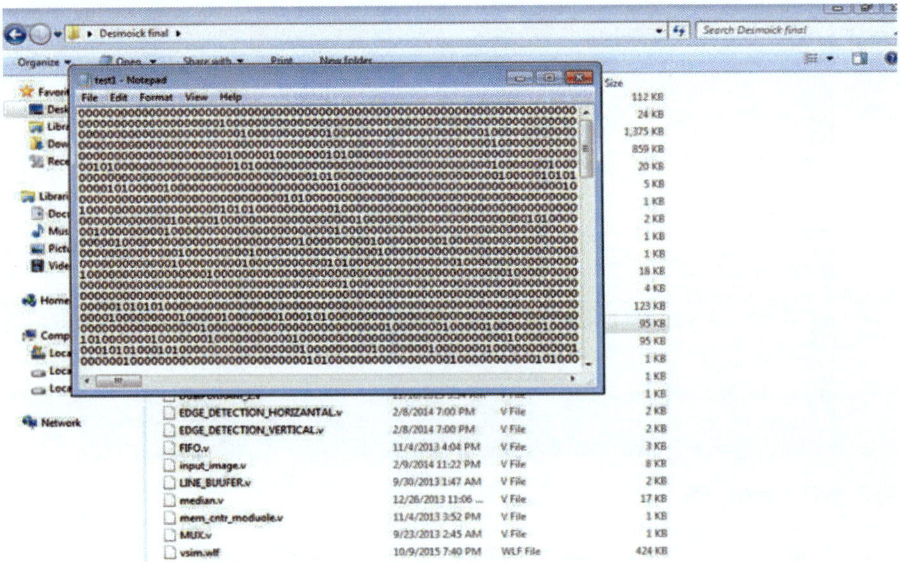

Fig. 8 Image converted into hexadecimal value

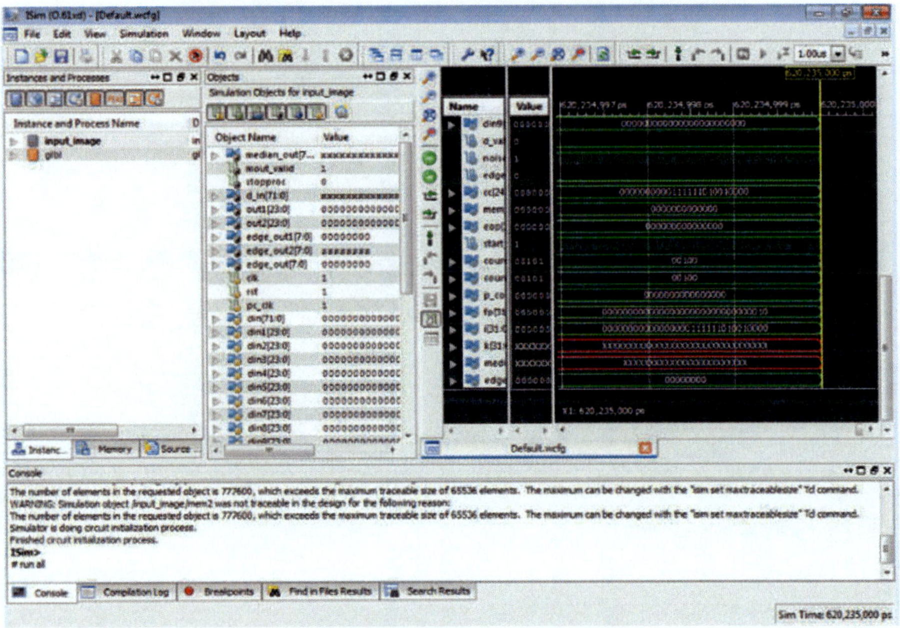

Fig. 9 Xilinx output for edge detection in horizontal and vertical directions

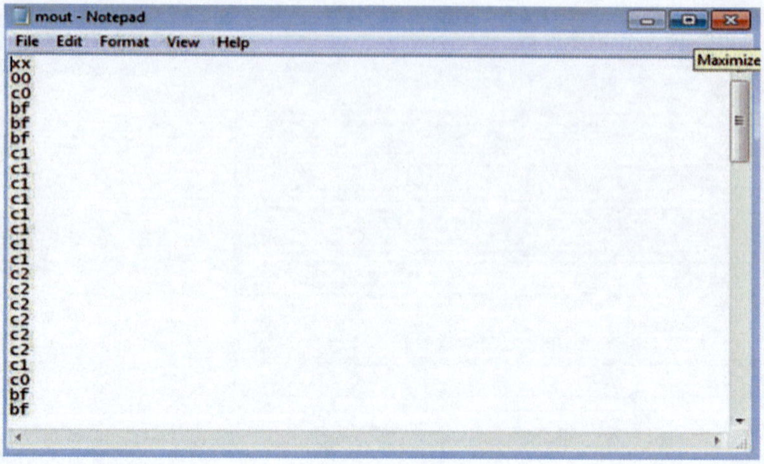

Fig. 10 Output text file

Fig. 11 Reconstructed image

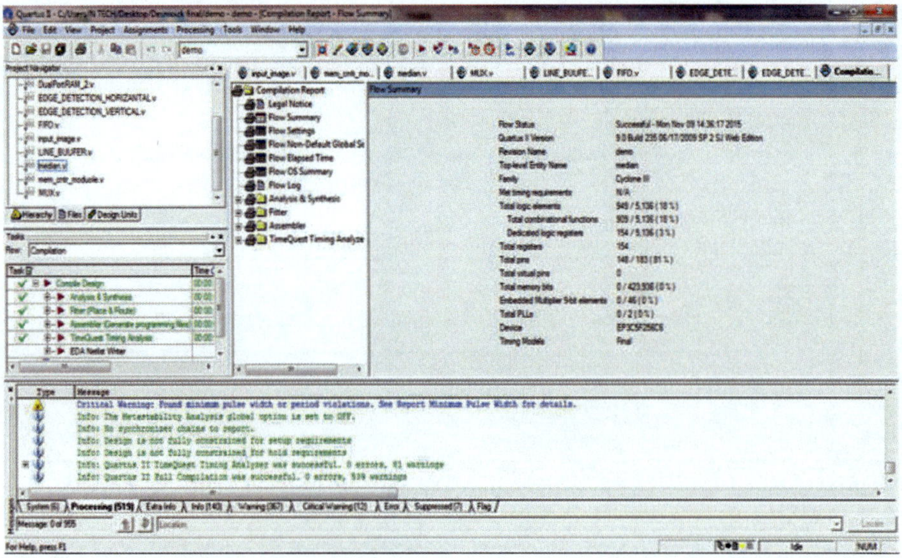

Fig. 12 Area analysis of output image

3.7 *Area Analysis of Output Image (Fig. 12)*

3.8 *Power Analysis*

The power analysis of output image is produced by Quartus II software. The objective of power analysis is to reduce countable register values simultaneously in different objects (Fig. 13).

3.9 *Graph of Core Power vs. Thermal Power Dissipation*

Figure 14 shows that the thermal power dissipation can slightly vary in between second interval and fourth interval but core power dissipation values are constant. In the proposed work we also discuss about image enhancement with clear analysis of power and area analysis and also find out signal-to-noise ratio value.

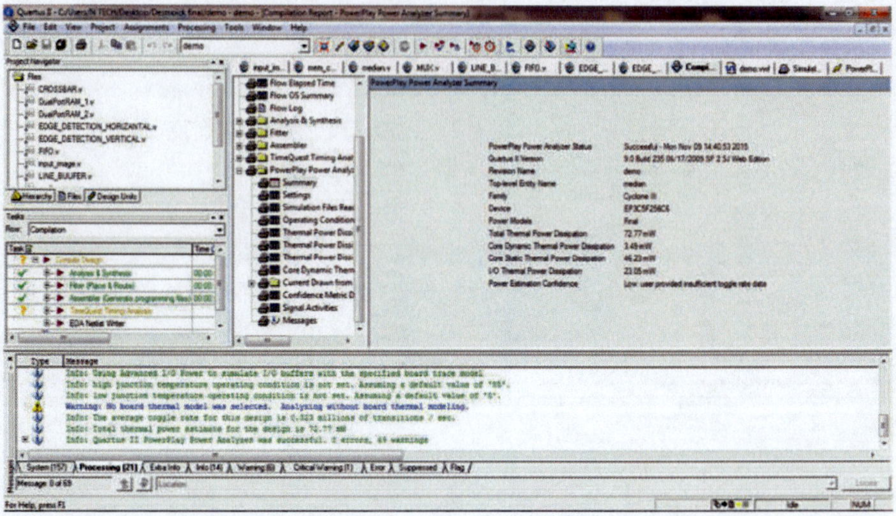

Fig. 13 Power analysis of output image

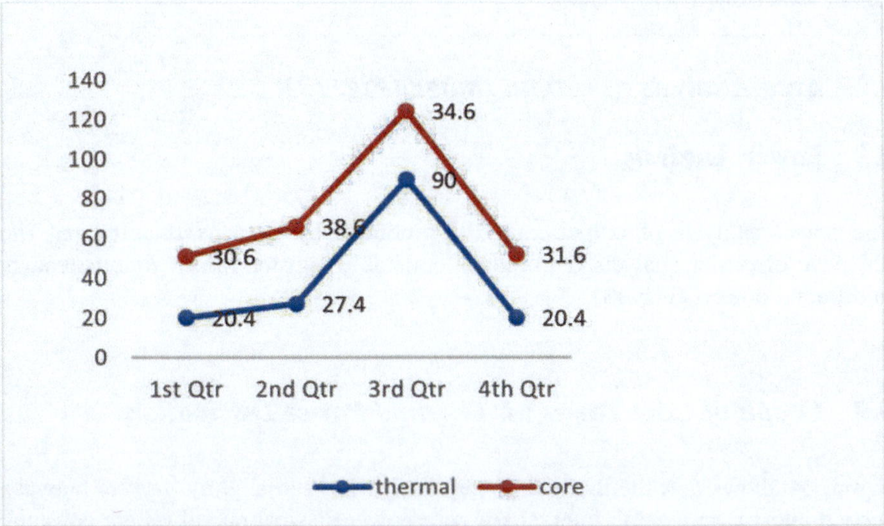

Fig. 14 Core vs. thermal power dissipation

3.10 *Area Analysis of Proposed Work*

The area analysis of output image is produced by Quartus II software. The objective of area analysis is to reduce countable register values simultaneously in different objects (Fig. 15).

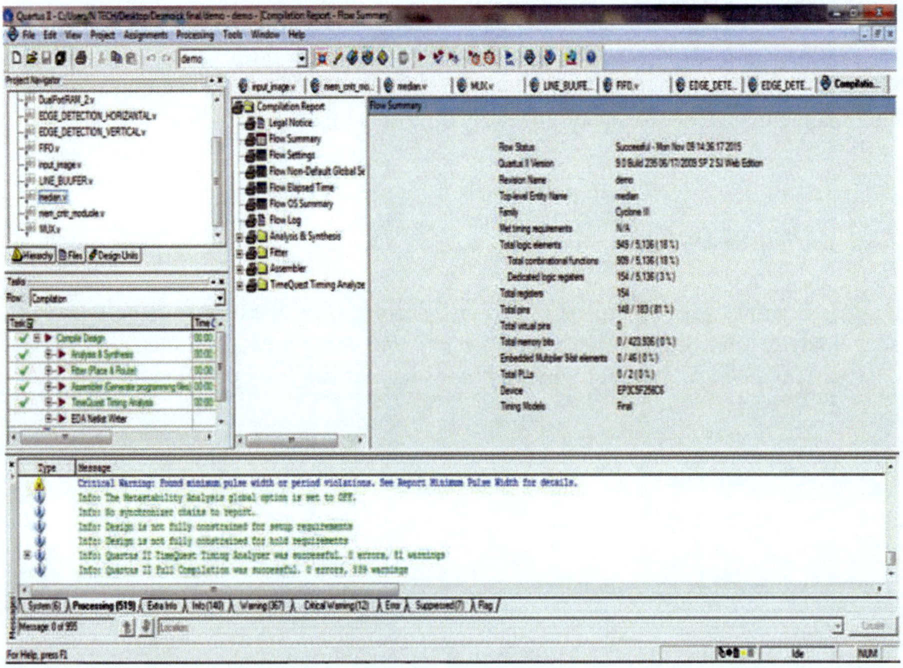

Fig. 15 Area analysis of proposed work

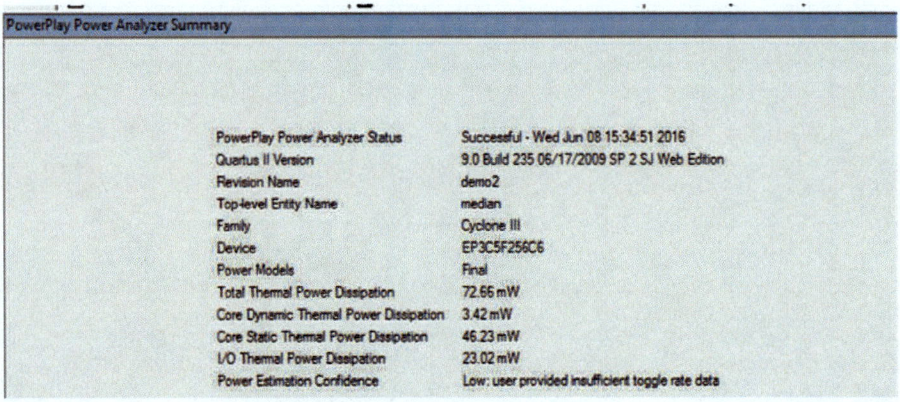

Fig. 16 Power analysis of proposed work

3.11 *Power Analysis of Proposed Work*

Figure 16 shows that energy consumption can be reduced linearly. Power optimization is achieved in proposed method.

3.12 *Signal-to-Noise Ratio Value*

Figure 17 illustrates that the signal-to-noise value is accurate for output image. So the image enhancement can be improved simultaneously.

3.13 *Output Image*

Figure 18 shows that corrupted image and corrected or enhanced image (see Fig. 18a) contain fully blurred image and some color components are missing in their position. It can be found out by green color pixels using fully pipelined method.

```
>> snr
  153.6632

>> snr
  153.6632

>> |
```

Fig. 17 SNR

Fig. 18 (**a**) Demosaicked image and (**b**) enhanced image

4 Conclusion

In this research chapter, demosaicing algorithm is proposed to improve a low-power, excessive overall performance for real-time photo applications. Here real time images are true input image. For real-time VLSI implementation, a low-power good-quality image is the main objective of our proposed work. Compared with previous techniques the power consumption can be linearly reduced by 8% or 90.6%. In future work we will add hybrid Gaussian filter interpolation for increasing the image enhancement, consuming less power, and considering the efficient solution for the quality of reconstructed image.

References

1. Jeon G, Anisetti M, Lee J, Bellandi V, Damiani E, Jeong J (2009) Concept of linguistic variable-based fuzzy ensemble approach: application to interlaced HDTV sequences. IEEE Trans Fuzzy Syst 17(6):1245–1258
2. Yun SH, Kim JH, Kim S (2008) Color interpolation by expanding a gradient method. IEEE Trans Consum Electron 54(4):1531–1539
3. Wu X, Zhang L (2006) Temporal color video demosaicking via motion estimation and data fusion. IEEE Trans Circ Syst Video Technol 16(2):231–240
4. Muresan DD, Parks TW (2005) Demosaicing using optimal recovery. IEEE Trans Image Process 14(2):267–278
5. Chang L, Tan YP (2004) Effective use of spatial and spectral correlations for color filter array demosaicking. IEEE Trans Consum Electron 50(1):355–365
6. Hirakawa K, Parks TW (2005) Adaptive homogeneity-directed demosaicing algorithm. IEEE Trans Image Process 14(3):360–369
7. Menon D, Andriani S, Calvagno G (2006) Demosaicing with directional filtering and a posteriori decision. IEEE Trans Image Process 16(1):132–141
8. Zhang L, Wu X (2005) Color demosaicking via directional linear minimum mean square-error estimation. IEEE Trans Image Process 14(12):2167–2178
9. Chen X, Jeon G, Jeong J (2013) Voting-based directional interpolation method and its application to still color image demosaicking. IEEE Trans Circ Syst Video Technol 24(2):255–262
10. Chen H, Cheng Y (2012) VLSI implementation of color interpolation in color difference spaces. In: IEEE international symposium on circuits and systems, pp 1680–1683
11. Chen SL, Chang HR (2015) Fully pipelined low-cost and high-quality color demosaicking VLSI design for real-time video applications. IEEE Trans Circ Syst II Express Briefs 62(6):588–592
12. Behrens A, Bommes M, Gross S, Aach T (2011) Image quality assessment of endoscopic panorama images. In: 18th IEEE international conference on image processing, pp 3113–3116
13. Devaraj SA, William BS (2017) Image steganography based on nonlinear chaotic algorithm. Int J Adv Res Innov Discov Eng Appl 2(2):1–8
14. Jasmine AM, Devaraj SA (2015) A novel processing chain for shadow detection and pixel restoration in high resolution satellite images using image imposing. Aust J Basic Appl Sci 9(16):216–223

DNN-Based Decision Support System for ECG Abnormalities

S. Durga, Esther Daniel, and S. Deepakanmani

1 Introduction

Medical tests results were waited on for days which then had to be interpreted based on the diagnosis. With the onset of IoT devices, there has been an introduction of machines based on artificial intelligence, which make use of continuous monitoring to aid disease detection, alerting the care holders or doctors through an alert system [1]. Moreover, these devices can also aid in decision-making through a decision support system (DSS). A major benefit from this transformation was the shift of tasks from a manual, hectic, and time-consuming methodology to a smarter, automated, and time-efficient one. Also, there were instances were medical practitioners weren't able to attend to patients due to lack of awareness about emergency cases, leading to fatal decisions, even death. These machines are trained using special AI algorithms, better known as machine learning and deep learning algorithms, where these algorithms once fed in the machines configure them, pushing them toward automation. Various innovative architectures have been brought forth into the picture through the usage of these algorithms. Successful prediction models with utmost precision and accuracy exist using these algorithms as their foundations.

There exist millions of patients across the world suffering from various diseases purely due to lack of availability and awareness of manpower in emergency cases [2–4]. People, now being busy with their everyday schedule, find it difficult to get proper heart monitoring and diagnosis for their loved ones in emergency calls. Also, in urban regions, quantity of patient outnumbers trained clinicians by a very huge margin, resulting in increased death rates due to unattended cases. So, an attempt

S. Durga (✉) · E. Daniel · S. Deepakanmani
Karunya Institute of Technology and Sciences, Coimbatore, India

A. Haldorai et al. (eds.), *2nd EAI International Conference on Big Data Innovation for Sustainable Cognitive Computing*, EAI/Springer Innovations in Communication and Computing, https://doi.org/10.1007/978-3-030-47560-4_27

to introduce a smarter and easier way to keep in touch with a doctor has been established through a generalized monitoring system framework. Initially, in our proposed system, we capture raw data from any available datasets and structure it. Once done, the trained system looks for the prediction of normal and abnormal conditions of the patient, meanwhile training itself to increase its accuracy rate. Once the prediction is done, the output is projected to the clinician staffs and patient's family through a website. Since healthcare equipment is being pushed toward automation and exposed to Internet connection for diverse data access, the proposed work tends to focus more on IoT healthcare devices.

This work aims at the examination of applying deep neural networks (DNN) [5] for the purpose of enhancing the sustainability in urban healthcare. It focuses on innovative architecture for a sustainable healthcare monitoring system which includes a decision support tool.

This system makes clinicians, staffs, patients, or other individuals aware of the patient's medical history, intelligently filtered or presented information at appropriate times, to improve healthcare systems. This tool includes computerized alerts and reminders to healthcare providers and patients. Apart from this, condition-specific order sets focused on patient data reports and summaries, documentation templates, and contextually relevant reference information are also made available among other tools.

Section 2 describes the literature review of the related work. Section 3 explains the detailed architecture of the proposed system. The details of the implementation and performance evaluations are given in Sect. 4. Section 5 concludes the paper with future research directions.

2 Literature Review

The electrocardiogram (ECG) signal is widely used to detect heart-related diseases. In 2017, a design was proposed to create a healthcare system in Internet of Things (IoT) using network layer system and computer software, which is the basic design of IoT healthcare network (IoThNet) [1, 6]. It talks about the introduction of several wearable devices for IoT and their applications in healthcare sectors. The authors [7] focus on the advanced and improved machine technologies which lead IoT to a new era, where remote control-based smart technology was introduced. A solution to diagnose the disease by analyzing the pattern found in the data using classification analyzing by decision tree and Naïve Bayes algorithms was found out in 2015 [6]. The risk level of the heart disease was predicted using an efficient clinical decision support system [8]. The empirical comparison of thyroid cancer data have been performed using C4.5 and C5.0 data mining algorithms [9]. The accuracy rate was computed to be 71.4% using the C4.5 algorithm. They also compare the performance of different types of algorithms. Deep learning-based cardiac arrhythmia detection based on long-duration electrocardiography (ECG) signal analysis has been proposed in the year 2018 [10].

Heart arrhythmia classification [11] using various machine learning techniques were discussed by Soman et al. The performance results of the machine learning- and deep learning-employed algorithms vary, depending on the dataset type, dataset size, and the impact of the skills of the algorithms. Recently, deep neural network (DNN) has witnessed the greater performance in the area of deep learning.

DNN uses large amounts of data to "train" the computer by labeling each case according to one of many predefined abnormalities, allowing the computer to discern what characteristics of ECGs are associated with any given abnormality [12]. This paper analyzes the performance of various machine and deep learning algorithms for classifying the ECG signal as abnormal and normal signals.

3 DNN-Based Decision Support System

This section describes the proposed DNN-based decision support system. The architecture of the DNN-based decision support system consists of four major modules, viz., (1) raw data input collection, (2) dataset structuring and manipulation section, (3) dataset training, and (4) testing. Figure 1 shows the architecture of DNN-based decision support system. The architecture describes a collection of processes that facilitate the designing, development, implementation, and maintenance of hospital data management systems. It includes the database systems that meet the

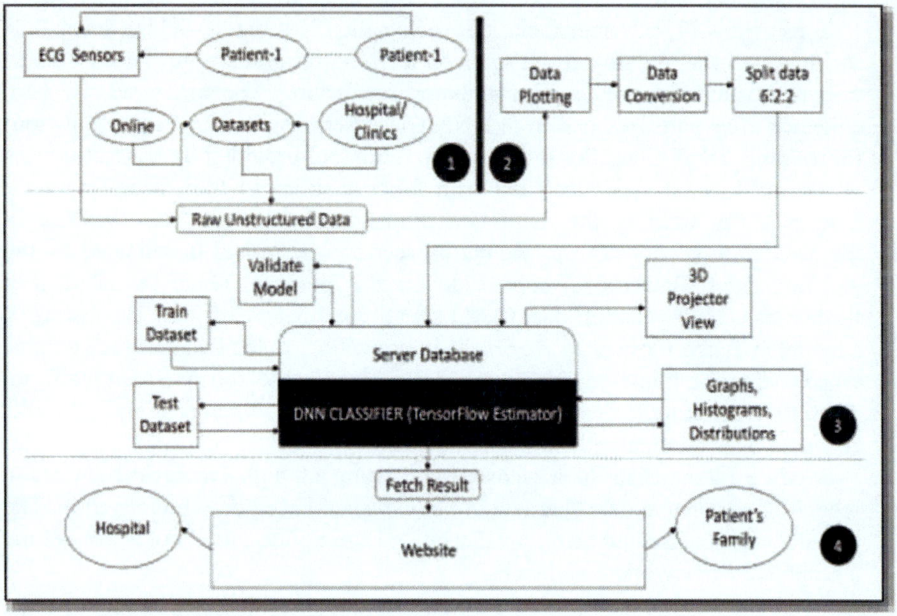

Fig. 1 Architecture of DNN-based decision support system

requirements of the hospitals and have high performance. The main objective of database designing is to produce physical design models of the proposed system.

The modules of the proposed decision support system are explained in detail below:

1. *Raw data input collection*—This module illustrates the process of preparing the dataset. Through ECG datasets (both normal and diseased) collected online from the hospital/general database, raw unstructured data is given as input to the server database for further rearrangement and model training. These datasets are generally in waveform format and hence are imported to the model via WFDB package.
2. *Dataset structuring and manipulation*—Module 2 consists of plotting the raw data collected graphically for visualization purposes. This ensures that the datasets collected are authentic and error-free. Once datasets are collected effectively, raw data is shuffled, converted, and structured as per the training model's framework. Once conversion is complete, dataset is split in the ratio of 6:2:2, for training (6), validation (2), and testing (2). The split data is stored separately in distinguished CSV files and is forwarded to module 3.
3. *Dataset training*—This is the most important section of our proposed system. Here, the server uses a deep learning algorithm to train the dataset for disease prediction/monitoring part. Deep neural networks consists of a set of algorithms, used to recognize a pattern modeled loosely after the human brain. DNN Classifier is a classification tool used to train neural networks in order to distinguish correlation between labels and data in a dataset. This is known as supervised learning.

   TensorFlow [7] provides a class of tools called Estimator—a high-level TensorFlow API that greatly simplifies deep learning programming. It encapsulates training, evaluation, prediction, and export for serving. The structured raw data, collected after partition, is sent to DNN Classifier provided by TensorFlow tool for training. DNN Classifier is a model training tool provided by Estimator class of TensorFlow—a class designed with functionalities to train neural network models. After training the dataset, a model is checked for its efficiency in the Validation section. In the Validation section, the model is validated by the partitioned test data (20%) kept aside for this purpose. Once validated, it is further tested using the test data (20%) kept aside during partition. The testing of a model in TensorFlow platform results in projection of detailed accuracy graphs. Once testing is completed, the outputs can be seen in the Tensor Board platform.
4. *Testing*—This module describes the execution section, which ensures the results accessed over the Internet using user devices.

   Module (4) consists of a website, consisting of both Doctor/Patient login, where the respective dashboards will be displayed based on who logged in. The website was constructed using JavaScript and a template called flot.js, for getting a graphical framework-based website.

The pseudo-code of the proposed DNN-based decision support system is given below.

Pseudo Code of DNN Based Decision Support System
Input: DNN Trained Model + Test Dataset
Output: Prediction Results

```
Step 1: data = Test dataset
Step 2: while (missing values)
    Step 2.1: if it is unstructured nominal then
            Step 2.1.1: dataset=Validation dataset
    Step 2.2: else
            Step 2.2.1: A=fetch (previous value of the specific missing value)
            Step 2.2.2: B=fetch (next value of the specific missing value)
            Step 2.2.4: result=round ((A+B)/2).
            Step 2.2.5: data=result.
            end if
    end while
Step 3: while (structured dataset)
  Step 3.1: dataset=DNN Classifier (Trained Model)
  Step 3.2: activate prediction class
Step 4: if expected values are not close to predicted values
  Step 4.1: total wrongs = total wrongs + 1
Step 5: Accurate = ((total values – total wrongs)*100/total values)
Step 6: if predict = diseased
  Step 6.1: Find disease class
    Step 6.2: Output=diseased
Step 7: else
  Step 6.1: Output=normal
Step 8: Final Output=Output, Accurate, Total Wrongs
Step 8: TensorBoard=Final Output
Step 9: Send to user (website)
```

The website is merged with TensorBoard in order to allow user navigation from the website to TensorFlow output window. Also, the website has been embedded with multiple other user interaction structures of features as follows:

Mailbox: For frequent communication with the doctor
Pulse Statistics: For pulse readings monitoring
Report Charts: For regular chart creation purpose
Profile Update: For updating user authentication and login details
Miscellaneous: Includes location, chatbox, maps, and other information

Doctor's Dashboard consists of statistical information of all the patients under their care and treatment. Statistical data includes patient's ID, name, age, contact info, diagnosed condition, monitoring stats, date, and present status. Every patient is directly hyperlinked to their own dashboard, enabling the doctor to easily access patient's dashboard and monitor their current status in detail. It also consists of mailbox and notification section to keep the doctor updated.

4 Performance Evaluation

The proposed system is analyzed using TensorFlow on a window system with core i5 processor 2.40 GHz and 4 GB RAM. The system is developed with Agile Software Development Model. Agile Software Development (ASD) is an umbrella term for a set of methods and practices based on the values and principles expressed in the Agile Manifesto. As discussed in the previous section, the proposed system includes the web portal for patient data monitoring and management. Figure 2 shows the doctor's home page to monitor the statistics of ECG signals.

The performance of the proposed DNN-based ECG classification system is compared with other state-of-the-art algorithms such as Naïve, SVM, 2DCNN, RNN, and KNN. The effectiveness of the proposed system is analyzed in terms of accuracy and sensitivity parameters. The performance parameters are defined as follows:

1. *Accuracy percentage*—no. of correct predictions made divided by the total number of predictions made, multiplied by 100 to turn it into a percentage.
2. *Sensitivity*—used for examining the partial derivatives of the outputs with respect to the inputs. Say the output vector $y \in R_i$ is given by $y = f(x)$, where $x \in R_j$ is the input vector and f is the function the network implements.

Figure 3 shows the performance comparisons in terms of accuracy percentage. The overall ECG classification assessment results are obtained by considering the 191 training, testing, and the whole dataset [13]. The results presented in Fig. 3 confirm that the proposed DNN-based system achieved superior performance than other available models.

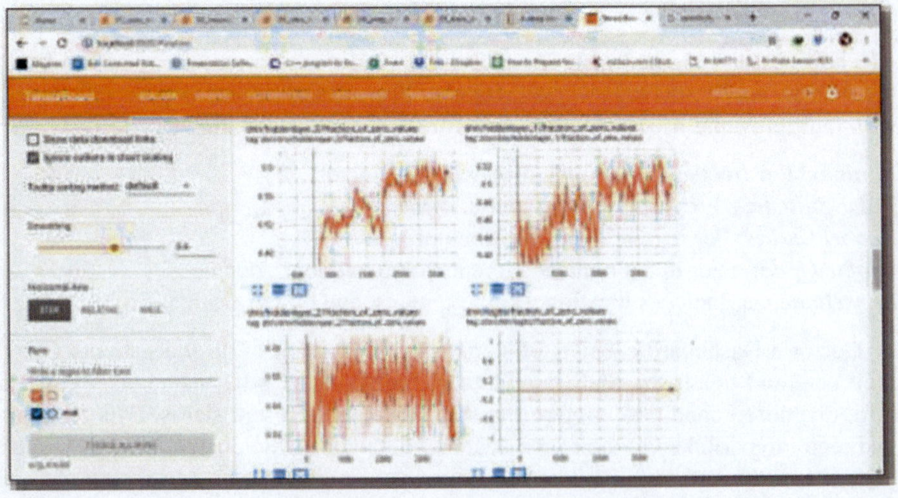

Fig. 2 ECG statistics

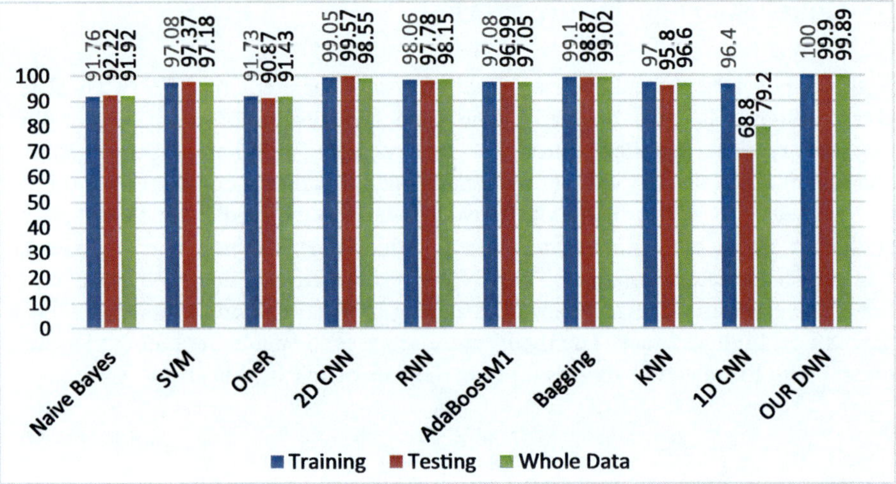

Fig. 3 Accuracy comparison between state-of-the-art model and proposed system for the ECG dataset

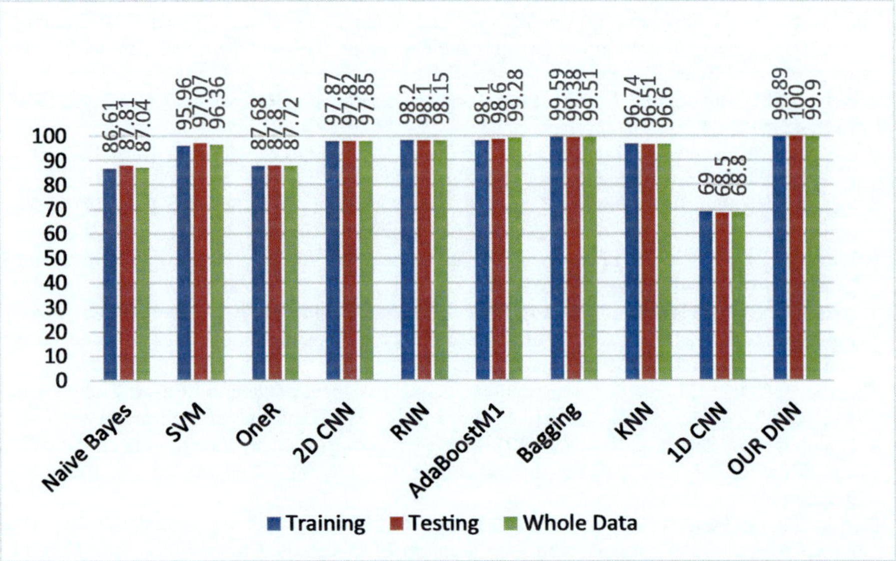

Fig. 4 Sensitivity comparison between state-of-the-art models and proposed system

The sensitivity comparisons between the proposed DNN-based system and other available models are depicted in Fig. 4. It is evident from the figure that the proposed system achieves 99.89% sensitivity compared to other state-of-the-art models. From the above comparisons with available models, it can be concluded that our DNN support system model attains the most efficient and accurate ECG classifications.

5 Conclusion and Future Work

In this paper, DNN-based decision-making system for predicting the ECG-related abnormalities has been analyzed. The proposed system is composed of four modules: raw data input collection, dataset structuring and manipulation, training, and testing. This system will surely benefit doctors, clinical companies, hospitals, patients, etc. and may provide vital breakthroughs to the medical and engineering students working in the field of deep learning tied along with practical implementations of sensor data acquisition. The DNN-based system was analyzed using tensor flow. The experiment results showed that the proposed system produces low error percentage, high accuracy. The main future scope is to enable the proposed system to deal with live data IoT data as input in the area of IoT healthcare.

References

1. Mutlag AA, Ghani MKA, Arunkumar N, Mohammed MA, Mohd O (2019) Enabling technologies for fog computing in healthcare IoT systems. Future Gener Comput Syst 90:62–78
2. Jayaratne M, Nallaperuma D, De Silva D, Alahakoon D, Devitt B, Webster K, Chilamkurti N (2019) A data integration platform for patient-centered e-healthcare and clinical decision support. Futur Gener Comput Syst 92:996–1008
3. Begum SFU, Begum I (2017) Smart health care solutions using IOT, vol 4(3), Mar 2017. ISSN: 2348-4845
4. Esteva A, Robicquet A, Ramsundar B, Kuleshov V, DePristo M, Chou K, Dean J (2019) A guide to deep learning in healthcare. Nat Med 25(1):24
5. Acharya UR, Oh SL, Hagiwara Y, Tan JH, Adam M, Gertych A, San Tan R (2017) A deep convolutional neural network model to classify heartbeats. Comput Biol Med 89:389–396
6. Iyer A, Sumbaly R (2015) Diagnosis of diabetes using classification mining techniques. Int J Data Mining Knowl Manag Proc 5:1–14
7. Abadi M, Barham P, Chen J, Chen Z, Davis A, Dean J, Kudlur M (2016) Tensorflow: a system for large-scale machine learning. In: 12th USENIX symposium on operating systems design and implementation (OSDI 16), pp 265–283
8. Anooj P (2012) Clinical decision support system: risk level prediction of heart disease using weighted fuzzy rules and decision tree rules. J King Saud Univ Comput Inform Sci 24:27–40
9. Upadhayay A, Shukla S, Kumar S (2013) Empirical comparison by data mining classification algorithms (C 4.5 & C 5.0) for thyroid cancer data set. Int J Comput Sci Commun Netw 3(1):64–68
10. Yıldırım Ö, Pławiak P, Tan RS, Acharya UR (2018) Arrhythmia detection using deep convolutional neural network with long duration ECG signals. Comput Biol Med 102:411–420
11. Soman T, Bobbie PO (2005) Classification of arrhythmia using machine learning techniques. WSEAS Trans Comput 4(6):548–552
12. Hung CY, Chen WC, Lai PT, Lin CH, Lee CC (2017) Comparing deep neural network and other machine learning algorithms for stroke prediction in a large-scale population-based electronic medical claims database. In: 39th Annual international conference of the IEEE Engineering in Medicine and Biology Society (EMBC), Jul 2017. IEEE, pp 3110–3113
13. Dataset Link: https://physionet.org/physiobank/database/

A Hybrid Artificial Bee Colony Algorithmic Approach for Classification Using Neural Networks

C. Mala, Vishnu Deepak, Sidharth Prakash, and Surya Lashmi Srinivasan

1 Introduction

Since artificial neural networks are utilised today in most fields of study, it is of utmost importance that they produce the best-quality solutions in the least amount of training time. In many problems, the achieved accuracy and reliability of the trained network in a given number of epochs depend heavily on the initialisation values chosen before training. As found by Du et al. [1], escaping saddle points is a major concern for efficient training of neural networks.

Also, as shown by Choromanska et al. [2], the probability on recovering a local minimum of poor quality for small neural networks is non-zero. This constraint can be alleviated by using approaches that incorporate global random search into the training process. This factor adds the capability to look beyond the immediate local solution concerning the process of training ANNs and gives room for a more comprehensive search of the solution space. The Artificial Bee Colony (ABC) algorithm [3] proposed by Karaboga and Basturk can be used for this purpose by setting the loss function of the neural network to be the objective function to be optimised by the bee colony while exploring. The algorithm mimics the foraging behaviour of honeybees that find the best quality of nectar using swarm intelligence properties. First, an initial population is sent out randomly to explore the search space. Then, these 'employee bees' return to the hive and notify the 'onlooker bees' of the quality of nectar they have found. The onlooker bees then seek to find better solutions in the vicinity of the best-quality nectar found by the employees, ensuring an effective investigation of the space in areas with a higher probability of finding a better solution. Once a set number of trials to find better-quality solutions

C. Mala (✉) · V. Deepak · S. Prakash · S. Lashmi Srinivasan
National Institute of Technology, Tiruchirappalli, Tamil Nadu, India

A. Haldorai et al. (eds.), *2nd EAI International Conference on Big Data Innovation for Sustainable Cognitive Computing*, EAI/Springer Innovations in Communication and Computing, https://doi.org/10.1007/978-3-030-47560-4_28

get over, the bee is converted into a scout bee which explores the solution space unconstrained, yet again. In the context of a neural net, this random search allows it to be free of the initial weight values and allows it to converge to the best results, within the least time. In this context, two algorithms are proposed (neural-modified ABC and layered-hybrid ABC) which have been tested on standard classification problems of increasing complexity and have outperformed the standard stochastic gradient methods and the base ABC algorithm in all cases. Further, an innovative approach to hyperparameter optimisation has also been proposed. Classification is one of the most useful tasks which can be performed by neural networks and presents a simple method to evaluate the efficiency of the proposed algorithms in the context of high dimensionality problems, and hence it has been chosen to test the algorithms. Classification problems also often suffer from getting stuck at local minima or saddle points and not being able to progress towards the global best solution, which is easily done by using the ABC algorithm. Conventional techniques often cannot distinguish between different minima, but the random global search capability of the ABC algorithm introduces a powerful modification to the way neural network classification works.

2 Related Works

Since its inception, there has been steady research conducted in the area of the artificial bee colony algorithm. Several variations with innovative modifications have been proposed to improve upon its performance.

Karaboga and Basturk [3] proposed the original ABC algorithm in 2007, which sparked renewed interest in the field of meta-heuristic optimisation. The basic algorithm was shown to outperform other algorithms such as genetic algorithm (GA) [4], particle swarm optimisation (PSO) [5] and others of its class.

The ABC algorithm when applied to several real-world problems showed enhanced results, and its performance in each context has been extensively studied. In [6] ABC was seen to provide the best optimum solution for the minimum spanning tree problem as compared to other methods. When applied to the travelling salesman problem [7], ABC produced results that were at par or slightly better than similar algorithms in almost all cases. The application of ABC to the general assignment problem was studied in depth in [8] by Baykosoglu et al. which provided good results. Liu and Tang [9] in 2018 utilised the ABC algorithm in the field of image processing by converting edge detection into an optimisation problem which was handled extremely well by the ABC algorithm. Chen et al. [10] applied the ABC algorithm to blind source separation of chaotic signals. The algorithm was successfully able to provide separation between the non-linear and non-Gaussian inputs as compared to the traditional independent component analysis method. In [11] Koylu has applied ABC successfully to data mining in the context of rule learning to facilitate online data streaming.

In 2011, Ozturk and Karaboga proposed to use ABC in the problem of clustering [12] and executed it successfully. In [13] Kiran et al. proposed using five different update equations for the bees in order to balance the global search capabilities of the colony with the local. However its effectiveness was not evaluated in the context of a neural network. In [14] Quan and Shi proposed using the approach of contractive mapping to enhance convergence speed at the cost of reduced global search capability. Banharnsakun et al. [15] devised a modification to ABC which biases towards the best-so-far found solution by transmitting this value globally to all the bees. However, this reduces the random search capability of the hive and hence defeats the purpose of usage in the context of neural networks.

Akay and Karaboga [16] proposed the concept of 'modification rate' or MR which is a parameter that is used to decide whether a certain component of the solution would be affected by the solution mutation function or not. However, in the case of neural networks, since almost all complex problems have relatively high dimensionality, having the mutation affect all components every time is the ideal choice. Karaboga and Gorkemli in 2014 formulated qABC [17] which proposes a different mutation function for onlooker bees by defining a 'neighbourhood' from which the mutation factor must be chosen. While this can increase convergence speed, it does so at the sacrifice of global search ability. In [18] Zhang et al. detailed a modified ABC algorithm to better design electromagnetic devices using an inheritance mechanism for solution mutation. Liu et al. [19] proposed an approach incorporating the concept of mutual learning to make sure that mutated solutions always have better fitness values.

This can be counterproductive in the case of neural networks, since the progression of fitness values in training need not always be a linear process while going from a local minima to a global minima.

Gao et al. in 2013 proposed an orthogonal learning strategy [20], built on top of the best-so-far framework in [15], which enhances the solution quality and converging speed but adds an excessive amount of computational overhead to the algorithm. The idea of parallelisation was explored by Narasimhan [21] in their formulation of PABC by using a shared memory architecture. However, the performance in the context of a neural network was not explored. In [22] Taspmar and Yildmm employed parallelisation of ABC on the peak-to-average power ratio problem, achieving exceptional results. The application of training neural networks was explored initially by Karaboga et al. [23] in 2007 and was successfully done so for low dimensionality problems. Next, a hybrid approach algorithm was applied to the complex problem to train neural networks in [24]. The approach used combined ABC with Levenberg-Marquardt (LM) algorithm which showed good results. However, it was only tested against problems of low dimensionality (3-bit parity problem and XOR problem).

Numerous improvements have been made consistently to effectively enhance the performance exhibited by the algorithm. However, most of these improvements do not apply when the problem complexity goes up (as in the case of neural networks). This paper aims to present a comprehensive study of the applicability and feasibility of using the ABC algorithm with neural network appropriate modifications (NMABC) and a novel layered-hybrid approach (LHABC) which can be extensively parallelised to speed up the search process. To test the proposed algorithms to their limits, the extremely complex problem of image colourisation has been chosen.

The proposed algorithms are tested on three different benchmark classification datasets of increasing complexity. Comparisons between the different algorithms are made on the basis of accuracy and loss for each application, and the final results and concluding remarks are presented along with an optimisation approach to hyperparameter tuning. The architecture proposed by Zhang et al. [25] has been chosen for this purpose over the ones proposed by Hu and Li [26] and Chen et al. [27] due to its non-requirement of human interaction and ease of evaluation (Table 1).

Table 1 Classifying conventional and proposed artificial bee colony algorithmic models (proposed algorithms)

Algorithmic model (year)	Enhancement achieved	Drawbacks/improvements to be made
ABC (2007)	Breakthrough work which surpassed similar algorithmic models like GA and PSO	Basic algorithm with no enhancements
Best-so-far ABC (2011)	Faster convergence by biassing towards the best-found solution so far	Reduces random global search capability
Modification rate-based ABC (2012)	More granular control over search space	Limiting search space is counterproductive when dealing with higher dimensions
qABC (2014)	Proposes specific neighbourhoods for each bee	Reduces random global search capability
Mutual learning-based ABC (2012)	Makes sure mutations always have better fitness values	Not applicable to neural networks since the loss functions need not decrease monotonically
Orthogonal learning-based ABC (2013)	Enhanced solution quality and convergence speed	Added extra computational overhead which becomes significant in the scale of large neural networks
PABC (2009)	Improvement achieved with shared memory architecture	Not tested on complex problems
NMABC	Specific neural network appropriate adjustments are made to mutation function, and gradual scale-up methodology is used for optimum performance	Not applicable to high dimensionality problems such as large neural networks
LHABC	Successfully incorporates the best of random search capability of ABC and gradient descent algorithms in a layered approach to offer a highly parallelizable solution	May not always offer a significantly better solution than standard gradient descent

3 Proposed Algorithms

This section begins with the explanation of the basic ABC algorithm and general classification and then proceeds to detail the two proposed modified versions.

3.1 *Algorithm: Base ABC*

The original algorithm is completely based on the foraging behaviour of honeybees. These bees are represented as solution vectors in our search space. The first step is to define the objective function (i.e.to be minimised or maximised) and the constraints of the search space (maximum and minimum limits that a bee can search). Then, the colony is initialised based on the *colony_size* parameter in which usually half the population is set as employee bees and the other half is set as worker bees. Each bee now evaluates the fitness of the solution it has found and remembers the best solution. A *max_trials* value is also set, which controls the number of times a bee checks around a given solution point before abandoning it and turning into a scout bee. Next, based on the number of iterations, each worker bee moves to a random location within the vicinity of an employee bee with a probability proportional to the quality of the solution (*fitness_value*). Each employee bee then mutates the current solution based on the following equation if the *max_trials* value for that bee has not been reached:

$$v_{mi} = x_{mi} + \varphi_{mi} (x_{mi} - x_{mk}) \quad (1)$$

where v_{mi} is the mutated solution for component i of bee m, x_{mi} is the original value of the solution, x_{mk} is a random component k of bee m and φ_{mi} is a random number between -1 and 1. If the *max_iteration* value has been crossed, that bee is converted into a scout bee which is re-initialised to a random location within the search space, and the whole process continues until the number of iterations is done. The process is represented in the form of a flow diagram in Fig. 1, detailing the iterative decision-making structure of the algorithm. The final food position represents the best function value found by the bees.

3.2 *Algorithm: Classification in Neural Networks*

Classification is a classic use case for neural networks which has been researched upon for several decades. The neural network in this case would represent a set of 'neurons' which are activated depending upon the trained weights acquired during the training phase. The neural network is set up as follows:

1. Decide on the number of hidden layers and number of neurons in each layer to set up the architecture of the neural network.

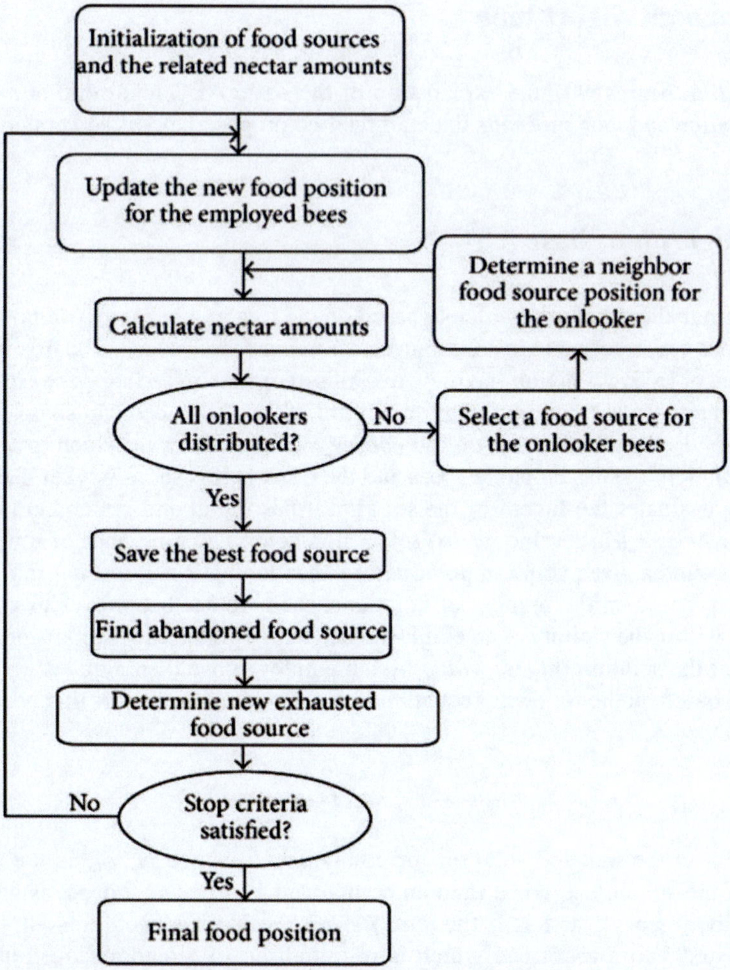

Fig. 1 Flowchart of ABC algorithm [28]

2. Train the neural network by using the method of backpropagation with training data.
3. Validate the accuracy of the neural network using the validation set, and run for more epochs until suitable accuracy is reached.

An example of the neural network architecture and class labels is given in Sect. 4.1 for the Iris dataset.

3.3 Algorithm: Neural-Modified ABC (NMABC)

The base algorithm does the job of searching and finding the global optima of the solution space very efficiently for generic problems. However, in the case of a neural network, the completely random initialisation and constraints imposed by the algorithm may work against finding the best solution in the lowest possible time. Hence the following modifications to the base algorithm are proposed:

- Have each bee initialise their values for the first time according to the normal neural network initialisation facilitated by the network compilation. This would end up having all weights initialised to very small values and will set biases to 0, which has been statistically proven to be the best method to start training.
- Unlike objective functions which have predetermined search spaces, the weights in neural networks completely depend on the problem type. This leads to situations where they can have very large or very small values. However, ABC performs better in smaller search spaces. To combine the best of both worlds, a search space-modifying algorithm is implemented as follows:

  $n_i = Num\_iterations,\ minf = Min\_function\_val$
  $maxf = Max\_function\_val,\ rf = Range\_factor$

  For i in range (n_i):

  - Carry out employee bees phase and onlooker bees phase
  - $minf = minf - rf$
  - $maxf = maxf + rf$

  Thus, the search space is widened from an initial value to a bigger range with each iteration that takes place, allowing the bees to focus on local solutions initially and then gradually scale up to a bigger search space.
- For high dimensionality solution spaces, modifying just one component of the solution vector is often not enough to converge to a solution in a smaller number of iterations. Hence the solution mutation equation is changed to:

$$v_m = x_m + \varphi_m (x_m - x_k) \quad (2)$$

  where

  - v_m refers to all the components of the mutated solution m
  - x_m refers to the current location of the bee
  - φ_m is a random value between -1 and $+1$
  - x_k is a randomly selected position of a bee from the colony

3.4 *Algorithm: Layered-Hybrid ABC (LHABC)*

The algorithm discussed above works very well in the case of low dimensionality problems to give fast convergence. However, as the problem complexity keeps increasing, the number of iterations required for the bee to find a high-quality solution vector goes up exponentially. It becomes mathematically infeasible to find solutions having the same or better quality than stochastic gradient descent. Hence, a hybrid approach algorithm that combines the best features of Artificial Bee Colony optimisation and stochastic gradient descent is proposed. A layered approach to the problem is adopted by adding stochastic gradient descent to the natural behaviour of the bees. Hence, each bee will compute its solution quality based on the metrics evaluated after applying the gradient descent algorithm to different solutions found by each bee and will then choose the best of them. The process continues until *Num_iterations* has been reached. The behaviour has been detailed as follows:

For *i* in range (*Num_iterations*):

1. Initialise/mutate bee position solution.
2. Run stochastic gradient descent on all the bees.
3. Evaluate the quality of solutions, and update the solution of each bee if the quality of the new solution is higher.
4. Add the *best_solution* found among all the bees to *optimal_solution_array* in the *i*th position.
5. *Num_iterations* = *Num_iterations* + 1.

In this manner, the bees first use their power of global search to find initialisation values, then apply gradient descent from all the different points initially found, evaluate their position quality and repeat until the number of iterations is satisfied. This approach effectively merges the global search capability of the bee colony algorithm together with the fast convergence of stochastic gradient descent to give better results in high dimensionality problems.

This assures that each iteration sees all weights (solution components) of the bee changing at once, which results in faster convergence and better solutions.

4 Simulation and Performance Analysis

A detailed analysis of the proposed algorithms is now presented based on problems with increasing levels of complexity formulated using standard benchmark datasets which are publicly available. All simulations were run using the Google Colab runtime environment which provides an NVIDIA Tesla K80 GPU, and all figures were made using Microsoft Excel.

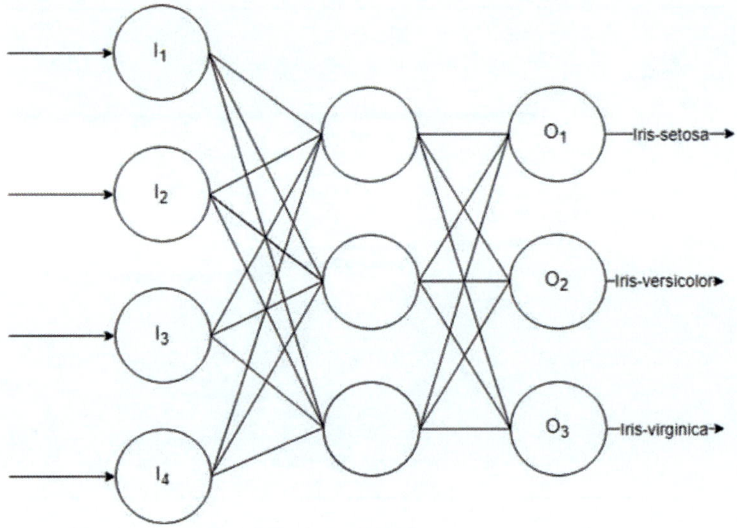

Fig. 2 The model used for classification for the Iris dataset

4.1 Iris Dataset

This set provides four measurements of a flower, based on which it has to be classified into one of three classes. The network was trained on 100 randomly selected samples and was tested on the 50 that remained. The network architecture for this neural net is constructed as per Fig. 2. The four inputs represent the length and width of petals and sepals of different Irises. This is connected to a three-neuron hidden layer which is in turn connected to a three-neuron output layer which is activated using the Softmax function to give class probabilities.

Testing ABC, NMABC and LHABC on Iris Dataset

The three optimisation techniques, discussed previously, are tested on the Iris network. The loss and accuracy are measured for the three optimisation techniques, and a graph is plotted. It is observed that NMABC and LHABC greatly outperform the basic implementation in Fig. 3. This is due to the availability of an increased search space and more appropriate initialisation values.

Similarly, in Fig. 4, both NMABC and LHABC reach higher values of accuracy faster than the basic algorithm. NMABC is able to achieve the highest peak accuracy owing to the fact that the hybrid approach compromises on the extent of random search capability of the bees in exchange for faster convergence. Also, LHABC requires far more parallel processing power to compute the results as shown in a similar time duration.

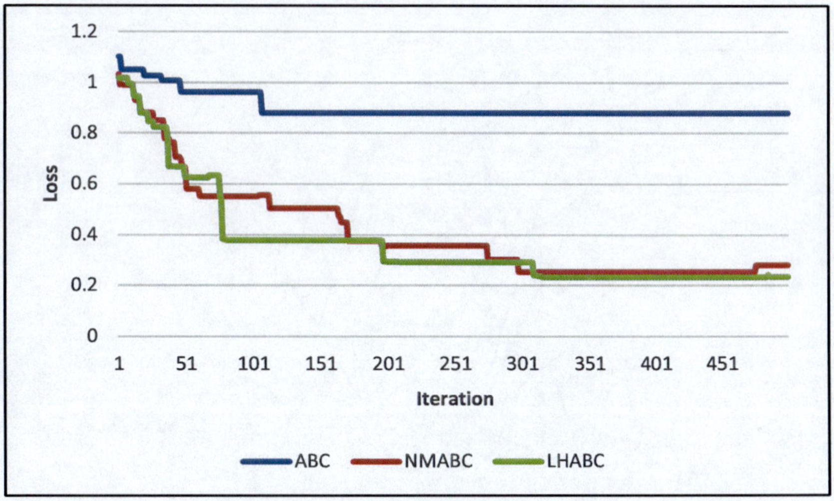

Fig. 3 Comparing loss from base and proposed algorithms on Iris dataset

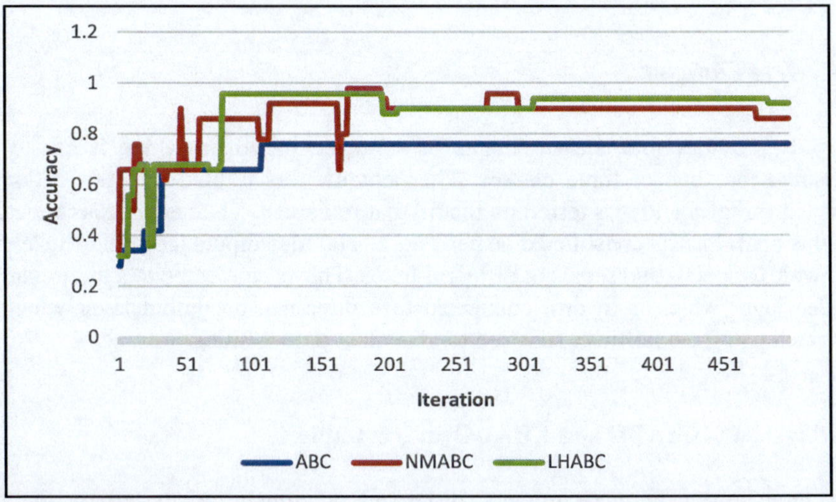

Fig. 4 Comparing accuracy from base and proposed algorithms on Iris dataset

The combined results are presented in Table 2, which summarises the behaviour of the algorithms. Hence, it can be concluded that NMABC is more suitable for problems of low dimensionality.

Table 2 Peak values for proposed algorithms compared to base algorithm for Iris dataset classification

Algorithm	Lowest loss	Peak accuracy (%)
ABC	0.8805	75.99
NMABC	0.2799	98
LHABC	0.2335	95.99

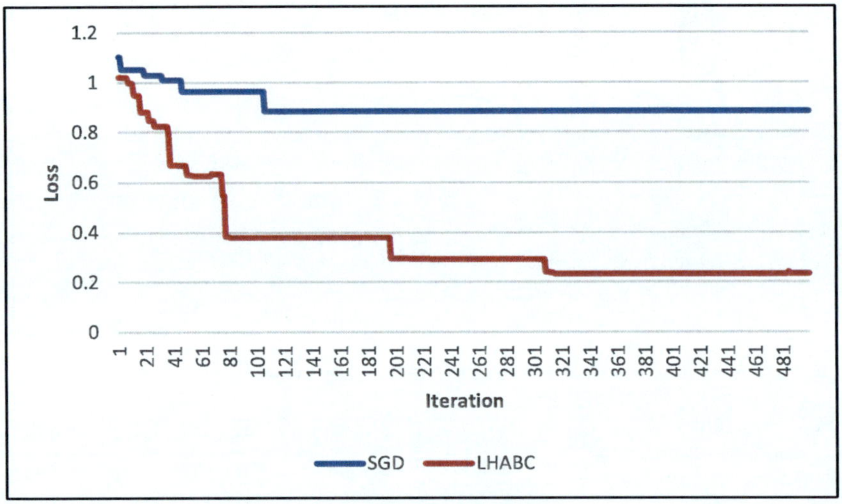

Fig. 5 Comparing loss from SGD and LHABC on Iris dataset

Testing SGD and LHABC on Iris Dataset

Next, LHABC is compared with the standard stochastic gradient descent (SGD) algorithm for training. Here, each iteration completed by the bee is equivalent to one epoch of the gradient descent algorithm since the gradient descent function has been incorporated into the behaviour of each bee.

Immediately, it is observed that in Fig. 5, LHABC is extremely effective at minimising loss values in very few iterations. This is owing to the swarm intelligence property of the bee colony working together to find the best possible location for gradient descent.

Again, LHABC greatly outperforms SGD in Fig. 6, especially in the early part of training where finding the best location in the n-dimensional solution space can be critical. LHABC also reaches a much higher peak accuracy with lesser number of iterations/epochs as compared to SGD. The peak obtained values show improvement of almost 10% in accuracy for less than a fifth number of epochs as seen in Table 3.

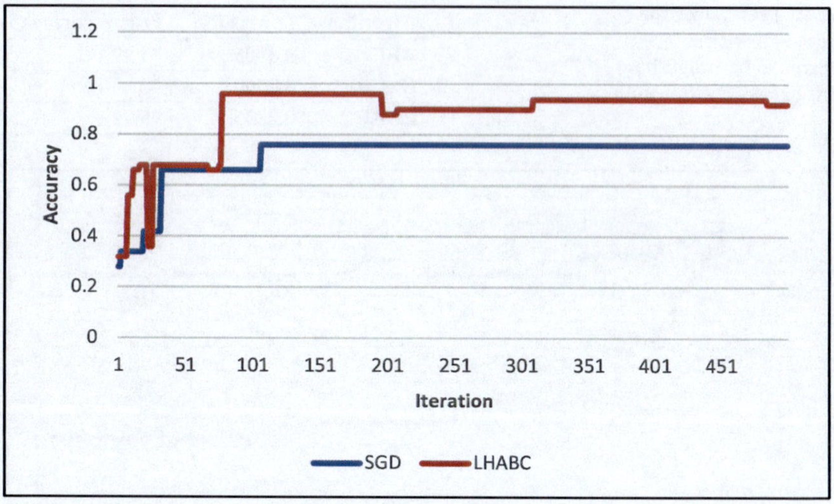

Fig. 6 Comparing accuracy from SGD and LHABC on Iris dataset

Table 3 Peak values for hybrid algorithm compared to gradient descent for Iris dataset classification

Algorithm	Lowest loss	Peak accuracy (%)	Epochs
SGD	0.4888	86	500
LHABC	0.2355	95.99	77

4.2 *MNIST Digit Classification*

MNIST is a popular handwritten digit database with digits 0–9 that are used to evaluate models for classification. As before, the performance of SGD, ABC, NMABC and LHABC is evaluated. The model used is a simple CNN network with one convolution layer followed by two fully connected layers giving 3510 dimensions in total to optimise.

Testing ABC, NMABC and LHABC on MNIST Dataset

Figures 7 and 8 clearly show the difference in capability of the hybrid algorithm as compared to the ones which do not incorporate gradient descent. ABC and NMABC fail to produce meaningful results as the dimensionality of the problem increases, as this results in exponential increases to computation time required to deliver similar results. These results are presented in Table 4, in which it is clearly seen that LHABC greatly outperforms the other algorithms.

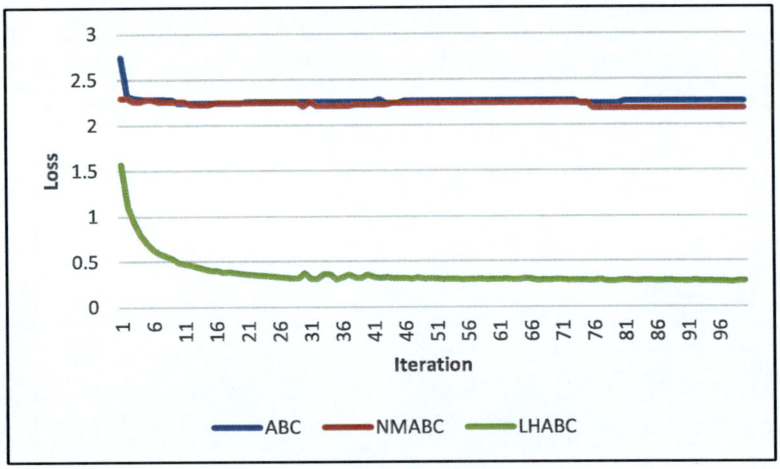

Fig. 7 Comparing loss from ABC, NMABC and LHABC on MNIST dataset

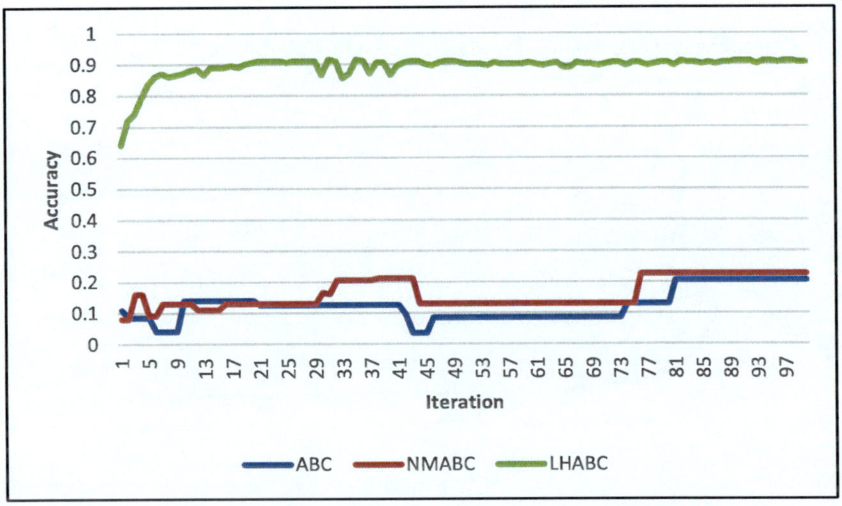

Fig. 8 Comparing accuracy from ABC, NMABC and LHABC on MNIST dataset

Testing SGD and LHABC on MNIST Dataset

When looking at the epochs-wise comparison between SGD and LHABC in Figs. 9 and 10, the proposed algorithm LHABC keeps up with or outperforms SGD at almost every point. The small inconsistencies in accuracy can be attributed to the global search incorporation which provides a higher value towards the end.

Hence, it is concluded that while ABC and NMABC are not viable for high dimensionality problems, LHABC can potentially do as well or even outperform

Table 4 Peak values for hybrid algorithm compared to gradient descent for MNIST dataset classification

Algorithm	Lowest loss	Peak accuracy (%)	Epochs
SGD	0.2444	92.5	79
LHABC	0.2390	93.5	94

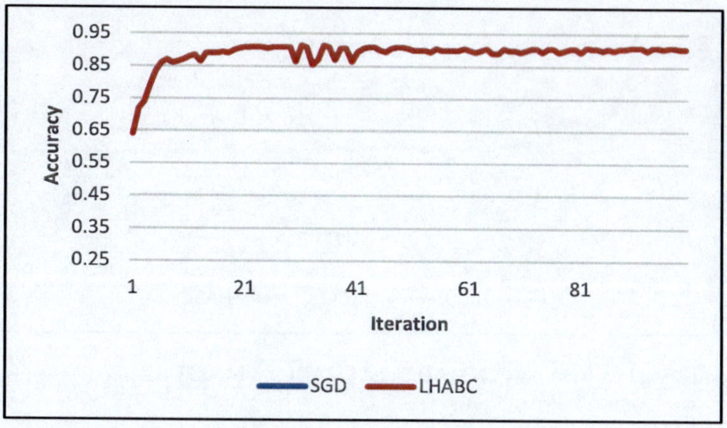

Fig. 9 Comparing accuracy from SGD and LHABC on MNIST dataset

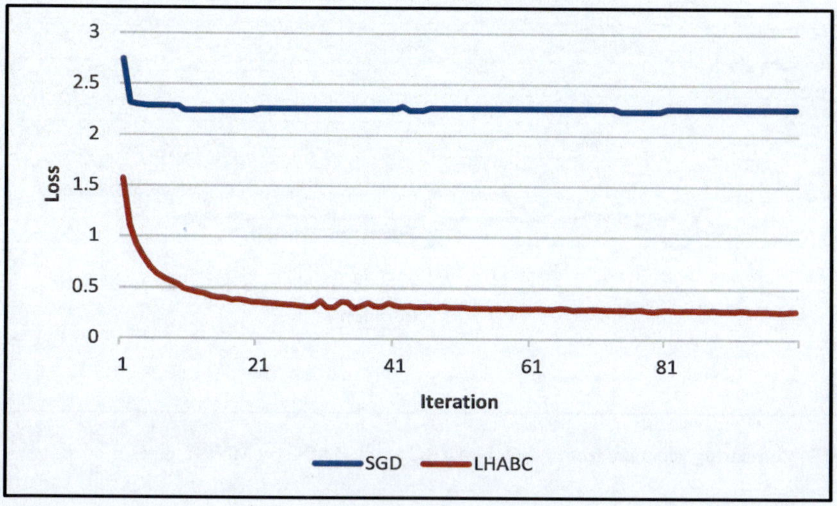

Fig. 10 Comparing loss from SGD and LHABC on MNIST dataset

SGD in almost all cases. In the peak values presented in Table 4, LHABC achieves marginally lower loss and a 1% higher accuracy value within the 100 epochs the experiment was run for. The reason for not witnessing a significant improvement as in the case of Iris is due to the dataset being more straightforward and not presenting many local minima pitfalls where SGD can get trapped.

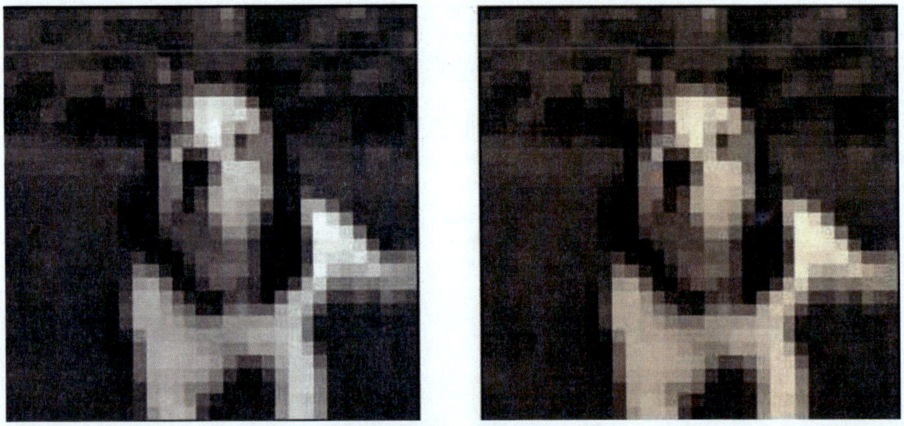

Fig. 11 Example of colourisation based on regression model

4.3 *CIFAR-10 Regression Colouriser*

The next problem the algorithms are applied to is the complex image colourisation task in which a greyscale image is fed in as input and the model predicts how best to colour the scene. The training set and validation set consist of 5000 and 1000 images of dogs, respectively, all sourced from the CIFAR-10 small image dataset. For training, the images are first converted to the CIE-*LAB* colour space, and the L (lightness) channel is separated out to represent the greyscale information. Here A and B are values ranging from -128 to $+127$. A represents the position in the gradient from green (negative) to red (positive), while B corresponds to the position between blue (negative) and yellow (positive). A and B combined act as the target values, which are recombined with L to retrieve the final colourised image. The traditional approach to the problem is to treat it as a regression task. However, this tends to give desaturated, brownish colours (as seen in Fig. 11) and fails to colourise the image properly.

As seen earlier, ABC and NMABC are infeasible when dealing with problems of high dimensionality; hence only the results for SGD and LHABC are compared.

Testing SGD AND LHABC on Regression Classification with CIFAR-10

In Figs. 12 and 13, LHABC reaches the same limiting values of loss and accuracy as SGD does but is able to do so in just one epoch/iteration as compared to SGD. The peak values are presented in Table 5, and it is seen that the number of epochs taken has been significantly reduced. While the quality of the final solution obtained remains the same, LHABC is successfully able to speed up the training process.

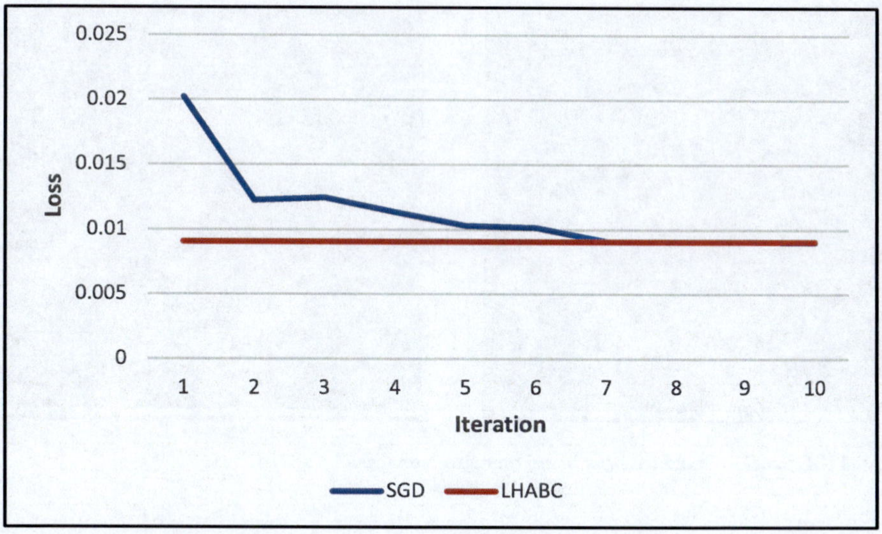

Fig. 12 Comparing loss from SGD and LHABC on CIFAR-10 dataset

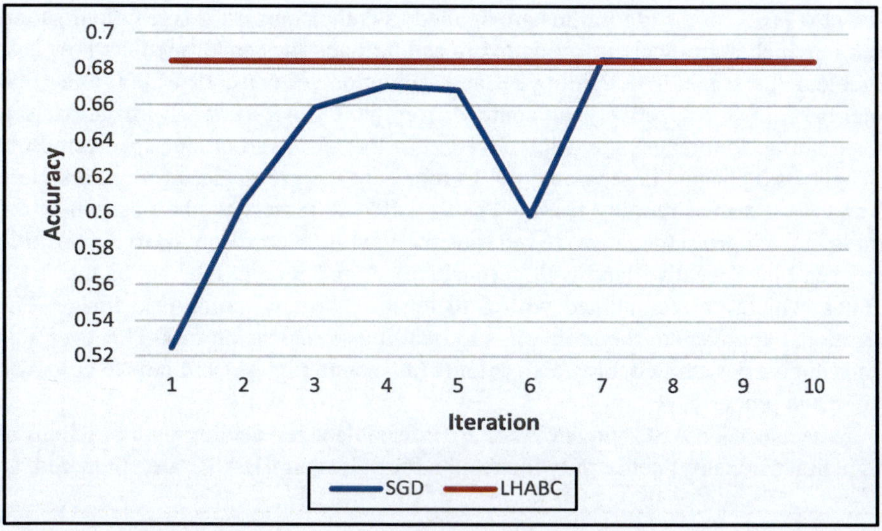

Fig. 13 Comparing accuracy from SGD and LHABC on CIFAR-10 dataset

4.4 CIFAR-10 Multinomial Classification Colouriser

Finding a suitable architecture to model the colourisation problem effectively is a strenuous task. Zhang et al. [25] proposed a multinomial classification approach to the problem which yielded very good results. The model used for this purpose

Table 5 Peak values for hybrid algorithm compared to gradient descent for CIFAR-10 dataset regression colourisation

Algorithm	Lowest loss	Peak accuracy (%)	Epochs
SGD	0.0090	68.45	6
LHABC	0.0091	68.44	1

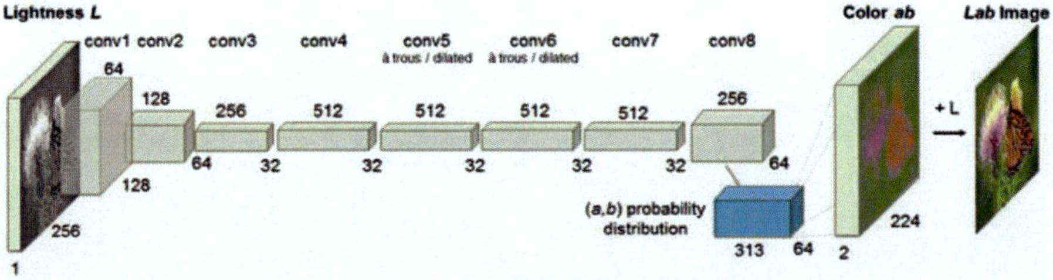

Fig. 14 Architecture of colourful image colourisation [25]

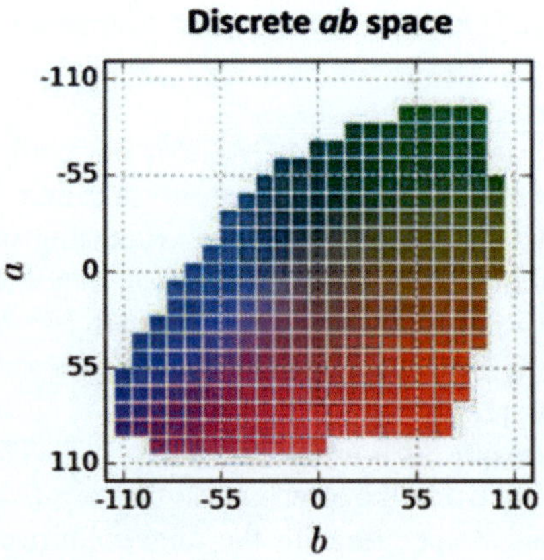

Fig. 15 Quantised *AB* colour space with a grid size of 10 [25]

is presented in Fig. 14. In this approach, the *A* and *B* components of the CIE-*LAB* colour space are quantised and divided into 313 uniform bins of size 10 (as seen in Fig. 15) which are in gamut. The likelihood for each of these bins (classes) is determined by creating a probability distribution for each class based on frequency of observation in a large set of sample photos. The separated *L* channel is fed into the model as input which goes through several blocks of convolution layers with ReLU activation functions. Finally, output values are compared based on Softmax classification probabilities for 313 classes for each pixel of the image. This distribution of *A* and *B* values is then combined with the original lightness (*L*) channel to produce the final colourised image in the post-processing phase.

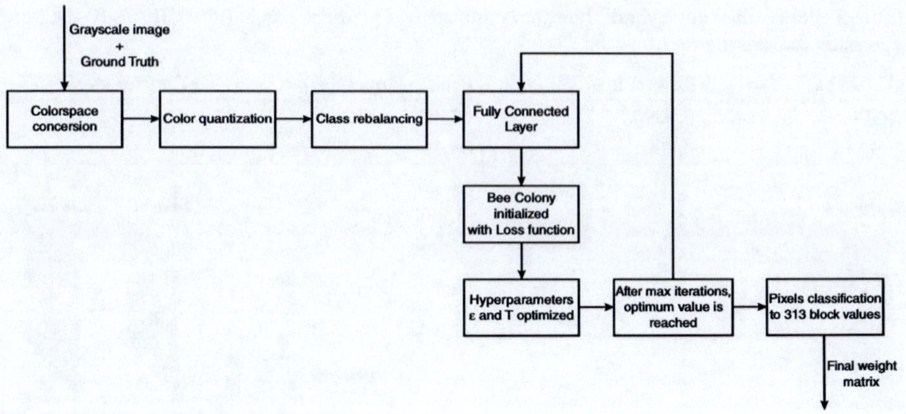

Fig. 16 Block diagram for hyperparameter tuning

Testing Hyperparameter Optimisation for Multinomial Classification with CIFAR-10

With over 16 million dimensions, even LHABC is not computationally feasible to be run in this case. Hence, the weights by themselves cannot be optimised using the meta-heuristic approach. However, in the post-processing phase, hyperparameters such as T (Softmax temperature), which have a major impact on the output produced, can be tuned using ABC. The block diagram for this process is presented in Fig. 16 which adds the hyperparameter optimisation module to the architecture proposed by Zhang et al. [25].

The effects of optimising T are shown in Fig. 17. The optimal value of T presented in Table 6 can hence be used as the value of the hyperparameter for obtaining the best colour temperature for the final colourised image. The baseline existing method is considered to be tuning of the parameter by trial-and-error basis which would take a lot of time and effort to zero in on the best value. The best output produced by the model based on this set of images will hence be obtained at the optimal value of T found by the ABC algorithm.

5 Conclusion

In this paper, modified versions of the artificial bee colony algorithm have been successfully implemented across problems of varying complexity and dimensionality. The base algorithm, ABC, was found to give similar or better results when compared to normal gradient descent for low dimensionality problems. The random search capability of the algorithm helped it find the global minimum in a relatively short amount of time for small problems. Our proposed algorithm, NMABC,

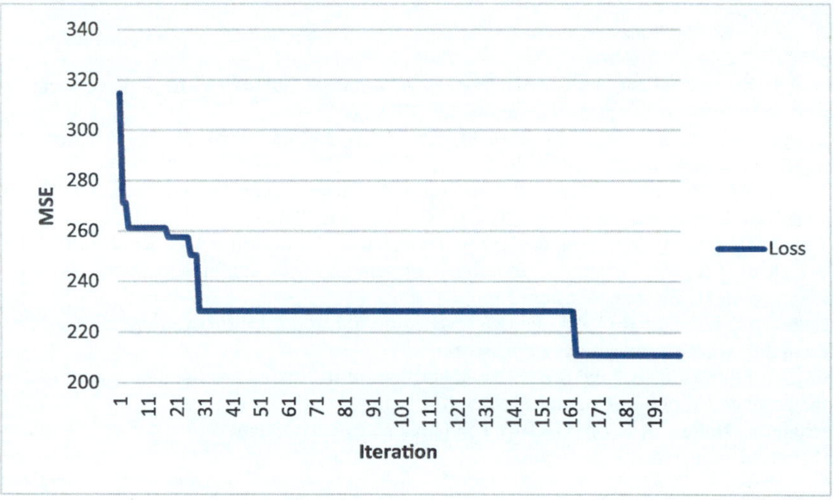

Fig. 17 Optimising value of T through ABC using MSE as objective function

Table 6 Optimal value of T and mean squared error for CIFAR-10 dataset classification colourisation

Optimal T	Lowest MSE
211	0.18386507

achieved better results with faster convergence for the same. This was due to careful adjustments made to the ABC algorithm to tweak it for the best performance in the neural network context. As the number of dimensions of a given optimisation problem increases, the average required time to find the global minimum increases exponentially. Therefore, NMABC was declared to not be suitable for high dimensionality problems, and hence we proposed the hybrid optimising algorithm, LHABC, which reached higher accuracy percentages with a significantly fewer number of epochs as compared to gradient descent, thereby increasing training efficiency. This behaviour can be completely parallelised for each independent bee to give even better results, which can be explored further. The applicability of the algorithms for hyperparameter tuning in the post-processing stage of image colourisation was also explored, resulting in images with more realistic levels of saturation. Hence, even in problems of extremely high dimensionality, NMABC can still be used as a valid method to improve the quality of solutions produced by targeting the hyperparameters of the problem instead of the weights. This extends the applicability of meta-heuristic techniques such as ABC to a wide array of problems like optimising learning rates, regularisation parameters and parameters in kernel functions for support vector machines as in [29], to name a few, which can be explored in the future work.

References

1. Du SS, Jin C, Lee JD, Jordan MI, Poczos B, Singh A (2017) Gradient descent can take exponential time to escape saddle points. In: NIPS 2017
2. Choromanska A, Henaff M, Mathieu M, Arous GB, LeCun Y (2014) The loss surfaces of multilayer networks. arXiv:1412.0233
3. Karaboga D, Basturk B (2007) A powerful and efficient algorithm for numerical function optimization: artificial bee colony (ABC). J Glob Optim 39:459
4. Holland JH (1984) Genetic algorithms and adaptation. In: Selfridge OG, Rissland EL, Arbib MA (eds) Adaptive control of ill-defined systems, NATO conference series (II systems science), vol 16. Springer, Boston, MA
5. Kennedy J, Eberhart R (1995) Particle swarm optimization. In: Proceedings of ICNN'95—international conference on neural networks
6. Singh A (2009) An artificial bee colony algorithm for the leaf-constrained minimum spanning tree problem. Appl Soft Comput J 9:625–631
7. Fenglei L, Haijun D, Xing F (2014) The parameter improvement of bee colony algorithm in TSP problem. Science Paper Online
8. Baykosoglu A, Ozbakir L, Tapkan P (2007) Artificial bee colony algorithm and its application to generalized assignment problem. In: Swarm intelligence: focus on ant and particle swarm optimization. Itech Education and Publishing, Vienna, pp 532–564
9. Liu Y, Tang S (2018) An application of artificial bee colony optimization to image edge detection. In: 13th International conference on natural computation, fuzzy systems and knowledge discovery (ICNC-FSKD), pp 923–929
10. Chen Y, Li Y, Li S (2014) Application of artificial bee colony algorithm in blind source separation of chaotic signals. In: IEEE 7th joint international information technology and artificial intelligence conference, pp 527–531
11. Koylu F (2017) Online ABC miner: an online rule learning algorithm based on artificial bee colony algorithm. In: 8th International conference on information technology (ICIT), pp 653–657
12. Ozturk C, Karaboga D (2011) A novel clustering approach: artificial bee colony (ABC) algorithm. Appl Soft Comput 11(1):652–657
13. Kiran MS, Hakli H, Gunduz M, Uguz H (2015) Artificial bee colony algorithm with variable search strategy for continuous optimization. Inf Sci 300:140,157
14. Quan H, Shi X (2008) On the analysis of performance of the improved artificial-bee-colony algorithm. In: 2008 4th international conference on natural computation
15. Banharnsakun A, Achalakul T, Sirinaovakul B (2011) The best-so-far selection in artificial bee colony algorithm. Appl Math Comput 11:2888–2901
16. Akay B, Karaboga D (2012) A modified artificial bee colony algorithm for real parameter optimization. Inform Sci 192:120–142
17. Karaboga D, Gorkemli B (2014) A quick artificial bee colony (qABC) algorithm and its performance on optimization problems. Appl Soft Comput 23:227–238
18. Zhang X, Zhang X, Yuen SY, Ho SL, Fu WN (2013) An improved artificial bee colony algorithm for optimal design of electromagnetic devices. IEEE Trans Magn 49:4811–4816
19. Liu Y, Ling X, Liang Y, Liu G (2012) Improved artificial bee colony algorithm with mutual learning. J Syst Eng Electron 23(2):265–275
20. Gao W-F, Liu S-Y, Huang L-L (2013) A novel artificial bee colony algorithm based on modified search equation and orthogonal learning. IEEE Trans Cybern 43(3):1011–1024
21. Narasimhan H (2009) Parallel artificial bee colony (PABC) algorithm. In: 2009 World congress on nature & biologically inspired computing (NaBIC)
22. Taspmar N, Yildmm M (2015) A novel parallel artificial bee colony algorithm and its PAPR reduction performance using SLM scheme in OFDM and MIMO-OFDM systems. IEEE Commun Lett 19(10):1830–1833

23. Karaboga D, Akay B, Ozturk C (2007) Artificial bee colony (ABC) optimization algorithm for training feed-forward neural networks. In: Torra V, Narukawa Y, Yoshida Y (eds) Modeling decisions for artificial intelligence. MDAI 2007, Lecture notes in computer science, vol 4617. Springer, Berlin
24. Ozturk C, Karaboga D (2011) Hybrid artificial bee colony algorithm for neural network training. In: 2011 IEEE congress of evolutionary computation (CEC)
25. Zhang R, Isola P, Efros AA (2016) Colorful image colourisation. In: 4th European conference on computer vision
26. Hu H, Li F (2018) Image colourisation by non-local total variation method in the CB and YIQ colour spaces. IET Image Process 12(5):620–628
27. Chen Y, Zong G, Cao G, Dong J (2017) Image colourisation using linear neighbourhood propagation and weighted smoothing. IET Image Process 11(5):285–291
28. Talatahari S, Mohaggeg H, Kh N, Manafzadeh A (2014) Solving parameter identification of nonlinear problems by artificial bee colony algorithm. Math Probl Eng 2014:1–6. https://doi.org/10.1155/2014/479197
29. Godinez-Bautista A, Padierna L, Rojas Dominguez A, Puga H, Carpio M (2018) Bio-inspired metaheuristics for hyper-parameter tuning of support vector machine classifiers. https://doi.org/10.1007/978-3-319-71008-2_10

Performance Analysis of Polar Codes for 5G Wireless Communication Network

Jyothirmayi Pechetti, Bengt Hallingar, P. V. N. D. Prasad, and Navin Kumar

1 Introduction

Fifth-generation New Radio (5G NR) is capable of fulfilling high-speed, low-latency, and high-reliability connections among mobile devices [1]. The goal of 5G is to provide availability of high throughput and high capacity per sector and at lower latency for some applications as compared to long-term evolution advanced (LTE-A) 4G LTE. The design of air interface would lower the latency and offer greater flexibility than 4G. The current development aims for achieving 20 Gbps speed and 1 ms latency [2]. Many new concepts and technologies at different layers are being explored to achieve and optimize some of these parameters. For example, some of the new error-correcting codes like Polar Codes and low-density parity check (LDPC) have been introduced in 5G NR as compared to 4G to address higher channel capacity requirements [3].

5G networks would use smaller cells and would be scalable and flexible. However, even with existing macro-cells, 5G would improve the capacity by multiple times [4] over current systems. This can be achieved by leveraging larger bandwidth and advanced antenna technologies.

Additionally, simple and efficient channel coding ensures reliability of the wireless channel. According to 3GPP [5], they have specified different coding techniques for transport channel and control channel. After many researches on the coding schemes suitable for 5G, an important innovation was achieved by

J. Pechetti · N. Kumar (✉)
Amrita School of Engineering, Bengaluru, Amrita Vishwa Vidyapeetham, Bengaluru, India
e-mail: navinkumar@ieee.org

B. Hallingar · P. V. N. D. Prasad
Tieto Sweden Support Services AB, Karlstad, Sweden

A. Haldorai et al. (eds.), *2nd EAI International Conference on Big Data Innovation for Sustainable Cognitive Computing*, EAI/Springer Innovations in Communication and Computing, https://doi.org/10.1007/978-3-030-47560-4_29

the Turkish scientist Erdal Arikan [6]. He suggested Polar Codes for achieving better capacity for binary discrete memoryless symmetric (B-DMS) channels. Both the encoder and the decoder processes help in the improvement of speed and complexity overhead, while the performance of decoders is tightly bounded to the characteristics of the Polar Code. These codes offer particularly strong error correction performance for short code blocks (CBs). Considering both complexity and performance, LDPC is found to be the most suitable for longer block code length in 5G development. These types of codes are normally used in shared channels, while Polar Codes are well suited for short block lengths which are typically used in control channels. This coding scheme seemed to have higher scope in yielding better performance.

Polar Code is normally designed as channel dependent. Authors in [7] have studied the performance of Polar Codes under different channels, that is, when channels are mismatched (i.e., optimized Polar Codes for one channel). They found to have a rate loss. In [8], authors identified a unique characteristic property of Polar Codes which is suboptimal and low complexity successive cancellation (SC) decoding algorithm. And, the Polar Codes are usually utilizing maximum likelihood (ML) decoding algorithm. Also, it was proved that under ML decoding, an optimized Polar Code for binary symmetric channel (BSC) achieves same capacity as any other binary input channel. It can be achieved at the cost of $O(N \log N)$ decoding algorithm. In yet another work [9, 10], Polar Coding methods were discussed that come at the cost of broadening regular Polar Code construction. Here it was noticed that finite length performance of Polar Codes was not good enough due to suboptimal nature of the standard SC decoding algorithm and relatively poor minimum distance properties of these codes. It was also noticed that the decisions made by successive cancellation decoder are sequential. The decoder processing time is proportional to the code length, which results in reduction of the throughput. Furthermore, a successive cancellation list (SCL) decoding algorithm was proposed [2] which has better performance than SC algorithm. SCL decoder provides better efficiency than SC decoder [11]. But it has very higher complexity compared to SC decoder due to the introduction of list at each cycle which leads to long processing time.

In this work, we studied and investigated different decoding algorithms. Performance comparison is performed, and a new methodology is proposed to overcome the long processing cycle times and larger memory size requirement. This can be achieved with proper selection of the algorithm in consideration of the standard deviation of the demodulated I&Q soft bits. According to the available literature, such study is not undertaken, and the proposed algorithm is also unique.

The remaining paper is organized as follows. Section 2 introduces overall system architecture. Section 3 focuses on design and analysis of different SCL decoding algorithm, while in Sect. 4, results are presented and discussed. Conclusion is presented in Sect. 5.

2 System Architecture

Polar Codes are based on channel polarization with lower encoding and decoding complexity for discrete memoryless channels. It is found that as the block length increases, the channels seen by individual bits through a certain transformation (called polar transformation) starts polarizing. This leads them to see either a pure noise channel or noiseless channel. Figure 1 shows a generic communication system using polar encoding and decoding. Before polar encoding, cyclic redundancy check (CRC) bits will be appended to the information bits. Transport block (TB) is segmented into multiple code blocks. CRC is calculated for the TB and each segmented code block. It allows error detection at the receiver side. If a code block is in error, there is no need to process remaining code blocks. The steps for construction of Polar Codes are as follows:

1. First, over a set of good channels (almost noiseless channels), information bits are sent, and zero is injected to the remaining bit channels.
2. The set of bit channels are sorted out from good to corrupt based on their corresponding reliability sequence.
3. Then the bit channels with reliability sequence having $(N - K)$ are called bad, while the rest are called good, where N is total block size and K is information bit size.
4. Therefore, this property is exploited to design codes for bit-interleaved coded modulation (BICM) channels. BICM channels can be modeled as a multi-channel comprising of several sets of binary input channels. These channels are used for transmitting coded bits.

Therefore, it is interesting to design efficient codes which can utilize different set of channels instead of single channel for designing Polar Codes. The coded bits are

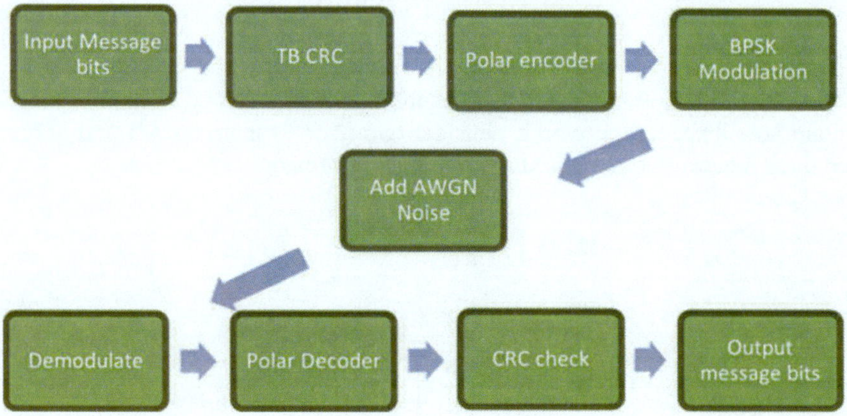

Fig. 1 Transmission chain

then passed through the channel, and at the receiver, the suitable decoder is used to bring back the original message bits.

3 Design and Analysis

3.1 *Frozen Bits and Its Significance*

If (N, K) bits are to be transmitted over a butterfly structure, then it is said that there will be $(N - K)$ number of frozen positions. So in these positions, frozen bits are placed which are basically zero. By doing this, a very high noisy node will be frozen to zero so that the next node will not get affected. The position of frozen nodes is decided according to reliability sequence [6]. In Polar Codes, poor channels are frozen which means channels with low reliability sequence. For example, if (16, 8) is to be transmitted where reliability sequence for 16 bits = 1 2 3 5 9 4 6 10 7 11 13 8 12 14 15 16, then the frozen channels will be 1 2 3 5 9 4 6 10 $(N - K)$, and information bits will be transmitted at these channels: 7 11 13 8 12 14 15 16 (K).

Polar Encoder

Encoding of Polar Codes can be achieved either with butterfly structure or in tree structure. In this paper, polar encoder is performed in butterfly structure. Kronecker product plays an important role in proceeding with polar encoder using butterfly structure. For a code which has the block length N, the generator matrix G_N is obtained from the base matrix G by taking the Kronecker product of $G^{\otimes n}$, where $n = (\log_2 N)$ and the base matrix is defined as:

$$G = \begin{bmatrix} 1 & 0 \\ 1 & 1 \end{bmatrix} \tag{1}$$

Operation of Kronecker product is different from matrix multiplication. To explain how its operation works, consider two (2×2) matrixes $M1$ and $M2$, and then the Kronecker product of $M1$ and $M2$ is performed.

$$M1 = \begin{bmatrix} a & b \\ c & d \end{bmatrix} \quad M2 = \begin{bmatrix} a^* & b^* \\ c^* & d^* \end{bmatrix}$$

$$A \otimes B = \begin{bmatrix} a\begin{bmatrix} a^* & b^* \\ c^* & d^* \end{bmatrix} & b\begin{bmatrix} a^* & b^* \\ c^* & d^* \end{bmatrix} \\ c\begin{bmatrix} a^* & b^* \\ c^* & d^* \end{bmatrix} & d\begin{bmatrix} a^* & b^* \\ c^* & d^* \end{bmatrix} \end{bmatrix} = \begin{bmatrix} aa^* & ab^* & ba^* & bb^* \\ ac^* & ad^* & bc^* & bd^* \\ ca^* & cb^* & da^* & db^* \\ cc^* & cd^* & dc^* & dd^* \end{bmatrix} \tag{2}$$

So, basically each element of matrix $M1$ is bit to bit multiplied with the complete $M2$ matrix. Therefore, the generator matrix can bc found at the nth product of base matrix G as in (3).

$$G = G^{\otimes n} \tag{3}$$

First, according to reliability sequence, least reliable channels are placed with frozen bits, and most reliable channels take information bits. Then these codewords are sent to polar encoder along with concatenation of CRC bits. In this work, we have chosen CRC24c according to 3GPP standard [5].

These codewords ($u1$, $u2$, ..., uN) are multiplied with generator matrix G as in Eq. (4).

$$y = u * G \tag{4}$$

After completing (4) the encoded codewords are sent to AWGN channel.

Beliefs. The rx_bits (received samples) undergo demodulation to get beliefs (r). They are obtained from Eq. (5).

$$r = \text{real}\,(\text{rxbits} * (1 - 1i)) \tag{5}$$

These bits can be both negative and positive. The larger the value of r, the stronger the beliefs. If the belief is positive, it is mapped to 0, and if it is negative, it is mapped to 1.

3.2 SC Decoder

Successive cancellation decoding has been specifically designed for Polar Codes with which they reach channel capacity at infinite code length. The block error rate (BLER) performance of Polar Codes using SC decoding is achieved with summation of the error probabilities of all the information over polarized channels [12]. The encoded bits (x) undergo binary phase shift keying (BPSK) modulation and are introduced to AWGN noise (r). These bits are sent to polar decoder.

Note that the butterfly units in the polar encoder present the correlation between the source bits. This ensures each coded bit with a given index to rely on all its preceding bits with smaller indices. Such correlation leads to a much better decoding performance because we use back tracing method for finding decoded bits and thus constitutes the central idea of a basic decoding algorithm. It is known as SC decoding [6]. Successive cancellation of the "interference" caused by the previous bits improves the reliability in the retrieval of source bits. Due to the regular structure of Polar Codes, the SC algorithm can be designed and discussed in terms of a trellis or a code tree structure.

SC decoder has a complexity of $O(N \log N)$ [12], where N is the code block length. The reliability sequence of the polarized channels can be tracked within this complexity. For better understanding, we use code length of $N = 8$ and message bits of $K = 4$. Also, the tree in looping algorithm is used in this work.

SC Decoding Algorithm

SC decoder will perform sequence of step by step decisions in butterfly structure. At every step the decision of the current step depends on the previous step decision. SC decoding can be observed as a soft/hard message passing algorithm. The tree (as shown in Fig. 2) consists of n stages as in (6) with a depth of d. Each stage includes 2^{n-1} nodes, and each depth contains a pair of check and variable nodes.

$$n = \log_2 N \tag{6}$$

SC decoder updates the nodes stage by stage and sequentially decides the estimation bit by bit which indicates that the bits are decoded step by step in a stated sequence. As in Fig. 3, it's done from top to bottom; start deciding on $\hat{u}_1$ and finally on $\hat{u}_N$, using hard decision decoding. One-bit decision $\hat{u}_i$ is made to find out where the next bit $\hat{u}_{i+1}$ starts. The confirmed decided bits will influence the decision of following bit decisions.

For a decision, all the associated likelihood ratios are calculated. All positions that influence the bit are related. Likelihood ratios on all the leaf nodes are calculated

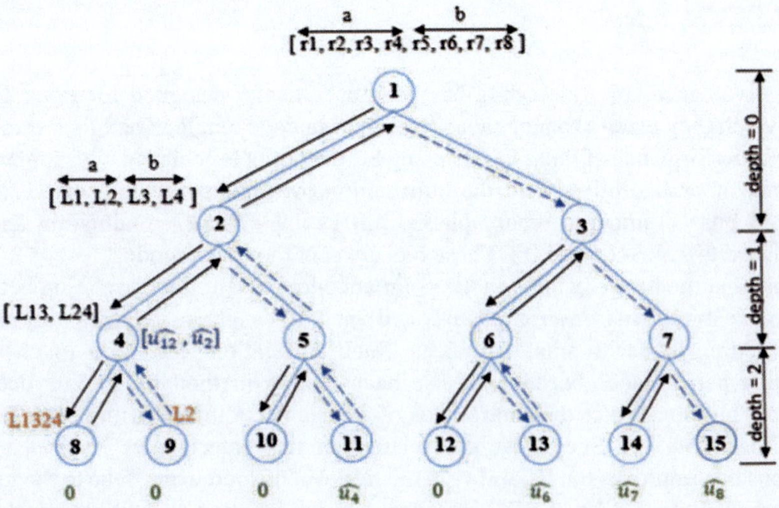

Fig. 2 SC decoder tree

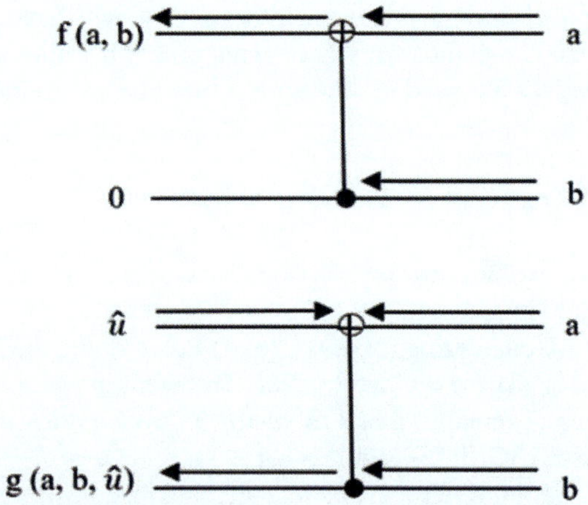

Fig. 3 Likelihood calculation significance

in the stated order (tree structure) before the first decision is made. The g-function is calculated as $(a + b)$ if the decision bit on position $\hat{u}$ is zero; else it is calculated as $(a - b)$ if the decision bit on position $\hat{u}$ is one. The decision made in the f-function decides if the g-function beliefs are calculated as $(a + b)$ or $(a - b)$.

Log Likelihood Ratio (LLR)

Likelihood ratio is designed in two different ways for Polar Codes depending on the leaf node position and is calculated as functions $f(r1, r2)$ for leaf node as in (8) and $(r1, r2)$ for right node as in (9). The f-function in (7) involves some transcendental functions that are complex for implementation in hardware; it can be approximated to (8).

$$f\ (a, b) = L = 2 * \tanh^{-1}\left(\tanh\left(\frac{a}{2}\right) * \tanh\left(\frac{b}{2}\right)\right) \tag{7}$$

$$\approx \text{sign}(a) * \text{sign}(b) * \min\left(|a|\,, |b|\right) \tag{8}$$

$$g\ \left(a, b, \hat{u}\right) = L = (-1)^{\hat{u}} * a + b \tag{9}$$

$$\hat{u} = \begin{cases} 0, \text{ if } L \geq 0 \\ 1, \text{ otherwise} \end{cases} \tag{10}$$

This can be associated to the butterfly structures in Fig. 2. As seen in the figure, LLR values from the f-function behave as XOR. In this way, it's easy to trace back the input bits. $\hat{u}$ is found from Eq. (10). The estimated $\hat{u}$ is forced to zero if the position of $\hat{u}$ is frozen; otherwise it will have the same value of $\hat{u}$. The value is saved together with all decided node bit values after a hard decision is made. These values are used in all future g-function calculations in the decoder.

Working of SC Decoder

To explain how the SC decoder works, this section describes the decoder in step by step process.

Considering $(N, K) = (8, 4)$ Polar Code, the input bits to the decoder ($r1$, $r2$, $r3$, ..., $r8$) are divided by $N/2$. The resulting first $N/2$ bits are considered to be vector a and remaining bits as vector b. Frozen positions are nodes 8, 9, 10, and 12 found from 3GPP [5] (Table 5.3.1.2-1).

All the left node beliefs are calculated using f-function from (8), and all the right node beliefs are found using g-function from (9). $\hat{u}$ is found at every leaf node using (10). If the position of $\hat{u}$ is at frozen node, then $\hat{u}$ is forced to zero as in (11). After computing $\hat{u}$ at each left leaf node from its belief (L1324), it is sent up to its mother node (node 4) as shown in Fig. 3.

$$\hat{u}_i = \begin{cases} 0, \text{ if it is frozen node} \\ \hat{u}, \text{ from (10)} \end{cases} \tag{11}$$

where i is node reference number.

Now, the beliefs and $\hat{u}$ from mother node (node 4) are sent to right leaf node (node 9) as shown in Fig. 3. g-function is computed at node 9 and respective $\hat{u}$ is found from (11). The procedure carries on at each mother node till the computation reaches $(2N - 1)$th node.

3.3 SCL Decoder

In this section successive cancellation list decoder is introduced. This decoder offers efficient output [11] compared with SC decoder. Complexity of SCL is $O(LN \log N)$ from [11]. The list decoder has a parameter L, called the list size. It is always in the powers of 2 and is generally said that larger values of L mean lower error rates but longer running times and larger memory usage. As far, best list size is considered as 32, and any list size above 32 wouldn't improve the performance significantly.

The SC decoder is very straightforward, easy to implement, and fast. However, it has some limitations; we will have to make a decision at each node. If its frozen node, then there is only one decision so there wouldn't be any problem. But slowly, when we come to the node which contains message bits, we will have to make a decision either 0 or 1 at $\hat{u}$. So, we are committing in one direction at that node as shown in Fig. 4 which could actually be an efficient choice. But the best part with

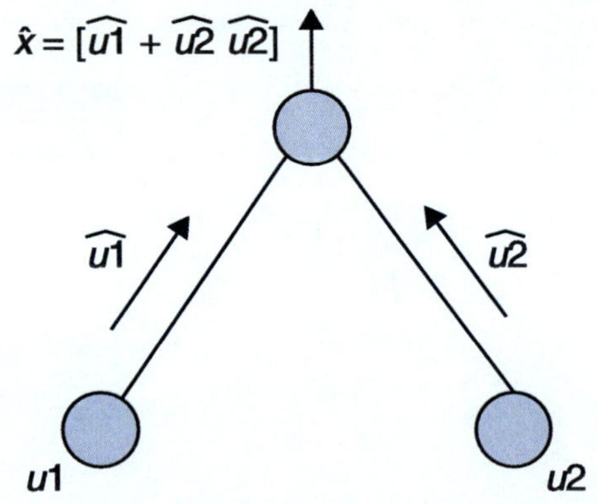

Fig. 4 Tree of SC decoder with two inputs

SC decoder is that even when the choice made at the leaf node is incorrect, SC decoder would eventually correct it.

However, the final codeword would have an error, because we made a single mistake at that node. So, when a mistake in decision like that is made, we are not allowed to go back and change the decision. So, to overcome this problem, "List" was introduced. In this algorithm, instead of making decision in one path which is according to the belief, we would make decision for both the paths, i.e., with $\hat{u} = 0$ and $\hat{u} = 1$. In this way, both the paths can be inspected.

SCL Decoder Algorithm

As discussed earlier, each node will be having two paths; one path is taken according to the belief, and the other path is opposite to the belief as in (12). Decision matrix is assigned at each split node.

$$\text{If } L(u_i) \geq 0, \quad \text{then for} \begin{cases} \hat{u}_i = 0, \text{DM}_i = 0 \\ \hat{u}_i = 1, \text{DM}_i = |L(u_i)| \end{cases}$$

$$\text{Else if } L(u_i) < 0, \quad \text{then for} \begin{cases} \hat{u}_i = 1, \text{DM}_i = 0 \\ \hat{u}_i = 0, \text{DM}_i = |L(u_i)| \end{cases} \tag{12}$$

So, when going against the belief, penalty is added. After making decision matrix, the decision at that node will get added to the next decision matrix as shown in Fig. 5. In the end, a set of decision matrixes will be found which will have final sum of the decision matrixes at that path.

However, when should the splitting stop at each node? As each split doubles the number of paths to be examined, they must be clipped, and the maximum number of paths allowable is the specified list size (L). This continues until the length of the list extents a limit L which is typically selected as a power of two. For a better

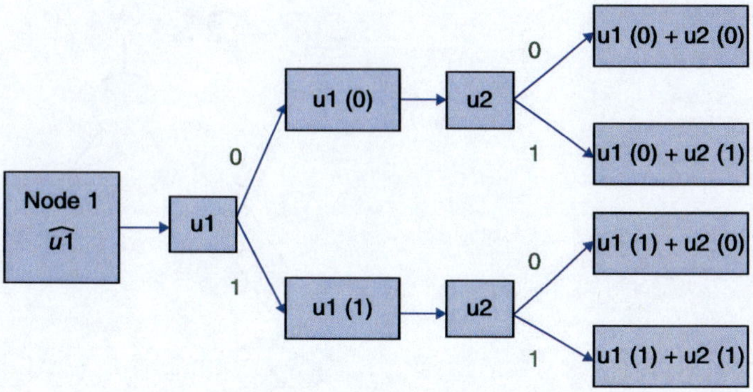

Fig. 5 Splitting a single node into two paths

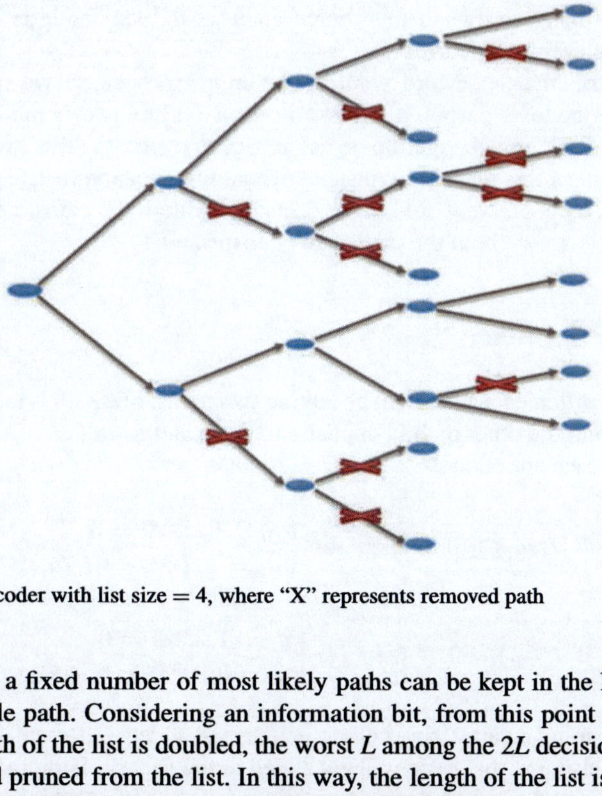

Fig. 6 SCL decoder with list size = 4, where "X" represents removed path

performance, a fixed number of most likely paths can be kept in the list instead of every probable path. Considering an information bit, from this point onward, each time the length of the list is doubled, the worst L among the $2L$ decision matrixes is identified and pruned from the list. In this way, the length of the list is preserved at L until the SCL decoding process completes attaining the last leaf node.

Example with list size $= 4$. Naturally, it is likely to keep the "best" paths at each stage and thus necessitates a pruning criterion. This pruning criterion will be to keep the most likely paths. Figure 6 shows splitting of a leaf node into two branches with a list size of $L = 4$. The first stage saves all possible decision matrixes since the list is not completed and so does the second stage. At stage 3, decision matrixes of all different paths are calculated as a sum of penalties at each step. In this case, since the number of decision matrixes is greater than L, only the four most likely paths are saved for the next list stage.

The crossed lines in the figure represent the least likely paths that are not saved; this indicates that those paths are added with high penalty. The full lines represent all paths saved in the list for that step. It can also be observed that at stage 4 also, we remove L number of decision matrixes. This splitting process continues till it reaches the next leaf node.

3.4 *Improved SCL Decoder*

As seen in previous section, SCL offers better performance compared to SC decoder but consumes more runtime. So, one of the main challenges faced by improved SCL is to obtain less runtime.

In improved SCL, the basic idea is to send the bits with higher noise to SCL decoder, as we have seen before that SCL decoder has high error correcting capacity. And, the bits with low noise should be sent to SC decoder, as it takes lesser runtime compared to SCL decoder. This idea would lead to decrease in overall runtime of the decoder. But the problem here is amount of AWGN in the bits is not known at the receiver side.

Standard Deviation

In statistics, the standard deviation is a measure that is used to quantify the amount of variation or dispersion of a set of data values. The formula for the sample standard deviation is as in Eq. (13) where $\overline{x}$ is mean and N is number of observations in the sample.

$$s = \sqrt{\frac{\sum_{i=1}^{N} (x_i - \overline{x})^2}{N - 1}} \tag{13}$$

So, to measure how much amount of AWGN has been added to the bits at the decoder, using standard deviation would be a better option. Standard deviation is done before sending the bits to the decoder as in Eq. (13). So, at each SNR standard deviation is performed.

Figure 7 contains simulation results of standard deviation performed with a random bit stream which are exposed to AWGN channel. It can be observed from the figure that at 5 dB, there is a drastic increase in standard deviation where standard deviation stands out to be approximately around 1.1–1.4. From the previous section observations, it is clear that at standard deviation of 1.1–1.4, there is high increase in noise. So, this is taken as threshold value to choose upon sending bits to SC or

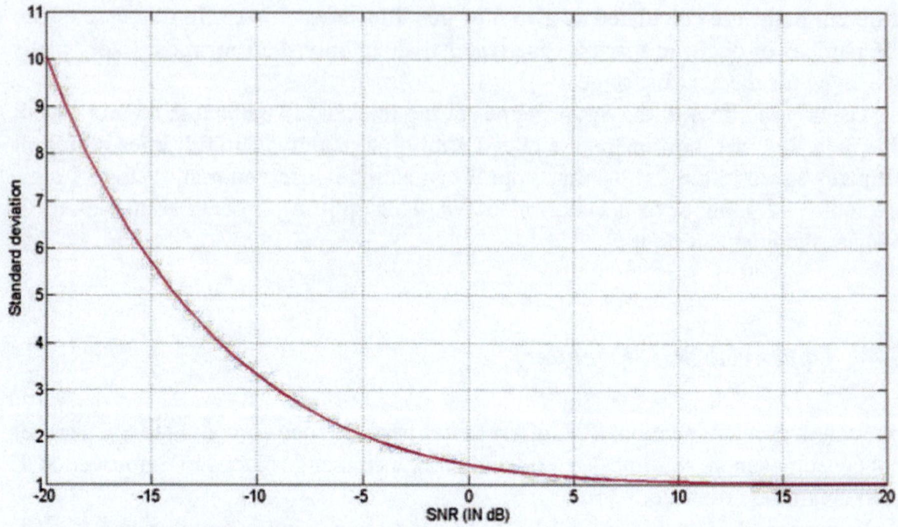

Fig. 7 Standard deviation results

SCL decoder as in Eq. (14).

$$\text{Modified decoder} = \begin{cases} \text{standard deviation} \geq \text{threshold, bits are sent to SCL decoder} \\ \text{standard deviation} < \text{threshold, bits are sent to SC decoder} \end{cases} \tag{14}$$

This way the beliefs to the decoder are decoded according to their exposure to noise. Simulation was done with this algorithm, and it is clearly observed that runtime has reduced compared to SCL decoder.

Advantage of Using Improved SCL

Complexity of SCL decoder is $O(LN \log N)$ where L is list size. SC complexity is $O(N \log N)$, so when bits divide into different decoders, their complexity would also reduce in the end. But in this work, complexity calculations are not taken into consideration; instead simulation runtime is calculated. However from the simulation runtime, it was found that runtime of improved SCL is almost reduced by 30% of SCL decoder which means the new algorithm is faster than SCL decoder.

4 Results and Discussion

In this section, simulation results are shown for SC decoder, SCL decoder, and improved SCL decoder. All the simulations are performed in MATLAB [13] over the additive white Gaussian noise with binary phase shift keying (BPSK) at 1000 frames.

4.1 *SC Decoder*

Figure 8 details about the performance characteristics of successive cancellation decoder with different sizes of the information block (*K*) and total block size (*N*) represented as (*N*, *K*). This figure shows that BLER for higher block size is having better coding gain compared to smaller block size. From the figure, there is improvement of 0.8 dB with block size of 1024 over 128 at 10^{-2} BLER.

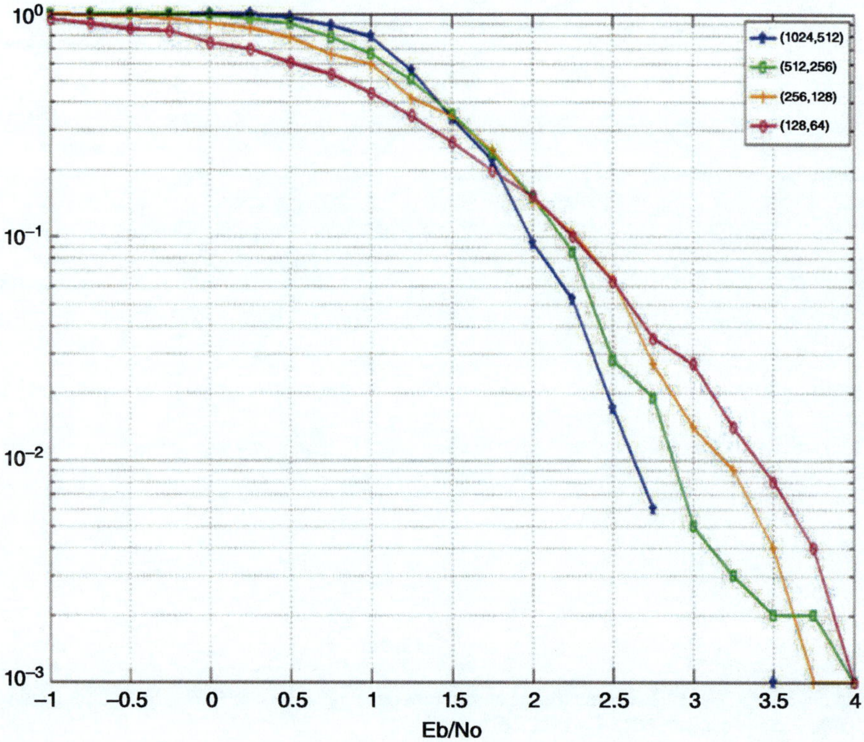

Fig. 8 BLER vs SNR in (dB) plot for rate = ½ simulated with 1000 subframes at an SNR of −1:0.25:4

As discussed in Sect. 3.1, frozen bits are known by the receiver which makes it easy to decode; this indicates that having more number of frozen bits gives smaller BLER. The bits other than information bits are frozen according to reliability sequence, which means for (1024, 512), it would have 512 frozen bits and, for (128, 64), it would have 64 frozen bits. So, it's clear that Polar Codes work best with higher block sizes.

Figure 9 shows the performance characteristics of successive cancellation decoder for block size (N) of 1024 bits with different rates (5/6, 1/4, 1/2) represented as (N, K) with information bits (K) obtained as per (15). It also shows that bits with lower rate perform better compared with bits with higher rates. From the figure, there is improvement of 0.5 dB with rate = 1/2 over rate = 5/6 at $\approx 2 \times 10^{-2}$ BLER.

$$K = N \times \text{Rate} \tag{15}$$

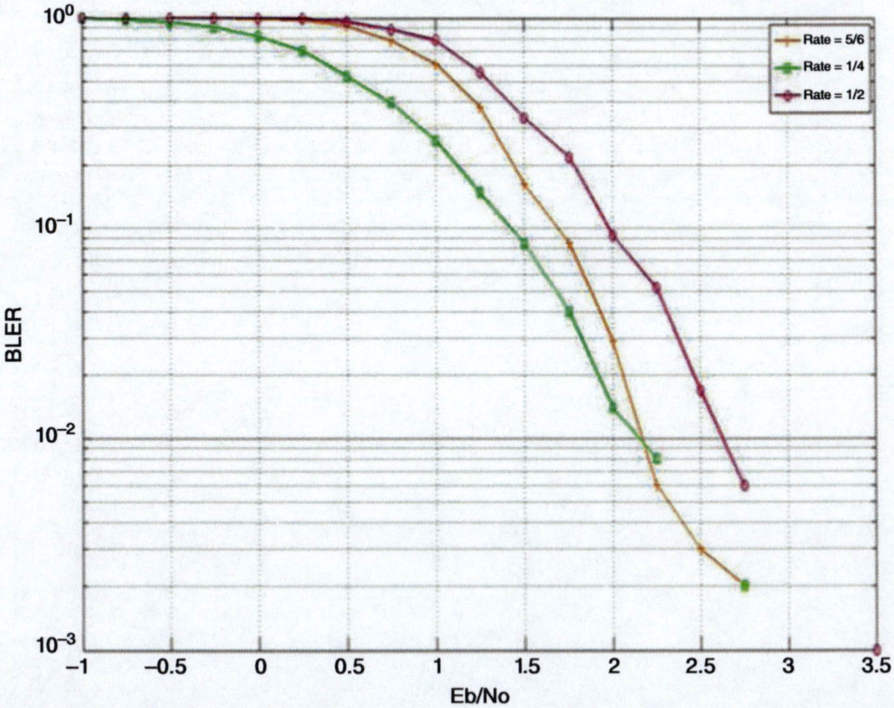

Fig. 9 SC decoder BLER comparisons at different rates for a block of (1024, 512). Simulated with 1000 subframes at an SNR of −1:0.25:4

4.2 SCL vs SC Decoder

Figure 10 details about the comparison of performance characteristics of successive cancellation decoder and successive cancellation list decoder for a block of (1024, 512) with different list sizes.

As seen from the Fig. 10, SCL decoder performance is better, and the higher the list size, the better the performance. It can be observed from the graph that after list size of 32, the performance is almost the same. This means, for any list size above 32, it wouldn't perform any better except that it consumes several numbers of cycles to run.

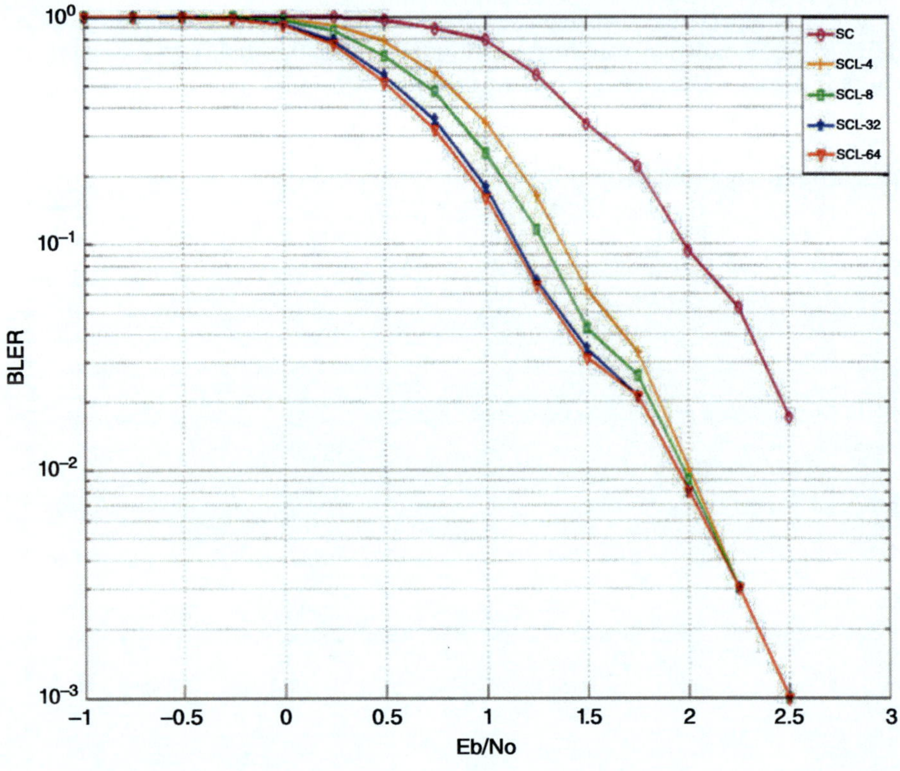

Fig. 10 BLER vs SNR (in dB) comparisons of SC and SCL decoder. Simulated with 1000 subframes at an SNR of −1:0.25:3

4.3 Improved SCL with Threshold 1.35

Figure 11 shows the comparison of performance characteristics of improved successive cancellation list decoder, successive cancellation decoder, and successive cancellation list decoder for a block of (1024, 512) with different list size of 4.

The above figure is simulated with CRC24c, rate = 1/2, subframes = 1000 with a block (1024, 512).

The simulation runtime for each decoder is as mentioned below:

SC: 1 h, 20 min
SCL: 4 h
Improved SCL (Modified): 3 h

It can be concluded that improved SCL reduces almost 1 h runtime of SCL decoder. So, simulation was performed with another threshold equal to 1.24, to check if there's a chance to get better performance.

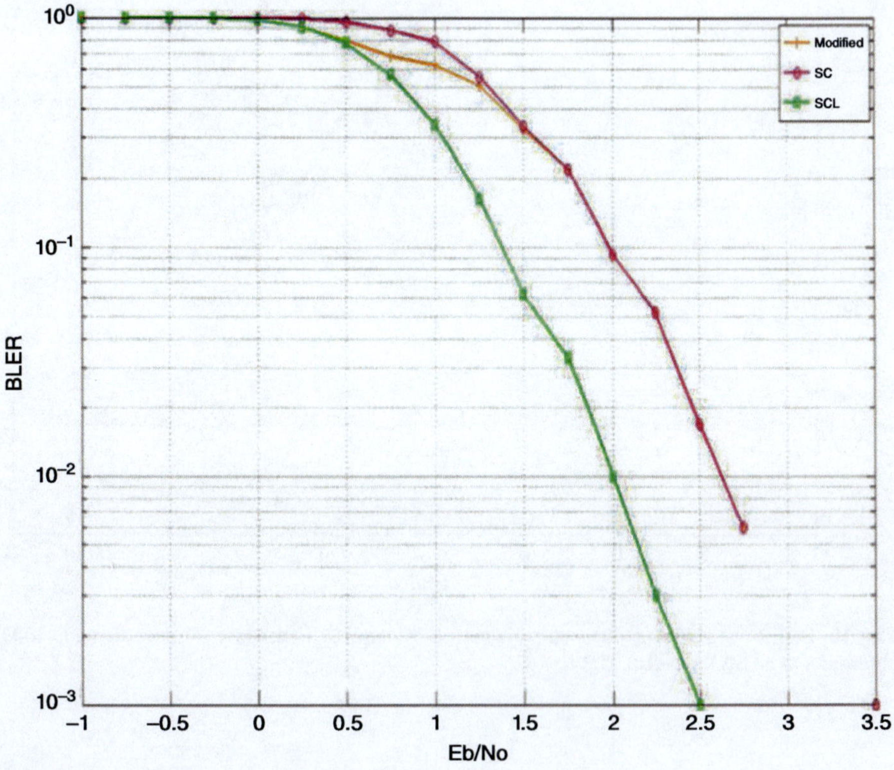

Fig. 11 Improved SCL with threshold 1.35

4.4 Improved SCL with Threshold 1.24

Figure 12 details about the comparison of performance characteristics of improved successive cancellation list decoder, successive cancellation decoder, and successive cancellation list decoder for a block of (1024, 512) with different list size of 4.

The above figure is simulated with CRC24c, rate = 1/2, subframes = 1000 with a block (1024, 512).

The simulation runtime for each decoder is as mentioned below.

SC: 1 h, 20 min
SCL: 4 h
Improved SCL (Modified): 2.5 h

It can be observed that with threshold of 1.24, it gives better runtime compared to threshold of 1.35. Simulations were performed with thresholds 1.2 and 1.4, but

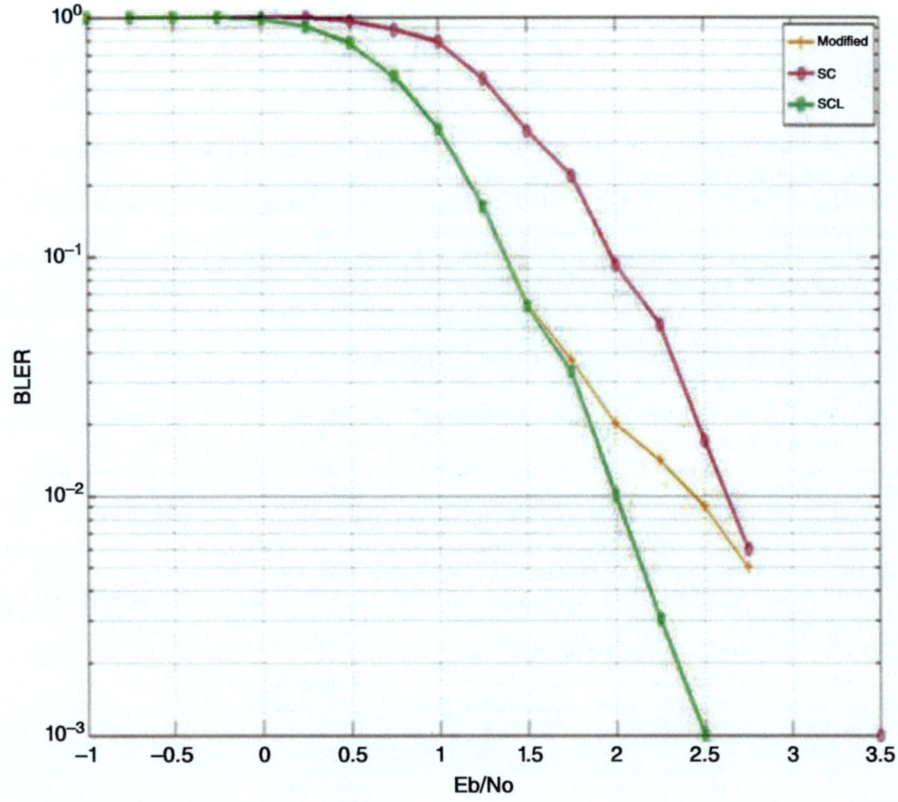

Fig. 12 Improved SCL with threshold 1.24

there was no change in the result. For threshold of 1.2, it obtains the same results as SC decoder, and for threshold with 1.4, it gives the same results as SC decoder.

5 Conclusion

In this work, SC and SCL decoders are analyzed and compared. Also, an improved SCL decoder has been implemented and compared with the results of SC and SCL decoder. In SC decoder, it can be noticed that channels with higher block size perform better compared to lower block size. Also, in SCL decoder, any list size above 32 would give almost the same results as with list size of 32, and also increasing the list size increases the performance of the decoder. For the improvised SCL decoder, performance was calculated based on total amount of time taken for simulation, and no calculations are made on memory utilization. But this idea can be extended further for hardware implementation.

References

1. Dahlman E, Parkvall S, Skold J (2018) 5G NR: the next generation wireless access technology, 1st edn. Academic Press, Chennai
2. Chen P, Xu M, Bai B, Wang J (2017). Design and performance of polar codes for 5G communication under high mobility scenarios. IEEE transaction paper
3. Mehta H, Patel D, Joshi B, Modi H (2014) 0G to 5G mobile technology: a survey. J Basic Appl Eng Res 1:56–60
4. EMF explained 2.0 how 5G works. http://www.emfexplained.info/?ID=259. Accessed Jan 2019
5. 3GPP TS 38.212 version 15.3.0 Release 15, ETSI TS 138 212 V15.3.0 (2018–10). Accessed Jan 2019
6. Arıkan E (2009) Channel polarization: a method for constructing capacity achieving codes for symmetric binary-input memoryless channels. IEEE Trans Inf Theory 55(7):3051–3073
7. Hassani SH, Korada SB, Urbanke R (2009) The compound capacity of polar codes. arXiv:0907.3291
8. Sasoglu E (2011) Polar coding theorems for discrete systems. PhD thesis, EPFL
9. Sasoglu E, Wang L (2013) Universal polarization. arXiv:1307.7495
10. Hassani SH, Urbanke R (2013) Universal polar codes. arXiv:1307.7223
11. Tal I, Vardy A (2015) List decoding of polar codes. IEEE Trans Inform Theory 61(5):2213–2226
12. Carlton A (2019) Computerworld "How polar codes work". IDG Communication, Inc., Boston, MA. https://www.computerworld.com/article/3228804/how-polar-codes-work.html
13. MATLAB (Matrix laboratory). https://www.mathworks.com/products/matlab.html. Accessed Jan 2019

Performance Analysis of Spectrum Sharing in mmWave Cellular Networks

Bandreddy Suresh Babu and Navin Kumar

1 Introduction

Millimeter wave (mmWave) technology is one of the emerging technologies for cellular mobile [1] and wireless local area networks. mmWave communications are characterized mainly by short range line of sight (LoS) communication and large bandwidth offering and directivity of antenna. Additionally, shadowing and atmospheric dependency are another characteristics [2, 3]. To ensure effectiveness and usefulness of mmWave, the network has been typically projected in the context of heterogeneous developments [4]. This means the part of connection is established with an anchor over traditional sub-6 GHz carrier and part over mmWave carrier. However, standalone mmWave systems can be developed. By using different bands, it may provide matching features, to ensure more efficient use of spectrum sharing paradigm [5].

Spectrum sharing [6] is a technique in which within a particular geographical area, there is a concurrent use of radio frequency band. Spectrum sharing is used to increase network capacity and also to maximize the usage of radio frequency spectrum with high data rate. Furthermore, it is used in communication services (data service, voice service, channel broadcasting). Additionally, it is used to minimize the radio spectrum pollution. Hybrid spectrum access [7] is a very efficient technology for mmWave networks. In hybrid access scheme, 1 GHz band at 28 GHz [8] is split over four operators (Base Station 1 (BS1), BS2, BS3, BS4). Spectrum pooling [9] is another spectrum management technique where the user can co-exist

---

B. Suresh Babu · N. Kumar (✉)
Department of Electronics and Communication Engineering, Amrita School of Engineering, Bengaluru, Amrita Vishwa Vidyapeetham, Bengaluru, India
e-mail: navinkumar@ieee.org

A. Haldorai et al. (eds.), *2nd EAI International Conference on Big Data Innovation for Sustainable Cognitive Computing*, EAI/Springer Innovations in Communication and Computing, https://doi.org/10.1007/978-3-030-47560-4_30

within a single allocation. In this, there exist primary and secondary users. Primary and secondary users are also called as licensed users and unlicensed users. Different types of spectrum classifications [10] are used such as open sharing and hierarchical access models, interweave, underlay, and overlay [11]. The open sharing model deals with those networks where each network will access with equal probability of the same spectrum. The hierarchical access model contains primary and secondary networks. Secondary networks access the spectrum without disturbing the primary network. Spectrum interweave represents the active information of users in the spectrum. Spectrum underlay deals with primary and secondary networks, while spectrum overlay deals with the secondary network access in the absence of primary network. Some of the existing models are dynamic range sharing model [12, 13], cooperative spectrum sharing model, and so on. In the dynamic range sharing model, the secondary users wait for the chance to access the spectrum. This model deals with channel occupancy and throughput calculation of different cases such as overlay and secondary user throughput. In cooperative spectrum sharing, the primary user chooses a lot of secondary users to serve the transmission.

Authors in [14] calculated some of the performance parameters of spectrum sharing of the different models such as no sharing, spectrum plus access, and spectrum plus interference. From all these models, the signal to interference noise ratio (SINR) is considered. Also, they calculated the coverage with SINR threshold with the base station density 30 and 60 BS/km^2 and UEs of 200 and 400. Furthermore, they analyzed the rate coverage for all the models. In [15], authors worked out for average achievable throughput with single user (SU) transmit power constraint over different types of cognitive radio networks. In [13, 16], authors have calculated the throughput per user, packet arrival rate and the spectrum occupancy probability when, the primary network shares the network with two other secondary users.

In this work, we present the efficient way of aggregation between licensed, hybrid, and pooled carrier. A millimeter wave-based spectrum sharing model is investigated among different types of spectrum baselines and different CR types. Comparative study for the cell association is also carried out. Additionally, we calculate the SINR for the three baselines. Finally, we obtained the throughput of CR types which vary with reference network (fully licensed) of different CR cases. At the same time, we also proposed the method to improve other techniques for spectrum sharing. Such exhaustive work is not reported in the literature.

Rest of the contents of this paper is presented as follows. In Sect. 2, the system framework model is presented and discussed. Analysis and mathematical details are presented in Sect. 3. In Sect. 4, results are discussed, while conclusion is presented in Sect. 5.

2 System Model

Fig. 1 shows a framework for the spectrum sharing based different types of cognitive radio networks and user equipments (UE).

- Let us consider j as the number of user equipments. Also, consider that different kinds of cognitive radio networks are present in a network area of $A \times A$, where A is the dimension of the geographical network.
- Assume that there are different base station densities per km^2 such as 30, 60, 90, and 120 BSs/km^2.
- Consider i as the number of base stations, and assume $i = (0–60)$, $j = (0–600)$.
- User equipment j is associated to the BSi with minimum path loss by using low-frequency carrier.
- Here, we consider three different spectrum sharing techniques (open sharing model): hybrid, pooled, and licensed.
- Two types of cell association techniques are used [1]:
  - Associating with the BS and the carrier
  - Associating with only the carrier
- Initially, the associated probability of the users which are associated with the low-frequency carrier is calculated.

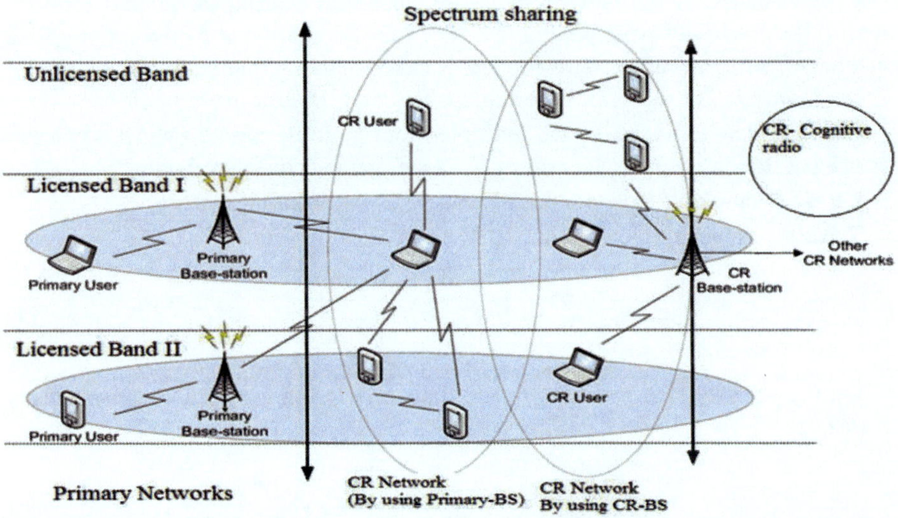

Fig. 1 Spectrum sharing based on user equipment (UEs) and cognitive radio (CR) network association with base station

- After calculating the association probability, different power constraint cases are investigated such as:
  - Case 1: Comparison of throughput for different types of spectrum sharing techniques (hybrid, pooled, licensed) with different percentiles (5, 50, 95) and base station density equal to 60 BS/km^2 for joint carrier and cell association.
  - Case 2: Same as case 1 but the association is done between the carrier only. In this case, BS is kept constant for entire simulation.
- The SINR for the three baselines (types of spectrum sharing, hybrid, pooled, licensed) with the BS density equal to 60 BS/km^2 is calculated.
- User equipments are associated to a base station. With different kinds of cognitive radio networks and with license spectrum sharing, the normalized ergodic throughput is calculated for two different cases:
  - Case 1: If the number of CR ≤ 100
  - Case 2: If the number of CR ≥ 100

3 Design and Analysis

3.1 *Association with Cell and Carrier*

The performance of the network over any spectrum sharing techniques (pooled, hybrid, licensed) will be depending on how the UE is allotted to the serving base station and with the carrier. This process is called association of cell and carrier.

Let i_m and j_m the subset of base station (BS) and UEs for operators M.

Let $H_{ij}^{(c)}$ denote matrix of the multiple input multiple output (MIMO) channel from base station i to UE j at carrier c. Channel gain is considered flat.

Let $w_{\mathrm{RX}_{ij}}^{(c)}$ and $w_{\mathrm{TX}_{ij}}^{(c)}$ denote the TX and RX beamforming vectors.

The single input single output (SISO) channel gain is given by:

$$G_{ij}^{(c)} = \left| w_{\mathrm{RX}_{ij}}^{(c)H} H_{ij}^{(c)} w_{\mathrm{TX}_{ij}}^{(c)} \right|^2 \tag{1}$$

Let $\overline{G}_{ijk}^{(c)}$ the average channel gain from the interfering base station BSk to user j of BSi.

$$\overline{G}_{ijk}^{(c)} = \frac{1}{N_k^{(c)}} \sum_j \left| w_{\mathrm{RX}_{ij}}^{(c)H} H_{ij}^{(c)} w_{\mathrm{TX}_{ij}}^{(c)} \right|^2 \tag{2}$$

where $N_k^{(c)}$ is the k number of base stations with the carrier c.

Table 1 Parameter setting for path loss model

Frequency	State	α	β	σ
28 GHz	NLOS	72	2.9	8.7 dB
	LOS	61.4	2	5.8 dB

The signal to interference noise ratio is expressed as:

$$\gamma_{ij}^{(c)} = \frac{\frac{P_{\mathrm{TX}_i}^{(c)}}{\mathrm{PL}_{ij}^{(c)}} G_{ij}^{(c)}}{\sum_{k \neq i} \frac{P_{\mathrm{TX}_k}^{(c)}}{\mathrm{PL}_{kj}^{(c)}} \overline{G}_{ijk}^{(c)} + W^{(c)} N_0} \tag{3}$$

where the $P_{\mathrm{TX}_i}^{(c)}$ is total transmitted power at carrier c from the base station i, $P_{\mathrm{TX}_k}^{(c)}$ is total transmitted power at carrier c from the base station k, N_0 is thermal noise power spectral density, and $\mathrm{PL}_{ij}^{(c)}$ is the path loss between BSi and UEj and $\mathrm{PL}_{kj}^{(c)}$ is the path loss between BSk and UEj.

The omnidirectional path loss is expressed as:

$$\mathrm{PL}(d)\ [\mathrm{dB}] = \alpha + \beta\ 10\log_{10}(d) + \varepsilon \tag{4}$$

where ε is the $N(0, \sigma^2)$ which is the log normal shadowing and parameters α, β, and σ are given in Table 1 [17].

3.2 *Different Association Techniques*

Associating with the BS and the Carrier

In this case, UE automatically selects the carrier and serving base station so as to maximize its data rate. The maximized value of cell-carrier assignment for UE is given as:

$$\left(c^*, i^*\right) = \arg \underbrace{\max}_{i \in \mathrm{Im}, c \in C} \left(\frac{W^{(c)}}{1 + N_i^{(c)}} \log_2 \left(1 + \gamma_{ij}^{(c)}\right)\right) \tag{5}$$

where the $N_i^{(c)}$ is number of UEs associated with cth carrier of ith base station, $W^{(c)}$ is the bandwidth of cth carrier, and $\gamma_{ij}^{(c)}$ is SINR between BSi and UEj with a carrier c.

If $N_{i*}^{(c)}$ is number of UEs associated with cth carrier of i^*th serving base station, the maximum association connection rate for UE is:

$$n_j = \frac{W^{(c^*)}}{1 + N_{i*}^{c^*}} \log_2\left(1 + \gamma_{i*j}^{(c^*)}\right) \tag{6}$$

Associating with Only Carrier

Now the base station and carrier are divided. That is, UE is assigned with a base station and is kept constant for the whole simulation, and the UE is permitted to select only carrier as a component of SINR. The updated carrier is given as:

$$\left(c^*\right) = \arg \underbrace{\max}_{c \in C} \left(\frac{W^{(c)}}{1 + N_{i*}^{(c)}} \log_2\left(1 + \gamma_{i*j}^{(c)}\right)\right) \tag{7}$$

3.3 *The Different Types of Cognitive Radio Networks and Its Signal to Noise Ratio*

Cognitive multi-access channel (C-MAC) [10] where we take K CRs which are denoted by CR1, CR2, ..., CRK transmits the unconventional messages to the cognitive radio network base station. h_k is the power gain of fading channel from kth CR to cognitive radio base station, where $k = 1, \ldots, K$. Likewise, p_k is transmission power from the kth cognitive radio networks, and P_k is the power constraint at kth cognitive radio network. Q_e is constant power, and e is a channel from primary radio network to cognitive radio network base station. The peak transmit power constraint is denoted as J. The primary radio network is protected by each and every cognitive radio network transmitter by applying Γ, which is represented as peak interference power constraint at primary radio network receiver. By combining both transmitter and interference power constraints, we get:

$$p_k \le \min\left(P_k, \frac{\Gamma}{g_k}\right), \quad \forall k, p \le \min\left(J, \frac{\Gamma}{g}\right) \tag{8}$$

The maximum achievable MAC signal to noise ratio of cognitive radio network k is then given as:

$$\gamma_k^{(\mathrm{MAC})} = \frac{h_k p_k}{1 + Q_e} \tag{9}$$

Cognitive broadcast channel also sends message to K cognitive radio networks. In between the cognitive radio multi-access channel and cognitive radio broadcast channel, we have assumed the channel reciprocity. For both the access channels, the channel gain is equal. The maximum achievable BC signal to noise ratio of cognitive radio network k is shown as:

$$\gamma_k^{(BC)} = \frac{h_k p}{1 + Q_{ek}} \tag{10}$$

In cognitive parallel access channel (C-PAC), there are K-CR transmitters denoted as CR-Tx_1, CR-Tx_2, ..., CR-Tx_k and K-CR receiver denoted as CR-Rx_1, CR-Rx_2, ..., CR-Rx_k; hk denotes the gain of the fading channels from Kth CR network to the receiver of the Kth cognitive radio network, whereas $k = 1, \ldots, K$, g_k, and e_k represents fading channel from Kth cognitive radio network to the Kth cognitive radio network receiver, respectively, and from primary radio network receiver. The peak transmitted power is represented as j at cognitive radio network base station in cognitive radio parallel access channel. p represents the transmit power from cognitive radio network base station. The maximum achievable PAC signal to noise ratio of cognitive radio network k is expressed as:

$$\gamma_k^{(PAC)} = \frac{h_k p_k}{1 + Q_{ek}} \tag{11}$$

From all the cognitive radios, we select one k cognitive radio network with a proper fading state which is having large SNR. h_k explains the power gain of fading channels from Kth CR network to receiver of Kth cognitive radio network. The fading state of particular users is represented as k^*. For each cognitive radio network, for example, for MAC, the k^* is given as:

$$k^*_{MAC} = \arg \max_{k \in \{1\in,\ldots,K\}} \frac{h_k \min\left(P_k, \frac{\Gamma}{g_k}\right)}{1 + Q_e} \tag{12}$$

where fading channel power gain is represented as g_k from the kth cognitive radio network to cognitive radio base station and Q_e is the received power of the primary radio network receiver to cognitive radio network base station. The fading state of particular users is represented as k^* for BC, which is expressed as:

$$k^*_{BC} = \arg \max_{k \in \{1\in,\ldots,K\}} \frac{h_k \min\left(J, \frac{\Gamma}{g}\right)}{1 + Q_{ek}} \tag{13}$$

h_k is the same as in MAC, but g is power gain of a fading channel from cognitive radio network base station to primary radio network transmitter, and e_k is the channel from PR-Tx to CRk. The fading state of particular users is represented as k^* for PAC; the k^* is given as:

$$k^*_{\text{PAC}} = \arg \max_{k \in \{1\epsilon, \ldots, K\}} \frac{h_k \min\left(p_k, \frac{\Gamma}{g_k}\right)}{1 + Q_{ek}} \tag{14}$$

where the h_k and p_k are power gains of fading channel and transmitted power.

4 Results and Discussions

4.1 Simulation Environment

For simulation, we have taken 60 BS/km^2 and 600 UEs within a geographical area $A \times A$ where A is dimension of the area.

Simulations are performed using MATLAB. Simulation results identify the better optimum values of the parameters. The values of the rest of the variables which are used in simulation are shown in Tables 1, 2, and 3.

4.2 Optimum Selection of Parameters

The values for modulation of spectrum sharing and multi-user interference are given below [1, 18]:

Figure 2 explains that it contains two phases of simulation where in the first phase, each and every UE is associated with high signal strength BS, such as BS which gives minimum path loss is selected by the UEs. In phase two, we pick one

Table 2 Throughput values measured in Gbps for the three cases with 30 BS/km^2 density

	Power constraint (i) value	Power constraint (ii) value	Power constraint (iii) value
Fully licensed 5%	0.0328	0.0362	0.0674
Hybrid 5%	0.0147	0.0190	0.0265
Fully pooled 5%	0.0003	0.0007	0.0007
Fully licensed 50%	0.3455	0.3848	0.4176
Hybrid 50%	0.3736	0.4492	0.5081
Fully pooled 50%	0.3143	0.4795	0.4878
Fully licensed 95%	0.9188	1.0770	1.0218
Hybrid 95%	1.9194	2.1810	2.0252
Fully pooled 95%	2.1926	2.6970	2.5461

Power limit for various power constraints are 30, 24.30 (GHz)

Table 3 Simulation parameters

Notation	Value	Description
A	0.3 km^2	Area of simulation
λ_{UE}	{300, 600}	UE density per km^2
λ_{BS}	{30, 60}	Base station density per km^2
F	28 GHz	Carrier frequency
P_{TX}	{24, 30} dBm	Transmit power
W	1 GHz	Total bandwidth
NF	7 dB	Noise figure
K	1	Cognitive radio BS
R_{TX}	{20, 26} dBm	Received power
M	4	Number of operators

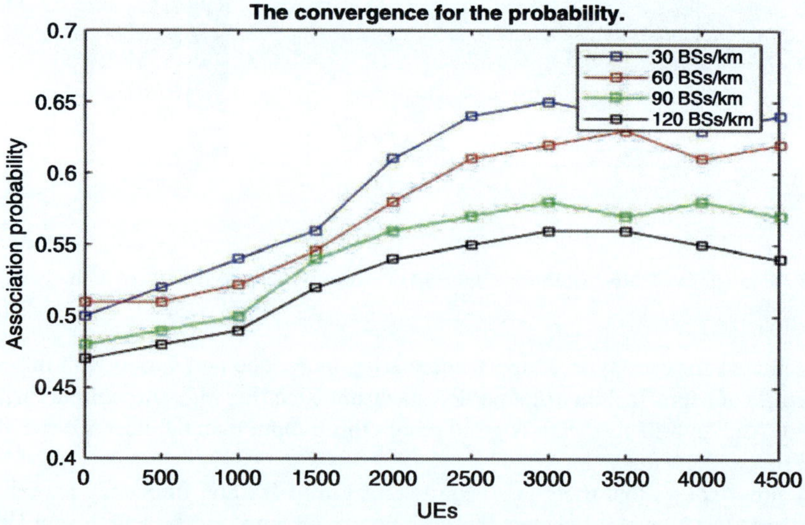

Fig. 2 Association probability of various BS densities with respect to UEs

UE at a time. By using Eqs. (5) and (7), we select the UE. Same process is followed for every UE. The numerical result in Fig. 2 uses this method. The association probability of base station densities (30, 60, 90, 120 BS/km^2) is calculated and varied with respect to various user equipment's association to the particular BS densities. The output is obtained by considering each BS density to the user equipment. The probability of user whether associated with low-frequency carrier or high-frequency carrier shows a stable value. The 30 BS/km^2 gives the high association probability as compared to other BS densities.

Figure 3 shows the different association algorithm performance when the selection of BS and frequency carrier is jointly selected by the UE. The throughput for different types of spectrum sharing techniques (licensed, pooled, hybrid) is calculated and varied with respect the percentile of users (5, 50, 95), i.e., where 5 percentile users indicate that they are using their data about 5%. The throughput

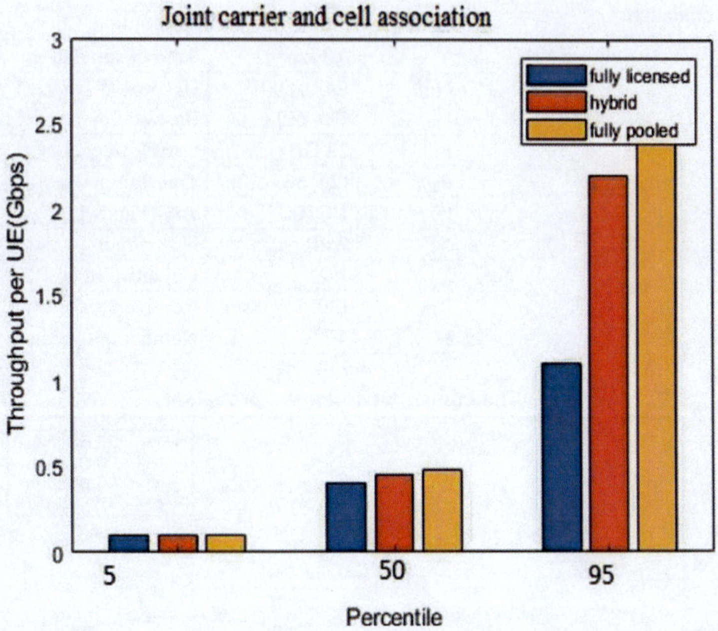

Fig. 3 Throughput of three baselines of 60 BS/km^2 with respect to percentile of users

is calculated for each type of spectrum sharing technique and varied with different percentile of users. It shows that pooled spectrum is getting high throughput than the hybrid. And, hybrid spectrum is getting high throughput than the licensed spectrum in all percentile values.

Figure 4 shows that there is no joint selection of BS and frequency carrier at a time. But the UE can only select the optimal carrier, and, for the whole simulation, the BS is kept constant. The throughput of different types for spectrum sharing techniques or three baselines (licensed, pooled, hybrid) is calculated and varied with respect to the percentile of users (5, 50, 95).

From these results, we can observe that for the best user such as 95th percentile, the throughput obtained by carrier only association is higher than the joint carrier and cell association. However, in medium case (50th percentile) and the worst users (5th percentile), the throughput obtained by carrier only association is lesser than that of the joint carrier and cell association. By definition of carrier only association, there is no user redistributing for balancing the BS load. Consequently, the association of the prime user to the lightly loaded BSs is done which obtain the higher data rate compared to the association of the cell and the carrier. Now for the same reason, the worst user gets the higher loaded BSs and gets poor performance. Therefore, the association of cell and carrier distributes the BSs and carriers most efficiently resulting in improved system.

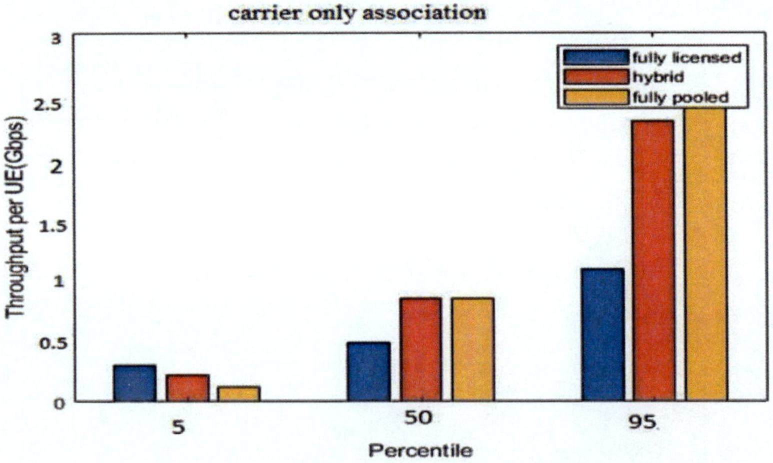

Fig. 4 Throughput of three baselines of 60 BS/km^2 with respect to percentile of users

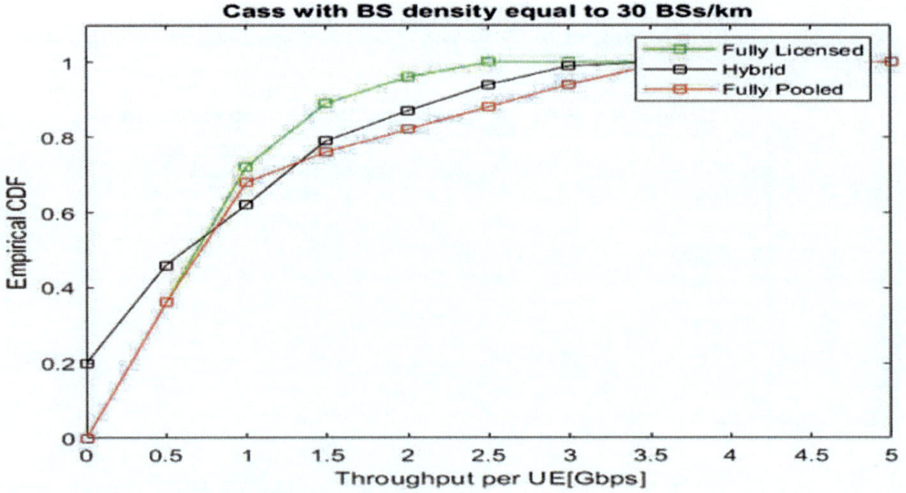

Fig. 5 Case with BS density equal to 30 BS/km^2

Figures 5 and 6 show that the hybrid case (blue line) is close to licensed curve (red line) for the worst user (bottom left of the graph), while the curve approaches the pooled (green curve) for the best user (top right), as usual. This shows that the hybrid scheme is close to the best performance in various situations (conditions).

Figure 7 shows that the cumulative distribution function (CDF) of each baseline is varied with the SINR and with the base station density equal to 60 BS/km^2. By increasing the bandwidth, there is decrease in SINR. All the baselines are considered and the SINR is calculated. By observing the SINR plot, we see that the three baselines give the different output.

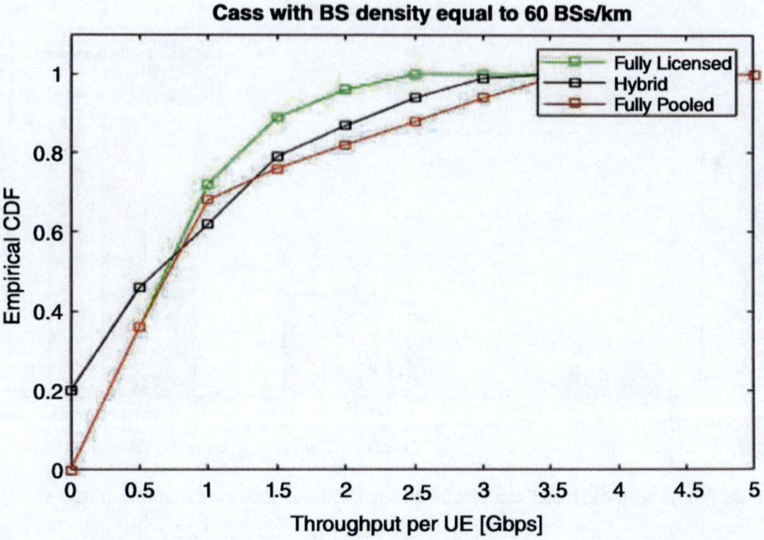

Fig. 6 Case with BS density equal to 60 BS/km^2. Compare these two cases with our three baselines

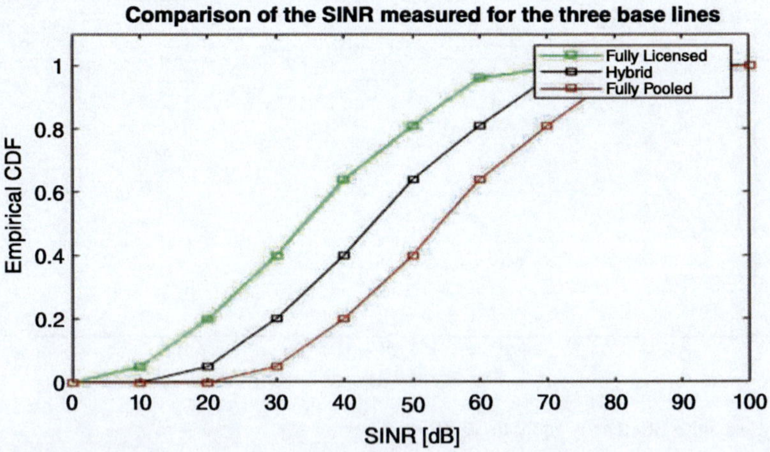

Fig. 7 SINR comparison for the three baselines

In Fig. 8 and from Table 2, the performance of fully pooled case is seen to have improved for the best user (95%) only and decreases for the medium and worst users (50%, 5%). As the bandwidth increases, the throughput of the fully licensed and hybrid is increased for all the users with the power constraint equal to 24 GHz.

Figure 9 shows that normalized ergodic throughput of dissimilar cognitive radio like multi-access channel, broadcast channel, parallel access channel, and our reference network varies with the number of CRs (k) [17]. This plot shows the

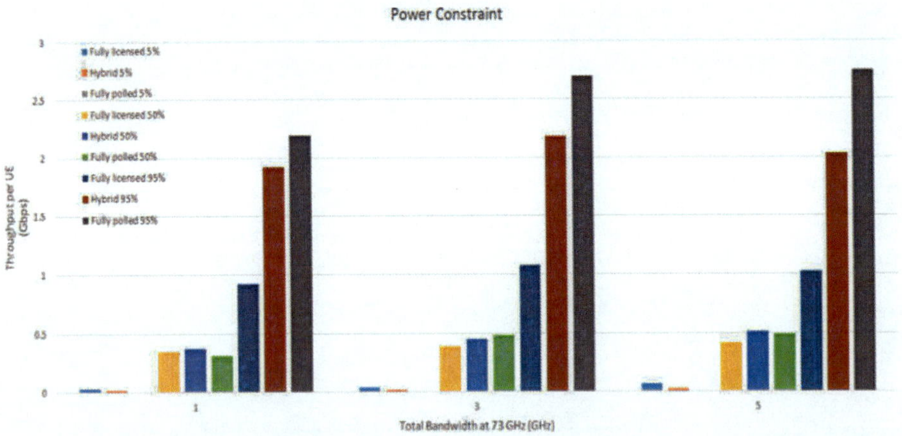

Fig. 8 Throughput comparison is done between three baselines and with varying total bandwidth 73 GHz, BS density equal to 60 BS/km^2, from Table 2

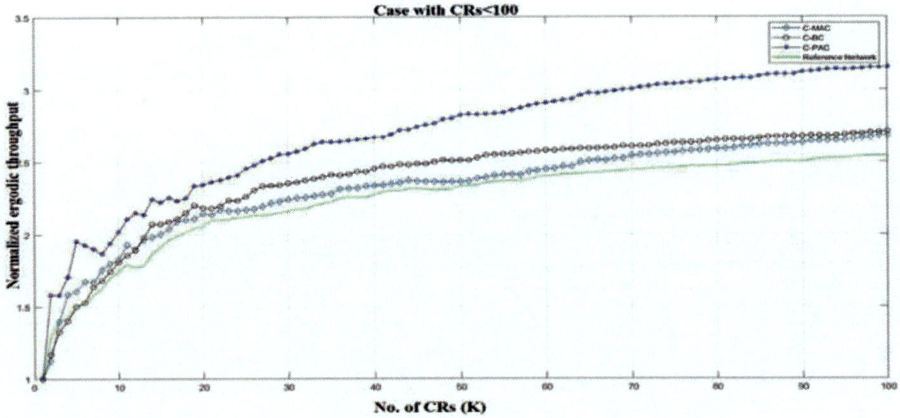

Fig. 9 CR ($K < 100$) networks and its normalized ergodic throughput (bps)

limited number of CRs ($k < 100$) where ($k = 1, \ldots, k$) CRs deployed with the UEs are already associated with the base station. The CRs which are additionally added to the network will help the UEs to quickly associate with the base station. Therefore, the throughput of CRs increases compared to the reference network.

Figure 10 describes ergodic throughput of Kth cognitive radio networks where ($K = 103$–106) CRs. In this, the ergodic throughput is calculated for large number of CRs associated with the base station. As the CR network number is increased, the interference between the CRs and UE is increased. From the plot, it is observed that the throughput of all CR types (MAC, BC, and PAC) decreases with increase in the CRs. We can observe that the throughput of reference network is higher than the other types.

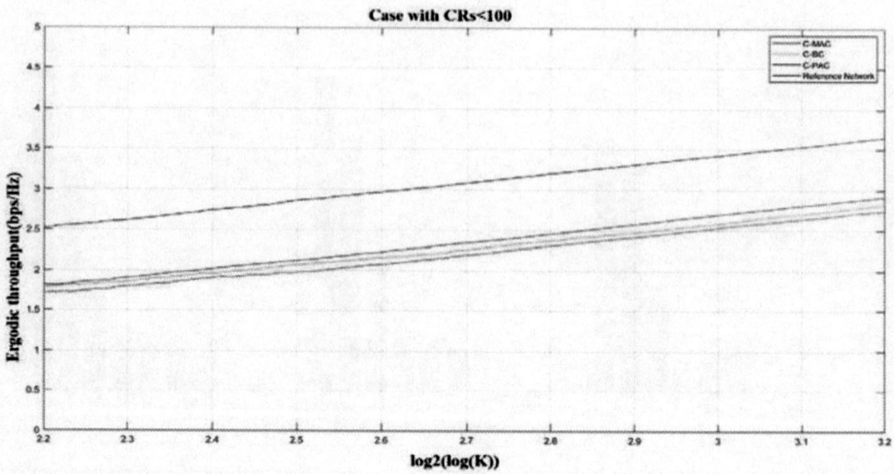

Fig. 10 CR ($K \geq 100$) networks and its asymptotic ergodic throughput

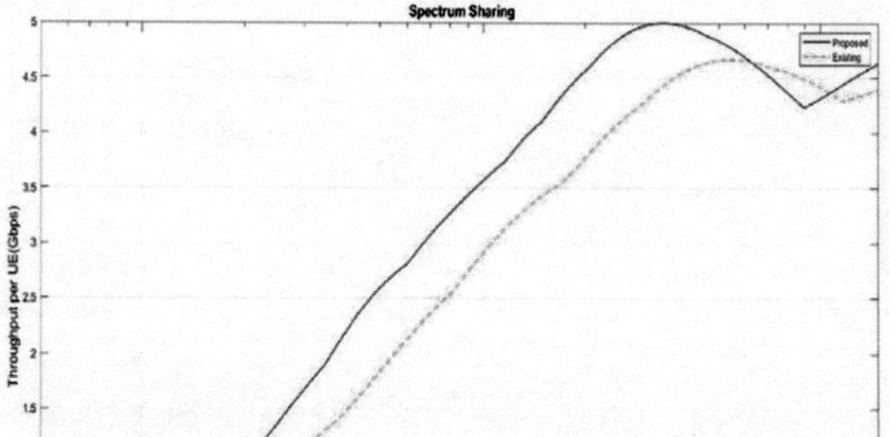

Fig. 11 Comparison of throughput between proposed and existing model

Figure 11 shows that the existing model (joint carrier and cell association) with the pooled case throughput is relatively lesser than the proposed model (different types of cognitive radio network model (MAC, BC, PAC)).

5 Conclusion

A millimeter wave-based different spectrum sharing model is investigated. Through analysis and simulation, we have shown the variation of baseline spectrum sharing (licensed, pooled, hybrid) with the number of user percentiles. We have also compared the cell association techniques like joint carrier and cell association and carrier only association. It is seen that the joint carrier and cell association offers higher throughput than the carrier only association. In this paper, we have also investigated the SINR for the three baseline techniques. Variations of SINR for number of CRs are also examined. Finally, we proposed a technique to obtain the high throughput from different CR types.

References

1. Rebato M, Boccardi F, Mezzavilla M, Rangan S, Zorzi M (2017) Hybrid spectrum sharing in mmWave cellular networks. IEEE Trans Cogn Commun Netw 3(2):155–168
2. Sheeba Kumari M, Rao SA, Kumar N (2019) Modeling and link budget estimation of directional mmWave outdoor environment for 5G. In: IEEE European conference on networks and communications (EuCNC 2019), June 18–21, Valencia, Spain
3. Sheeba Kumari M, Rao SA, Kumar N (2017) Outdoor millimeter-wave channel modeling for uniform coverage without beam steering. In: LNICST, vol 218. Springer, Cham, pp 233–244
4. Jorswieck EA et al (2014) Spectrum sharing improves the network efficiency for cellular operators. IEEE Commun Mag 52(3):129–136
5. Li G, Irnich T, Shi C (2014) Coordination context-based spectrum sharing for 5G millimeter-wave networks. In: Proceedings of the international conference on cognitive radio oriented wireless networks and communications, June 2014, pp 32–38
6. Lackpour A, Hamilton C, Jacovic M, Rasheed I, Rivas Rey X, Dandekar KR (2017) Enhanced 5G spectrum sharing using a new adaptive NC-OFDM waveform with reconfigurable antennas. In: 2017 IEEE international symposium on dynamic spectrum access networks (DySPAN), Piscataway, NJ, pp 1–2
7. Rebato M, Boccardi F, Mezzavilla M, Rangan S, Zorzi M (2010) Hybrid spectrum access for mmWave networks. In: 2016 Mediterranean ad hoc networking workshop (Med-Hoc-Net), Vilanovai la Geltru, pp 1–7
8. Zhao H et al (2013) 28 GHz millimeter wave cellular communication measurements for reflection and penetration loss in and around buildings in New York city. In: 2013 IEEE international conference on communications (ICC), Budapest, pp 5163–5167
9. Boccardi F et al (2016) Spectrum pooling in MmWave networks: opportunities, challenges, and enablers. IEEE Commun Mag 54(11):33–39
10. Weiss TA, Jondral FK (2004) Spectrum pooling: an innovative strategy for the enhancement of spectrum efficiency. IEEE Commun Mag 42(3):S8–S14
11. Yingxiao W, Zhen Y (2010) Spectrum sharing under spectrum overlay and underlay with multi-antenna. In: 2010 Ninth international conference on networks, Menuires, pp 237–242
12. Kim C, Ford R, Rangan S (2014) Joint interference and user association optimization in cellular wireless networks. In: 2014 48th Asilomar conference on signals, systems and computers, Pacific Grove, CA, pp 511–515
13. Nair SS, Schellenberg S, Seitz J, Chatterjee M (2013) Hybrid spectrum sharing in dynamic spectrum access networks. In: International conference on information networking 2013 (ICOIN), Bangkok, pp 324–329

14. Rebato M, Mezzavilla M, Rangan S, Zorzi M (2016) Resource sharing in 5G mmWave cellular networks. In: 2016 IEEE conference on computer communications workshops (INFOCOM WKSHPS), San Francisco, CA, pp 271–276
15. Zhang R, Cui S, Liang Y (2009) On ergodic sum capacity of fading cognitive multiple-access and broadcast channels. IEEE Trans Inform Theory 55(11):5161–5178
16. Sorayya R, Suryanegara M (2017) The model of spectrum sharing between a primary and two secondary operators. In: 2017 International conference on smart technologies for smart nation (SmartTechCon), Bangalore, pp 518–522
17. MacCartney GR, Samimi MK, Rappaport TS (2014) Omnidirectional path loss models in New York City at 28 GHz and 73 GHz. In: 2014 IEEE 25th annual international symposium on personal, indoor, and mobile radio communication (PIMRC), Washington, DC, pp 227–231
18. Samimi MK, Rappaport TS, MacCartney GR (2015) Probabilistic omnidirectional path loss models for millimeter-wave outdoor communications. IEEE Wirel Commun Lett 4(4):357–360

Performance Analysis of VLC Indoor Positioning and Demonstration

S. Rishi Nandan and Navin Kumar

1 Introduction

Visible light communication (VLC) is an emerging technology with many indoor and outdoor applications, and many are yet to be explored [1]. Because of certain unique features such as simultaneous dual functionality, low-cost design, and no interference with radio frequency (RF), the technology has attracted significant research and development (R&D) in recent years. Additionally, the development in Internet of Things (IoT) is helping push this technology for many use cases. Furthermore, as VLC uses unlicensed spectrum (light wave—430–770 THz), more and more opportunities are expected in the future. In the indoor applications, it can provide spectrum relief to the widely used Wi-Fi network [2].

Localization/positioning in the indoor as well as outdoor environment is becoming very significant, almost a necessity based on requirements [3]. While the popular global positioning system (GPS) [4] works relatively well outdoor with an accuracy of few hundreds of centimeter, it does not show this accuracy in the indoor environment. Many other technologies such as Wi-Fi-based localization [5] and cellular system-based localization are being explored for the indoor environment, which give different accuracies. Precise indoor positioning systems (IPS) could play an important role in many sectors. Retail sectors, industries, warehouses, hypermarkets, etc. have shown interest in the indoor positioning systems. Since, it is capable of providing improved navigation which in turn helps them avoid unrealized sales when customers find it difficult to search for the items they need. As targeted

S. Rishi Nandan · N. Kumar (✉)
Department of Electronic and Communication Engineering, Amrita School of Engineering, Bengaluru, Amrita Vishwa Vidyapeetham, Bengaluru, India
e-mail: navinkumar@ieee.org

A. Haldorai et al. (eds.), *2nd EAI International Conference on Big Data Innovation for Sustainable Cognitive Computing*, EAI/Springer Innovations in Communication and Computing, https://doi.org/10.1007/978-3-030-47560-4_31

advertising is possible using these systems, the revenue automatically increases. The increasing popularity of the indoor positioning systems is confirmed by the anticipated growth in IPS market from $5.22 billion in 2016 to a massive $40.99 billion by 2022 [6]. Few of the major applications would be real-time navigation, improving user experience, visually impaired navigation, and optimizing the layout. Despite the ever-increasing marketability, IPS remains a "grand challenge," as existing systems fail to provide precise location. Wi-Fi- and other RF-based approaches deliver accuracies measured in meters, and failure to provide information about the orientation makes them a poor choice for many applications like retail navigation and shelf-level advertising.

There are many IPS approaches such as infrared-based (IR), ultrasonic, Bluetooth, RFID, ultrawide band (UWB), and WLAN. IPS based on IR [7] gives us a highly accurate location, but it suffers from vulnerability from indoor lighting or sunlight interference. Also, the cost will be high if we desire a really accurate system as we need more receivers. Furthermore, using IR indoors may affect the human health. Ultrasonic-based systems [8] have a very simple system structure and give us high accuracy, but it suffers from problems such as multipath effects. The key to the Bluetooth-based systems [9] is the connection between the user's Bluetooth device and the Bluetooth anchor nodes in the building. This technology is low cost and consumes very less power, and almost every mobile device comes with this technology. It suffers from low accuracy compared to other systems and it's also susceptible to noise interference. Systems based on RFID [10] have high accuracy and low cost but are susceptible to the environmental conditions. Also, for both Bluetooth- and RFID-based systems, the propagation distance is short. All the above systems including UWB [11] require special devices to transmit signals making it expensive to deploy. WLAN-based system [12] uses wireless network signal making it cheaper to deploy. It has a reasonable accuracy, and the ease of integrating with the mobile device makes it a good choice. All of these systems have several shortcomings such as short distances, expensive, limited bandwidth, and multipath effects. VLC-based localization is another option which is found to be cost-effective and of high accuracy. Since, VLC can be built on existing infrastructure or with slight modifications, the technology is cost-effective.

In this work, various positioning algorithms have been analyzed and discussed. The algorithms have been simulated and compared for their error accuracy. At the same time, we have hardware that implemented two of the popular algorithms and demonstrated their working. The measured data in a small area is considered and error in localization is calculated. We have used four sets of LED emitters, used time-division multiple access scheme, and received signal strength indicator in properly divided and defined coordinates as grid for the receiver to calculate the positioning. It is found that we obtain an accuracy as close as a couple of centimeter. The study is unique of its kind in hardware development and demonstration of their capabilities with a different concept introduced in the algorithm.

The rest of the paper has four different sections. Section 2 describes the overall system configuration where the layout and other scenario are presented. Section 3 discusses different algorithms and their mathematical characteristics and analysis.

Section 4 presents an overview of the hardware design and demonstration as proof of concept. Demonstration of the hardware and system is also presented as result. Finally, Sect. 5 presents the conclusion.

2 System Configuration

A visible light communication based positioning system is analyzed in the indoor environment in a small area. Its description and layout is illustrated in this section. As shown in Fig. 1, four LED emitters (VLC transmitters) marked as T1, T2, T3, and T4 have been fixed at a constant height of 82 cm from the ground surface in a model of dimensions $70 \times 60 \times 82$ cm. Their coordinates are defined as T1(0,0,82), T2(70,0,82), T3(70,60,82), and T4(0,60,82) as shown in the figure. Such placement ensures efficient overlap of the coverage area, facilitating ease of receiving light by the receiver circuit. The receiver circuit can be moved around anywhere on the *XY* plane for it to successfully determine its own position after receiving the coordinate data from the transmitters.

The system in fact comprises of four transmitters and a single receiver or two receivers depending on the algorithm. Figure 2 describes a general VLC [13] transmitter and receiver for a simple understanding. The transmitter is a combination of LEDs and the LED driver. As the popular characteristic of VLC [14], LED performs dual function for lighting in general and helps in determining position.

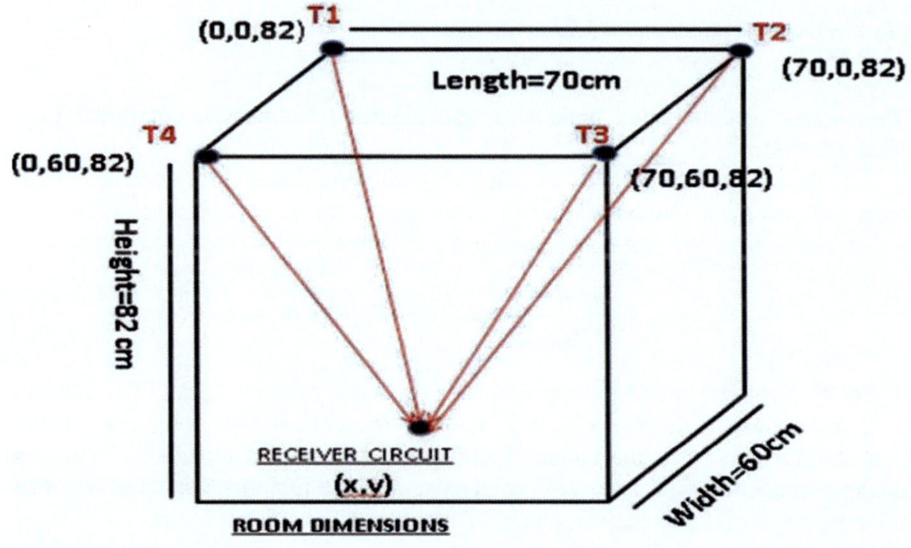

Fig. 1 System configuration

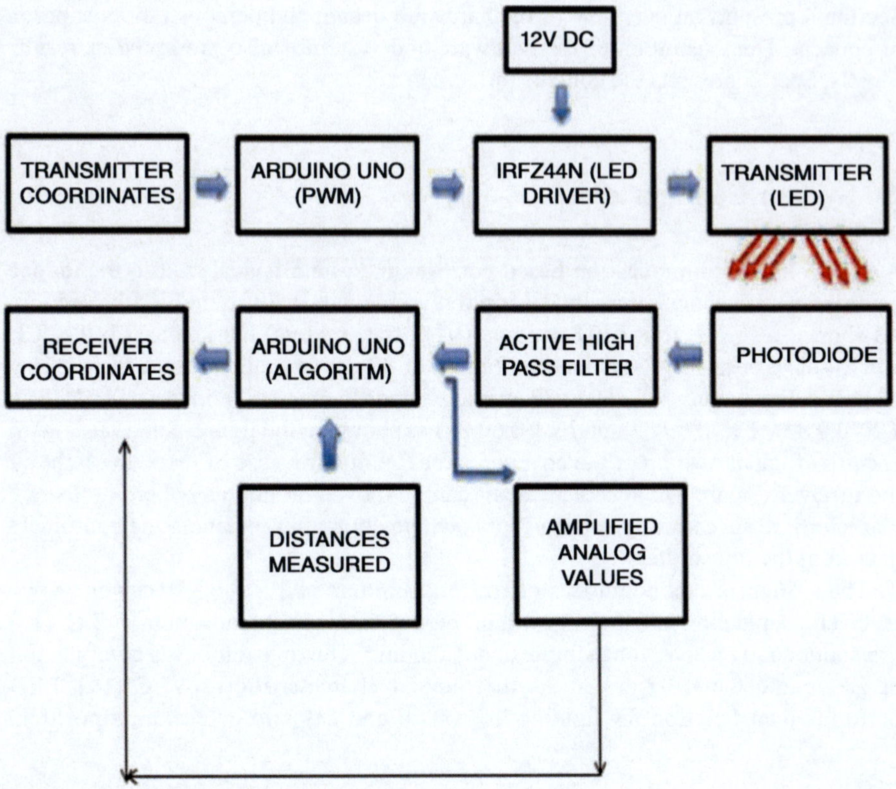

Fig. 2 VLC-based positioning system architecture

The receiver on the other hand is a light sensor (photodiode) connected to a microcontroller.

A LED light usually follows a Lambertian radiation pattern [15], and the optical received power is given by:

$$P_{\text{LOS}} = P_{\text{t}} \frac{(m+1)\,A}{2\pi {d_0}^2} \cos(\Phi)^m T_{\text{s}}(\theta)\, g(\theta) \cos(\theta) \tag{1}$$

where P_{t} is the transmitted power and m is the mode number of the LED pattern. A is the photodiode's physical area, $T_{\text{s}}(\theta)$ is the optical filter gain, $g(\theta)$ is the concentrator gain, Φ is the radiation angle w.r.t. transmitter normal axis and θ is angle of incidence w.r.t. receiver normal axis. d_0 is the Euclidean distance between the transmitter and the receiver, and FOV is the receiver's field of view.

The transmitter sends its own coordinates by switching on and off at a very high frequency undetected by the human eye. The receiver receives this modulated light and processes it with the help of a microcontroller to calculate the coordinates.

Knowledge about the distance or angle of arrival between the individual transmitters and receiver is essential. Once we have all the transmitter coordinates and the distance/angle between each transmitter and the receiver, we can perform a positioning algorithm to estimate the coordinates.

3 Localization Algorithms

In this section, we discuss the algorithm for positioning. Three different positioning algorithms based on triangulation are discussed. Triangulation uses the geometrical properties of triangles to estimate the absolute position. It's further subdivided into two types—angulation and lateration. Lateration methods determine the target object by evaluating the distances from all the transmitters. This method uses a variety of techniques to calculate the distances mathematically between the individual transmitters and the receiver such as received signal strength (RSS), time of arrival (TOA), and time difference of arrival (TDOA). Angulation works on the principle of measuring the angle of arrival (AOA) from every transmitter in the system.

3.1 Circular Lateration

We have discussed RSS-based circular lateration. This system measures the received signal power and calculates the propagation losses the signal has encountered. Using the path loss model, the distances are estimated.

Assume (M_i, N_i) to be the coordinates of the transmitters and (x, y) the coordinate of the receiver. R_i is the distance between the ith LED system and the receiver. Then each circle drawn can be expressed as:

$$(M_i - x)^2 + (N_i - y)^2 = R_i^2 \tag{2}$$

Theoretically, the intersection of three or more circles should yield the position of the receiver. However, this is not a realistic solution as noises will have a huge role to play:

$$\begin{aligned} R_i^2 - R_1^2 &= \left(x - M_i^2\right) + \left(y - N_i^2\right) - \left(x - M_1^2\right) - \left(y - N_1^2\right) \\ &= M_i^2 + N_i^2 - M_1^2 - N_1^2 - 2x\,(M_i - M_1) - 2y\,(N_i - N_1) \end{aligned} \tag{3}$$

where $i = 1, 2, \ldots, n$, and $n = 4$ (in our case) and R_1 denotes the distance between transmitter 1 and the receiver. The equations characterizing the system can be modified into:

$$AX = B \tag{4}$$

where

$$X = [x \;\; y]^T \tag{5}$$

$$A = \begin{bmatrix} (M_2 - M_1) & (N_2 - N_1) \\ \vdots & \vdots \\ (M_n - M_1) & (N_n - N_1) \end{bmatrix} \tag{6}$$

and

$$B = 0.5 \begin{bmatrix} (R_1{}^2 - R_2{}^2) + (M_2{}^2 + N_2{}^2) - (M_1{}^2 + N_1{}^2) \\ (R_1{}^2 - R_n{}^2) + (M_n{}^2 + N_n{}^2) - (M_1{}^2 + N_1{}^2) \end{bmatrix} \tag{7}$$

The least squares solution of the system is given as:

$$X = \left(A^T A\right)^{-1} A^T B \tag{8}$$

3.2 *Hyperbolic Lateration*

Hyperbolic lateration methods employ time difference of arrival (TDOA) measurements. The difference in time at which signals from two distinct transmitters arrived is calculated. The signals have to be relayed exactly at the same time, as transmitter synchronization is of utmost importance in TDOA-based IPS systems. Once the signal is received from two reference points, the difference in arrival time can be used to calculate the difference in distances between the receiver and the two reference points. The difference can be calculated using the equation: $\Delta d = c_* \Delta t$.

Every hyperbola can be expressed as:

$$D_{ij} = R_i - R_j = \sqrt{(M_i - x)^2 + (N_i - y)^2} - \sqrt{(M_j - x)^2 + (N_j - y)^2} \tag{9}$$

where $i! = j$ (i is not equal to j)

$$(R_1 + D_{i1})^2 = R_i{}^2$$

$$M_i{}^2 + N_i{}^2 - M_1{}^2 - N_1{}^2 - 2x\,(M_i - M_1) - 2y\,(N_i - N_1) - D_{i1} - 2D_{i1}R_1 = 0 \tag{10}$$

where $i = 1, 2, \ldots, n$. Equation (9) mentioned above can be transfigured into the form:

$$AX = B$$

$$X = \begin{bmatrix} x \; y \; R_1 \end{bmatrix}^T \tag{11}$$

$$A = \begin{bmatrix} (M_2 - M_1) & (N_2 - N_1) & D_{21} \\ \vdots & \vdots & \vdots \\ (M_n - M_1) & (N_n - N_1) & N_{21} \end{bmatrix} \tag{12}$$

and

$$B = 0.5 \begin{bmatrix} \left(M_2{}^2 + N_2{}^2\right) - \left(M_1{}^2 + N_1{}^2\right) - (D_{21})^2 \\ \left(M_n{}^2 + N_n{}^2\right) - \left(M_1{}^2 + N_1{}^2\right) - (D_{n1})^2 \end{bmatrix} \tag{13}$$

The least squares solution of the system is given as in Eq. (8):

$$X = \left(A^T A\right)^{-1} A^T B$$

In this work, we have not used the TDOA to estimate the distance between the receivers and the transmitters as the height of our system is very small and the time difference of arrival of signals from two reference points would be negligible. Since we have only simulated the position of the receiver for hyperbolic lateration, we have calculated the distances between the transmitters and the receiver using geometry as we have simulated an ideal case for positioning.

3.3 *Angulation*

The angles of arriving signals from every transmitter are measured by the receiver in an angle of arrival-based system. AOA-based systems do not require synchronization, and this feature could be a blessing in disguise as it reduces the complexity of the system. Also, most smartphones currently in use possess front-facing cameras which are inherently imaging receivers.

Let a_j denote the angle measured with respect to the jth transmitter:

$$\tan a_j = \frac{(y - N_j)}{(x - M_j)} \tag{14}$$

where $j = 1, 2, \ldots, n$, we can rewrite the equation as:

$$(x - M_j)\sin a_j = (y - N_j)\cos a_j \tag{15}$$

The equations representing the angulation-based IPS system can be modified into a matrix form:

$$AX = B$$

where

$$\begin{bmatrix} x\ y \end{bmatrix}^T$$

$$A = \begin{bmatrix} -\sin a_1 & \cos a_1 \\ \vdots & \vdots \\ -\sin a_n & \cos a_n \end{bmatrix} \tag{16}$$

and

$$B = \begin{bmatrix} N_1 \cos a_1 - M_1 \sin a_1 \\ \vdots \\ N_n \cos a_n - M_n \sin a_n \end{bmatrix} \tag{17}$$

The least squares solution of the system is given as in Eq. (8):

$$X = \left(A^T A\right)^{-1} A^T B$$

Since we have only simulated angulation algorithm for an ideal positioning case, we calculated the distance between the transmitters and the receiver using geometry.

3.4 Differential Detection

The transmitters suffer from illumination fluctuations. The fluctuation of LED's illumination will increase the error in positioning. The positioning error must be reduced to ameliorate the accuracy, and we satisfy this desired condition by minimizing the impact of LEDs' flickering. The transmitter power P_t is not fixed. We can use $P_t(t)$ to represent it and calculate the received power in multiple time stamps.

For a photodiode, two samples of the received power have a relationship as:

$$\frac{P_{rA,t1}}{P_{rA,t2}} = \frac{P(t1)_0}{P(t2)_0} \tag{18}$$

We derive

$$R = \frac{P(t1)_0 - P(t2)_0}{P(t1)_0} \tag{19}$$

where R denotes the change rate of the signal intensity. R is unrelated to the detector's position; it only depends on the received power/intensity in various time stamps. For the two photodiodes taken into consideration, we designate one to be the base detector as we are aware of the received power in different time stamps. As both the photodiodes receive light concurrently from the transmitters, the base photodiode can compensate the variations in received intensity values of the other mobile detector [16].

We have assumed the base detector (A) to be in the middle of the room and the mobile detector (B) intensity has to be compensated. The signal strength measured by the base detector from a single transmitter for continuous time stamps can be recorded as $A1, A2, A3, \ldots$ Then the change rate of signal intensity can be written as $R_1 = \frac{A2-A1}{A1}$, $R_2 = \frac{A3-A1}{A1}$, ..., $R_{n-1} = \frac{An-A1}{A1}$. For the mobile photodiode B, the received signal power B_n can be calculated as superposition of varying light illumination and original received signal strength:

$$B_n = B_1\,(1 + R_{n-1}) \tag{20}$$

where B_1 is the received power at any particular coordinate of the mobile photodiode and R is calculated using the received power of the base detector at different time stamps.

This value is not governed by the flickering of the light source. The discussion above is restricted to one transmitter and the intensities received by two detectors. We can extend this idea to our system which uses four LEDs. After we obtain the measured received signal strength of each transmitter by the mobile photodiode (B), we can perform circular lateration algorithm using the compensated intensities.

4 Hardware Design and Results

We discuss about the hardware design and the results in this section. Both sets of results are presented; the simulations and the hardware demonstration.

4.1 *Hardware Description*

The hardware mainly consists of two separate crucial systems. In the transmitter system, we have used a high-speed power MOSFET (IRFZ44N) as the LED driver. The LED driver is capable of alternating between PWM signal input and the DC input. We are able to achieve a transmission frequency of 10 kHz using the Arduino Uno.

The system consists of four transmitters as shown in Fig. 3. Each transmitter is a closely wired network of four LEDs and a high-speed MOSFET connected to a 12 V DC power supply. The receiver we have designed in our implementation is a combination of a photodiode and an active high pass filter. While photodiode detects the light, the active high pass filter blocks DC current and amplifies the current generated due to modulated light.

For this demonstration, we have assumed a noise-free environment. For lux measurement, we have used an android application, light meter.

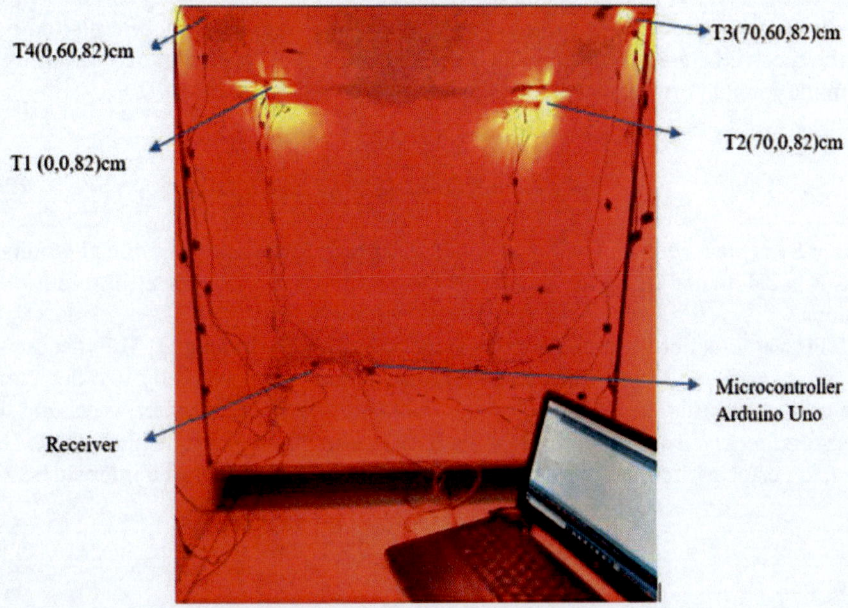

Fig. 3 Hardware demonstration (circular lateration)

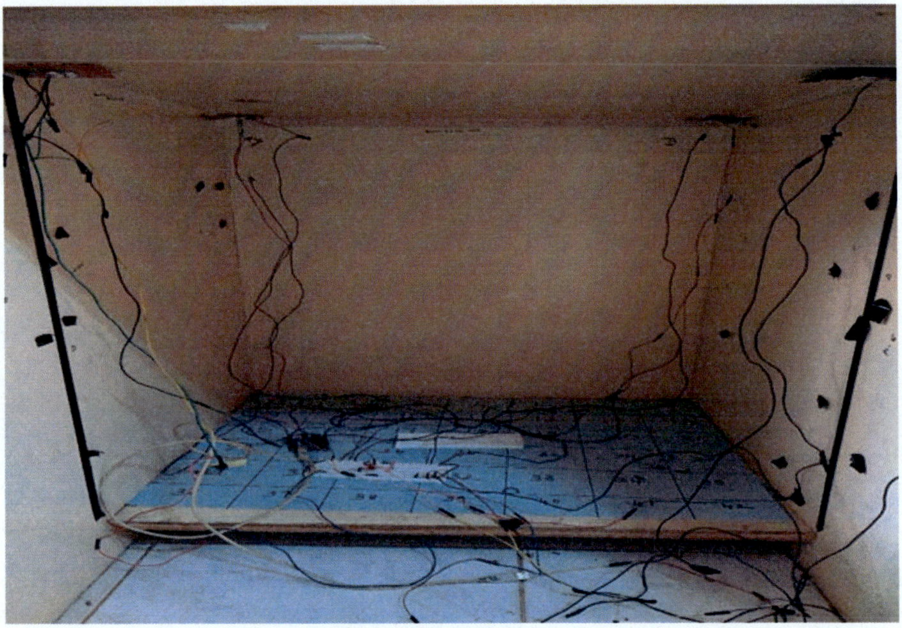

Fig. 4 Hardware demonstration (differential detection)

As shown in Fig. 3, we use a single photodetector for localization. The area under consideration is lit up by four transmitters powered by a DC supply. The data is transmitted using time-division multiple access (TDMA).

The receiver (photodiode + Active high pass filter) can be moved around anywhere in the given area, and upon receiving the coordinates of the transmitters, the receiver with a help of a microcontroller (Arduino) finds its own coordinates in our system model.

4.2 *Results*

Figure 5 shows the plot of the coordinates of the algorithms simulated in MATLAB for circular lateration, hyperbolic lateration, and angulation algorithms. We see that all algorithms map to the defined coordinates. This is an important step before we measure the position of an object using different algorithms in our system model.

The error plot is shown in Fig. 6. The error obtained from circular lateration is very small and can be neglected. However, the hyperbolic lateration and angulation show small error. It is found that the small error is inherent in these algorithms compared to the ideal theoretical value. Although the error is small, in order to compare with the actual position measured with hardware, we have to consider them as reference.

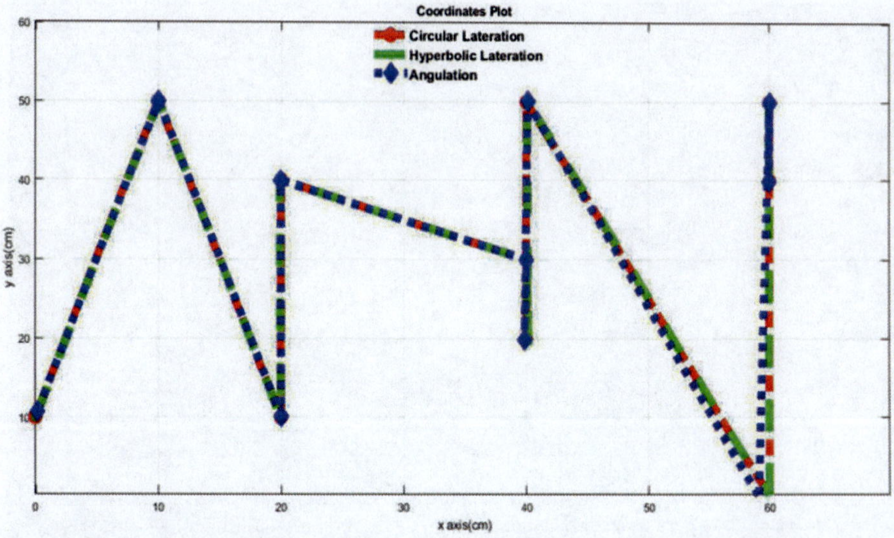

Fig. 5 Coordinate plots for algorithms

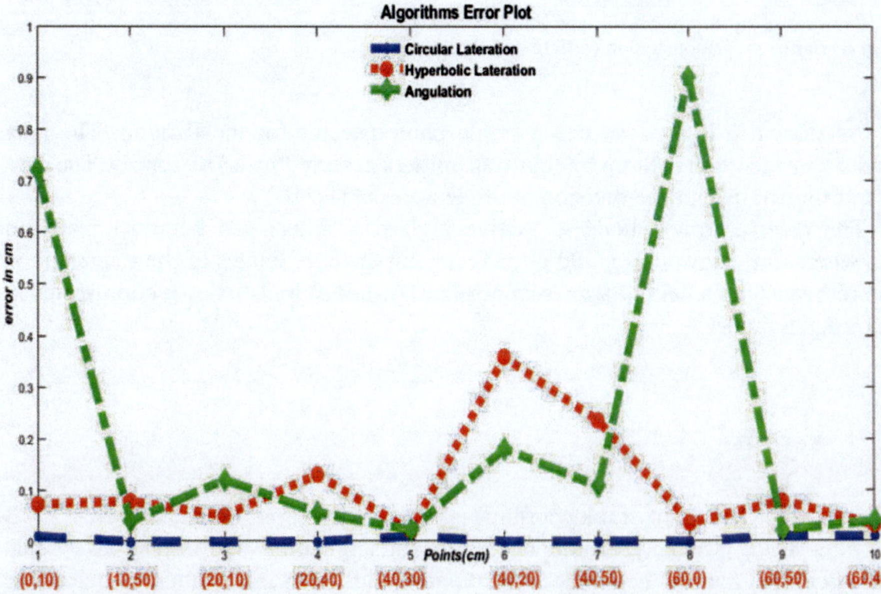

Fig. 6 Algorithm error plot

Figure 7 is the plot for measurement of positioning by demonstrating circular lateration and differential detection in a hardware model. *X*-axis has the same ten sample points as in simulation. The *Y*-axis gives the error in measurement compared

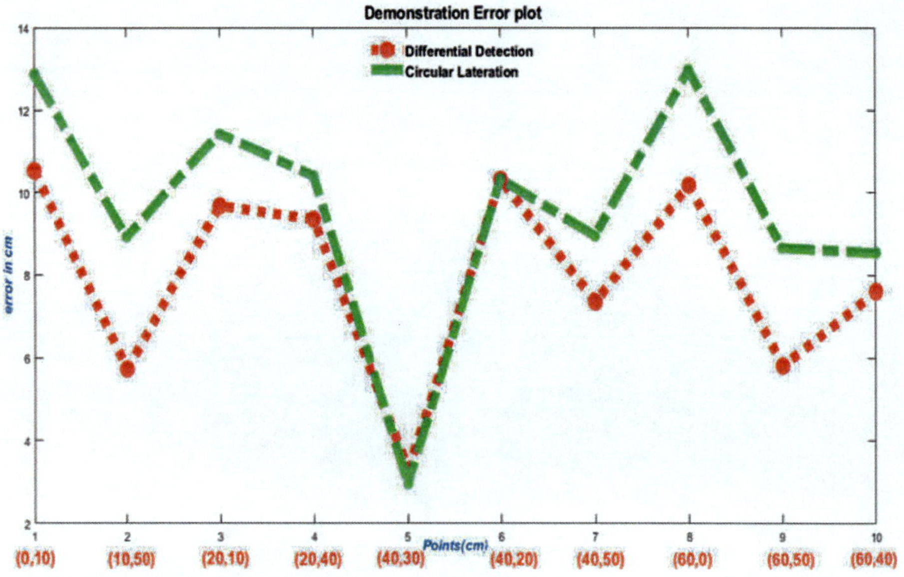

Fig. 7 Demonstration error plot

to the actual position. The maximum error observed in the demonstration using circular lateration is 13 cm, whereas for differential detection, the maximum error is 10 cm. The increased accuracy in differential detection is due to the use of two photodetectors instead of one. The error we have obtained in this demonstration is mainly due to the use of a non-precise lux meter to calculate the received power. Hence, the error in calculating distance between the LEDs and receiver plays a huge role in the precision of our system.

Figure 8 shows the 3D intensity/received power distribution of the four transmitters in the defined area. We can estimate the received power at each point in the defined area using this figure, and we can conclude the power distribution of each transmitter. It can help us to calculate the signal-to-noise ratio if the system is also used for data communication simultaneously with positioning

4.3 Summary of Measurement

In Table 1, the coordinates obtained from simulation and hardware demonstration as well as actual coordinates are shown. Using this, the error can be calculated. It is seen that the most accurate algorithm among the three is circular lateration. The distance between the transmitters and the receiver considered for the simulations is exact and was calculated by geometry.

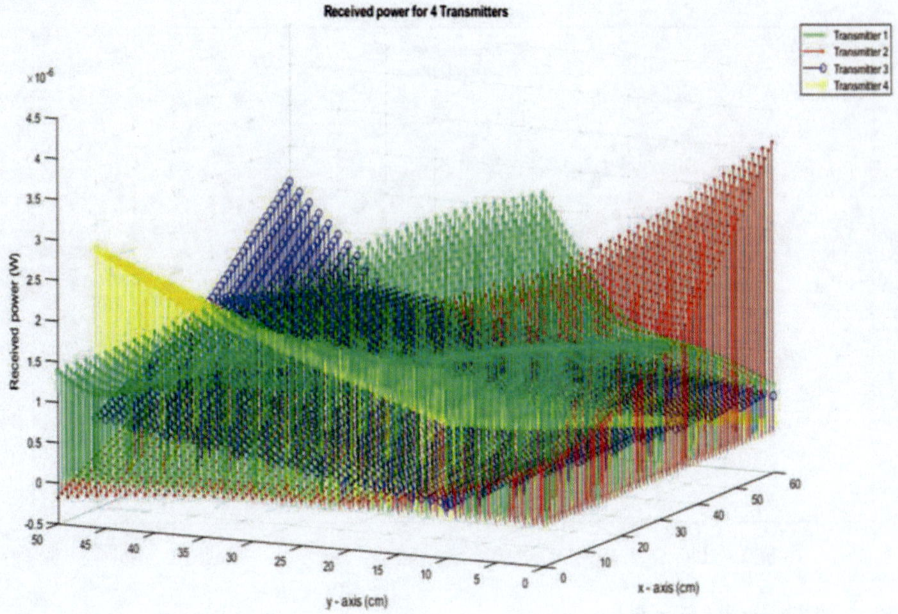

Fig. 8 Received power distribution for four transmitters

The last two columns give us the location obtained through hardware demonstration. In this, the received signal strength was used to calculate the distances between the transmitters and the receiver.

5 Conclusion

A comparative study of indoor positioning system using VLC is performed for widely popular algorithms. It is found that the hyperbolic lateration and angulation algorithms inherit small error. However, circular lateration is found to give the best accuracy. We have also implemented the system in hardware and demonstrated two of them and compared the performance with theoretical results. Hardware implementation, as expected, gives small error of a couple of cm which is still precise in indoor positioning applications. Using a good microcontroller and an accurate lux meter would further reduce the error. The area considered for our demonstration of system model is small and hence we could not validate all the algorithms, but the performance can be visualized using the demonstrated one. It would be possible to use visible light communication based positioning system in the indoor environment.

Table 1 Comparison of coordinates obtained with the actual value

No.	Actual position	Circular lateration algorithm position (MATLAB simulation)	Hyperbolic lateration algorithm position (MATLAB simulation)	Angulation algorithm position (MATLAB simulation)	Circular lateration (distance between Tx and Rx calculated using RSS) (demonstration)	Differential detection (demonstration)
1	(60,0)	(60,0)	(59.98,0.03)	(59.10,0.04)	(50.46,8.77)	(52.60,6.97)
2	(0,10)	(0,10.01)	(0.06,10.04)	(0.17,10.70)	(10.85,16.89)	(7.51,17.36)
3	(20,10)	(20,10)	(19.97,9.96)	(20.01,10.12)	(24.53,20.48)	(23.08,19.16)
4	(40,20)	(40,20)	(40.16,19.68)	(39.85,19.90)	(32.52,27.10)	(32.17,26.70)
5	(40,30)	(39.99,30)	(40.02,30)	(40.02,29.98)	(39.31,32.89)	(39.91,33.29)
6	(20,40)	(20,40)	(19.89,40.07)	(19.96,39.96)	(28.44,33.93)	(27.53,34.47)
7	(60,40)	(60.01,40)	(59.97,39.98)	(60,39.96)	(52.22,43.48)	(54.61,45.35)
8	(10,50)	(10,50)	(10.06,49.95)	(10.04,50)	(16.72,44.15)	(14.19,46.10)
9	(40,50)	(40,50)	(40.05,50.23)	(40.10,50.04)	(37.76,41.34)	(38.14,42.90)
10	(60,50)	(60,49.99)	(59.94,49.95)	(60.02,49.98)	(51.89,47.01)	(54.23,49.36)

References

1. Rahman MS, Haque MM, Kim K-D (2011) Indoor positioning by LED visible light communication and image sensors. Int J Electr Comput Eng 1(2):161–170
2. Varshney V, Goel RK, Qadeer MA (2016) Indoor positioning system using Wi-Fi and bluetooth low energy technology. In: 2016 13th International conference on wireless and optical communication networks
3. Wang L, Guo C, Luo P, Li Q (2017) Light localization algorithm based on received signal strength ratio with multi-directional LED array. In: ICC: WS01—The 3rd workshop on optical wireless communications (OWC)
4. Cherntanomwong P, Chantharasena W (2015) Indoor localization system using visible light communication. In: 7th International conference on information technology and electrical engineering (ICITEE)
5. Lashkari AH, Parhizkar B, Mike NG, Ngan AH (2010) WiFi-based indoor positioning. In: 2010 2nd International conference on computer and network technology
6. https://www.geospatialworld.net/blogs/indoor-positioning-indoors-gps-stops-working/
7. Cahyadi WA, Chung YH, Adiono T (2019) Infrared indoor positioning using invisible Beacon. In: 2019 11th International conference on ubiquitous and future networks (ICUFN)
8. Majchrzak J, Michalski M, Wiczynski G (2009) Distance estimation with a long range ultrasonic system. IEEE Sens J 9(7):767–773
9. Kalbandhe AA, Patil SC (2016) Indoor positioning system using bluetooth low energy. In: 2016 International conference on computing, analytics and security cast (CAST)
10. Saab SS, Nakad ZS (2011) A standalone RFID indoor positioning system using passive tags. IEEE Trans Ind Electr 58(5):1961–1970
11. Ling RWC, Gupta A, Vashistha A, Sharma M, Law CL (2018) High precision UWB-IR indoor positioning system for IoT applications. In: 2018 IEEE 4th world forum on internet of things (WF-IoT)
12. Chen C, Chen Y, Lai H-Q, Han Y, Ray Liu KJ (2016) High accuracy indoor localization: a WiFi-based approach. In: 2016 IEEE international conference on acoustics, speech and signal processing (ICASSP)
13. Anand M, Kumar N (2014) New effective and efficient dimming and modulation technique for visible light communication. In: 2014 IEEE 79th Vehicular technology conference (VTC Spring)
14. Jha MK, Addanki A, Lakshmi YVS, Kumar N (2015) Channel coding performance of optical MIMO indoor visible light communication. In: 2015 International conference of advances in computing, communications and informatics, ICACCI 2015. Institute of Electrical and Electronics Electronics Engineers Inc., pp 97–102
15. Shawky S, El Shimy MA, El-Sahn ZA, Rizk MRM (2017) Improved VLC—based indoor positioning system using a regression approach with conventional RSS techniques, 978-1-5090-4372-9/17. 13th IEEE international conference on wireless communications and mobile computing conference (IWCMC)
16. Lv H, Feng L, Yang A, Guo P, Huang H, Chen S (2017) High accuracy VLC indoor positioning system with differential detection. IEEE Photon J 9(3):7903713

Ranking of Educational Institutions Based on User Priorities Using Multi-criteria Decision-Making Methods

A. U. Angitha and M. Supriya

1 Introduction

In this age of science and technology, India holds a significant place in the education industry. Moreover, India has one of the largest populations in the age range of 5–24 years. This results in huge demand for technical or professional education today. The education industry has grown nominally in recent years, so does the number of students. Today, students prefer higher education and also are going abroad to take up various courses. The Government of India is focused on getting universal recognition for its education system. Many colleges are trying to get NBA (National Board of Accreditation) accreditation. Due to this, a number of ranking schemes have been introduced to ensure education standards in India. Some of the ranking systems followed for educational institutions along with the parameters they emphasize on are presented in Table 1.

The existing ranking systems grade the educational institutions based on some of the predefined parameters listed in Table 1. With the aim of helping people know the best and the worst among them, a number of factors were taken into consideration, and accordingly the institutions were ranked. Several ranking systems were proposed and their evaluation was based on several criteria. In order to handle such multiple criteria, multi-criteria decision-making (MCDM) was introduced. A number of MCDM methods exist such as MAUT, ELETRE, PROMTHEE, OMDM and AHP.

A. U. Angitha (✉) · M. Supriya
Department of Computer Science and Engineering, Amrita School of Engineering, Bengaluru, Amrita Vishwa Vidyapeetham, Bengaluru, India
e-mail: m_supriya@blr.amrita.edu

A. Haldorai et al. (eds.), *2nd EAI International Conference on Big Data Innovation for Sustainable Cognitive Computing*, EAI/Springer Innovations in Communication and Computing, https://doi.org/10.1007/978-3-030-47560-4_32

Table 1 Various ranking schemes available in India

National Institutional Ranking Framework (NIRF)	Quacquarelli Symonds (QS)	Times Higher Education (THE)	National Assessment and Accreditation Council (NAAC)
Teaching, learning and resources (TLR)	Academic reputation	Teaching (the learning environment)	Curricular aspects
Research and professional practice (RP)	Employer reputation	Research (volume, income and reputation)	Teaching, learning and evaluation
Graduation outcomes (GO)	Faculty/student ratio	Citations (research influence)	Research, innovations and extension
Outreach and inclusivity (OI)	Citations per faculty	International outlook (staff, students and research)	Infrastructure and learning resources
Peer perception	International faculty ratio	Industry income (knowledge transfer)	Student support and progression
			Governance, leadership and management
			Institutional values and best practices

This paper focusses on ranking of educational institutions based on user-defined preferences. This work considers few of the contributing attributes from the NAAC parameters and models a ranking system using MCDM methods AHP and PROMETHEE. The attributes include resources related to teaching and learning such as strength of the students (includes doctoral students too) and ratio of faculty to students, graduation outcomes such as combined percentage for placements rate, higher education and entrepreneurship and many more.

The existing work and the problem description are discussed in Sects. 2 and 3. The implementation and the performance details are mentioned in the subsequent section. The paper is concluded in Sect. 5.

2 Literature Survey

Multi-criteria decision-making is a sub-field of operations research that deals with conflicting criteria. It generally involves choosing the best alternative among several potential candidates, subject to a number of criteria or attribute that may or may not be concrete. The goal is to choose an optimal solution which precisely analyses a number of criteria and helps in making decisions. Figure 1 gives an overview of how a decision or evaluation would be carried out.

MCDM has been used in helping librarians for ranking materials and in digitization and renewal of resources based on PROMETHEE [1]. It has also been

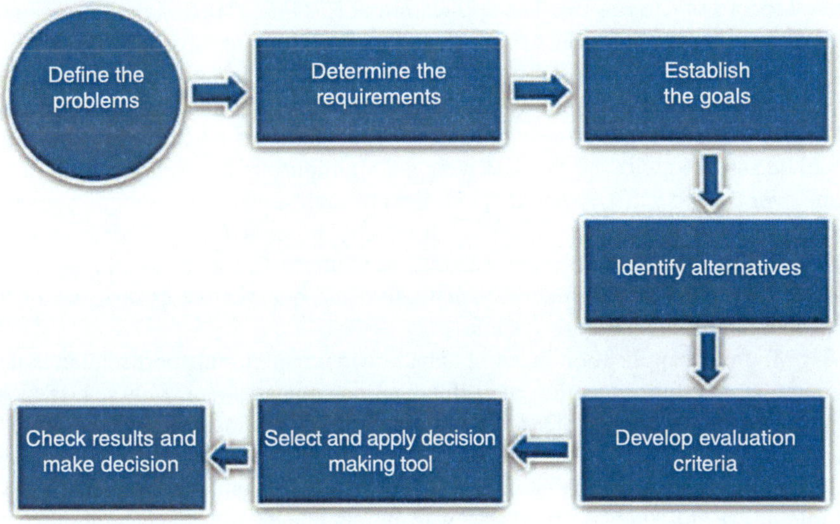

Fig. 1 Decision-making process

implemented in the educational sector for evaluating scholarly performance of the students based on several factors such as academic portfolios, participation in the extracurricular activities, biographical details, etc. [2]. MCDM also finds its use in evaluating software reliability growth models (SRGMs) using entropy distance-based approach (EDBA) in the software reliability domain [3]. Unfortunately, in real situations, decisions regarding optimal solution would involve certain conflicts and dissatisfaction.

To support multiple criteria decision-making for uncertain data, fuzzy-based MCDM has been introduced in literature [4]. It permits inputs which are not precise and allows to solve problems of greater complexity using a minimal set of rules. The disadvantage of fuzzy systems is that it sometimes is difficult to develop and may demand for a number of simulations to cope up with a real scenario. Fuzzy set theory finds its application in various other areas such as economic, social, medical, engineering and management. In 2013, a model for Trust Management built upon Fuzzy Logic was used to compare the cloud service providers offered within the market [5].

PROMETHEE helps in finding a best suitable alternative that suits the goal and is relevant to the gist of the problem rather than narrowing down to a "right" decision. A variation of PROMETHEE is PROMETHEE II, which enables the application of the model in situations where the criteria to be considered are more in number and the goal seems to be the same as in the case of any decision-making approaches. This method works on the basis of similarity measure wherein all the alternatives are compared to highlight the similar and the dissimilar features among them to determine the net flow scores. Based on these scores, the method gives a total ordered record of choices [6]. The application of PROMETHEE II in digital library management system helps the user to rank the collection areas for faster access [1].

The current paper centres around PROMETHEE and carries out the ranking of the educational institutions using multiple criteria. As the system emphasizes on the user preferences, the criteria considered for ranking need to be evaluated and should be assigned with appropriate weights, highlighting their relative importance of one over the other. So, to quantify the level of importance, a systematic approach considered in this work is analytical hierarchy process (AHP). AHP makes a pairwise comparison of the criteria based on mathematics and psychology [7]. It is a tool that helps to transform the qualitative and quantitative evaluations of the decision-maker into a ranking system.

AHP found its application in supply chain management, transport-related issues, e-commerce, portfolio selection and safety and risk management [8–11]. AHP along with VIKOR method of MCDM formed a hybrid MCDM system and was used to rank institutions in [12]. To evaluate the performance of nearly 20 National Institutes of Technology, an approach that integrated PROMETHEE and geometrical analysis for interactive aid (GAIA) methods was implemented considering nine pivotal criteria [13]. Internet service providers were also ranked according to the user preferences using AHP and technique for order of preference by similarity to ideal solution (TOPSIS) methods [14]. A hierarchical trust model was proposed in 2015 to rank service providers along with numerous plans for their infrastructure [15]. A comparative study between AHP-based and fuzzy-based MCDM techniques was conducted to rank the cloud computing services in which fuzzy AHP seemed to give a clearer set of values compared to the AHP mechanism [16]. In the subsequent year, another combination of analytical and fuzzy MCDM methods was used to rank cloud service providers based on trust [17].

The structure of AHP is depicted in Fig. 2. In this paper, the selection of the alternative is corresponding to the selection of the educational institution, and the various parameters are the criteria. Each criterion, constituted as a vector, is multiplied by its weight and results in the score of the alternative with respect to the criterion.

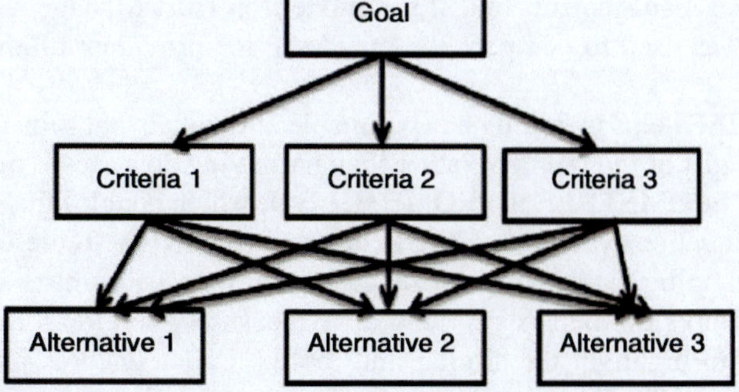

Fig. 2 Structure of AHP

3 Problem Description

Ranking of the institutions is based on certain parameters which are of great priority to the users. The association between the criteria, alternatives and the ultimate result is depicted in Fig. 3 at distinct levels of AHP. This figure represents the criteria/attributes and the alternatives enlisted in Table 2. The alternatives correspond to the colleges being considered. The values corresponding to each attribute are taken from the Annual Quality Assurance Report submitted by the colleges every year to the NAAC committee which is generally accessible from the college website. The colleges considered for the work are among the top colleges and universities in India. Henceforth, in this work, the attributes considered are renamed as A1, A2, etc., and the alternatives are represented as college1, college2, etc.

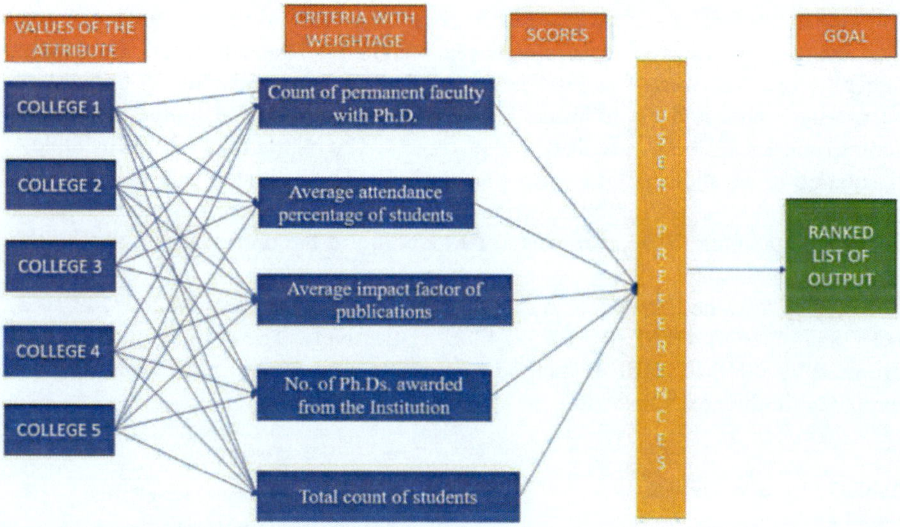

Fig. 3 Relationship between the parameters, alternatives and the result

Table 2 Attributes and contributing parameters

	No. of permanent faculty with PhD (A1)	Average percentage of attendance of students (A2)	Avg. impact factor of publications (A3)	No. of PhDs awarded by faculty from the institution (A4)	Total no. of students (A5)
College1	503	95	2.726	38	19, 213
College2	239	97.5	1.67	85	14, 325
College3	537	80	1	112	23, 804
College4	535	87.5	3	80	18, 232
College5	1219	88.14	0.896	282	33, 836

The user can assign scores to each attribute based on his/her preference. These scores are taken up by AHP to assign weights to the attributes as described in the previous section. After the weights are assigned, PROMETHEE method compares the relative importance of each institution and provides the customized ranked list of colleges that can be considered while selecting an appropriate institution.

So, this piece of work addresses ranking of the educational institutions based on the user preferences. AHP and PROMETHEE are the MCDM methods which will estimate the rank.

4 Results and Discussions

4.1 AHP-Based Weight Derivation

Initially, every attribute is assigned a weight by the user based on their preference. The weights are derived from the scores assigned to the attributes. The scores generally range between 1 and 9, 9 being the most preferred attribute. A test case of score assignment is listed in Table 3. AHP performs a pairwise comparison of the scores for each attribute as represented in Table 4. The matrix thus formed in Table 4 is squared in iterations till it matches a threshold. The next step is to calculate the row sum of the matrix in Table 5 to normalize the data. These normalized values denote the weights to the attributes corresponding to the preferences given by the user in Table 3.

Weights for the attributes A1, A2, A3, A4 and A5 are 0.22857143, 0.2, 0.14285714, 0.17142857 and 0.25714286, respectively. The weights thus derived are used by PROMETHEE II method to rank the educational institutions which is explained in the next subsection.

Table 3 Scores assigned to the attributes

Count of permanent faculty with PhD	8
Average attendance percentage of students	7
Average impact factor of publications	5
No. of PhDs awarded from the institution	6
Total count of students	9

Table 4 Pairwise comparison of the scores available for the five attributes

	C1	C2	C3	C4	C5
C1	1	1.142857	1.6	1.333333	0.888889
C2	0.875	1	1.4	1.166667	0.777778
C3	0.625	0.714286	1	0.833333	0.555556
C4	0.75	0.857143	1.2	1	0.666667
C5	1.125	1.285714	1.8	1.5	1

Table 5 Squared matrix formed after two iterations

	C1	C2	C3	C4	C5
C1	125	142.8571	200	166.6667	111.1111
C2	109.375	125	175	145.8333	97.22222
C3	78.125	89.28571	125	104.1667	69.44444
C4	93.75	107.1429	150	125	83.33333
C5	140.625	160.7143	225	187.5	125

Table 6 Values obtained on applying (1)

	A1	A2	A3	A4	A5
C1	0.269388	0.857143	0.869772	0	0.250525
C2	0	1	0.367871	0.192623	0
C3	0.304082	0	0.04943	0.303279	0.485829
C4	0.302041	0.428571	1	0.172131	0.200246
C5	1	0.465143	0	1	1

4.2 *PROMETHEE Calculations*

PROMETHEE II method promises completeness in ranking unlike PROMETHEE I. Using the weights calculated from AHP technique, PROMETHEE ranks the institutions in multiple steps as described below:

Step 1: The evaluation matrix (decision matrix) in Table 1 is normalized using the formula (1), and the results of this matrix are listed in Table 6.

$$R\,[i]\,[j] = \frac{\text{matrix}\,[i]\,[j] - \text{minInColumns}\,[j]}{\text{maxInColumns}\,[j] - \text{minInColumns}\,[j]} \tag{1}$$

Step 2: Determine the estimated difference in the alternatives with reference to the other alternatives.

So, we perform a pairwise comparison of every institution with every other institution to form a difference matrix R of size (rowcount $*$ rowcount $-$ 1, column count). The number of columns remains the same. However, the number of rows increase by (row $-$ 1) times. In this case, we get a matrix of size R(20, 5) as depicted in Table 7.

Step 3: Determine the preference Function P.

Further we modify the matrix so obtained to form a preference matrix P by using the logic (2)

$$\text{if } R\,[i]\,[j] < 0, \quad P\,[i]\,[j] = 0$$

$$\text{else} \quad P\,[i]\,[j] = R\,[i]\,[j] \tag{2}$$

Table 7 Pairwise comparison of every institution with others

	A1	A2	A3	A4	A5
C1–C2	0.269388	−0.14286	0.501901	−0.19262	0.250525
C1–C3	−0.03469	0.857143	0.820342	−0.30328	−0.2353
C1–C4	−0.03265	0.428571	−0.13023	−0.17213	0.050279
C1–C5	−0.73061	0.392	0.869772	−1	−0.74947
C2–C1	−0.26939	0.142857	−0.5019	0.192623	−0.25053
C2–C3	−0.30408	1	0.318441	−0.11066	−0.48583
C2–C4	−0.30204	0.571429	−0.63213	0.020492	−0.20025
C2–C5	−1	0.534857	0.367871	−0.80738	−1
C3–C1	0.034694	−0.85714	−0.82034	0.303279	0.235303
C3–C2	0.304082	−1	−0.31844	0.110656	0.485829
C3–C4	0.002041	−0.42857	−0.95057	0.131148	0.285582
C3–C5	−0.69592	−0.46514	0.04943	−0.69672	−0.51417
C4–C1	0.032653	−0.42857	0.130228	0.172131	−0.05028
C4–C2	0.302041	−0.57143	0.632129	−0.02049	0.200246
C4–C3	−0.00204	0.428571	0.95057	−0.13115	−0.28558
C4–C5	−0.69796	−0.03657	1	−0.82787	−0.79975
C5–C1	0.730612	−0.392	−0.86977	1	0.749475
C5–C2	1	−0.53486	−0.36787	0.807377	1
C5–C3	0.695918	0.465143	−0.04943	0.696721	0.514171
C5–C4	0.697959	0.036571	−1	0.827869	0.799754

On doing so, all the negative values in the matrix become zero forming the preference matrix.

Step 4: Determine the aggregated preference AggrPref.

After the preference matrix is formed, an aggregated preference matrix AggrPref[][] is calculated using Eq. (3), taking into consideration the criteria weights, which was obtained from the AHP process:

$$\text{AggrPref}\,[i]\,[j] = \frac{\sum_0^n \text{weight}\,[j] * P\,[i]\,[j]}{\sum_0^n \text{weight}\,[j]} \tag{3}$$

The AggrPref is represented in Table 8.

Step 5: Depending on the values of the alternatives, determine the ranks of the alternatives.

Leaving (positive) flow for the *a*th alternative and entering (negative) flow for the *a*th alternative:

$$\frac{1}{m-1}\sum_{b=1}^{m} \text{AggrPref}\,(a, b) \quad (a \neq b)$$

$$\frac{1}{m-1}\sum_{b=1}^{m} \text{AggrPref}\,(b, a) \quad (a \neq b)$$

Table 8 Aggregated preference matrix after applying (3)

	C1	C2	C3	C4	C5
C1	0	0.197695	0.28862	0.098643	0.202653
C2	0.061593	0	0.245492	0.117799	0.159524
C3	0.120427	0.213401	0	0.096384	0.007061
C4	0.055576	0.210834	0.22151	0	0.142857
C5	0.531148	0.624122	0.503749	0.514419	0

Table 9 Entering flow and leaving flow values

	C1	C2	C3	C4	C5	Leaving flow
C1	0	0.197695	0.28862	0.098643	0.202653	0.262537
C2	0.061593	0	0.245492	0.117799	0.159524	0.194802
C3	0.120427	0.213401	0	0.096384	0.007061	0.145758
C4	0.055576	0.210834	0.22151	0	0.142857	0.210259
C5	0.531148	0.624122	0.503749	0.514419	0	0.724479
Entering flow	0.256248	0.415351	0.41979	0.275748	0.170699	

Table 10 Net outranking flow values

	Leaving flow	Entering flow	Net outranking flow
C1	0.262537	0.256248	0.00629
C2	0.194802	0.415351	−0.22055
C3	0.145758	0.41979	−0.27403
C4	0.210259	0.275748	−0.06549
C5	0.724479	0.170699	0.553781

Table 11 Ranked list of institutions

	Net outranking flow	Rank
C1	0.00629	2
C2	−0.22055	4
C3	−0.27403	3
C4	−0.06549	5
C5	0.553781	1

Table 9 represents the Entering flow and Leaving flow calculated using the above formula.

Step 6: Calculate the net outranking flow for each alternative.

$$\text{LeavingFlow}(a) - \text{EnteringFlow}(a)$$

The calculated net outranking flows in shown in Table 10.

Step 7: Determine the ranks of all the alternatives depending upon the values of the net outranking flow as shown in Table 11. According to the user preferences considered in the initial table, the ranked list is C5, C1, C3, C2, and C4. However, the weights are subjective and can vary from user to user to derive a new set of ranks as per the requirement.

On analysing the complexity of the above implementation, it can be found that AHP mechanism does calculations involving power iterations of certain values which are of order $O(n^3)$, where n represents the number of institutions. As the number of parameters or the institutions increase, computation cost also increases. Also, if an attribute is added or removed, the entire computation from the leaf level goes through repetition. PROMETHEE has a quadratic time complexity, i.e. $O(qn^2)$, where q and n represent the count of criteria and alternatives, respectively. Hence the overall complexity of this ranking process is $O(n^3 + qn^2)$.

5 Conclusion

This study proposes a mechanism for evaluating and ranking the educational institutions on the grounds of certain criteria by implementing AHP and PROMETHEE II. The results based on the scores provided by the users for the data values presented in the institutions' website help the users retrieve the rank according to their preferences. The weights for the criteria generated using AHP result in a more fairly objective and guaranteed choice. A change in the user preference leads to a new set of weights and new set of ranking. The data pertaining to the attributes have been taken from the institutional websites; however, in the future, the data can also be retrieved using web crawling techniques. The attributes considered can also be increased and the same can be evaluated in multiple levels for a more effective ranking system.

References

1. Hemili M, Laouar MR (2018) Use of multi-criteria decision analysis to make collection management decisions. In: 2018 3rd International conference on pattern analysis and intelligent systems (PAIS), Tebessa, pp 1–5. https://doi.org/10.1109/PAIS.2018.8598495
2. Wati M, Novirasari N, Pakpahan HS (2018) Evaluation of scholarly performance student using multi-criteria decision-making with objective weight. In: 2018 International electronics symposium on knowledge creation and intelligent computing (IES-KCIC), Bali, Indonesia, pp 56–61
3. Gupta A, Gupta N, Garg R, Kumar R (2018) Evaluation, selection and ranking of software reliability growth models using multi criteria decision making approach. In: 2018 4th International conference on computing communication and automation (ICCCA), Greater Noida, India, pp 1–8
4. Perzina R, Ramik J, Mielcova E (2018) FDA—fuzzy decision analyzer. In: 2018 IEEE international conference on fuzzy systems (FUZZ-IEEE), Rio de Janeiro, pp 1–7. https://doi.org/10.1109/FUZZ-IEEE.2018.8491455
5. Supriya M, Sangeeta K, Patra GK (2013) Comparison of cloud service providers based on direct and recommended trust rating. In: 2013 IEEE international conference on electronics, computing and communication technologies, Bangalore, 2013, pp 1–6
6. Eppe S, De Smet Y (2014) Approximating Promethee II's net flow scores by piecewise linear value functions. Eur J Oper Res 233(3):651–659

7. Saaty TL (1980) The analytic hierarchy process. McGraw-Hill, New York
8. Kong F, Liu H (2005) Applying fuzzy analytic hierarchy process to evaluate success factors of e-commerce. Int J Inform Syst Sci 1(3–4):406–412
9. Oguztimur S (2011) Why fuzzy analytic hierarchy process approach for transport problems? ERSA conference papers ersa, vol 11. European Regional Science Association, p 438
10. Tiryaki F, Ahlatcioglu B (2009) Fuzzy portfolio selection using fuzzy analytic hierarchy process. Inf Sci 179(1–2):53–69. https://doi.org/10.1016/j.ins.2008.07.023
11. Ozkan B, Bashgil H, Sahin N (2011) Supplier selection using analytic hierarchy process: an application from Turkey. In: Proceedings of the world congress on engineering, vol II, July 2011, UK
12. Wu H-Y, Chen J-K, Chen I-S (2012) Hsin-Hui Zhuo, ranking universities based on performance evaluation by a hybrid MCDM model, measurement
13. Ranjan R, Chakraborty S (2015) Performance evaluation of Indian technical institutions using PROMETHEE-GAIA approach. Inform Educ 14:105–127
14. Dheeraj M, Teja GS, Yathendra N, Supriya M (2017) Ranking of Internet service providers considering customer priorities using multiple criteria decision-making methods. In: 2017 International conference on smart technologies for smart nation (SmartTechCon), Bangalore, pp 1213–1220
15. Supriya M, Sangeeta K, Patra GK (2016) A fuzzy based hierarchical trust framework to rate the cloud service providers based on infrastructure facilities
16. Supriya M, Sangeeta K, Patra GK (2015) Comparison of AHP based and fuzzy based mechanisms for ranking cloud computing services. In: 2015 International conference on computer, control, informatics and its applications (IC3INA), Bandung, pp 175–180
17. Supriya M, Sangeeta K, Patra GK (2016) Trustworthy cloud service provider selection using multi criteria decision making methods. Eng Lett 24:1–10

Evaluation of Graph Algorithms for Mapping Tasks to Processors

Sesha Kalyur and G. S. Nagaraja

1 Introduction

Parallel machines are characterized by two main traits, namely the processor count and the type of communication links that connect them. This offers a wide latitude in the choice of architectures for constructing the parallel machine. Some of the popular links include the Shared Bus, Fully Connected Network, Linear Array, Ring, Mesh, Tori, Grid, Cube, Tree, Fat Tree, Benes Network, and Hypercube among others. Independent of the link types, parallel machines can also be classified as shared memory multi-processor where all processors have equal and similar access to a shared memory pool, and distributed memory multi-processor, where each processor has access only to its private memory pool and sharing happens via explicit data exchange. We have uniform memory access when all of the memory in the pool are accessed with the same latency and non-uniform memory access when memory is hierarchical when different layers have different access times. Accordingly we have Uniform Memory Access (UMA) parallel machines and Non-Uniform Memory Access (NUMA) parallel machines [5, 7].

When distributing tasks to the processors of a parallel machine, one has to pay attention and ensure that it happens uniformly in which case the parallel machine is said to be load balanced. Formally, load balancing is a process, where the computing resources of a parallel machine are distributed evenly, from an execution standpoint. Load balancing can be broadly classified as static and dynamic. Static methods carry out the distribution or the mapping algorithm prior to task deployment and

S. Kalyur (✉) · G. S. Nagaraja
Department of Computer Science and Engineering, R. V. College of Engineering, VTU, Bangalore, India
e-mail: nagarajags@rvce.edu.in

A. Haldorai et al. (eds.), *2nd EAI International Conference on Big Data Innovation for Sustainable Cognitive Computing*, EAI/Springer Innovations in Communication and Computing, https://doi.org/10.1007/978-3-030-47560-4_33

the assignment of tasks to processors, and it remains fixed for the duration of their execution [19]. On the contrary, the dynamic schemes are complex, deploy the mapping algorithms at run time with the additional option of remapping and triggering task migrations, to keep the system balanced [14]. Most of the existing mapping methods are heuristic in nature [2, 3, 11].

Earlier research work involved studying, and developing a methodology, to map parallel tasks of any given program, to a suitable processor topology in linear or near-linear time. The outcome was seven algorithms of varying complexity that uses the graph model to solve the mapping problem. As part of the research, it was argued using examples that any set of tasks and their communication dependencies as well as the topology details of the parallel machine can be captured in directed graphs. These direct graphs serve as input to the mapping algorithms which then use the data in the graphs for making mapping decisions. This is a continuation of the work where we evaluate the algorithms in an empirical fashion using cost equations by deploying the algorithms for various permutations of the processor and task counts. The results of the evaluation study are captured in the form of tables and graphs and reported. We also elaborate on the process and Topmap the tool we use for this work.

In Sect. 2 of the paper, we look at existing work in the related areas of research. In Sect. 3 of the paper, we provide a quick review of the task mapping problem and introduce the readers to the various proposed algorithms to solve them. In Sect. 4 of the paper, we take a tour of the Topmap tool which is used extensively in this evaluation and analysis work. Section 5 is devoted to a discussion of the cost equations which are the core components of our evaluation portfolio. In Sect. 6 of the paper we present the results of the measurement and the analysis of the results. In Sect. 7 the final section of the paper, where we conclude the paper by revisiting the motivations for taking up our research work, and briefly summarizing our findings and contributions.

2 Related Works

Graphs are very popular tools to represent data and have been used to store processor and network topologies in the past[1, 2, 21]. Graph aggregation, summarization, and inference are well studied [8, 9, 18, 20].

Load balancing and task assignment have been studied by several researchers [2–4, 6, 14, 16, 17, 19]

Task placements on various machine architectures such as MPI, HPC, UMA, NUMA, etc. using user supplied placement strategy hints have been studied by [10, 11]. Others have used profiling to gather such data [15].

Existing mapping solutions are mostly approximate solutions and fail to provide deterministic and close to optimal solutions in linear or near-linear time. The algorithms presented in a related paper attempt to do just that. Here in this work we assess the performance of the algorithms with the help of cost equations to see how well they utilize the resources of the system both execution and bandwidth and at the same time keeping the load balancing requirement alive.

3 A Quick Review of the Task Assignment Problem (TAP)

As the name suggests Task Assignment Problem (TAP) is an activity where the tasks which are produced as a result of parallelization are assigned to the processors of a suitable parallel machine, without which the parallelization process would be incomplete. This process requires two graphs as inputs, one capturing the topology information of the parallel machine called here as the Processor Topology Graph (PTG) and the other holding the task details called as the Task Communication Graph (TCG).

PTG could hold processor details such as cycle speed in the processor nodes. Edges could hold the bandwidth parameter for the connection. In a similar vein, TCG holds at a minimum, the task cycles data in the nodes and the edge could hold the communication volume if any between the tasks concerned.

The result of running the TAP process produces an augmented graph called the Task Assignment Graph (TAG). Besides the pairing information, it has all the data necessary, to compute the performance metrics and complete the algorithm evaluation process.

We propose the following algorithms, to solve the Task Assignment Problem. The inputs to all of the algorithms are supplied as two graphs, namely the PTG and TCG. The result of running the algorithms is returned in the form of TAG which has the placement information plus all the parameter data necessary for calculating the performance metrics such as the mapping efficiency both computation and communication based.

1. **Minima Strategy algorithm:** This algorithm uses a very simple strategy by randomly mapping tasks to processors. It may not yield best results for all process task count permutations but could serve as a baseline number for analysis.
2. **Maxima Strategy algorithm:** This algorithm aims to provide the best topology for the given task set by providing direct connections between each pair of communicating tasks with the links at the highest bandwidth chosen from the minima strategy. This algorithm should produce the best results for all permutations of processor and task counts.
3. **Dimenx Strategy algorithm:** This algorithm has two variants, one uses the execution cycles of the tasks and another the communication bandwidth between the tasks to drive placement.
4. **Dimenxy Strategy algorithm:** This algorithm classifies tasks as computation bound or bandwidth bound, and then maps them accordingly.
5. **Graphcut Strategy algorithm:** This algorithm creates subgraphs of PTG and TCG and matches the subgraphs based on shape similarities. Once the subgraphs

are matched, it is trivial to match individual processors with tasks from the matched subgraphs.

6. **Optima Strategy algorithm:** This algorithm uses a two-step approach to mapping. In the first phase, it maps the tasks to a virtual topology with no limitations and providing direct connections, edge-by-edge. In the second phase the processors and tasks are moved from a virtual to an actual physical domain. It handles inconsistencies such as duplicate mappings for one task by keeping the map with lowest overhead and getting rid of other low yield mappings.
7. **Edgefit Strategy algorithm:** This algorithm maps tasks to processors edge-by-edge after sorting so that the best edges are mapped. Inconsistencies are handled by a backtracking post-phase followed by a load-balance phase which is optional.

Algorithm details and complexity analysis are provided in a separate related publication. These algorithms could be graded, based on their complexity and their performance yield. Any one of these algorithms can be deployed in any given scenario, involving a particular combination of tasks and processor topology. However it is possible to provide some advice, on the viability of these algorithms to a particular situation. For instance, Minima strategy is the simplest to implement and may provide good results in some cases such as a scenario that involves a few processors and a large number of tasks with little or no communication between them. Maxima strategy is useful as a benchmark configuration, to compare with an actual topology, by providing an upper limit on performance, for a given combination of tasks. Dimenx and Dimenxy are strategies to employ in situations, which demand optimizing both processor cycles and bandwidth. Graphcut, Optima, and Edgefit strategies are complex both in terms of implementation and runtime demands. They are useful for scenarios involving big and complex topologies and a large set of tasks, with large bandwidth and cycle requirements, where careful placement of tasks is well worth the effort.

4 Topmap: A Topology Mapper Tool for NUMA Architectures

Topmap derived from the phrase Topology Mapper is a tool that reads the topology of the machine in Comma Separated Values (CSV) format, called as the Processor Topology Table (PTT). Similarly the task dependency information is also fed as a CSV formatted file referred to as the Task Communication Table (TCT). There is an accompanying CSV formatted file referred to as the Task Computation Table (TMT) that stores the computation profile of the tasks. PTT is organized as a two dimensional table with each row and column corresponding to the processor numbers, which start from 0 and go up to N-1, where N is the size of the machine, in terms of the number of processors. Location PTT(I, J) stores the network reachability information, going from processor I to J. A value of 1.0 means a network hop of 1, and it actually means they are directly connected. A positive value greater than

Table 1 Conceptual machine topology table of 16 processors

	1	2	3	4	5	6	7	8	9	10	11	12	13	14	15	16
1	0	1	1	1	2	2	2	2	3	3	3	3	4	4	4	4
2	1	0	1	1	2	2	2	2	3	3	3	3	4	4	4	4
3	1	1	0	1	2	2	2	2	3	3	3	3	4	4	4	4
4	1	1	1	0	2	2	2	2	3	3	3	3	4	4	4	4
5	2	2	2	2	0	1	1	1	2	2	2	2	3	3	3	3
6	2	2	2	2	1	0	1	1	2	2	2	2	3	3	3	3
7	2	2	2	2	1	1	0	1	2	2	2	2	3	3	3	3
8	2	2	2	2	1	1	1	0	2	2	2	2	3	3	3	3
9	3	3	3	3	2	2	2	2	0	1	1	1	2	2	2	2
10	3	3	3	3	2	2	2	2	1	1	1	1	2	2	2	2
11	3	3	3	3	2	2	2	2	1	1	0	1	2	2	2	2
12	3	3	3	3	2	2	2	2	1	1	1	0	2	2	2	2
13	4	4	4	4	3	3	3	3	2	2	2	2	0	1	1	1
14	4	4	4	4	3	3	3	3	2	2	2	2	1	0	1	1
15	4	4	4	4	3	3	3	3	2	2	2	2	1	1	0	1
16	4	4	4	4	3	3	3	3	2	2	2	2	1	1	1	0

1 means, they are not directly connected, and it would require multiple hops to go from I to J. A higher value of hop means it is a slower connection and vice versa. The values in the table only depend on the architecture of the target machine. It should be noted the PTT(I, J) may not be equal to PTT(J, I), although in most practical cases it is usually the case. Table 1 on page 427 provides conceptual topology data for a machine configuration of 16 processors.

Similarly the Task Communication Table (TCT) stores the communication profile, between the parallel tasks of a program. Thus it is specific to the given program, and only depends on the communication pattern, of all the parallel tasks with the others in the mix. Location TCT(I, J) stores the communication coefficient, between any two tasks I and J. A higher value of the coefficient means there is tighter coupling between the tasks involved, from a communication perspective, which is proportional to the total number of bytes transferred, in either direction between the tasks involved. As such in all cases TCT(I, J) is always equal to TCT(J, I), and one can look into storing just the significant half of the table, for space saving purposes. The values actually stored in each cell can be imagined to be a normalized multiple of the actual communication in bytes.

The accompanying Task Computation Table (TMT) (not shown here but will be later with actual generated values) stores the execution cycles required for each of the tasks. So TMT(I) is the execution cycles consumed by task I. Table 2 on page 428 provides conceptual communication data for a configuration of 8 tasks. Each cell data can be imagined as a normalized float value, that is generated from the count of instructions for the task.

Table 2 Conceptual task communication table of 8 tasks

	1	2	3	4	5	6	7	8
1	0.00	0.73	0.52	0.81	0.30	0.91	0.81	0.65
2	0.73	0.0	0.13	0.50	0.91	0.77	0.06	0.50
3	0.51	0.13	0.00	0.84	0.62	0.39	0.93	0.42
4	0.81	0.50	0.84	0.00	0.35	0.40	0.48	0.41
5	0.30	0.91	0.62	0.35	0.00	0.11	0.24	0.20
6	0.91	0.77	0.39	0.40	0.11	0.00	0.76	0.86
7	0.81	0.057	0.93	0.48	0.24	0.76	0.00	0.09
8	0.65	0.50	0.42	0.41	0.20	0.86	0.09	0.00

Table 3 Processor topology table of 16 processors [cols 0–7] generated by Topmap

	0	1	2	3	4	5	6	7
0	0.0	68.0	0.0	0.0	0.0	0.0	0.0	0.0
1	68.0	0.0	93.3	28.9	0.0	0.0	0.0	0.0
2	0.0	93.3	0.0	0.0	0.0	0.0	82.7	0.0
3	0.0	28.9	0.0	0.0	900.2	74.6	592.0	0.0
4	0.0	0.0	0.0	900.2	0.0	0.0	0.0	0.0
5	0.0	0.0	0.0	74.6	0.0	0.0	0.0	0.0
6	0.0	0.0	82.7	592.5	0.0	0.0	0.0	954.2
7	0.0	0.0	0.0	0.0	0.0	0.0	954.2	0.0
8	0.0	0.0	0.0	0.0	0.0	0.0	75.6	0.0
9	0.0	0.0	0.0	0.0	0.0	0.0	1.6	0.0
10	0.0	0.0	0.0	0.0	54.6	0.0	771.1	0.0
11	0.0	0.0	0.0	0.0	0.0	0.0	0.0	0.0
12	0.0	0.0	0.0	0.0	0.0	0.0	0.0	0.0
13	0.0	0.0	0.0	0.0	0.0	0.0	0.0	0.0
14	0.0	0.0	0.0	0.0	0.0	0.0	0.0	0.0
15	0.0	0.0	0.0	0.0	0.0	86.2	0.0	97.7

Topmap generates these tables for any user specified configuration, which is a combination of the processor and task counts. However the values in PTT and TCT are actual bytes, in terms of bandwidth in the former case and volume of communication in the latter case. In case of TMT, it is the execution cycles for each task in the mix. The current version of the tool generates these values randomly or the users can use the CSV editor that is integrated in to the tool, to enter the values (from a profile data maybe) manually, and this is a limitation at present. We plan to plug in the data from our auto-parallelization framework [12, 13], so that the next version of Topmap has access to the actual inter-task communication profiles of real programs.

Tables 3 and 4 on pages 428 and 429, respectively, display topology data for a machine configuration of 16 processors generated by Topmap. The numbers represent the bandwidth in MB, between any two directly connected processors. A zero value denotes a lack of direct connection between any two processors, but are

Table 4 Processor topology table of 16 processors [cols 9–16] generated by Topmap

	9	10	11	12	13	14	15	16
1	0.0	0.0	0.0	0.0	0.0	0.0	0.0	0.0
2	0.0	0.0	0.0	0.0	0.0	0.0	0.0	0.0
3	0.0	0.0	0.0	0.0	0.0	0.0	0.0	0.0
4	0.0	0.0	0.0	0.0	0.0	0.0	0.0	0.0
5	0.0	0.0	54.6	0.0	0.0	0.0	0.0	0.0
6	0.0	0.0	0.0	0.0	0.0	0.0	0.0	86.2
7	75.6	1.6	771.1	0.0	0.0	0.0	0.0	0.0
8	0.0	0.0	0.0	0.0	0.0	0.0	0.0	97.7
9	0.0	0.0	0.0	0.0	0.0	0.0	0.0	0.0
10	0.0	0.0	0.0	0.0	0.0	0.0	0.0	0.0
11	0.0	0.0	0.0	4.2	70.6	52.1	77.1	21.1
12	0.0	0.0	4.3	0.0	0.0	0.0	0.0	0.0
13	0.0	0.0	70.6	0.0	0.0	0.0	0.0	0.0
14	0.0	0.0	52.1	0.0	0.0	0.0	0.0	0.0
15	0.0	0.0	77.1	0.0	0.0	0.0	0.0	0.0
16	0.0	0.0	21.6	0.0	0.0	0.0	0.0	0.0

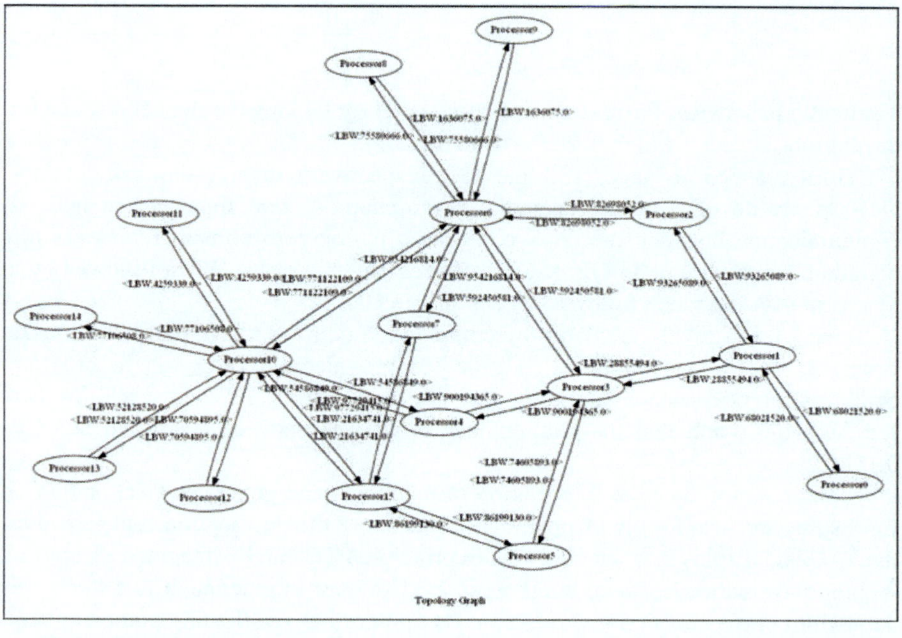

Fig. 1 Processor topology graph of 16 processors

however indirectly connected by intermediate processors at the cost of extra hops. The connections are bidirectional in nature and exhibit the same bandwidth in either direction.

Figure 1 on page 429 displays the corresponding PTG for the machine configuration of 16 processors tabulated above. In the graph we can see what the line

Table 5 Task communication table of 8 tasks (cols: 0–3) generated by Topmap

	0	1	2	3
0	0.0	79.4G	0.0	0.0
1	79.4G	0.0	0.23M	0.26M
2	0.0	0.24M	0.0	0.0
3	0.0	0.26M	0.0	0.0
4	0.0	0.0	0.0	0.44M
5	0.0	0.0	0.0	45.1G
6	0.0	0.0	0.1M	0.19M
7	0.0	0.0	0.0	0.0

Table 6 Task communication table of 8 tasks (cols: 4–7) generated by Topmap

	4	5	6	7
0	0.0	0.0	0.0	0.0
1	0.0	0.0	0.0	0.0
2	0.0	0.0	0.1M	0.0
3	0.44M	45.1G	0.19M	0.0
4	0.0	0.0	0.0	0.0
5	0.0	0.0	0.0	0.0
6	0.0	0.0	0.0	0.15M
7	0.0	0.0	0.15M	0.0

bandwidth is between Processor1 and Processor3 by looking for the LBW value on the appropriate edge which is 28.9 MB in this case.

Tables 5 and 6 on pages 430 and 430, respectively, display the task communication profile of 8 tasks generated by Topmap. A zero represents a lack of communication between the tasks concerned. A non-zero number represents the communication volume in GB or MB between any two tasks. When followed by a M it is in MB and when followed by a G it is in GB.

Table 7 on page 431 tabulates the computation data for a task profile of 8 tasks generated by Topmap. Figure 2 on page 431 provides the TCG, for the program configuration of 8 tasks assimilating the information from TCT and TMT. We can see from the graph that the communication volume between task6 and task7 is 0.15MB.

Topmap assimilates the information in the tables and generates PTG and TCG for display and also for the mapper algorithms to use the information, and generates the TAG and display it when its ready. In principle PTG can be imagined as, storing the hop information between the processors. However in practice, it just stores the bandwidth of the connections, in GB or TB or a normalized floating point number.

Table 8 on page 433 tabulates the task assignment data, for a simulation profile of 16 processors and 8 tasks, generated with Topmap using the Edgefit Strategy mapping algorithm. We can see from the table, that, for instance, on Row-8, processor PRC_7 has been assigned task TSK_4.

Figure 3 on page 432 illustrates the TAG that is generated by the Edgefit mapping algorithm, for the configuration of 16 processors and 8 tasks presented earlier. The

Table 7 Task computation table of 8 tasks

	Tasks	Cycles
1	0	8673
2	1	3, 042, 027
3	2	9, 781, 752
4	3	56, 203, 253
5	4	4, 099, 408
6	5	2, 887, 313
7	6	7744
8	7	6, 802, 327

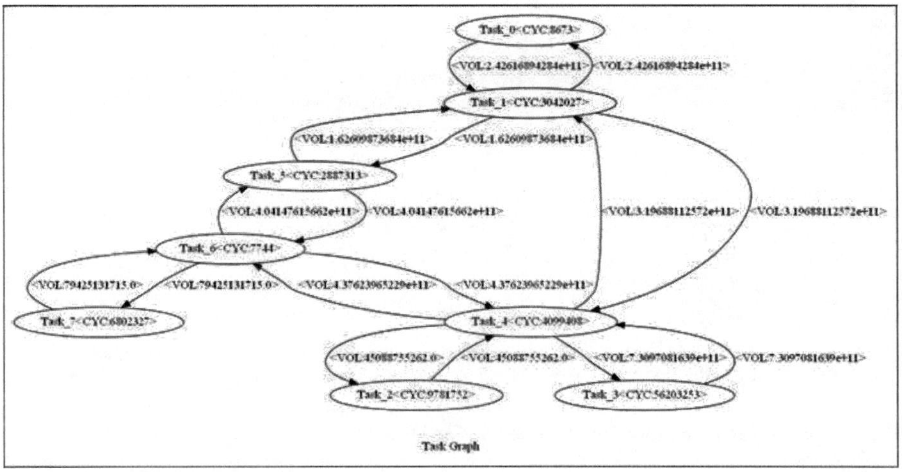

Fig. 2 Task communication graph for 8 tasks

Table 8 Task assignment table (cols: 1–8) for Edgefit placement

	Processor	Tasks
1	PRC_0	None
2	PRC_1	TSK_0
3	PRC_2	TSK_7
4	PRC_3	TSK_1
5	PRC_4	TSK_2
6	PRC_5	None
7	PRC_6	TSK_3
8	PRC_7	TSK_4

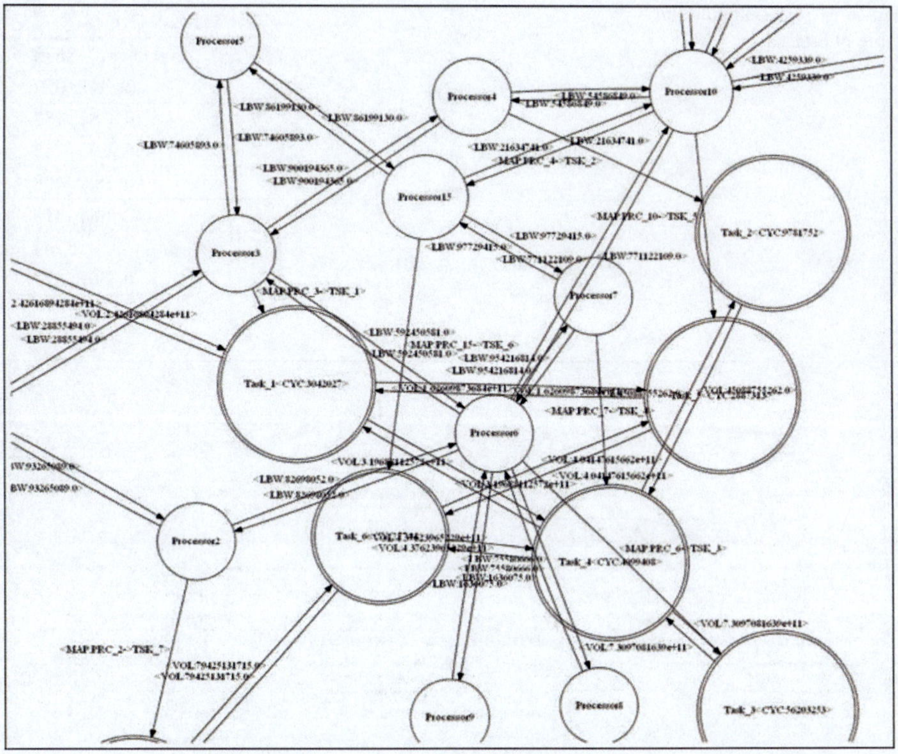

Fig. 3 Task assignment graph for 16 processors and 8 tasks

smaller nodes in the graph denote the processors, and the larger nodes represent the tasks. All the information that was part of PTG and TCG are there in TAG, plus a directed edge from the processor node to the task node, with a parameter that highlights the mapping. If the processor maps multiple tasks, there would be one edge per mapping. For instance, from the graph, we can see that there is an edge going from Processor_7 to Task_4 which is the same information we saw from the table. But this is an augmented graph that contains all processor and task nodes and edges along with all the parameter data, including volume data on edges connecting two task nodes (Table 9).

Topmap also comes with a CSV editor that can be used to create the Processor Topology Table, the Task Computation and the Task Communication Tables. Figures 4, 5, and 6 on pages 433, 433, and 434, respectively, display the steps involved in the creation of a Task Communication Table manually using the CSV editor. This is the same table referred to earlier which was generated randomly by Topmap.

Table 9 Task assignment table (cols: 9–16) for Edgefit placement

	Processor	Tasks
9	PRC_8	None
10	PRC_9	None
11	PRC_10	TSK_5
12	PRC_11	None
13	PRC_12	None
14	PRC_13	None
15	PRC_14	None
16	PRC_15	TSK_6

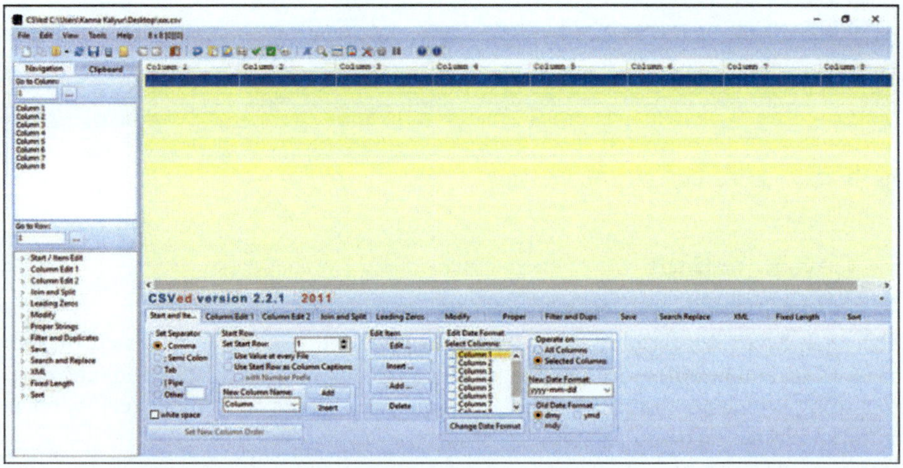

Fig. 4 CSV editor at the start of TCT creation

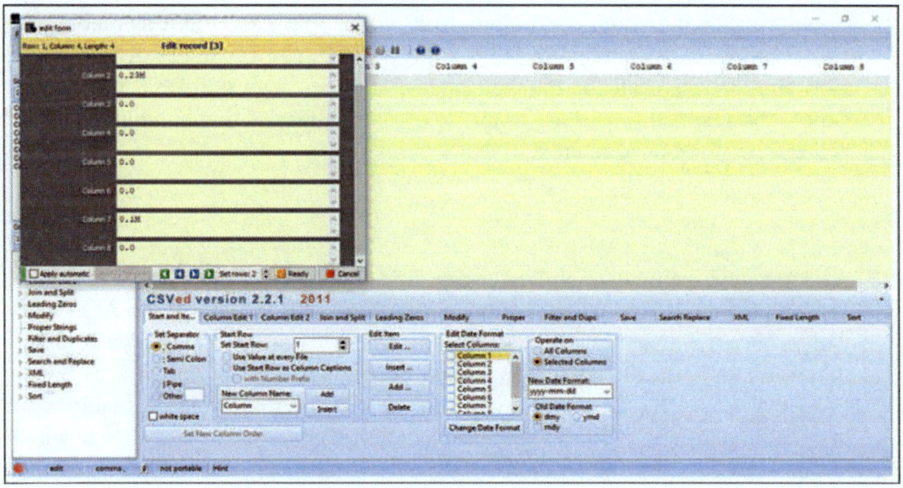

Fig. 5 CSV editor at the point of adding a row/record during TCT creation

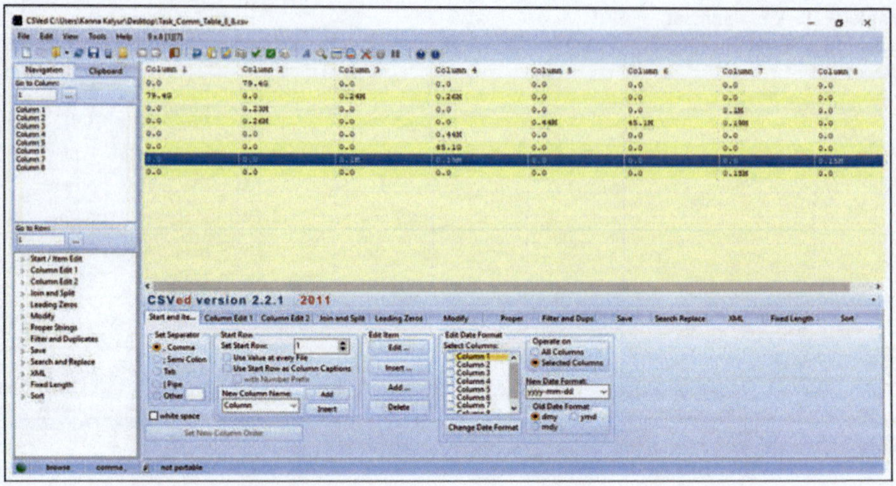

Fig. 6 CSV editor at the end of TCT creation

5 Cost Equations

The Task Mapping algorithms presented in a related paper are evaluated on the basis of several performance metrics, which are calculated using the following cost equations, which are discussed in detail below. Consult Table 10 on page 435, for a definition of the metrics.

1. **Average line bandwidth** (lbw_a)**:** Average line bandwidth is the average bandwidth available at any mapped processor in the topology. The algorithm that provides a higher value of lbw_a for a given set of tasks and topology is superior. In the equations below, lbw_i is the line bandwidth available at a particular processor node i in the topology graph, lbw_t is the total bandwidth after summation at all the mapped nodes, and the range is from 0 to n which denote the mapped processor count, lbw_a is the average line bandwidth which is the main number of interest.

$$lbw_t = \Sigma_0^{n-1} lbw_i \tag{1}$$

2. **Average line volume** (vol_a)**:** Average line volume is the average communication volume at any of the mapped processors in the topology. A strategy that achieves a low vol_a by careful placement of tasks is performing better than the others. In the equations below, vol_i is the line volume at a particular processor node i in the topology graph, vol_t is the total volume requirement after summation at all the mapped nodes and the range is from 0 to n which

Table 10 Column definitions

	Column	Definition	Interpretation
1	SPT	Specifies an Unique Configuration, an experimental step, which is a string concatenated from the Strategy, Processor count and Task count	–
2	LBW	Specifies the Average Bandwidth that exists between any two processors.	High value is better.
3	VOL	Specifies the Average Volume of network traffic that exists between the tasks.	Low value is better.
4	CYC	Signifies the Average Processor Cycles, a measure of computation derived from the sum of the instruction counts for all the tasks on the processor.	Low value is better.
5	XBW	Specifies the Average Bandwidth Excess, the difference between the traffic volume and the bandwidth offered by the link.	Low value is better.
6	OVC	Signifies the Average Overhead from a cycles perspective, incurred by mapping a specific task to a specific processor.	Low value is better.
7	OVN	Signifies the Average Overhead from a bandwidth perspective, incurred by mapping a specific task to a specific processor.	Low value is better.
8	UTC	Signifies the Average System Utilization from the task placement perspective.	High value is better.
9	UTN	Signifies an Average System Utilization from a link bandwidth perspective.	High value is better.
10	EFC	Signifies Average Efficiency from a network bandwidth perspective.	High value is better.
11	EFN	Signifies an Average Efficiency from a computation perspective.	High value is better.
12	TPP	Specifies the Average Number of Tasks assigned to each processor.	Low value is better.
13	RAT	Specifies the Average Computation to Communication Ratio, a measure of useful work done.	High value is better.

denote the mapped processor count, vol_a is the average line bandwidth which is the main number of interest.

$$vol_t = \Sigma_0^{n-1} vol_i \tag{2}$$

$$vol_a = \frac{vol_t}{n} \tag{3}$$

3. **Average processor cycles** (**cyc_a**)**:** Average processor cycles are the cycle count at any of the mapped processors in the topology. For a given task set, the objective is to consume minimum processor cycles, so an algorithm that provides the lowest number is deemed to be superior. In the equations below,

cyc_i is the line bandwidth available at a particular processor node i in the topology graph, cyc_t is the total bandwidth after summation at all the mapped nodes, and the range is from 0 to n which denote the mapped processor count, cyc_a is the average line bandwidth which is the main number of interest.

$$cyc_t = \Sigma_0^{n-1} cyc_i \tag{4}$$

$$cyc_a = \frac{cyc_t}{n} \tag{5}$$

4. **Average bandwidth deficit** (xbw_a)**:** Average bandwidth deficit is the difference between the bandwidth requirement of the tasks and the actual value provided by the topology. A strategy that carefully pairs a task with a processor by considering the bandwidth requirements will exhibit lower values of xbw_a and is considered to be outperforming its rivals. In the equations below, xbw_i is the line bandwidth deficit at a particular processor node i in the topology graph, xbw_t is the total bandwidth deficit after summation at all the mapped nodes, and the range is from 0 to n which denote the nodes, xbw_a is the average line bandwidth which is the main number of interest.

$$xbw_t = \Sigma_0^{n-1} xbw_i \tag{6}$$

$$xbw_a = \frac{xbw_t}{n} \tag{7}$$

5. **Average tasks per processor** (tpp_a)**:** Average tasks per processor is the average number of tasks mapped at any of the processors in the topology. An important aspect of mapping tasks is how well the tasks are distributed, on the topology. So a good algorithm will not overload a processor with tasks, when processing resources are available. In the equations below, tpp_i is the number of tasks at a particular processor node i in the topology graph, tpp_t is the total tpp after summation at all the mapped nodes and the range is from 0 to n which denote the mapped processor count, tpp_a is the average tasks per processor which is the main number of interest.

$$tpp_t = \Sigma_0^{m-1} tpp_i \tag{8}$$

$$tpp_a = \frac{tpp_t}{n} \tag{9}$$

6. **Average computation overhead** (ovc_a)**:** The average computation overhead is the average of the total overheads experienced by each task from a computation perspective in the system. This is a number that considers how quickly the set of tasks are processed by the topology, and so an algorithm that returns a lower number is performing well. It is computed as follows in the equations below where m is the number of tasks mapped to processor p, n is the number of tasks in the system, cyc_t is the total cycles consumed by one task. cyc_p is the

total cycles consumed by the processor to which it is mapped, ovc_t is the total computation overhead, and ovc_a is the average overhead.

$$cyc_p = \Sigma_0^{n-1} cyc_t \tag{10}$$

$$ovc_t = 1 - \frac{cyc_t}{cyc_p} \tag{11}$$

$$ovc_a = \frac{ovc}{n} \tag{12}$$

7. **Average communication overhead** (ovn_a)**:** The average communication overhead is the average of the total overheads experienced by each task from a communication perspective in the system. The communication overheads are an important factor, in limiting the turnaround time of tasks. So a strategy that keeps this number low is doing well against the competition. It is computed as follows in the equations below where m is the number of tasks mapped to processor p, n is the number of tasks in the system, xbw_t is the total excess bandwidth consumed by one task. xbw_p is the total excess bandwidth consumed by the processor to which it is mapped, ovn_t is the total communication overhead, and ovn_a is the average overhead.

$$xbw_p = \Sigma_0^{m-1} xbw_i \tag{13}$$

$$ovn_t = 1 - \frac{xbw_p}{xbw_t} \tag{14}$$

$$ovn_a = \frac{ovn_t}{n} \tag{15}$$

8. **Aggregate computation utilization** (utc_a)**:** Average computation utilization is a measure of how well the resources are utilized from a computation perspective. An algorithm that provides a higher number is doing well compared to its peers.

   In the equations below, prc_u is the unused processor count which is actually a count of processors that do not have tasks assigned, prc_t is the total processor count in the topology, cyc_p is the cycle count for a mapped processor p, cyc_i is the cycle count for a task i out of a total of m mapped to p, cyc_j is the cycle count for a processor j, cyc_a is the average cycles across all processors, ovc_p is the overhead at a particular processor p, utc_p is the utilization at a particular processor p, $penp$ is the penalty for not using unused processors in the topology, $pent$ is the penalty for crowding tasks at a processor in the topology, utc_e is the effective utilization that factors in the penalty, utc_t is the sum of utilization across the mapped processors, m is the total count of tasks mapped at processor p, n is the total count of processors mapped to tasks, utc_a is the average utilization for the entire topology.

$$cyc_p = \sum_{0}^{m-1} cyc_i \tag{16}$$

$$cyc_a = \frac{\Sigma_0^{n-1} cyc_j}{n} \tag{17}$$

$$ovc_p = cyc_p / cyc_a \tag{18}$$

$$utc_p = \frac{1.0}{ovc} \tag{19}$$

$$penp = \frac{proc_t}{(proc_t + proc_u)} \tag{20}$$

$$pent = \frac{tcnt}{tcnt + tpp)} \tag{21}$$

$$utc_e = utc_p * penp * pent \tag{22}$$

$$utc_t = \Sigma_0^{n-1} utc_e \tag{23}$$

$$utc_a = \frac{utc_t}{n} \tag{24}$$

9. **Aggregate network utilization** (utn_a)**:** Average network/bandwidth utilization is a measure of how well the bandwidth resources are utilized in the system. An algorithm that returns a higher number is obviously doing well.

   In the equations below, pec_u is the unused processor count which is actually a count of processors that do not have tasks assigned, pec_t is the total processor count in the topology, xbw_p is the bandwidth deficit for a mapped processor p, xbw_i is the bandwidth deficit for a task i out of a total of m mapped to p, xbw_j is the bandwidth deficit for a processor j, xbw_a is the average bandwidth deficit across all processors, ovn_p is the overhead at a particular processor p, utn_p is the utilization at a particular processor p, pec_t is the total processor edge count in the system, pec_u is the total processor edge count unused in the system, tec_t is the total task edges in the system, $tec_p p$ is the total task edges per processor in the system, $penp$ is the penalty for not using available processor edges for mapping when available, and $pent$ is a penalty for overloading a processor edge with task edges when free processor edges are available. ovn_n and utn_n are values for a particular mapped processor, utn_t is the total utilization for mapped processors in the topology, and utn_a is the average utilization.

$$xbw_p = \sum_{0}^{m-1} xbw_i \tag{25}$$

$$xbw_a = \frac{\Sigma_0^{n-1} xbw_j}{n} \tag{26}$$

$$penp = \frac{pec_t}{(pec_t + pec_u)} \tag{27}$$

$$pent = \frac{tec_t}{(tec_t + tec_p p)} \tag{28}$$

$$ovn = \frac{xbw}{xbw_a vg} \tag{29}$$

$$utn_p = \frac{1.0}{ovn_p} \tag{30}$$

$$utn_n = utn_p * penp * pent \tag{31}$$

$$utn_t = \Sigma_0^{m-1} utn_n \tag{32}$$

$$utn_a = \frac{utn_t}{m} \tag{33}$$

10. **Average computation efficiency** (efc_a)**:** Average computation efficiency is a measure of how well the algorithms perform in terms of their effectiveness from a cycles perspective so high numbers reflect the suitability of an algorithm to a particular scenario.

    The average computation efficiency is the average of the total efficiency experienced by each task from a computation perspective in the system.
    In the equations below, n is the total number of tasks in the system, ovc_i is the computational overhead for a particular task i (equation shown earlier), efc_i is the computational efficiency of the mapping, efc_t is the cumulative across all the mapped processors, and efc_a is the average for the topology.

$$efc_i = \frac{1.0}{ovc_i} \tag{34}$$

$$efc_t = \Sigma_0^{n-1} efc_i \tag{35}$$

$$efc_a = \frac{efc}{n} \tag{36}$$

11. **Average communication efficiency** (efn_a)**:** Average network/bandwidth efficiency is a measure of how well the algorithms perform in terms of their effectiveness from a bandwidth perspective. A strategy that exhibits a larger number is suited to a situation that demands higher bandwidths from processor connections.

    The average communication efficiency is the average of the total efficiency experienced by each task from a communication/bandwidth perspective in the system.

    In the equations below, n is the total number of tasks in the system, ovn_i is the communication overhead for a particular task i (equation shown earlier), efn_i is the communication efficiency of the mapping, efn_t is the cumulative across all the mapped processors, and efn_a is the average overhead for the topology

$$efn = \frac{1.0}{ovn} \tag{37}$$

$$efn_t = \Sigma_0^{n-1} efn_i \tag{38}$$

$$efn_a = \frac{efn}{n} \tag{39}$$

12. **Average computation to communication ratio (rat_a):** The computation to communication ratio measures the effectiveness of the mapping from both the computation cycles and communication bandwidth perspective. Higher numbers mean better mapping. In the equations below, ovc is the computation overhead, ovn is the communication overhead, ovc_avg is the average computation overhead, ovn_avg is the average communication overhead, $xrat$ is the computation part of the ratio, $crat$ is the communication part of the ratio, rat is the computation to communication ratio, rat_i is the computation to communication ratio factoring the overhead for an individual task. rat_t and rat_a are the total and average ratios, respectively.

$$xrat = \frac{ovn}{(ovn + ovn_avg)} \tag{40}$$

$$crat = \frac{ovc}{(ovc + ovc_avg)} \tag{41}$$

$$rat = \frac{xrat}{crat} \tag{42}$$

$$rat_t = \Sigma_0^{n-1} rat_i \tag{43}$$

$$rat_a = \frac{rat_t}{n} \tag{44}$$

6 Results

This section presents the performance results of the various mapping algorithms, for a variety of processor and task configurations. In this study we have included topologies with 4, 8, 16, 32, 64, and 128 processors and task configurations of 4, 8, 16, 32, 64, and 128 tasks. All the possible combinations of these configurations are used. The results of all of the experiments, where each unique experiment is for a particular configuration, are stored in a CSV formatted file. The main performance number for each configuration, is a measure of its efficiency from both the network and the processor utilization perspective. Each row of a table corresponds to a unique configuration that is identified by the string representing the strategy, and the number of processors and tasks involved. For instance, the configuration Minima-4–8 refers to the Minima strategy with 4 processors and 8 tasks. There are totally 12 columns for each configuration, and each one denotes a particular result, of running the experiment for that configuration. The following table defines the meaning of each column. It should be noted that in a table when reported the value corresponding to a particular cell is normalized and raw score is not reported.

Table 10 on page 435 provides a concise definition of each of the column, from the results table presented further down.

Due to space limitations we are not presenting the values in table format but two sets of plots are presented next. The first set is the Top_20 plots for each of the metrics which is the Top 20 numbers for each and the Agg_20 for each of the metrics which is the aggregate scores for each algorithm. However the aggregate values are also reported in table format also.

Figures 7a, b (page 441), 8a, b (page 441), 9a, b (page 442), 10a, b (page 442), and 11a, b (page 442) provide a graphical view of the Top-20 scores, for all configurations for all algorithms.

From the Top-20 plots of LBW it can be seen that configurations which are not stressing the algorithms, namely those with higher or equal number of processors as the tasks, show higher numbers. On the other hand, in the case of VOL plot it is seen that configurations with higher task counts show bigger numbers because bandwidth is being squeezed in these cases. Similarly in the case of XBW plot,

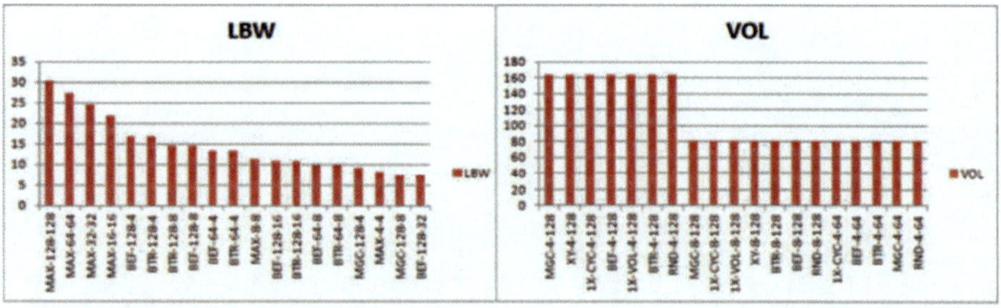

(a) (b)

Fig. 7 TOP-20 scores LBW and VOL. **(a)** TOP-LBW. **(b)** TOP-VOL

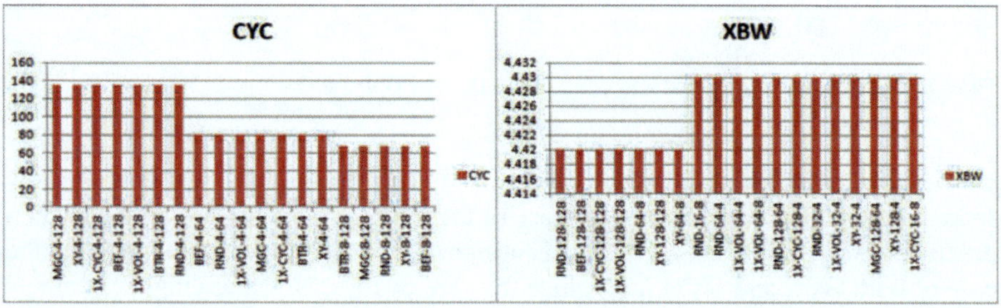

(a) (b)

Fig. 8 TOP-20 scores CYC and XBW. **(a)** TOP-CYC. **(b)** TOP-XBW

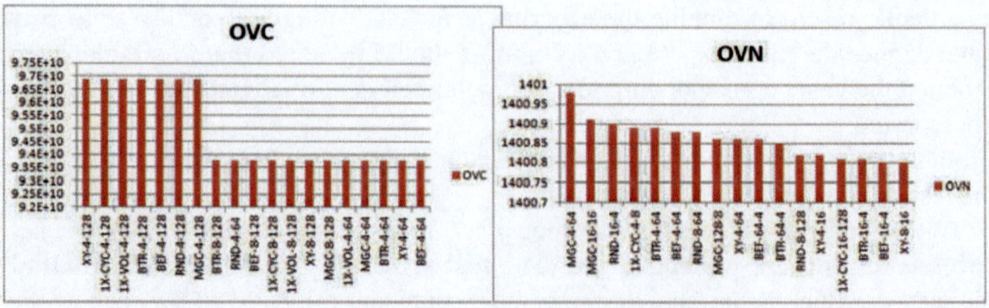

Fig. 9 TOP-20 scores OVC and OVN. (**a**) TOP-OVC. (**b**) TOP-OVN

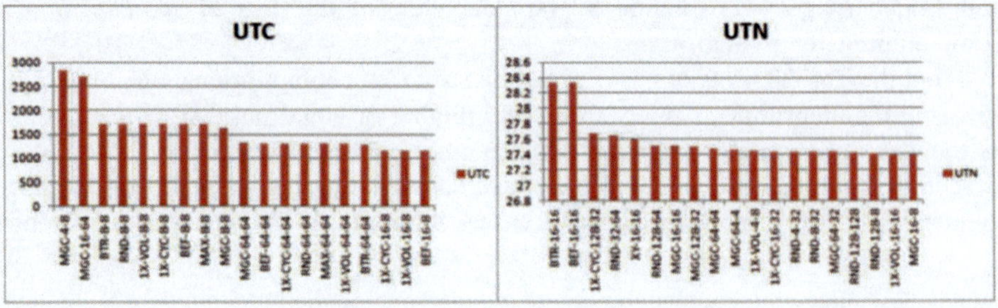

Fig. 10 TOP-20 scores UTC and UTN. (**a**) TOP-UTT. (**b**) TOP-UTH

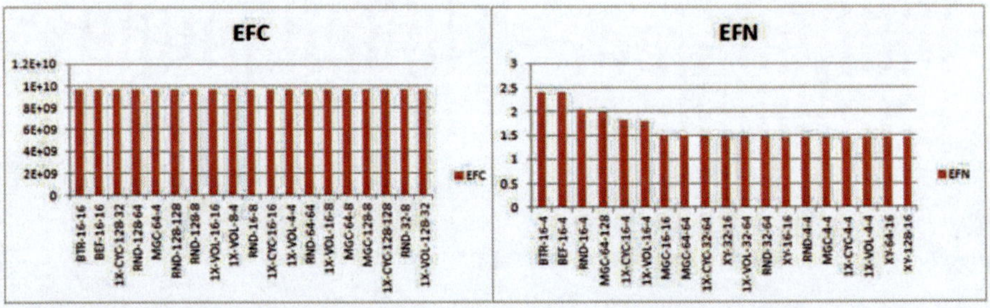

Fig. 11 TOP-20 scores EFC and EFN. (**a**) TOP-EFC. (**b**) TOP-EFN

we see that configurations with processor counts greater than task counts show spare bandwidth as expected. In the case of CYC plot we see higher numbers when processors have more than one task executing which is also expected behavior. In the case of both OVC and OVN plots since they are overheads, we see higher numbers for configurations which have more tasks than processors. UTC and UTN are indicators of effective system utilization, and we see higher UTC for configurations that have equal or almost equal numbers of processors and tasks, for UTN we

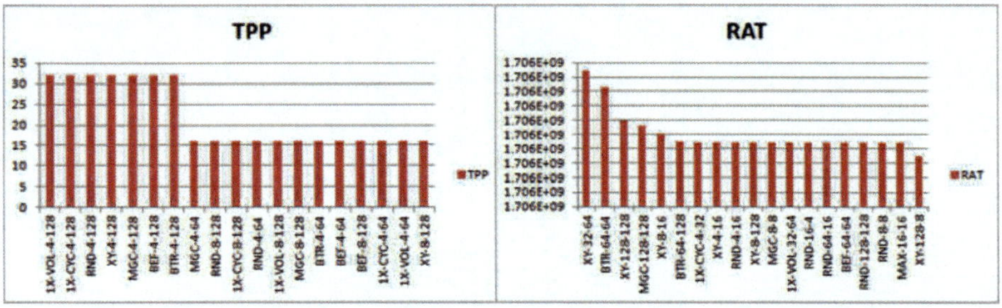

(a) (b)

Fig. 12 TOP-20 scores TPP and RAT. (**a**) TOP-TPP. (**b**) TOP-RAT

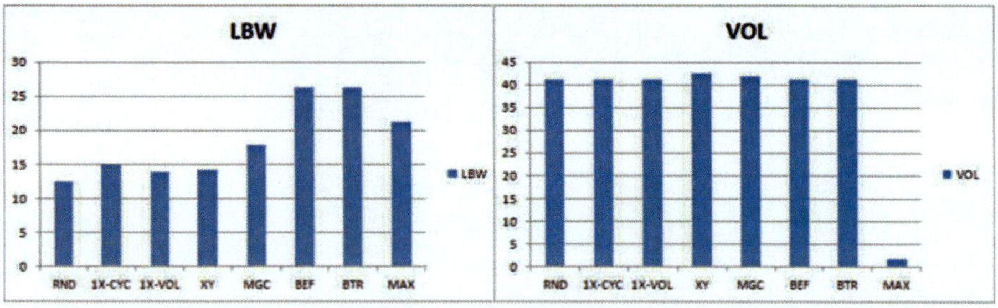

(a) (b)

Fig. 13 Aggregate scores LBW and VOL. (**a**) AGG-LBW. (**b**) AGG-VOL

see good numbers when processor count outnumbers task counts. Since EFC and EFN are performance numbers we see higher numbers when processors outnumber tasks. TPP plot is just useful to ensure that the algorithms are doing a good job of distributing tasks around the processors in the topology. The RAT plot is interesting we see a good number of configurations from both categories, namely processor outnumbering and tasks outnumbering configurations showing up. Please note that the legends on the plots are a little different than what we refer to the algorithms here. RND is same as Minima, 1X-CYC is same as Dimenxc, 1X-VOL is same as Dimenxv, XY is same as Dimenxy, MGC is same as Graphcut, BEF is same as Edgefit, BTR is same as Optima, and MAX is same as Maxima (Fig. 12).

Table 11 on page 444 provides a listing of aggregate scores for each of the algorithms.

Figures 13a, b, 14a, b, and 15a, b on page 445 and Figs. 16a, b and 17a, b on page 446 provide a graphical view of the aggregate scores, for all configurations for all algorithms.

From the aggregate LBW plot we see that the Edgefit, Optima, and Maxima algorithms all outperforming others which is indicative of good bandwidth management through careful processor to task mapping. From the VOL plot we see almost similar numbers for all the algorithms except Maxima. Maxima shows a very

Table 11 Aggregate scores

	SPT	LBW	VOL	CYC	XBW	OVC	OVN	UTC	UTN	EFC	EFN	TPP	RAT
1	Minima	12.44	41.08	17.23	−28.63	10.97	−1106.31	9932.63	26.48	2.1E+11	−11.41	135	1064.57
2	Dimenxc	14.94	41.08	17.23	−26.14	10.97	−1006.34	9395.34	53.09	2.1E+11	−9.92	135	1096.79
3	Dimenxv	13.91	41.08	17.23	−27.16	10.97	−835.49	10, 370.44	47.71	2.1E+11	−9.33	135	1367
4	Dimenxy	14.12	42.64	17.94	−28.52	15.67	−761.29	2276.93	18.19	1.22656E+11	−7.3	141.56	8,293,467,908
5	Graphcut	17.75	41.91	17.56	−24.16	12.8	−1060.71	14, 439.7	26.56	1.83047E+11	−11.35	137.52	9,458,103,810
6	Edgefit	26.19	41.08	17.23	−14.89	10.97	256.8	10, 370.44	16.04	2.1E+11	−14.69	135	2032.59
7	Optima	26.24	41.13	17.26	−14.88	11.12	256.8	10, 975.8	43.21	2.08594E+11	−14.69	135.16	−682,507,625.8
8	Maxima	21.23	1.7	0.65	19.53	0	1932.06	3901.54	8.19	60,000,000,000	0.9	6	0.89

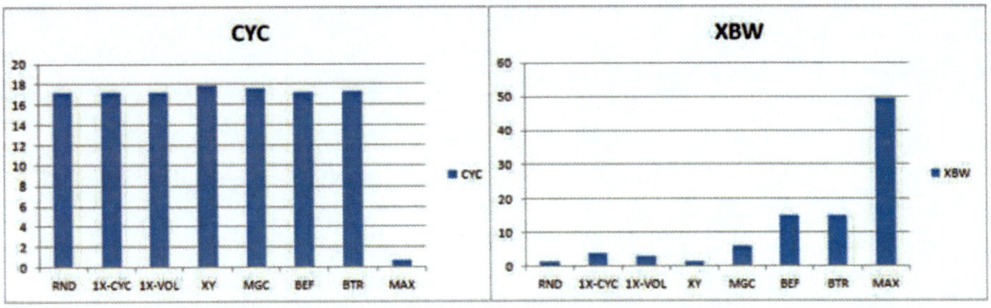

(a) (b)

Fig. 14 Aggregate scores CYC and XBW. (**a**) AGG-CYC. (**b**) AGG-XBW

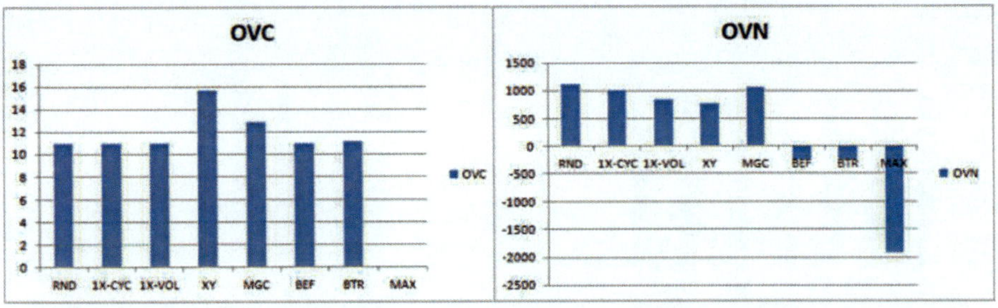

(a) (b)

Fig. 15 Aggregate scores OVC and OVN. (**a**) AGG-OVC. (**b**) AGG-OVN

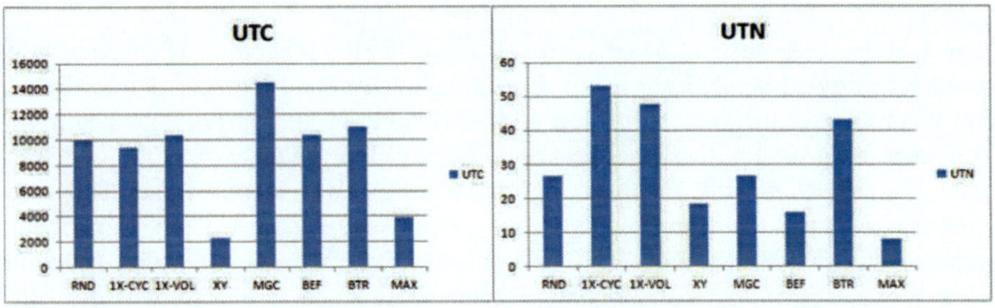

(a) (b)

Fig. 16 Aggregate scores UTC and UTN. (**a**) AGG-UTC. (**b**) AGG-UTN

small number since it creates a topology that produces the best result for the taskset including creating processors to match the task count. For the same reason Maxima outperforms the rest by posting low cycle counts as seen in the CYC plot. The XBW plot once again confirms good bandwidth management from Edgefit, Optima, and Maxima with Maxima staying well ahead of the other two. OVC plot shows high overhead posted by Dimenxy and Graphcut, and the rest showing comparable numbers but Maxima clearly outperforms all with no overhead. As before we see

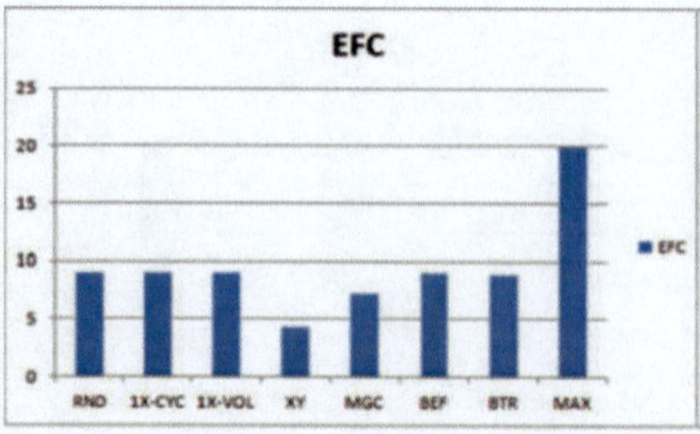

(a)

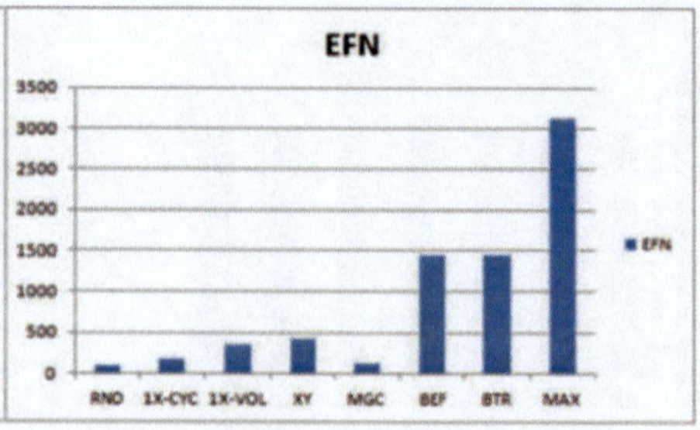

(b)

Fig. 17 Aggregate scores EFC and EFN. (**a**) AGG-EFC. (**b**) AGG-EFN

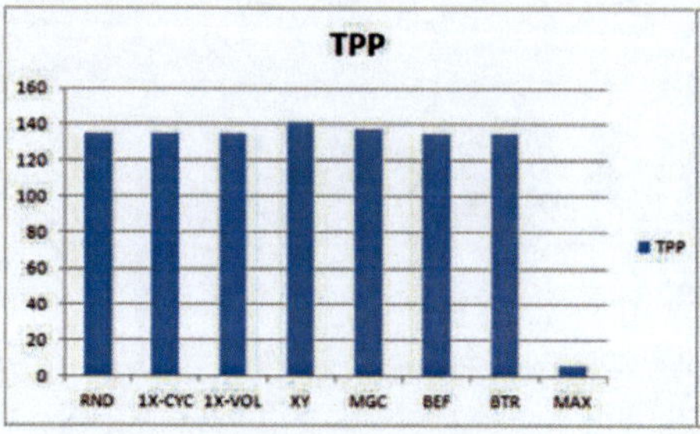

(a)

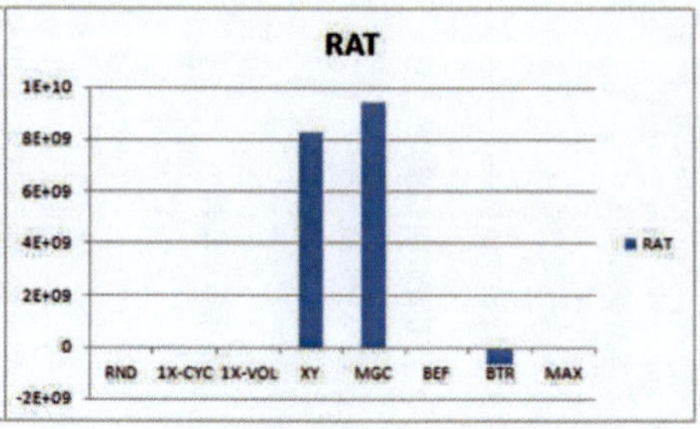

(b)

Fig. 18 Aggregate scores TPP and RAT. (**a**) AGG-TPP. (**b**) AGG-RAT

that Edgefit, Optima, and Maxima doing really well by keeping OVN low with good handling of bandwidth. From the UTC plot we see Graphcut doing a good job of producing the best utilization with Dimenxy doing poorly. Once again we see some surprises, with Dimenxc and Dimenxv doing well with UTN which is probably understandable since the algorithm is built with this in mind. They use direct connections between processors only when needed otherwise they just focus on distribution of tasks. With EFC we see all of them doing well with Maxima clearly outperforming. With EFN we see the same trend as before, with Edgefit, Optima, and Maxima clearly the leaders. TPP numbers for all the algorithms are similar indicating that all of them do a good job of distributing the tasks around. Maxima shows a low number because it uses processor count to match the task count always and so keeps the TPP count low. The RAT score is supposed to check if both computation and communication are managed well, which seems to be the case for all but Dimenxy and Graphcut seem to excel the others which is indicative of good computation and communication management at the same time (Fig. 18).

7 Conclusion

In this research work, we studied the problem of mapping parallel tasks of a program, to the processors of a multi-processor machine. The problem is interesting and challenging, since effective mapping depends on two criteria, namely the total execution cycles consumed by each processor, and the overall bandwidth provided to the tasks by the topology. We presented the algorithms and their complexity in an earlier paper. The focus of this paper was to evaluate the algorithms and study their performance from an empirical standpoint. We developed cost equations that take the graph parameters as input which are changed by the mapping of specific tasks to specific processors in the topology. This helps us to determine which algorithm and configuration combination performs better in comparison. What we have observed is that the Maxima algorithm performs best since we provide custom topology choices for a given taskset. From a bandwidth perspective, Edgefit and Optima configurations do well for most configurations from a computation perspective. We see that Graphcut and Dimenxy does well from both perspectives for many configurations due to their design choices. Dimenxc and Dimenxv do reasonably well in computation weighted and communication weighted situations, respectively, since by design they just focus on one criteria. Minima as expected is not spectacular due to its simple strategy of random selection but posts reasonable scores often.

References

1. Abbas W, Egerstedt M (2012) Robust graph topologies for networked systems. IFAC Proc Vol 45(26):85–90. 3rd IFAC workshop on distributed estimation and control in networked systems. https://doi.org/https://doi.org/10.3182/20120914-2-US-4030.00052, http://www.sciencedirect.com/science/article/pii/S147466701534814X
2. Ahmad I, Kwok YK (1998) On exploiting task duplication in parallel program scheduling. IEEE Trans Parallel Distrib Syst 9(9):872–892. https://doi.org/10.1109/71.722221
3. Ahmad I, Dhodhi MK, Ghafoor A (1995) Task assignment in distributed computing systems. In: Proceedings international phoenix conference on computers and communications, March 1995, pp 49–53. https://doi.org/10.1109/PCCC.1995.472512
4. Chou TCK, Abraham JA (1982) Load balancing in distributed systems. IEEE Trans Softw Eng 8(4):401–412
5. Culler D, Singh JP, Gupta A (1998) Parallel computer architecture: a hardware/software approach. Morgan Kaufmann, San Francisco
6. e Silva EDS, Gerla M (1991) Queueing network models for load balancing in distributed systems. J Parallel Distrib Comput 12(1):24–38
7. Feng TY (1981) A survey of interconnection networks. Computer 14(12):12–27
8. Gkantsidis C, Mihail M, Zegura E (2003) Spectral analysis of internet topologies. In: INFOCOM 2003. Twenty-second annual joint conference of the IEEE computer and communications. IEEE societies, vol 1. IEEE, Piscataway, pp 364–374
9. Grieco LA, Alaya MB, Monteil T, Drira K (2014) A dynamic random graph model for diameter-constrained topologies in networked systems. IEEE Trans Circuits Syst II Express Briefs 61(12):982–986. https://doi.org/10.1109/TCSII.2014.2362676

10. Hursey J, Squyres JM, Dontje T (2011) Locality-aware parallel process mapping for multi-core Hpc systems. In: 2011 IEEE international conference on cluster computing, Sept 2011, pp 527–531. https://doi.org/10.1109/CLUSTER.2011.59
11. Jeannot E, Mercier G (2010) Near-optimal placement of MPI processes on hierarchical NUMA architectures. In: D'Ambra P, Guarracino M, Talia D (eds) Euro-Par 2010 - parallel processing. Springer, Berlin, pp 199–210
12. Kalyur S, Nagaraja GS (2016) ParaCite: auto-parallelization of a sequential program using the program dependence graph. In: 2016 International conference on computation system and information technology for sustainable solutions (CSITSS), Oct 2016, pp 7–12. https://doi.org/10.1109/CSITSS.2016.7779431
13. Kalyur S, Nagaraja GS (2017) Concerto: a program parallelization, orchestration and distribution infrastructure. In: 2017 2nd international conference on computational systems and information technology for sustainable solution (CSITSS), Dec 2017, pp 1–6. https://doi.org/10.1109/CSITSS.2017.8447691
14. Loh PKK, Hsu WJ, Wentong C, Sriskanthan N (1996) How network topology affects dynamic load balancing. IEEE Parallel Distrib Technol 4(3):25–35. http://dx.doi.org/10.1109/88.532137
15. Pilla LL, Ribeiro CP, Cordeiro D, Bhatele A, Navaux PO, Méhaut JF, Kalé LV (2011) Improving parallel system performance with a NUMA-aware load balancer. Technical report
16. Shirazi BA, Kavi KM, Hurson AR (1995) Scheduling and load balancing in parallel and distributed systems. IEEE Computer Society Press, Washington
17. Tantawi AN, Towsley D (1985) Optimal static load balancing in distributed computer systems. J ACM 32(2):445–465
18. Tian Y, Hankins RA, Patel JM (2008) Efficient aggregation for graph summarization. In: Proceedings of the 2008 ACM SIGMOD international conference on management of data. ACM, New York, pp 567–580
19. Wang YT et al (1985) Load sharing in distributed systems. IEEE Trans Comput 100(3):204–217
20. Xu Y, Salapaka SM, Beck CL (2014) Aggregation of graph models and Markov chains by deterministic annealing. IEEE Trans Automat Contr 59(10):2807–2812
21. Zegura EW, Calvert KL, Donahoo MJ (1997) A quantitative comparison of graph-based models for internet topology. IEEE/ACM Trans Netw 5(6):770–783

Smart Trash Segregator Using Deep Learning on Embedded Platform

Namrata Anilrao Mahakalkar and Radha D.

1 Introduction

As per the 2011 census, India's population raised expeditiously than expected [1]. The amount of waste production per capita per day is increasing every year [2]. Waste generations per capita per day in kilograms in major cities of India are 0.35, 0.45, 0.26, 0.52, 0.44, 0.54, and 0.360 in Mumbai, Delhi, Kolkata, Chennai, Bangalore, Hyderabad, and Ahmedabad, respectively [2]. Maharashtra generates 15,364–19,204 metric tons of waste in a day followed by Uttar Pradesh, West Bengal, Pondicherry, and Tamil Nadu which generate 11,522–15,363 metric tons of waste per day. Andhra Pradesh and Kerala generate 7683–11,522 metric tons. Karnataka, Goa, Gujarat, Rajasthan, Madhya Pradesh, and Tripura generate 3842–7682 metric tons. Other states of India generate less than 3841 metric tons per year. From such giant figures of waste generation, only 69% of waste is collected. Only 28% of total collected waste gets treated as all types of waste are collected in the same bin. Mixed waste collection at the centralized level makes the segregation process worse.

Nagendra Bhargavi et al. [3] studied the solid waste collection and its management crisis in Bangalore city. After 9 months of analysis, the report states the reason why only 20% of waste is getting picked up by pourakarmikas (refuse collectors) in Bangalore city of India. These reasons are non-segregation of waste, irregular waste collection, traffic and spillover waste on the road while waste collection, failure in formal requirement with transport. This analysis was carried out by the

N. A. Mahakalkar (✉) · Radha D.
Department of Computer Science and Engineering, Amrita School of Engineering, Bengaluru, Amrita Vishwa Vidyapeetham, Bangalore, India
e-mail: nammahakalkar@gmail.com; d_radha@blr.amrita.edu

A. Haldorai et al. (eds.), *2nd EAI International Conference on Big Data Innovation for Sustainable Cognitive Computing*, EAI/Springer Innovations in Communication and Computing, https://doi.org/10.1007/978-3-030-47560-4_34

Public Affairs Centre (PAC) waste tracker app which makes data available on the cloud. Waste segregation awareness is also a key component when it comes to solid waste management issues. Lois Eyram Agbefe et al. [4] focuses on the municipal solid waste (MSW) management crisis in Ghana. Insight of waste collection, waste traders and common man perspective about waste segregation are noted by interviewing key stakeholders. To overcome the crisis, the paper [4] proposed two effective ways. First is awareness and training about waste segregation among traders and people. The second one is proper infrastructure by municipal for waste segregation at source. Technical guidance committee of solid waste management (SWM), Bruhat Bengaluru Mahanagar Palika (BBMP) has an opinion that wet, dry and sanitary waste segregation at door-step is neither practical nor economical [5]. The waste needs to be segregated automatically to avoid human intervention and the segregation classes should be recycled friendly so that the system will be more useful. If the garbage is classified accurately, then the recycling efficiency can be improved as compared to the existing waste management ecosystem. More infrastructural and technological development is required to overcome the waste management crisis.

The traditional method of segregation at the industrial level includes many steps. The very first step is manual segregation followed by a process of the large rotational drum with small size holes to filter small size waste. Then metallic objects are separated by electromagnetic or eddy current based separator. Different types of plastic wastes are separated by near field infrared scanner. After that, at some places, X-ray is used for remaining waste segregation. This giant mechanical arrangement demands more labor power, time and unavoidably invites human health hazards. The health hazards to the labor who work for trash collection, the industrial trash segregation, and people living nearby area to the landfill sites reduce their life expectancy. Due to the financial crisis, they have to accept this work irrespective of knowing its consequences. Garbage segregation using sensor mechanism and supporting hardware in paper [6–10] has limitations of a limited number of segregation classes, bulky hardware, structural design issues and is less accurate. So few experimenters have chosen a deep learning-based approach for segregation. Carlos Vrancken et al. [11] demonstrated step by step approach to create a database. This approach helps improving classification accuracy. Experimentation on less and more number of images with varying factors (angle, height, width, size, tilting, cropping) of images was carried out. The images were classified into two classes such as paper and cardboard with a total of 24 training images and 20 testing images. Further literature such as Common garbage classification using MobileNet [12] and Capsule neural network (CapsulNet) [13] focusing on deep learning-based approach achieves remarkable accuracy but for less number of classes. In the case of mixed waste, image detection can be done as shown in paper [14].

The proposed Smart Trash Segregator completely automates the waste segregation system at an organizational level (airport, railway stations, etc.). This helps to segregate six practically possible garbage classes, especially for Indian waste. The proposed system achieves superior accuracy for multiple waste classes. The segregation process of waste management can be tested using this prototype before the actual implementation of the system. This model is tested with a newly created dataset of Indian wastes. The model can be tuned for any particular area like industrial area or city area waste, hospital waste, etc. It will reduce waste management related problem up to some extent. It will also upgrade the quality of life of refuse collectors and segregation. It can be an add-on to waste management to solve the current problem. This will enable sustainable waste management. The novelty of the proposed method is that it segregates more than two types of waste and tested up to six classes and achieved accuracy ranges from 85% to 96%. The Smart Trash Segregator (STS) is proposed as a first step to deal with sustainable trash recycling.

2 System Architecture

System architecture consists of software model and hardware model.

2.1 Software Model

Software model in Fig. 1 shows the flow for training the model. Checking system software availability and installation requirements are as essential as dataset creation. Dataset is a very important asset as well as a requirement to start the software design. Following Table 1 gives information about the available dataset and created dataset. In created dataset section, Picam-dataset is a dataset containing images that captured by pi-camera of 5 mega pixels focusing over STS system trash tray. Whereas Nam-dataset is a dataset captured by 5 mega pixels Sony camera on white background.

Tensorflow has widely used framework for deep learning applications [15] and it has a large community base. Keras gives the user-friendly libraries for image pre-processing, classification, etc. Tensorflow and Keras libraries are supported by Raspberry Pi and can also run on cloud-based Google Colabs. So the model uses Keras with Tensorflow backend for training the model.

The prebuilt Convolution Neural Network (CNN) model performs well on the garbage image classification problem [16]. The transfer learning approach is used to train the model. The general idea of transfer learning is to take the pre-trained weights on the ImageNet dataset for earlier layers. Because earlier layers learn the general features such as edge, texture, background, etc. Then train the weights of the deeper layers with a custom dataset, where an object is identified by the layer [17]. There are many prebuilt CNN models such as ResNet, DenseNet, ImageNet, InceptionNet, Inception ResNet, etc. Here four CNN models with pre-trained weights on ImageNet dataset are used to extract the features at the starting layers of the network. The last three fully connected layers are trained on customized data which are in Trashnet and Nam-dataset. The model is trained for hundred

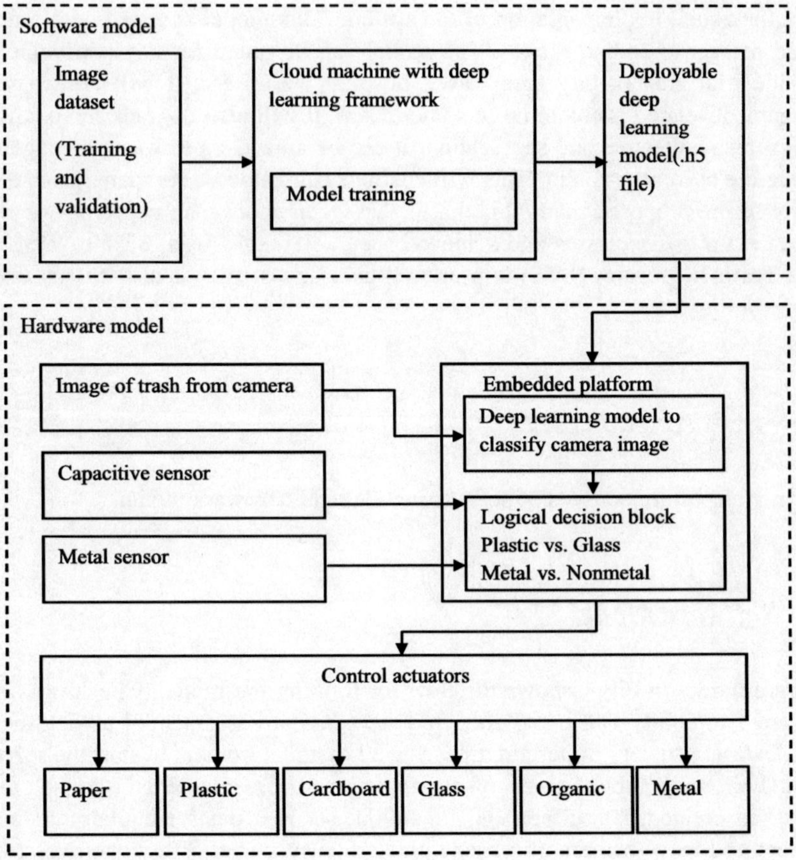

Fig. 1 High level system design

Table 1 Available and created dataset table

	Available dataset		Created dataset	
Class	Trashnet	FDM	Nam-dataset	Picam-dataset
Paper	594	100	198	289
Plastic	482	100	350	355
Cardboard	403	–	23	347
Glass	501	100	42	373
Metal	410	100	24	338
Organic/foliage	–	100	500	260
Trash	137	–	–	–
Total	2527	500	1137	1962

epochs with no data augmentation. However, each model exhibits different accuracy performance for the given dataset. Hence the performance of all the models is evaluated for the trash dataset to get maximum accuracy as discussed in Sect. 3.3. It is observed that the ResNet50 with Nam-dataset outperforms the other model.

At a higher level, the prebuilt CNN model is trained over the custom dataset using "Google Colab." The system DNN model has weights of the ResNet50 model trained on the ImageNet dataset. On top of it, three more fully connected layers are stacked to train over custom data. That makes this model a custom ResNet50 model for Nam-dataset. Transfer learning Keras ResNet50 custom model gives the result as shown in Table 3. The trained deep learning model is generated in .h5 file format. This trained model (.h5) will then be deployed into the Raspberry Pi for further operations of the hardware module as shown in Fig. 1.

2.2 *Hardware Model*

Hardware model incorporates the structural hardware design, electronic component mounting, and interconnections to ensure the workability of the system. Figure 2 shows the hardware parts of the system.

System Hardware Making Process Step 3 of Fig. 2 shows the complete system hardware without trash bins. During the making process of hardware individual models have been tested. Individual modules referred to as follows: (a) An image capturing on keypress, (b) Classification of captured trash image to appropriate class using DNN, (c) Sensor module for classification of organic and metal waste. (d) Circular/horizontal movement of the tray at each trash bin. (e) 90-degree up-down movement of the tray for dropping the trash to the bin and coming back at the vertical position.

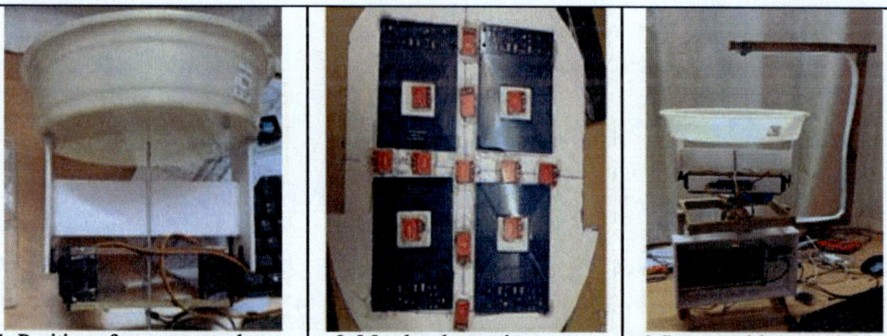

1. Position of servos to trash tray | 2. Metal and capacitor sensor arrangement | 3.System without trash bins

Fig. 2 Hardware parts of the system

Other Electronic Board Requirements Interconnection details of the hardware are shown in Fig. 1. The PWM expander board PCA9685 is used because the Raspberry Pi 3 B+ model has only two PWM pins and needs to drive more than two servo motors referred to as control actuators. PCA9685 is a 16 PWM channel with a 12 bit I2C interfaced servo driver board. Before using I2C, the PI needs to configure I2C with SMBus support and I2C-tools. The Raspberry Pi communicates to the PWM expander PCA9685 via an I2C interface. The metal sensor gives analog values. ADS1115 ADC is used to convert its output to digital. ADS1115 is a 16 bit 4 channel analog to digital converter. Four metal sensors can be interfaced with the ADS1115 board at A0, A1, A2, and A3 pins. GPIO expander is also required to make the system design less complicated and to manage the interconnections between Raspberry Pi and rotating arm. All three boards are I2C so they can work on the same I2C bus and have different addresses.

Implementation As soon as the user keeps the trash on to the trash tray and presses the "y" button from the keyboard, the system loads the trained (ported on Raspberry Pi) deep learning model and the camera captures the image of trash placed in the tray. Trained model (custom ResNet50 trained over Nam-dataset, Picam-dataset) in the Raspberry Pi classifies the captured trash image into an appropriate class. Model gives the output in the form of class index such as 0 for "metal," 1 for "glass," 2 for "paper," 3 for "cardboard," 4 for "plastic," and 5 for "organic." According to the class index, the first servo motor which performs the horizontal movement of the tray at each trash bin positions the trash tray to the specific index bin. For example, if the model classifies the trash as 4 which is the class index of plastic, then the tray rotates horizontally for 240° clockwise and stops at the plastic bin. Each bin is placed 60° apart. Once the tray is horizontally rotated, the next two servo motors (which are meant for 90° up-down movements of tray) help to drop the trash into the bin. After dropping trash into the bin, the tray moves back to the vertical position. When a different or same type of trash is placed onto the tray, the system captures the image then classifies into the appropriate class and drops the garbage into the corresponding bin. Figure 3 shows the implementation flowchart, whereas Figs. 4 and 5 show the front and top view of the complete system. System introduction, working and inference videos are available at https://sites.google.com/site/namrataprojects/home/smart-trash-segregator-using-deep-learning-on-embedded-platform.

The system performs proper identification/classification but in some cases, it mystifies and gives the wrong class index. In the case of identification of plastic and glass, the model classifies glass as plastic or plastic as glass in some cases. Hence redundant mechanism for the segregation is used. Capacitive sensors and deep learning classification output are used to make the decision. This hybrid approach makes the system more accurate. Another case of misclassification happens between metal and plastic. This misclassification is resolved by using a metal sensor and a deep learning-based approach for garbage segregation. Capacitive and metal sensors are embedded in the system to make the system more accurate and reliable. The sensor module/plate is positioned in such a way that thirteen capacitive sensors

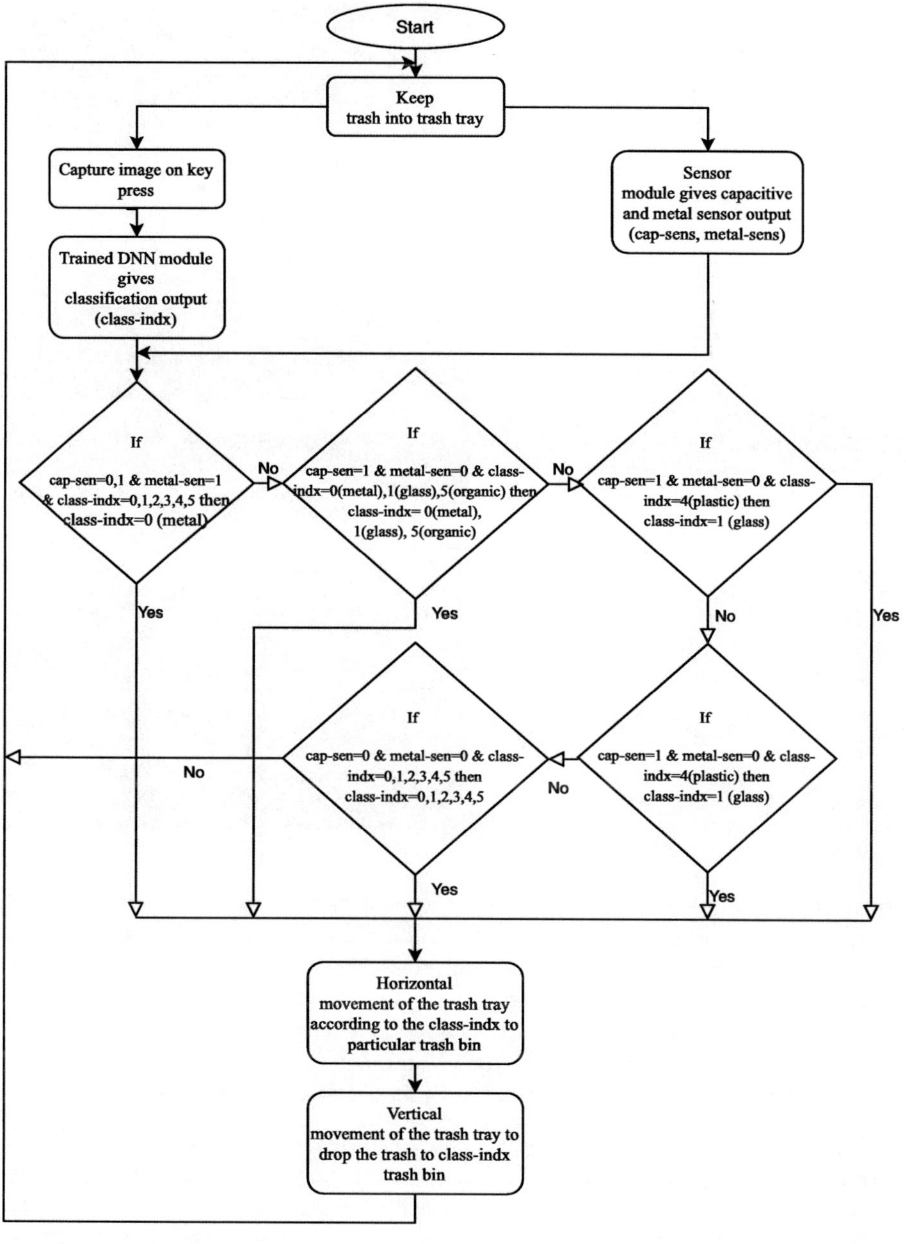

Fig. 3 System flowchart

Fig. 4 Front view

Fig. 5 Top view

touch the sensing surface and the magnetic field of four metal sensors faces the bottom of the trash tray. The sensitivity of the capacitive sensor is maximum for wet/organic trash followed by metal and glass because of the relative dielectric constant (ε) of the materials. The capacitive sensor is not sensitive to cardboard, paper, and plastic. So the glass versus plastic identification is achieved as the capacitive sensor is not sensitive to the plastic material. The metal sensor is also embedded into the sensor module to classify among metal, plastic, and others.

Figure 3 shows the system implementation flowchart, it constitutes a sensor module as well as a DNN module. As soon as the trash is kept onto the trash bin, the sensor module gives output for the capacitive sensor and the metal sensor and camera module captures the image. DNN module gives the output in the form of class index. DNN module output and sensor module output are analyzed and the final output of the class index depending on the conditions is given by the logical block to the actuators. The logical block conditions are as follows:

- If the metal sensor output is positive, then irrespective of other sensor outputs the trash is classified as a metal class.

- If the capacitive sensor output is positive, the metal sensor output is negative and class index output of the DNN module is one, then the trash is classified as glass.
- If the capacitive sensor output is positive, the metal sensor output is negative and the class index output of the DNN module is four, i.e. plastic class index, then the trash is classified as glass.
- If the capacitive sensor output is positive, the metal sensor output is negative and the class index output of the DNN module class index is two or three that is of paper or cardboard, respectively, then the trash is classified as the maximum probability of DNN from organic, glass, and metal.
- If the capacitive sensor output is negative, the metal sensor output is negative and class index output of the DNN module is one or two or any other class index, then the classification is based on DNN classification.

3 Simulation Results and Discussion

Multiple models and methods were evaluated to check the classification performance of the trash dataset. Sections 3.1 and 3.2 explain the result of the ten layer CNN model and sixteen layer CNN model, respectively. Sections 3.3 and 3.4 use transfer learning approach. The following section shows the results of each method in detail.

3.1 Results of Ten Layer CNN Model for Two Class Classification

Using Tensorflow and Keras, ten layers of convolution neural network are built and trained over Nam-dataset. This network achieves approximately 20% accuracy for six class classification. But achieves approximately 88–94% accuracy for two types (plastic and paper, organic and paper) of trashes. Dataset of 1200 images is used and 400 images of each class are trained for 15 epochs. It has hyperparameters like dropout = 0.5, Adam optimizer and ReLu activation at each convolution layer. This model is validated over a total of 400 images containing 200 images of each class. Data augmentation is done as a part of dataset pre-processing. Data is augmented with a rotational range of 40, width and height shift of 0.2, horizontal flip and a zoom range of 0.2. The same hyperparameter and data augmentation are applied for the results shown in Figs. 6, 7, 8, 9, 10, and 11. These figures are of loss and accuracy of various waste classes.

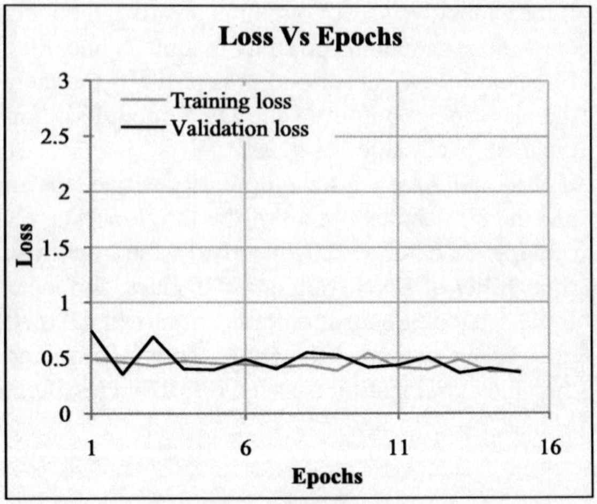

Fig. 6 Loss graph for plastic and paper class on training and validation set

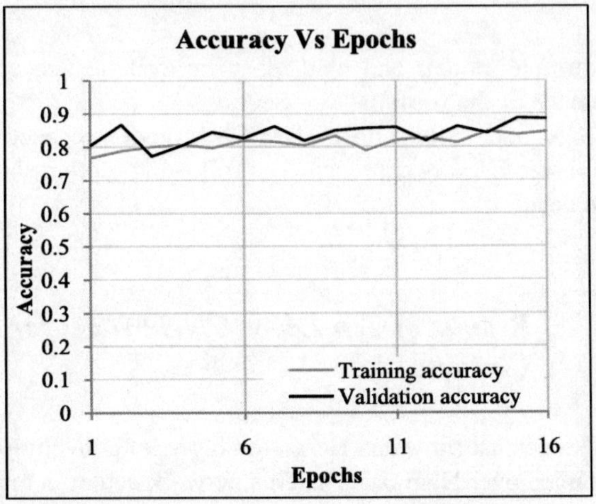

Fig. 7 Accuracy graph for plastic and paper class on training and validation set

3.2 *Sixteen Layer CNN Model for Six Class Classification*

Improved network with 16 layers with the same data augmentation parameter mentioned in the 10 layer CNN network [Sect. 3.1] works well for all six class classification. The dataset is having 1800 training images (300*6) and 600 validation images (100*6), which are taken from Nam-dataset and Trashnet dataset. Model is trained for 50 epochs and achieves 84% and 80% as training and validation accuracy, respectively, as shown in Figs. 10 and 11.

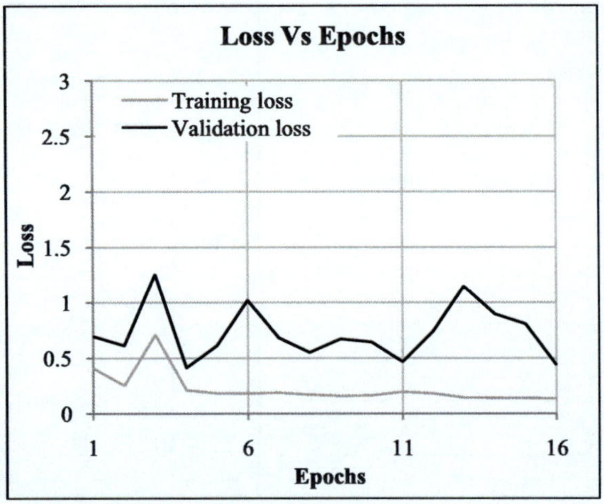

Fig. 8 Training and validation loss graph for organic and paper class

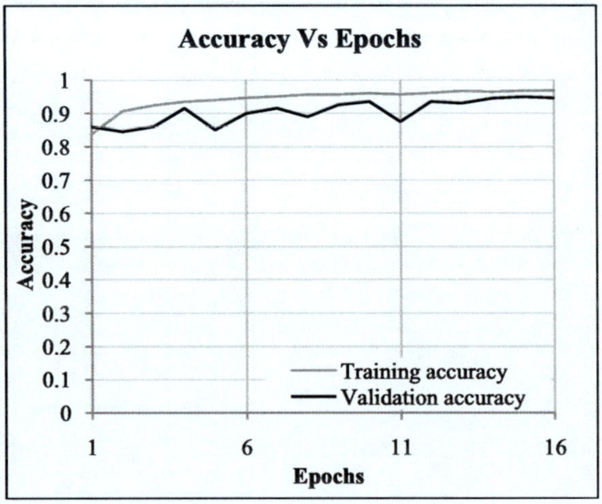

Fig. 9 Training and validation accuracy graph for organic and paper class

3.3 *Comparison of CNN Models Using Fastai and Keras Frameworks and Selection of Best Model for Custom Dataset*

In the framework and model identification process, experimentation on Fastai using ResNet34, ResNet50, and ResNet101 are carried out over the Trashnet dataset. In Table 2, models are trained from scratch with 25 epoch using GPU. It has 1262 training images, 630 validation images, and 635 test images. Table 2 shows the validation accuracy percentage from confusion matrix. Validation accuracy of ResNet50 and ResNet101 is the same as 94% but the classification percentage of each class is different. Glass and paper class confusion percentage is more in

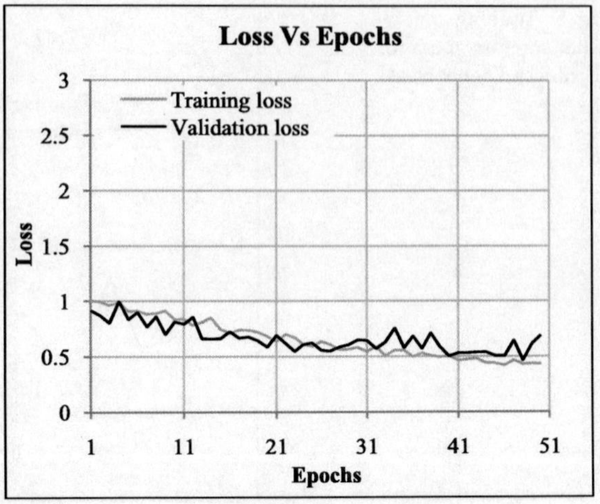

Fig. 10 Training and validation loss graph for six classes

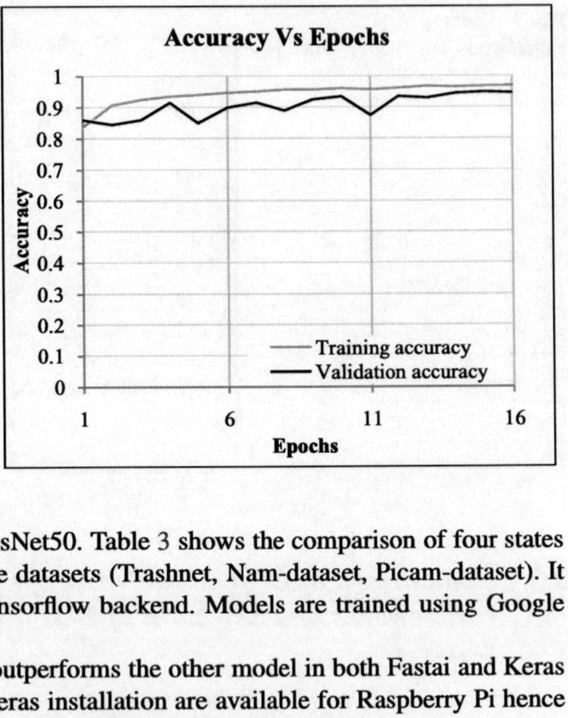

Fig. 11 Training and validation accuracy graph for six classes

ResNet101 as compared to ResNet50. Table 3 shows the comparison of four states of the art CNN model on three datasets (Trashnet, Nam-dataset, Picam-dataset). It uses Keras running on the Tensorflow backend. Models are trained using Google Colabs with GPU.

Tables 2 and 3, ResNet50 outperforms the other model in both Fastai and Keras framework. Tensorflow and Keras installation are available for Raspberry Pi hence the ResNet50 with Keras using Tensorflow backend is a better choice.

Table 2 Fastai Model experimentation results

Model	Confusion matrix: Actual	Predicted: cardboard	Predicted: glass	Predicted: metal	Predicted: paper	Predicted: plastic	Predicted: trash	Validation accuracy
ResNet34	cardboard	91	0	0	9	1	0	84%
	glass	2	98	13	0	12	0	
	metal	2	4	91	2	2	1	
	paper	7	0	1	137	1	2	
	plastic	0	8	9	3	91	3	
	trash	1	1	4	5	2	27	
ResNet50	cardboard	98	0	0	2	0	1	94%
	glass	0	122	1	0	2	0	
	metal	0	2	99	0	0	1	
	paper	2	0	0	164	0	0	
	plastic	0	1	3	2	109	5	
	trash	1	1	0	3	0	29	
ResNet101	cardboard	99	0	0	2	0	0	94%
	glass	0	118	1	0	6	0	
	metal	0	1	98	0	3	0	
	paper	2	0	0	142	0	4	
	plastic	0	1	2	2	110	5	
	trash	1	1	1	1	0	30	

3.4 *Trash Segregation in Hardware Module Using Custom ResNet50 Model*

Figure 12 shows the images with its actual class and predicted class. The actual class index of the images is like 0:"metal," 1:"glass," 2:"paper," 3:"cardboard," 4:"plastic," 5:"organic." Almost all the images are correctly classified in Fig. 12 except one image of metal at position (3, 3) with class index zero. Image is of a metal spoon but the model has predicted it as cardboard with class index 3. So such kind of misclassification problem can be easily avoided by using a metal sensor. The predicted class system takes the action of dropping the trash into the correct bin as shown in Fig. 13. The accuracy of such classification and dropping is identified as 95%.

3.5 *Trash Segregation in Hardware Model Using Sensor Module and DNN Module*

Raspberry pi is loaded with the ResNet50 model trained over Picam-dataset which achieves 85% accuracy. An experiment is conducted to know the improvement of the system after integrating a sensor network to the system. Figures 14 and 15 show the confusion matrix using DNN module only and confusion matrix using DNN with sensor module, respectively. Hundred trash images are considered, of which 24 are of cardboard, 13 are of glass, 18 are of metal, 14 are of paper, 14 are of

Table 3 Comparison table of CNN models

Model	Dataset	Training vs. Validation loss	Training vs. validation accuracy	Comment
ResNet50	Trashnet	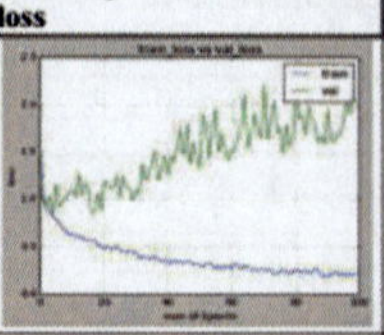	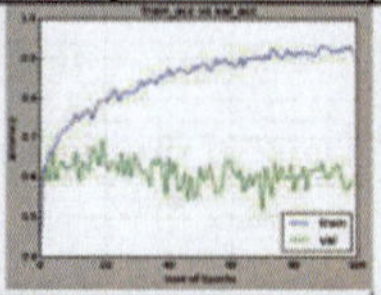	loss=1.2760, accuracy: 63.0435%
ResNet50	**Nam-dataset**	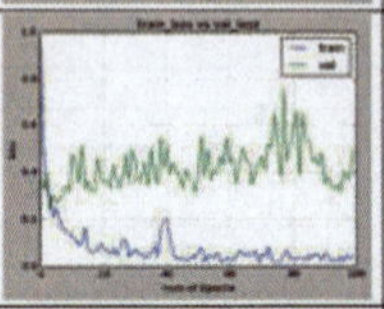		**loss=0.4982, accuracy: 95.0877%**
InceptionNet V2	Trashnet	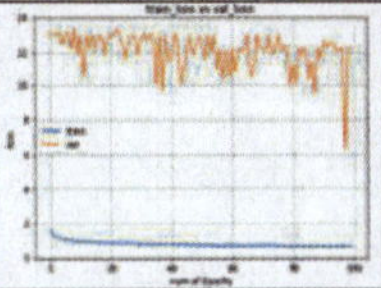	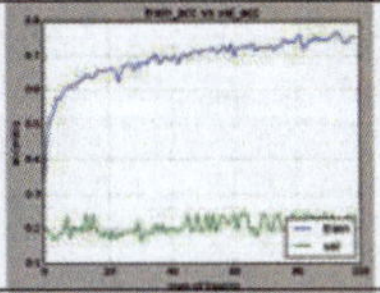	loss=12.2368, accuracy: 24.0803%
InceptionNet V2	Nam-dataset	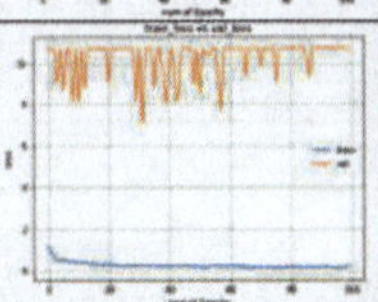	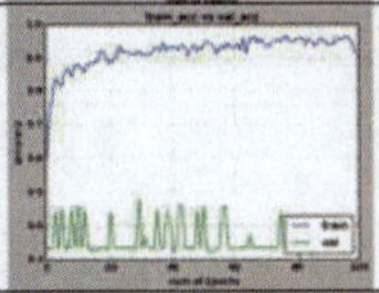	loss=10.7454, accuracy: 33.3333%
MobileNetV2	Trashnet	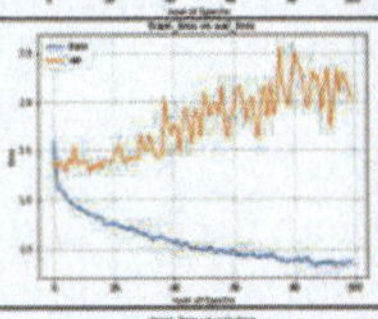	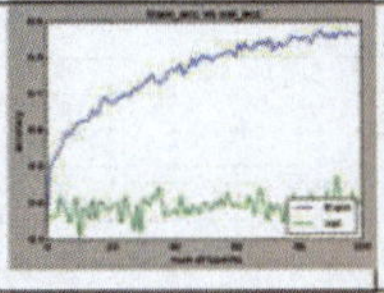	loss=2.0192, accuracy: 41.0156%
MobileNetV2	Nam-dataset	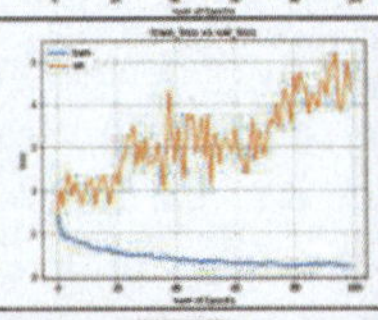	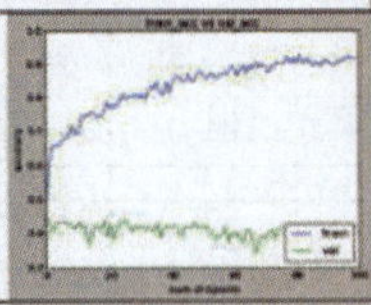	loss=4.3533, accuracy: 42.1053%
DenseNet121	Trashnet	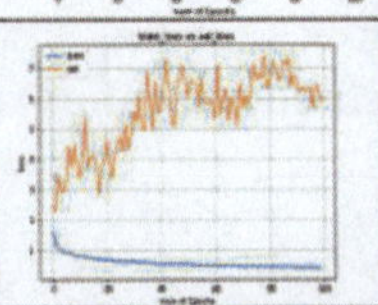	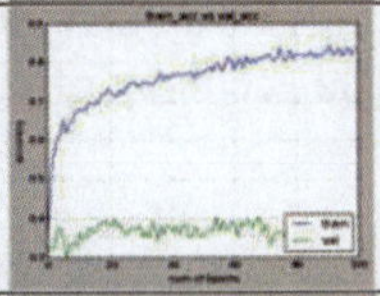	loss=5.8906, accuracy: 37.6254%
DenseNet121	Nam-dataset	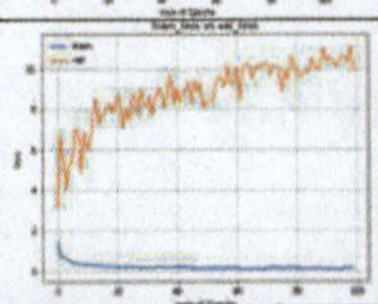	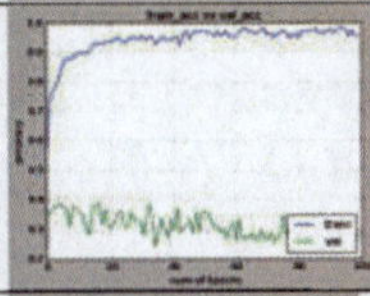	loss=9.9433, accuracy: 29.1228%
ResNet50	**Picam-dataset**	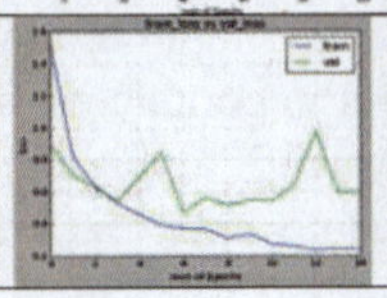	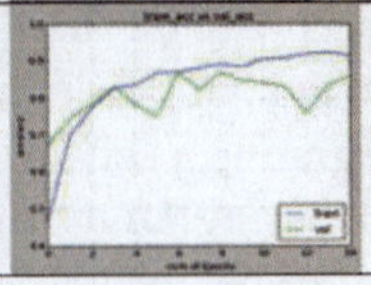	**loss=0.5967, accuracy: 85.7434%**

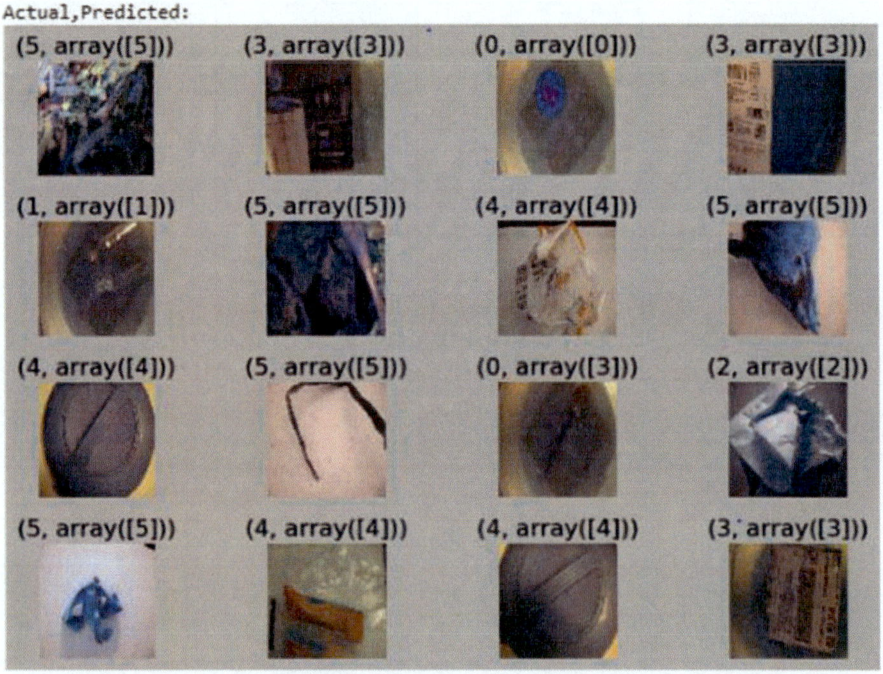

Fig. 12 Actual and predicted class index images

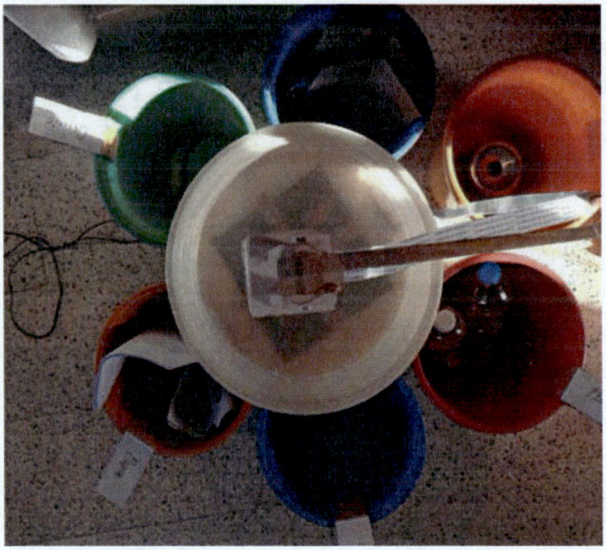

Fig. 13 Segregated trash by the system

Actual	cardboard	22	0	0	2	0	0
	glass	0	13	0	0	0	0
	metal	0	5	13	0	0	0
	paper	0	0	0	14	0	0
	plastic	0	4	0	0	10	0
	organic	0	1	1	1	0	14
		cardboard	glass	metal	paper	plastic	organic
Predicted by DNN							

Fig. 14 Confusion matrix for DNN module without sensor module

Actual	cardboard	24	0	0	0	0	0
	glass	0	13	0	0	0	0
	metal	0	2	16	0	0	0
	paper	0	0	0	14	0	0
	plastic	0	0	0	0	14	0
	trash	0	2	0	0	0	15
		cardboard	glass	metal	paper	plastic	trash
Predicted by DNN and Sensor module together							

Fig. 15 Confusion matrix for DNN and Sensor module together

plastic, and 17 are of organic. DNN module gives 86% accuracy and DNN with sensor module gives 96% accuracy.

4 Conclusion

The proposed system is to segregate waste into different classes which will be useful for recycling. The traditional waste management system is not efficient for the segregation of wastes. The proposed prototype model of Smart Trash Segregator at source level can be efficient in segregating the waste into six classes. The system uses a sensor module and DNN module which achieves the maximum possible accuracy of 96%, average accuracy as 91%. It is used to segregate the trash autonomously. A scaled prototype version with robust material and high-speed capacity enable the system to be a sustainable system for waste management. The classification is done for one trash at a time.

5 Future Scope

If mixed waste is kept into the trash tray, the system will take the majority trash class decision and drop it into the corresponding class bin. This problem can be overcome by designing a robotic arm that will be capable of picking the trash one by one and dropping into the appropriate bin. The robotic arm is one of the future scopes for making the system more practically sound to achieve mixed garbage classification. IoTisation of the system can be done to make recycling more efficient. The idea of IoTisation includes a common interface through which all trash traders/recyclers and generators can be connected. Traders/recyclers can get online data regarding trash availability. This includes type, quantity, and GPS location. An interface can be developed to know the live information about trash with its location. Trash generator can sell their trash to the traders/recyclers and traders/recyclers can buy trash from the trash generators. To route to the location, GPS data can be provided through the application itself. Trash segregation and communication to corresponding traders for recycling eventually improve the trash management activity.

References

1. Census of India (2011) Office of Registrar general and census commissioner, India. http://www.dataforall.org/dashboard/censusinfoindia-pca/. Accessed 30 June 2019
2. Samar L (2019) India's challenges in waste management. down toearth.org.in, 8 May 2019. https://www.downtoearth.org.in/blog/waste/india-s-challenges-in-waste-management-56753. Accessed 30 June 2019
3. Bhargavi N, Arvind L, Priyanka A (2019) Mobile application in municipal waste tracking: a pilot study of "PAC waste tracker" in Bangalore city, India. J Mater Cycles Waste Manage 21:705–712
4. Agbefe LE, Lawson ET, Yirenya-Tawiah DJ (2019) Awareness on waste segregation at source and willingness to pay for collection service in selected market in Ga West Municipality, Accra, Ghana. J Mater Cycles Waste Manage 21:905. https://doi.org/10.1007/s10163-019-00849-x
5. Reporter staff (2019) In a first, BBMP drafts by-lows for solid waste management. thehindu.com, 4 August 2019. https://www.thehindu.com/news/cities/bangalore/in-a-first-bbmp-drafts-by-laws-for-solid-waste-management/article28810097.ece. Accessed 4 Aug 2019
6. Pereira W, Parulekar S, Phaltankar S, Kamble V (2019) Smart bin (waste segregation and optimisation). In: 2019 Amity international conference on artificial intelligence (AICAI), Dubai, 2019, pp 274–279
7. Chandramohan A, Mendonca J, Shankar, NR, Baheti NU, Krishnan NK, Suma MS (2004) Automated waste segregator. In: 2014 Texas instruments India educators' conference (TIIEC), Bangalore, pp 1–6
8. Kumar BRS, Varalakshmi N, Lokeshwari SS, Rohit K, Manjunath, Sahana DN (2017) Eco-friendly IOT based waste segregation and management. In: 2017 international conference on electrical, electronics, communication, computer, and optimization techniques (ICEECCOT), Mysuru, 2017. IEEE, New York, pp 297–299

9. Madankar A, Patil M, Khandait P (2019) Automated waste segregation system and its approach towards generation of ethanol. In: 2019 5th international conference on advanced computing and communication systems (ICACCS), Coimbatore. IEEE, New York, pp 571–573
10. Hassan H, Saad F, Mohd Raklan MS (2018) A low-cost automated sorting recycle bin powered by Arduino microcontroller. In: 2018 IEEE conference on systems, process and control (ICSPC), Melaka, pp 182–186
11. Vrancken C, Longhurst P, Wagland S (2019) Deep learning in material recovery: development of method to create training database. Expert Syst Appl 125:268–280. ISSN: 0957-4174
12. Rabano SL, Cabatuan MK, Sybingco E, Dadios EP, Calilung EJ (2018) Common garbage classification using MobileNet. In: 2018 IEEE 10th international conference on humanoid, nanotechnology, information technology, communication and control, environment and management (HNICEM), Baguio City, pp 1–4. https://doi.org/10.1109/HNICEM.2018.8666300
13. Sreelakshmi K, Akarsh S, Vinayakumar R, Soman KP (2019) Capsule neural networks and visualization for segregation of plastic and non-plastic wastes. In: 2019 5th international conference on advanced computing and communication systems (ICACCS), Coimbatore, pp 631–636
14. Vishwakarma S, Radha D, Amudha J (2018) Effectual training for object detection using eye tracking data set. In: 2018 international conference on inventive research in computing applications (ICIRCA), 2017, pp 225–230
15. Jeff H (2018) Deep learning framework power scores 2018. towarddatascience.com, 20 September 2018. https://towardsdatascience.com/deep-learning-framework-power-scores-2018-23607ddf297a. Accessed 5 Aug 2019
16. Bircanoğlu C, Atay M, Beşer F, Genç Ö, Kízrak MA (2018) RecycleNet: intelligent waste sorting using deep neural networks. In: 2018 innovations in intelligent systems and applications (INISTA), Thessaloniki. IEEE, New York, pp 1–7
17. Ankit P (2018) Understanding your convolution network with visualization. towarddatascience.com. 1 October 2018. https://towardsdatascience.com/understanding-your-convolution-network-with-visualizations-a4883441533b. Accessed Sept 2019

Efficient Graph Algorithms for Mapping Tasks to Processors

Sesha Kalyur and G. S. Nagaraja

1 Introduction

Parallel machines are characterized, by the number of processing elements, and the different ways these processors can be interconnected. Since the prime reason for using a parallel machine is to share the software workload, the configuration which includes both the processors and the interconnection network mainly determines the performance one can expect from the parallel machine. This ultimately dictates how the machine scales, in relation to larger problem sizes and the resulting addition to the processor count.

The interconnection networks are diverse in nature, dictated by performance considerations, physical design, and implementation, and are collectively referred to as the Processor Interconnect Topology. Historically the published literature presents several references to processor interconnect topologies [5, 7]. Some of the popular ones include the Shared Bus, Fully Connected Network, Linear Array, Ring, Mesh, Tori, Grid, Cube, Tree, Fat Tree, Benes Network, and Hypercube among others. The following sections look at some of them in detail, and also provide a mathematical abstraction for representing them, and using them in analysis.

The computation in a higher level language program can be represented internally in a compiler, by a representation such as a three address form [15]. The dependencies that exist between statements or instructions of the program, manifest in two forms namely, Data Dependence and Control Dependence [17, 29]. Data dependence exists between two instructions, when one of them reads a datum

S. Kalyur (✉) · G. S. Nagaraja
Department of Computer Science and Engineering, R.V. College of Engineering, VTU, Bangalore, India
e-mail: nagarajags@rvce.edu.in

A. Haldorai et al. (eds.), *2nd EAI International Conference on Big Data Innovation for Sustainable Cognitive Computing*, EAI/Springer Innovations in Communication and Computing, https://doi.org/10.1007/978-3-030-47560-4_35

that the other has written. Control dependence exists between two instructions, when the execution of one of them is conditional on the results produced by the execution of the other. There are several mathematical representations available to represent a program internally, but the Program Dependence Graph (PDG) is a popular technique that captures these program artifacts [8]. PDG is a convenient tool to partition a sequential program into parallel tasks [11, 12, 15, 16, 28]. While detection of parallelism in software is a popular area of research, it is not central to the topic we have chosen for our research work here. We start with the assumption that the concurrent tasks are available and focus on the related problem of mapping them to available processors. The main research goal is to find out if the choice of a processor for a particular task makes a difference to the overall efficiency of the mapping process.

The problem of assigning parallel tasks of a program, to processors of a distributed system efficiently, is referred to here as the Task Assignment Problem. An effective solution to this problem normally depends on the following criteria namely, Load Balancing and Minimization of Communication Overhead. Load Balancing is a process, where the computing resources of a distributed system are uniformly loaded, by considering the execution times of the parallel tasks. Existing load balancing strategies can be broadly classified as static and dynamic, based on when the load balancing decision is made. In static methods, the load balancing decision is taken at the task distribution time and the assignment of tasks to processors remains fixed for the duration of their execution [27]. On the other hand, the dynamic schemes are adaptive to changing load conditions and tasks are migrated as necessary, to keep the system balanced. The latter scheme is more sensitive to changing topological characteristics of the machine, especially the communication overheads [20].

Minimization of communication overhead involves clustering the tasks carefully on the target machine, so as to minimize the inter-task communication demands. There are two factors influencing this decision namely, the communication granularity and the topological characteristics of the underlying distributed machine [14]. Task assignment is not an easy problem to solve, since load balancing and communication minimization are inter-related, and improving one adversely affects the other. This is believed by many to be an NP Complete problem [2]. There are solutions proposed in the literature based on Heuristics and other search techniques [3]. However missing are efficient algorithms that are scalable, complete, and deterministic.

Typically, mapping problems that involve binary relations could be represented and studied by creating a model based on graphs. Topology representation is a problem that can be modeled as a graph, with processors as nodes and edges representing the processor connections. The tasks could also be represented as a graph, with tasks as nodes and edges denoting the communication between them. The solution would then be the mapping of the task nodes to the suitable nodes of the processors. This would require solutions to several sub-problems such as the following: How to gather the various properties of a graph, such as the node and edge count, individual connections, clustering, weights, etc.? How do we capture graph similarities? The obvious way of course is by visual inspection, which works for manageable sizes, but not for real world large graphs. Is it possible to solve this problem, in a deterministic fashion? [26] is a good survey paper, on the topic of graph comprehension and cognition.

This research work involves studying, and developing a methodology, to map parallel tasks of any given program, to a suitable processor topology in linear or near linear time. Subsequent sections provide more details of the findings, including several algorithms, that produce task assignments, that are designed to be progressively more efficient than the earlier algorithms.

In Sect. 1, we look at existing solutions to the task mapping problem, highlight the deficiencies in the current solutions, and elaborate on the motivations for pursuing our research work. In Sect. 2, we discuss the various popular ways of connecting the processors of a distributed machine and a suitable mathematical representation for each. Section 3 defines the Task Assignment Problem, which is the topic of investigation in this work. In Section 4 we present the various mapping algorithms proposed in this paper. In Sect. 5, we theoretically examine the fitness of each algorithm, in terms of its run time complexity. In Sect. 6, the final section of the paper, we conclude the paper by revisiting the motivations for taking up our research work, and briefly summarizing our findings and contributions with some ideas for extending the work for the future.

2 Related Works

Historically, researchers have used graphs to represent processor and network topologies [31]. Such a graph has been used to compute robustness of a network, towards corruptions related to noise and structural failures [1].

How do we capture both the topology and task details in a single graph? We found several references on the topic of graph aggregation, and incremental graph construction, subject to certain constraints. One such idea is a topological graph constrained by a virtual diameter, signifying properties such as communication [10]. However they do not provide implementation details. Researchers have used the spectral filtering techniques for analyzing network topologies, using Eigen vectors to group similar nodes in the topology, based on geography or other semantic properties [9]. Comparing directed graphs, including those of different sizes by aggregating nodes and edges, through deterministic annealing has been studied [30]. Graph aggregation techniques based on multi-dimensional analysis for understanding of large graphs have been proposed [25].

Graph Summarization is a process of gleaning information out of graphs for understanding and analysis purposes. Most of the methods are statistical in nature which use degree distributions, hop-plots, and clustering coefficients. But statistical methods are plagued by the frequent false positive problem and so are other methods. Analytical methodologies on the other hand are immune to such limitations [25].

Load Balancing on Distributed Systems has been studied extensively for a long time, and a plethora of papers exist on this topic [4, 6, 20, 23, 24, 27]. However load balancing is just one of the criteria that determines efficient performance of a distributed application. It needs to be complemented with minimization of communication overheads to see positive performance results.

Several researchers have tackled the task assignment problem before with varying degree of success. One such solution to the problem, based on the genetic algorithm technique in the context of a Digital Signal Processing (DSP) system has been proposed [3]. Researchers have used duplication of important tasks among the distributed processors to minimize communication overheads and generate efficient schedules [2]. Methods using the Message Passing Interface (MPI), on to a High Performance Computing (HPC) machine with Non-Uniform Memory Access (NUMA) characteristics, using a user supplied placement strategy has been tried by several groups with effective results [13, 14]. A few researchers have used both load balancing and communication traits as criteria to drive the mapping decisions [22]. They use dynamic profiling, to glean performance behavior of the application. However dynamic profiling is heavily biased on the sample data used and in our opinion, static analysis of the communication characteristics is a better technique and should yield optimum results in most situations.

Besides the domain of Parallel Processing, are there other fields where the task mapping problem has been explored? Assignment of Internet services based on the topology and the traffic demand information is one such domain [21]. Mapping tasks to processing nodes has also been studied at length, by researchers working in the Data Management domain. Tasks in a Data Management System, are typically characterized by data shuffling and join operations. This demands extra care in parallelization besides static partitioning, such as migrating tasks to where data is to realize maximum benefit [19]. Query processing is another area, where static partitioning of tasks runs into bottlenecks, and the authors solve this by running queries on small fragments of input data, whereby the parallelism is elastically changed during execution [18]. However our solution to the mapping problem is a general technique, and is not specific to the Data Management problem or the Internet domain and should be easily adaptable here.

In this research work, we propose a mathematical representation, based on directed graphs, to represent both the machine topology and the parallel task profiles. These graphs are then read by our task mapper, to map the processors and tasks. The following sections look at the problem and solutions in greater detail.

3 Processor Interconnection Topologies

Processors in a Multiprocessor machine can be interconnected in several interesting ways that mainly affect how the resulting machine scales as processors are added. It is important both from a problem representation and solution perspective, that we study these topologies in some detail and understand them. We next discuss several popular processor interconnection topologies found in published literature, and show with the help of diagrams, how each of these topologies could be represented mathematically as directed graphs.

At one end of the spectrum is the Shared Bus, where at any given time, a single communication is in progress. At the other end lies the Fully Connected Network, where potentially at any given time, all the processors could be involved in private communication. If the number of processors in the network is N, with a shared bus we can only realize a bandwidth of $O(1)$, but with a fully connected network, we could extract a bandwidth of $O(N)$ [5, 7]. There are several other topologies that fall in between, and we will look at a few of them in the following paragraphs.

Figure 1a, b on page 471 illustrates a simple bus topology, for connecting processors of a machine and its graph representation. Likewise a Linear Array is a simple interconnection of processor nodes, connected by bidirectional links. Figure 2a, b on page 471 represents linear topology and its corresponding graph.

A Ring or a Torus is formed from a Linear Array, by connecting the ends. There is exactly one route from any given node to another node. Figure 3a on page 472 is an illustration of a topology organized in the form of a ring, which allows communication in one direction between any pair of nodes. The associated Fig. 3b on page 472 illustrates how the topology can be represented as a graph. The average

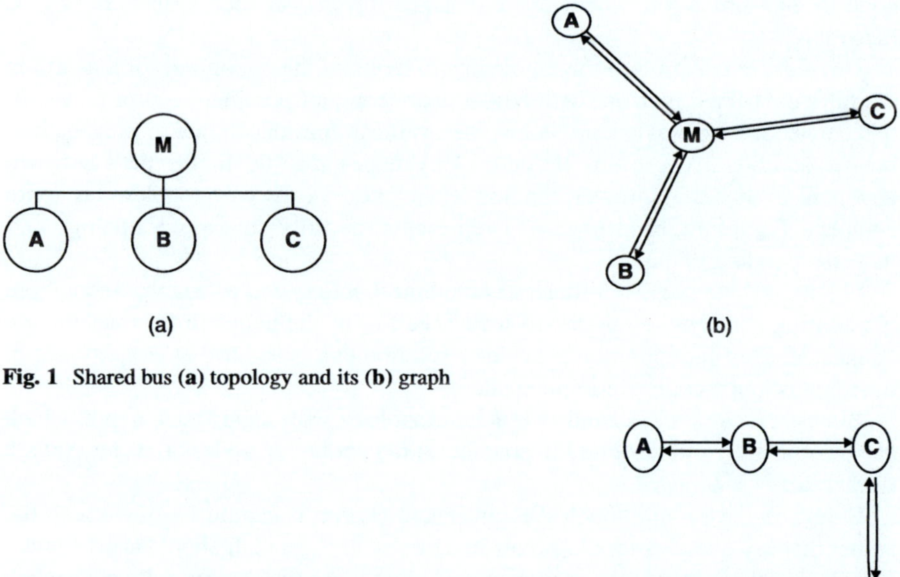

Fig. 1 Shared bus (a) topology and its (b) graph

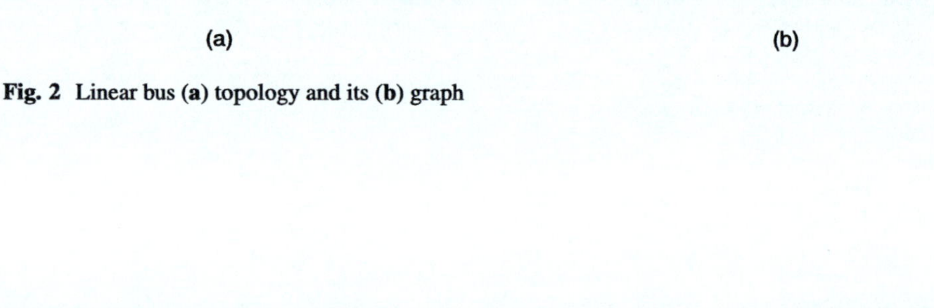

Fig. 2 Linear bus (a) topology and its (b) graph

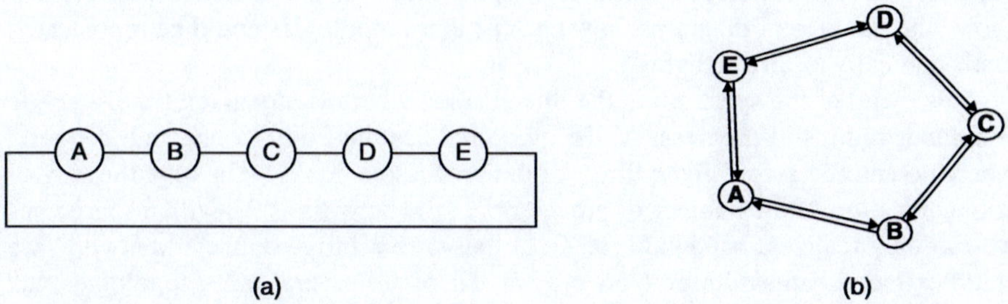

Fig. 3 Ring bus (a) topology and its (b) graph

distance between any pair of nodes in the case of a ring is $N/3$ and in the case of a Linear Array it is $N/2$, where N is the number of nodes in the network. We could potentially realize a bandwidth of $O(N)$, from such a connection.

Grid, Tori, and Cube are higher dimensional network configurations, formed out of Linear arrays and Rings. Specifically they are K-ary, D-$cube$ networks with K nodes, in each of the D dimensions. These configurations provide a practical scalable solution packing more processors in higher dimensions. To travel from any given node to another, one crosses links across dimensions, and then to the desired node in that dimension. The average distance traveled in such a network is $D * (2/3) * K$.

The most efficient of all the topologies in terms of the communication delays is the fully connected network with edges, connecting all possible pairs of nodes. In this connection topology, there is just the overhead introduced due to propagation, but no additional overheads introduced by nodes that lie in the path between any pair of nodes. However the implementation of such a topology is quite complex. Figure 4a, b on page 473 represents the fully connected topology and its corresponding graph.

Figure 5a, b on page 473 illustrates the Star topology and its graph, where there is a routing node in the middle, to which the rest of the nodes in the machine are connected. This configuration provides a solution that is somewhat less efficient in time, but is less complex to implement.

Binary trees represent another efficient topology with logarithmic depth, which can be efficiently implemented in practice. A network of N nodes offers an average distance of $N * Log(N)$.

Based on the sample topologies presented earlier, it should be obvious to the reader that any complex topology can be modeled as a graph. It should also be noted that all graphs representing topologies, including those that are not fully connected, should allow a path between any pairs of nodes, even though not directly connected, by a route that passes through other intermediary nodes.

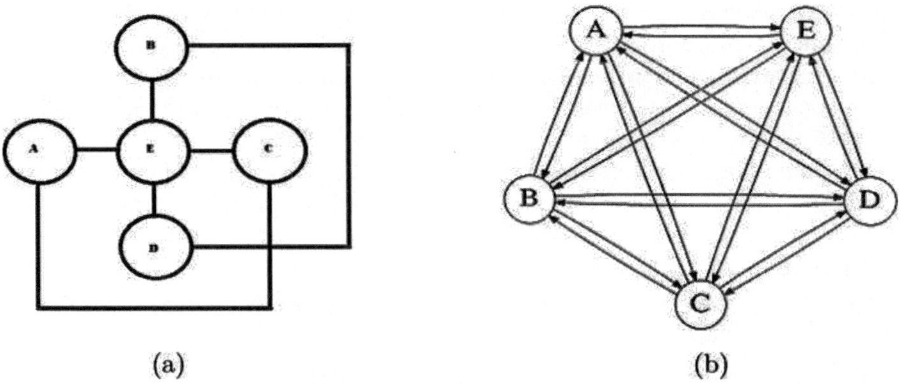

Fig. 4 Fully connected (**a**) topology and its (**b**) graph

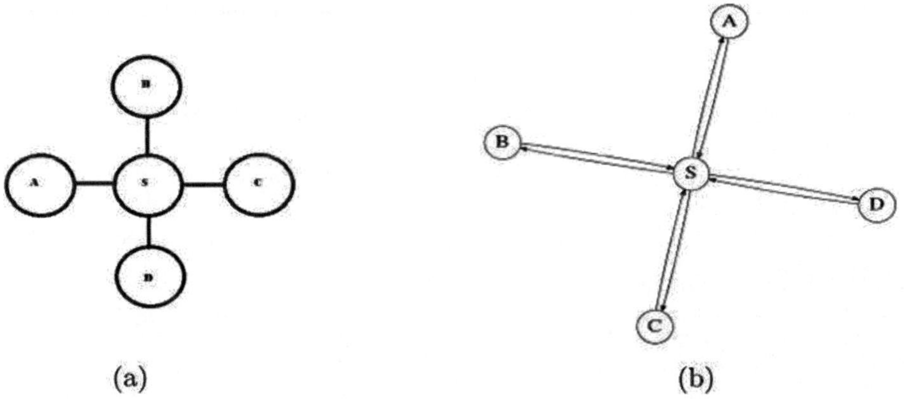

Fig. 5 Star (**a**) topology and its (**b**) graph

4 Task Assignment Problem

The problem of assigning parallel tasks of a program to the processing elements of a suitable machine topology is referred to here as the Task Assignment Problem. The characteristics of the machine relevant for the assignment, is captured in a graph, which we refer to here, as the Processor Topology Graph (PTG). The nodes of such a graph represent the processors and the edges represent the connections between the processors. The nodes could include parameters that capture processor characteristics, such as the clock rate. Similarly, the edge parameters could capture the bandwidth details of the connection, both of which could serve to drive the placement decisions. The communication details pertaining to the tasks can be captured in another graph, which we refer to as the Task Communication Graph (TCG). The nodes of the TCG represent the tasks and node parameters could represent the computation cycles for the task. The edges denote the communication that exists between any pair of tasks and the edge parameters could represent the volume of communication if any between the tasks concerned. Both the node and edge parameters could provide input for the placement algorithms. A Task Assignment Graph (TAG) captures the augmented information, from both the PTG and TCG, that mainly conveys the task to processor mappings.

In summary, the task assignment problem could be defined as a selective process whereby a particular processor is chosen among a list of available processors to act as a host for executing a particular task. For clarification purposes, a task is just a collection of executable instructions grouped together for the purpose of convenience. We propose the following algorithms, to solve the Task Assignment Problem:

1. **Minima_Strategy:** Uses a random mapping strategy
2. **Maxima_Strategy:** Provides a topology with direct connections between processors
3. **Dimenx_Strategy:** Focuses on either the cycle or bandwidth requirement of tasks
4. **Dimenxy_Strategy:** Considers both the task cycles and bandwidth requirement
5. **Graphcut_Strategy:** Creates subgraphs out of the topology and task graphs and maps them
6. **Optima_Strategy:** Maps tasks to a virtual topology and then remaps to an actual physical topology
7. **Edgefit_Strategy:** Sorts the edges of the topology and task graphs and maps the best edges

The following subsections provide the details of each one of these algorithms.

5 Task Mapping Algorithms

This section provides algorithmic details of each of the algorithms which follows a summary or objective of the algorithm.

1. **Minima_Strategy Algorithm:** This is a very simple scheme where in the list of processors in PTG, as well as the list of tasks in the TCG are randomly shuffled as a first step. Then the tasks in the shuffled TCG are assigned to processors, in the shuffled PTG in a random fashion one-on-one. While this scheme does not guarantee efficient task placements, we can definitely use the result in comparisons, for measuring the effectiveness of other strategies.

   Algorithm 1 on page 475 provides the steps required to implement the Minima Strategy algorithm.
2. **Task Mapping by Maxima_Strategy:** The objective of this algorithm is to provide a topology that produces maximum benefits to the set of tasks from a topology standpoint. It achieves this by providing direct connections to the tasks

Algorithm 1 Minima_Strategy algorithm

```
procedure MINIMA_STRATEGY(PTG, TCG)                ▷ The Minima strategy routine
    TAG ← MERGE_GRAPH(PTG, TCG)
    TAG ← MINIMA_MAPPER(TAG)
    return TAG
end procedure
procedure MINIMA_MAPPER(TAG)                       ▷ Minima mapper routine
    proc_list ← GET_PROC_NODE_LIST(TAG)
    task_list ← GET_TASK_NODE_LIST(TAG)
    map_list ← NEW_LIST()
    len ← LIST_SIZE(task_list)
    while len ≠ 0 do
        map_list ← NEW_LIST()
        proc ← CHOOSE_RANDOM(proc_list)
        task ← CHOOSE_RANDOM_UNIQUE(task_list)
        MAP_TASK(map_list, proc, task)
        len ← LIST_SIZE(task_list)
    end while
    TAG ← MAP_TASKS_TO_PROCS(TAG, map_list)
    TAG ← MERGE_NODE_AGGREGATES(TAG)
    return TAG
end procedure
procedure MAP_TASK(map_list, proc, task)   ▷ Store the proc, task tuple in the map_list
    tuple ← MAKE_TUPLE(proc, task)
    ADD_TO_LIST(map_list, tuple)
end procedure
```

so that the communicating tasks need not experience router and other network delays.

Algorithm 2 on page 476 provides the steps required to implement the Maxima_Strategy algorithm.

3. **Dimenx_Strategy algorithm:** There are actually two sub-strategies here. One that takes into consideration the task execution cycles and another that considers the task communication bandwidth. Accordingly, the algorithm uses a depth-first-listing of processors to achieve the first objective where the underlying assumption is that processor connection is not important so we can afford to map to processors that are farther apart in the topology. To achieve criteria two, the algorithm maps tasks to a breadth-first-listing of processors since the direct connection between processors would be required.

   Algorithm 3 on page 477 provides the details necessary to implement Dimenx_Strategy of mapping tasks to processors.
4. **Dimenxy_Strategy algorithm:**
   The Dimenxy strategy algorithm tries to achieve two demands of the tasks at the same time, by maximizing the computation part, and controlling the communication overheads, by minimizing the communication to computation ratio, through effective mapping of tasks to processors. Tasks are grouped into clusters

Algorithm 2 Maxima_Strategy algorithm

```
procedure MAXIMA_STRATEGY(PTG, TCG)            ▷ Maxima strategy entry interface
    TAG ← MATCH_TASK_GRAPH_SHAPE(PTG, TCG)
    TAG ← MAXIMA_MAPPER(TAG)
    return TAG
end procedure
procedure MATCH_TASK_GRAPH_SHAPE(PTG, TCG)     ▷ Create a machine topology to match the task profile
    TAG ← COPY_GRAPH(TCG)
    graph_profile ← GET_GRAPH_NODE_EDGE_PROFILE(TCG)
    max_lbw ← GET_MAX_LBW(PTG)
    ADD_TO_PROFILE(graph_profile, max_lbw)
    PTG ← UPDATE_GRAPH_NODE_EDGE_PROFILE(PTG, graph_profile)
    TAG ← RECONSTRUCT_GRAPH(TAG, PTG)
    return TAG
end procedure
procedure MAXIMA_MAPPER(TAG)                   ▷ Mapper for the Maxima configuration
    task_list ← GET_TASK_LIST(TAG)
    proc_list ← GET_PROC_LIST(TAG)
    map_list ← NEW_LIST()
    for i ← 1, n − 1 do   ▷ All processor nodes and edges are maximal here, so pairing a task to a process is trivial
        task ← task_list[i]
        proc ← proc_list[i]
        pair ← MAKE_PAIR(task, pair)
        ADD_TO_LIST(map_list, pair)
    end for
    TAG ← MAP_TASKS_TO_PROCS(TAG, map_list)
    TAG ← MERGE_NODE_AGGREGATES(TAG)
    return TAG
end procedure
```

referred to here as segments, based on their inherent nature, whether computation or communication biased, and then mapped accordingly segmentwise.

Algorithm 5 on page 479 describes the Dimenxy_Strategy Algorithm for generating the TAG.

5. **Graphcut_Strategy algorithm:**
Graphcut Strategy algorithm, creates subgraphs out of PTG and TCG that are matching or similar from a shape perspective, and so can be easily mapped. It is important to slice a graph only to the extent that we get subgraphs that are similar in shape, and more amenable for mapping. The subgraphs are similar in shape, in terms of the number of nodes constituting the subgraphs, and the number of edges, and the number of edges incident and leaving the nodes of the concerned subgraphs. One should be careful not to take this slicing too far, in which case we can end up with a graph of just nodes, with no shape or edge information, making the mapping decisions difficult.

Algorithm 7 on page 481 describes the Graphcut_Strategy Algorithm for generating the TAG.

Algorithm 3 Dimenx_Strategy algorithm

```
procedure DIMENX_STRATEGY(PTG, TCG)                          ▷ strategy entry point
    TAG_C ← MERGE_GRAPH(PTG, TCG))                ▷ Create TAG based on task cycles
    TAG_V ← MERGE_GRAPH(PTG, TCG))  ▷ Create TAG based on task communication volume
    TAG_C ← DIMENX_MAPPER_CYC(TAG_C, PTG, TCG)  ▷ Create mappings based on cycles
    TAG_V ← DIMENX_MAPPER_VOL(TAG_V, PTG, TCG)  ▷ Create mappings based on volume
    pair ← MAKE_PAIR(TAG_C, TAG_V)
    return pair
end procedure
procedure DIMENX_MAPPER_CYC(TAG, PTG, TCG)▷ Mapper uses the execution cycles of the tasks as the basis for mapping decisions
    proc_list_sorted ← GENERATE_DFS_LIST(PTG)  ▷ Create a depth first listing (DFS) of processor nodes so that they are spaced apart
    for i ← 0, n − 1 do
        cyc ← GET_TASK_CYC(task_list[i])
        ADD_TO_LIST(cyc_list, cyc)
    end for
    cyc_list_sorted ← RSORT_LIST(cyc_list)       ▷ Create an inverted sorted list of cycles
    for i ← 0, n − 1 do                                  ▷ Create a sorted list of tasks
        cyc ← cyc_list_sorted[i]
        for j ← 0, n − 1 do
            task ← task_list[j]
            task_cyc ← GET_TASK_CYC(task)
            if cyc = task_cyc then
                ADD_TO_LIST(task, task_list_sorted)
            end if
        end for
    end for
    map_list ← NEW_LIST()
    for i ← 0, n − 1 do    ▷ Create the map with the best processor in the proc_list_sorted list
        task ← task_list_sorted[i]
        proc ← proc_list_sorted[i]
        pair ← MAKE_PAIR(task, proc)
        ADD_TO_LIST(map_list, pair)
    end for
    TAG ← MAP_TASKS_TO_PROCS(TAG, map_list)              ▷ map tasks to processors
    TAG ← MERGE_NODE_AGGREGATES(TAG) ▷ Compute node property values for fitness calculation purposes
    return TAG
end procedure
```

Algorithm 4 Dimenx_Strategy algorithm (cont…)

```
procedure DIMENX_MAPPER_VOL(TAG, PTG, TCG)    ▷ Mapper uses the communication
volume of the tasks as the basis for mapping decisions
    proc_list_sorted ← GENERATE_BFS_LIST(PTG)    ▷ Create a breadth first listing (BFS)
of processor nodes so that they are bunched together
    for i ← 0, n − 1 do
        vol ← GET_TASK_VOL(task_list[i])
        ADD_TO_LIST(vol_list, vol)
    end for
    vol_list_sorted ← RSORT_LIST(vol_list)    ▷ Create an inverted sorted list of volumes
    for i ← 0, n − 1 do    ▷ Create a sorted list of tasks
        vol ← vol_list_sorted[i]
        for j ← 0, n − 1 do
            task ← task_list[j]
            task_vol ← GET_TASK_VOL(task)
            if vol = task_vol then
                ADD_TO_LIST(task, task_list_sorted)
            end if
        end for
    end for
    map_list ← NEW_LIST()
    for i ← 0, n − 1 do    ▷ Create the map with the best processor in the proc_list_sorted list
        task ← task_list_sorted[i]
        proc ← proc_list_sorted[i]
        pair ← MAKE_PAIR(task, proc)
        ADD_TO_LIST(map_list, pair)
    end for
    TAG ← MAP_TASKS_TO_PROCS(TAG, map_list)    ▷ map tasks to processors
    TAG ← MERGE_NODE_AGGREGATES(TAG) ▷ Compute node property values for fitness
calculation purposes
    return TAG
end procedure
```

6. **Optima_Strategy algorithm:**
   The Optima_Strategy algorithm follows a two step process where the sorted edges of the TCG are mapped to a virtual topology with no processor limitations. Then the second mapping step is carried out using node sorted graphs of topology and task graphs.
7. **Edgefit_Strategy algorithm:**
   Edgefit_Strategy involves mapping the task edge with the highest communication volume, with the topology edge with the highest bandwidth, and so on. This seems like an easy task to achieve, but practically poses consistency issues, because a task can only be mapped, to a single processor at any time. So an extra post processing step is required, whereby inconsistent mappings have to be sorted out.

   Algorithm 11 on page 484 describes the Edgefit_Strategy Algorithm for generating the TAG.

Algorithm 5 Dimenxy_Strategy algorithm

```
procedure DIMENXY_STRATEGY(PTG, TCG) ▷ The strategy entry point into the Dimenxy
    TAG ← MERGE_GRAPH(PTG, TCG)
    TAG ← DIMENXY_MAPPER(TAG, PTG, TCG)    ▷ Create mappings based on cycles and volume
    return TAG
end procedure
procedure DIMENXY_MAPPER(TAG, PTG, TCG)    ▷ Mapper that uses both the execution cycles, and volume of communication of the tasks as the basis for mapping decisions
    proc_list_dfs ← GET_DFS_LIST(PTG)    ▷ Create a depth first listing (DFS) of processors so that they are spaced apart
    proc_list_bfs ← GET_BFS_LIST(PTG)    ▷ Create a depth first listing (BFS) of processors so that they are bunched closer
    task_list ← GET_TASK_LIST(TCG)
    map_list ← NEW_LIST()
    proc_dfs_idx ← 0
    proc_bfs_idx ← 0
    for task ← task_list[0], task_list[n − 1] do    ▷ Create the map list

        task_type ← GET_TASK_TYPE(TCG, task)
        if task_type =′ MAP′_X then
            proc ← proc_list_bfs[proc_bfs_idx]
            edge ← MAKE_EDGE(task, proc)
            ADD_TO_LIST(map_list, edge)
            proc_bfs_idx ← proc_bfs_idx + 1
            flag ← IS_LAST_INDEX(proc_list_bfs, proc_bfs_idx)
            if flag ≠ 0 then
                proc_bfs_idx ← 0
            end if
        else
            proc ← proc_list_dfs[proc_bfs_idx]
            edge ← MAKE_EDGE(task, proc)
            ADD_TO_LIST(map_list, edge)
            proc_dfs_idx ← proc_dfs_idx + 1
            flag ← IS_LAST_INDEX(proc_list_dfs, proc_dfs_idx)
            if flag ≠ 0 then
                proc_dfs_idx ← 0
            end if
        end if
    end for
    TAG ← MAP_TASKS_TO_PROCS(TAG, task_proc_map)    ▷ Map tasks to processors
    TAG ← MERGE_NODE_AGGREGATES(TAG)    ▷ Compute node property values for fitness calculation purposes

    return TAG
end procedure
```

Algorithm 6 Dimenxy_Strategy algorithm (cont...)

```
procedure CYC_TO_INS(TCG, node)    ▷ Converts cycles to an equivalent number of instructions
    IPC ← 1.2    ▷ IPC number is chosen based on the modern processor trends
    cyc ← GET_CYC(TCG, node)
    ins ← cyc * IPC
    return ins
end procedure
procedure VOL_TO_INS(TCG, edge_list)    ▷ Converts volume to an equivalent number of instructions
    vol ← 0.0
    for edge ← edge_list[0], edge_list[n − 1] do
        vol ← vol + GET_VOL(TCG, edge)
    end for
    ins ← vol/128.0    ▷ For normalization purposes we assume 128 bytes are equal to one instruction
    return ins
end procedure
procedure GET_TASK_TYPE(TCG, task)    ▷ Advice if the task is computation or communication dominant
    adv ← EMPTY_STRING()
    node_list ← GET_NODE_LIST(TCG)
    for node ← node_list[0], node_list[n − 1] do
        if node = task then
            edge_list ← GET_EDGE_LIST(TCG, node)
            cyc ← CYC_TO_INS(TCG, node)
            vol ← VOL_TO_INS(TCG, edge_list)
            if cyc > vol then
                adv =' MAP'_X
            else
                adv = 'MAP'_Y
            end if
        end if
    end for
    return adv
end procedure
```

We presented seven algorithms with varying degree of complexity in terms of implementation and accordingly offer varying levels of performance. So which algorithm offers maximum benefit for a particular scenario? Minima strategy is simple to implement and may work well in many situations, especially when there are few processors in the topology and a large number of tasks. Since Maxima strategy uses direct connections between processors, it provides an upper limit on the maximum performance level possible, for any combination of processors and tasks. Dimenx and Dimenxy are good strategies to employ in situations, when both the tasks cycles and bandwidth dictate performance. Graphcut, Optima, and Edgefit are complex strategies to implement and run, and should be employed for complex processor topologies and large task counts, where communication volumes and bandwidth expectations are high. In such scenarios the extra time spent, in carefully

Algorithm 7 Graphcut_Strategy algorithm

```
procedure GRAPHCUT_STRATEGY(PTG, TCG)
    TAG ← MERGE_GRAPH(PTG, TCG)
    TAG ← GRAPHCUT_MAPPER(TAG) returnTAG
end procedure
procedure GRAPHCUT_MAPPER(TAG, PTG, TCG)                ▷ The Graphcut mapper
    subgraph_pair ← GRAPHCUT_CUT(PTG, TCG)
    subgraph_list_ptg ← subgraph_pair[0]
    subgraph_list_tcg ← subgraph_pair[1]
    map_list ← GRAPHCUT_FIT(subgraph_list_ptg, subgraph_list_tcg)
    for map_list ← list_of_map_list[0], list_of_map_list[n − 1] do
        TAG ← MAP_TASKS_TO_PROCS(TAG, map_list)
        TAG ← MERGE_NODE_AGGREGATES(TAG)
    end for
    return TAG
end procedure
procedure GRAPHCUT_CUT(PTG, TCG)        ▷ Create subgraphs and sort them based on
connected edges
    subgraph_list_ptg ← MIN_CUT_GRAPH(PTG)
    subgraph_list_tcg ← MIN_CUT_GRAPH(TCG)
    subgraph_list_ptgs ← DEGREE_SORT_GRAPH(subgraph_list_ptg)
    subgraph_list_tcgs ← DEGREE_SORT_GRAPH(subgraph_list_tcg)
    pair ← MAKE_PAIR(subgraph_list_ptgs, subgraph_list_tcgs)
    return pair
end procedure
```

Algorithm 8 Graphcut_Strategy algorithm (cont...)

```
procedure GRAPHCUT_FIT(subgraph_list_ptg, subgraph_list_tcg)      ▷ Fit the task and the
topology sub-graphs
    map_list ← NEW_LIST()
    for i ← 0, n − 1 do
        subgraph_tcg ← subgraph_list_tcg[i]
        subgraph_ptg ← subgraph_list_ptg[i]
        map_list ← GRAPHCUT_FIT_SUBGRAPH(map_list, subgraph_ptg, subgraph_tcg)
    end for
    return map_list
end procedure
procedure GRAPHCUT_FIT_SUBGRAPH(map_list, subgraph_list_ptg, subgraph_list_tcg)  ▷
Map nodes of the subgraph
    tcg_nodes ← GET_NODE_LIST(subgraph_tcg)
    ptg_nodes ← GET_NODE_LIST(subgraph_ptg)
    for i ← 0, n − 1 do
        pair ← MAKE_PAIR(tcg_nodes[i], ptg_nodes[i])
        ADD_TO_LIST(map_list, pair)
    end for
    return map_list
end procedure
```

Algorithm 9 Optima_Strategy algorithm

```
procedure OPTIMA_STRATEGY(PTG, TCG)                    ▷ Optima strategy entry
    TAG ← ADD_GRAPH(PTG, TCG)
    TAG ← OPTIMA_MAPPER(TAG, PTG, TCG)
    return TAG
end procedure
procedure OPTIMA_MAPPER(TAG, PTG, TCG)        ▷ Maps tasks to a virtual machine
topology followed by a remap to an actual topology
    VAG ← MAP_VIRTUAL(TAG, PTG, TCG)
    TAG ← MAP_PHYSICAL(TAG, VAG, PTG, TCG)
    TAG ← MAP_SOLO_TASKS(TAG, PTG)
    TAG ← UPDATE_NODE_VALUES(TAG)
    return TAG
end procedure
procedure MAP_VIRTUAL(VAG, PTG, TCG)      ▷ Map tasks to a virtual topology where
there is no shortage of optimal processors
    VAG ← INIT_DIGRAPH
    psl ← EDGE_SORT_GRAPH(PTG,'LBW','DSC')   ▷ descending sort based on LBW
param
    tsl ← EDGE_SORT_GRAPH(TCG,'VOL','DSC')   ▷ descending sort based on VOL
param
    for i ← 0, n − 1 do
        ADD_TASK_NODE(VAG)
    end for
    for e ← 0, n − 1 do
        ADD_TASK_EDGE(VAG)
    end for
    for i ← 0, n − 1 do
        ADD_TOPOLOGY_NODE(VAG)
        ADD_TOPOLOGY_NODE(VAG)
        proc_edge ← psl[i]
        task_edge ← tsl[i]
        ADD_TOPOLOGY_EDGE(VAG, proc_edge)
        VAG ← MAP_TASK_EDGE_TO_PROC_EDGE(VAG, task_edge, proc_edge)
    end for
    return VAG
end procedure
```

mapping tasks to the appropriate processors, translates to measurable performance benefits and is definitely worth the effort.

6 Complexity Analysis of the Algorithms

In this section, we look at the complexity of the algorithms proposed in this paper. We use the standard $Big - O$ notation, which defines the upper bounds, on the scalability of algorithms in general. The O on the left hand side in each of these equations stands for the $Big - O$ measure. The purpose of providing these equations

Algorithm 10 Optima_Strategy algorithm (cont...)

```
procedure MAP_PHYSICAL(TAG, VAG, PTG, TCG)   ▷ Replace virtual processors with real processors
    PTG ← UPDATE_EDGE_VALUES(PTG)   ▷ Calculate aggregate bandwidth, volume and deficit
    TCG ← UPDATE_EDGE_VALUES(TCG)
    proc_sort_list ← NODE_SORT_GRAPH(PTG,'LBW','DSC')   ▷ Sort processors and tasks in the topology

    task_sort_list ← NODE_SORT_GRAPH(TCG,'VOL','DSC')
    pair_list ← NEW_LIST()
    for index ← 0, n − 1 do   ▷ Create the pair list
        proc ← proc_sort_list[index]
        task ← task_sort_list[index]
        pair ← MAKE_PAIR(task, proc)
        LIST_ADD(pair_list, pair)
    end for
    VAG ← GRAPH_READ_PAIR_LIST(VAG, pair_list)   ▷ Add the mappings to VAG
    TAG ← MERGE_GRAPH(TAG, VAG)   ▷ Merge the virtual and the assignment graph
    TAG ← UPDATE_NODE_VALUES(TAG)   ▷ Aggregate values for fitness calculation
    return TAG
end procedure
```

is mainly to acquaint the readers about the complexity of the algorithms presented earlier and provide mathematically an expectation in terms of performance. In this study we have focused on the time complexity of the algorithms using the $Big - O$ notation. Since we believe that the algorithms are moderate in their use of memory, we chose not to discuss or analyze their space complexity behavior at this point.

1. **Minima_Strategy:** The algorithm is simple. It just maps every task in the set randomly with a processor in the topology. So there is just one loop which picks a task in sequence from the list of tasks and maps with a processor it has randomly chosen from the list of processors. If you ignore the work done in generating the lists from their corresponding graphs PTG and TCG, the complexity is just $O(N)$ where N is the number of tasks.

$$O(Minima) = O(N) \tag{1}$$

   where N is the number of tasks.

2. **Maxima_Strategy:** Initialization involves finding the maximum value of bandwidth which means looping through the list of processor edges which translates to a complexity of $O(E)$ where E is the number of edges in the original topology. It also involves creating a processor topology that produces the best results for the given task set. This involves creating processor nodes and edges mimicking the task graph. This means an additional complexity of $O(N)$ where N is the number of nodes in the task set. Another order of $O(N^2)$ where we are setting

Algorithm 11 Edgefit_Strategy algorithm

```
procedure EDGEFIT_STRATEGY(PTG, TCG)                    ▷ Edgefit strategy entry
    TAG ← ADD_GRAPH(PTG, TCG)
    TAG ← EDGEFIT_MAPPER(TAG, PTG, TCG)
    return TAG
end procedure
procedure EDGEFIT_MAPPER(TAG, PTG, TCG)                 ▷ Edgefit mapper entry
    TAG ← MAP_GREEDY(TAG, PTG, TCG)
    TAG ← BACKTRACK(TAG)
    TAG ← BALANCE(TAG, PTG, TCG)
    TAG ← ASSIGN_SOLO(TAG, PTG)
    TAG ← UPDATE_NODE_VALUES(TAG)
    return TAG
end procedure
procedure MAP_GREEDY(TAG, PTG, TCG)          ▷ Pair up the best graph edges in greedy
manner
    proc_sort_list ← EDGE_SORT_GRAPH(PTG,'LBW','DSC')
    task_sort_list ← EDGE_SORT_GRAPH(TCG,'VOL','DSC')
    pair_list ← NEW_LIST()                              ▷ Create the pairings
    for i ← 0, n − 1 do
        proc_edge ← proc_sort_list[i]
        task_edge_edge ← task_sort_list[i]
        pair ← MAKE_PAIR(task_edge, proc_edge)
        LIST_ADD(pair_list, pair)
    end for
    TAG ← GRAPH_READ_PAIR_LIST(TAG, pair_list)
    TAG ← UPDATE_NODE_VALUES(TAG)
    return TAG
end procedure
```

up direct edges between any pair of tasks for realizing the best communication results.

$$O(Maxima) = O(N) + O(N^2) + O(E) \tag{2}$$

where N is the number of tasks, E is the number of topology edges.

3. **Dimenx_Strategy:** Finding the Dfs listing of all the processors in the topology which involves looping through the nodes and for each node follow each direct edge and visit them in turn which translates to an order of $O(EN)$ where E is the number of processor edges and N is the number of processor nodes. Finding a descending order listing of all task nodes based on the execution cycles translates to $O(M^2)$ when using a simple bubble sort where M is the number of task nodes. Mapping processors to tasks is $O(M)$ like before where M is the number of tasks. So totally we have,

$$O(Dimenx_C) = O(EN) + O(M^2) + O(M) \tag{3}$$

Algorithm 12 Edgefit_Strategy algorithm (cont. . .)

```
procedure BACKTRACK(TAG)                                  ▷ Ensure pairing consistency
    map_list ← GET_TASK_MAP_LIST(TAG)
    for i ← 0, n − 1 do
        for j ← 0, n − 1 do
            mapi ← map_list[i]
            mapj ← map_list[j]
            if mapi ≠ mapj then
                taski ← GET_TASK(mapi)
                taskj ← GET_TASK(mapj)
                if taski = taskj then
                    xbwi ← GET_XBW(mapi)
                    xbwj ← GET_XBW(mapj)
                    if xbwi ≤ xbwj then
                        ADD_MAP(TAG, mapi)
                    else
                        ADD_MAP(TAG, mapj)
                    end if
                end if
            end if
        end for
    end for
end procedure
procedure MIGRATE(TAG, task, current, proc_list)          ▷ Move task to a lightly loaded
processor
    for target ← proc_list[0], proc_list[n − 1] do
        ldt ← GET_TASK_LOAD(TAG, target)
        ldc ← GET_TASK_LOAD(TAG, current)
        if ldt < ldc then
            MOVE_TASK(TAG, current, target)
            return TAG
        end if
    end for
end procedure
procedure BALANCE(TAG)                                    ▷ Ensure load balance in the topology
    proc_list ← GET_PROC_LIST(TAG)
    for proc ← proc_list[0], proclist[n − 1] do
        task_list ← GET_TASK_LIST_FOR_PROC(TAG, proc)
        flag ← IS_LOAD_BALANCED(TAG, task_list) ▷ Tasks per processor is used as the
load metric
        if flag ≠ 1 then
            TAG ← MIGRATE(TAG, task, proc_list)
        end if
    end for
    return TAG
end procedure
```

where M is the number of tasks and N is the number of processors and E is the number of processor edges.
Similarly when we use Volume as the reference we have a task edge sorting step which is the only step that is different from above. So in total we have,

$$O(Dimenx_V) = O(EN) + O(F^2) + O(M) \tag{4}$$

where M is the number of tasks and N is the number of processors, E is the number of processor edges, and F is the number of task edges.

4. **Dimenxy_Strategy:** Involves the generation of both the Dfs and Bfs listing of the processors in the topology. As explained earlier, this translates to a complexity of $O(EN)$ each. Computing aggregate volumes is aggregating edge weights in the task nodes which is of complexity $O(F)$, where F is the number of edges in the task graph. Sorting the task nodes based on their communication volumes is a $O(M)$ where M is the number of task nodes. Mapping is a $O(M)$ order where M is the number of task nodes.

$$O(Dimenxy) = O(EN) + O(F) + 2 * O(M) \tag{5}$$

where M is the number of tasks, N is the number of processors, E is the number of processor edges, and F is the number of edges in the task graph.

5. **Graphcut_Strategy:** Graph cut step which is a $O(EM)$ where E is the number of edges and M is the number of nodes in the processor topology graph. Similarly for the task graph this step is a $O(FN)$ order of complexity. The sort of the list of subgraphs of PTG is a worst case order of $O(M^2)$ where M is the number of nodes in the processor topology. Similarly for tasks it is a $O(N^2)$ where N is the number of task nodes. Matching the subgraphs and the nodes within corresponding subgraphs is a $O(N^2)$ order where N is the number of task nodes. So in total we have,

$$O(Graphcut) = O(EM) + O(FN) + O(M^2) + 2 * O(N^2) \tag{6}$$

where N is the number of tasks and M is the number of processors, E is the number of processor edges and F is the number of task edges.

6. **Optima_Strategy:** Map virtual step involves building a virtual graph that has the same number of processor nodes and number of edges equal to the number of edges in the task graph. The mapping step involves going through the list of task edges and mapping the virtual processors one on one. This translates to the following, $O(M)+2*O(F)$ where M is the number of processor nodes and F is the number of task edges. Mapping to a physical topology involves aggregating edge properties to nodes in the topology and task graphs which translates to $O(E)+O(F)$ where E is the number of edges in PTG and F in TCG. Two node sort operations of the PTG and TCG graphs are of order $O(M^2)$ and $O(N^2)$ respectively, where M is the number of nodes in PTG and N the corresponding number in TCG. Mapping edge is a $O(NF)$ in the worst case where N is the

number of task edges and F is the edges that involve a particular task node. Totally this translates to

$$O(Optima) = O(M)+3*O(F)+O(E)+O(M^2)+O(N^2)+O(NF) \quad (7)$$

where N is the number of task nodes and M is the number of processor nodes, E is the number of PTG edges and F the number of TCG edges.

7. **Edgefit-Strategy:** The edge sort steps of PTG and TCG are of order $O(E^2)$ and $O(F^2)$ as discussed earlier, where E is the number of processor edges and F the number of task edges. Mapping of edges in greedy fashion is of order $O(F)$, where F is the number of edges in TCG. Backtrack which involves comparison of one edge with another is a $O(F^2)$ order where F is the number of edges in the task graph. The $Task\text{-}per\text{-}processor$ (for load balancing) values gathering step is a $O(M)$, where M is the number of processors. Together they are $3 * O(M)$. Migrate step involves finding a suitable processor for the migrating task and its worst case order is $O(M)$, where M is the number of processors. Balance step involves sorting the processor graph once and looping till all processors are balanced, and each time in the loop balance the load-balance checks are called. This translates to $O(M^2) + O(M * (4 * M))$ and we have the following equation:

$$O(Edgefit) = O(E^2) + O(F) + O(F^2) + 5 * O(M^2) \quad (8)$$

7 Preliminary Results

This section gives some preliminary results on the performance of the various algorithms, for a configuration that consists of a topology of 512 processors and 1024 tasks. This configuration was generated by a tool, where in the processors connections were randomly chosen, as well as the bandwidth of the connections. The topology is not a fully connected one, but all processors are connected, with some of them with direct connections and others with indirect connections, enabled by one or more intermediate processors. Similarly, the communicating tasks and non-communicating tasks were randomly chosen, as also the volume of the communication in the former case. The bandwidths across processor boundaries were determined through simulation. The result has been captured in the form of a table and a plot below. While the actual simulation values are not relevant for this discussion, both the line bandwidth and task communication volumes were randomly generated for this experiment and they were generated in the form of a comma-separated-value files and stored as parameters in the appropriate graphs such as the PTG or the TCG as relevant. The algorithms used these bandwidths to guide the task placements and at the end they were evaluated based on the overall bandwidth overheads they were able to achieve in the topology. Less overheads mean better placement and translates to a better network/bandwidth efficiency.

Detailed analysis, results, and plots are planned to be presented in a separate work which is an extension to the present research.

Table 1 on page 488 lists the bandwidth overheads experienced by the algorithms for an example configuration that involves a topology of 512 processors and 1024 tasks. As seen from the table, we see that MAXIMA configuration has achieved the best/lowest overhead as expected along with OPTIMA. We also see that the highest overhead for this particular experiment was grabbed by GRAPHCUT which needs further study to determine the causes. The middle values are grabbed by the others.

Figure 6 on page 488 lists the bandwidth overheads experienced by the algorithms for an example configuration that involves a topology of 512 processors and 1024 tasks. The plot basically conveys the same information as the table in a graphical format. And we can see that MAXIMA algorithm has done better than the others in limiting the bandwidth overheads by mapping tasks and processors effectively.

Table 1 Bandwidth overheads experienced by the algorithms

Sl. No.	Algorithm	Bandwidth overhead
1	MINIMA-512-1024	3.16
2	DIMENX-C-512-1024	3.16
3	DIMENX-B-512-1024	3.16
4	DIMENXY-512-1024	3.16
5	GRAPHCUT-512-1024	3.45
6	EDGEFIT-512-1024	3.16
7	OPTIMA-512-1024	2.45
8	MAXIMA-512-1024	2.45

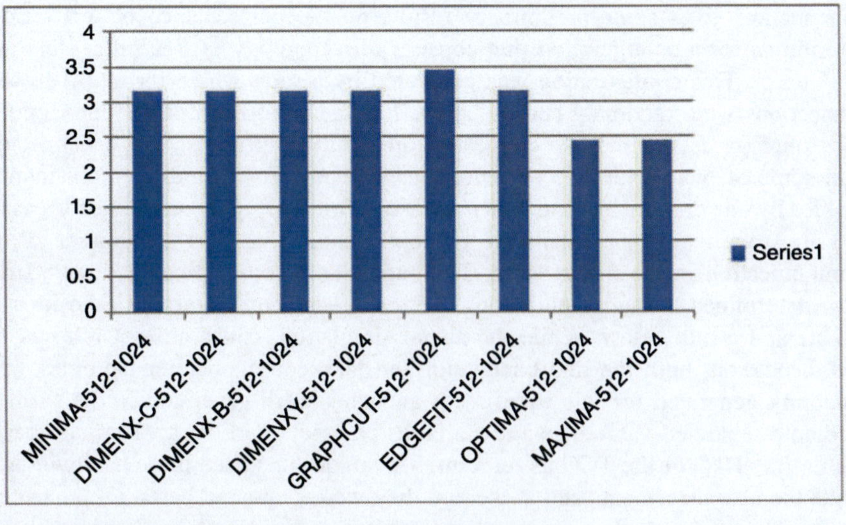

Fig. 6 Bandwidth overheads experienced by the algorithms

8 Conclusion

In this research work, we studied the problem of mapping parallel tasks of a program to the processors of a multiprocessor machine. The problem is interesting and challenging, since effective mapping depends on two criteria, namely the total execution cycles consumed by each processor and the overall bandwidth provided to the tasks by the topology. Effective use of processing resources of machine is possible by distributing the tasks across multiple processors which also serves the load balancing cause. To effectively use the bandwidth resources of a topology, more work is required to effectively choose processors for the tasks. We characterized this problem as the task assignment problem. We presented the following seven algorithms to solve the problem, based on the mathematical abstraction of a graph: Minima_Strategy, Maxima_Strategy, Dimenx_Strategy, Dimenxy_Strategy, Graphcut_Strategy, Optima_Strategy, and Edgefit_Strategy. These algorithms read the topology and task profiles, in the form of two weighted directed graphs namely, the Processor Topology Graph (PTG) and the Task Communication Graph (TCG), and generate a Task Assignment Graph (TAG), also a directed graph as output with the required task to processor mappings. These algorithms are general and are applicable to a wide range of machine architectures, including distributed multiprocessors such as NUMA. Future work involves characterizing these algorithms based on their performance and development of a tool or infrastructure to study and manage topologies and task profiles, as well as design custom topologies for optimum performance.

References

1. Abbas W, Egerstedt M (2012) Robust graph topologies for networked systems. IFAC Proc Vol 45(26):85–90 (2012). https://doi.org/https://doi.org/10.3182/20120914-2-US-4030.00052, http://www.sciencedirect.com/science/article/pii/S147466701534814X. 3rd IFAC workshop on distributed estimation and control in networked systems
2. Ahmad I, Kwok YK (1998) On exploiting task duplication in parallel program scheduling. IEEE Trans Parallel Distrib Syst 9(9):872–892. https://doi.org/10.1109/71.722221
3. Ahmad I, Dhodhi MK, Ghafoor A (1995) Task assignment in distributed computing systems. In: Proceedings international phoenix conference on computers and communications, March 1995, pp. 49–53. https://doi.org/10.1109/PCCC.1995.472512
4. Chou TCK, Abraham JA (1982) Load balancing in distributed systems. IEEE Trans Softw Eng 8(4):401–412
5. Culler D, Singh JP, Gupta A (1998) Parallel computer architecture: a hardware/software approach. Morgan Kaufmann Publishers Inc., San Francisco
6. e Silva EDS, Gerla M (1991) Queueing network models for load balancing in distributed systems. J Parallel Distrib Comput 12(1):24–38
7. Feng TY (1981) A survey of interconnection networks. Computer 14(12):12–27
8. Ferrante J, Ottenstein KJ, Warren JD (1984) The program dependence graph and its use in optimization. In: Proceedings of the 6th colloquium on international symposium on programming. Springer, London, pp 125–132. http://dl.acm.org/citation.cfm?id=647326.721811

9. Gkantsidis C, Mihail M, Zegura E (2003) Spectral analysis of internet topologies. In: INFOCOM 2003. Twenty-second annual joint conference of the IEEE computer and communications. IEEE societies, vol 1. IEEE, New York, pp 364–374
10. Grieco LA, Alaya MB, Monteil T, Drira K (2014) A dynamic random graph model for diameter-constrained topologies in networked systems. IEEE Trans Circ Syst II: Express Briefs 61(12):982–986. https://doi.org/10.1109/TCSII.2014.2362676
11. Horwitz S, Reps T (1992) The use of program dependence graphs in software engineering. In: International conference on software engineering, 1992, pp 392–411. https://doi.org/10.1109/ICSE.1992.753516
12. Horwitz S, Reps T, Binkley D (1990) Interprocedural slicing using dependence graphs. ACM Trans Program Lang Syst 12(1):26–60. http://doi.acm.org/10.1145/77606.77608
13. Hursey J, Squyres JM, Dontje T (2011) Locality-aware parallel process mapping for multi-core HPC systems. In: 2011 IEEE international conference on cluster computing, September, pp 527–531. https://doi.org/10.1109/CLUSTER.2011.59
14. Jeannot E, Mercier G (2010) Near-optimal placement of MPI processes on hierarchical NUMA architectures. In: D'Ambra P, Guarracino M, Talia D (eds) Euro-Par 2010 - parallel processing. Springer, Berlin, Heidelberg, pp 199–210
15. Kalyur S, Nagaraja GS (2016) ParaCite: auto-parallelization of a sequential program using the program dependence graph. In: 2016 international conference on computation system and information technology for sustainable solutions (CSITSS), October, pp. 7–12. https://doi.org/10.1109/CSITSS.2016.7779431
16. Kalyur S, Nagaraja GS (2017) Concerto: a program parallelization, orchestration and distribution infrastructure. In: 2017 2nd international conference on computational systems and information technology for sustainable solution (CSITSS), December, pp 1–6. https://doi.org/10.1109/CSITSS.2017.8447691
17. Kennedy K, Allen JR (2002) Optimizing compilers for modern architectures: a dependence-based approach. Morgan Kaufmann Publishers Inc., San Francisco
18. Leis V, Boncz P, Kemper A, Neumann T (2014) Morsel-driven parallelism: a NUMA-aware query evaluation framework for the many-core age. In: Proceedings of the 2014 ACM SIGMOD international conference on Management of data. ACM, New York, pp 743–754
19. Li Y, Pandis I, Mueller R, Raman V, Lohman GM (2013) NUMA-aware algorithms: the case of data shuffling. In: CIDR
20. Loh PKK, Hsu WJ, Wentong C, Sriskanthan N (1996) How network topology affects dynamic load balancing. IEEE Parallel Distrib Technol 4(3):25–35. http://dx.doi.org/10.1109/88.532137
21. Pantazopoulos P, Karaliopoulos M, Stavrakakis I (2014) Distributed placement of autonomic internet services. IEEE Trans Parallel Distrib Syst **25**(7):1702–1712
22. Pilla LL, Ribeiro CP, Cordeiro D, Bhatele A, Navaux PO, Méhaut JF, Kalé LV (2011) Improving parallel system performance with a NUMA-aware load balancer. Tech. rep.
23. Shirazi BA, Kavi KM, Hurson AR (1995) Scheduling and load balancing in parallel and distributed systems. IEEE Computer Society Press, Los Alamitos (1995)
24. Tantawi AN, Towsley D (1985) Optimal static load balancing in distributed computer systems. J ACM 32(2):445–465
25. Tian Y, Hankins RA, Patel JM (2008) Efficient aggregation for graph summarization. In: Proceedings of the 2008 ACM SIGMOD international conference on management of data. ACM, New York, pp. 567–580
26. Von Landesberger T, Kuijper A, Schreck T, Kohlhammer J, van Wijk JJ, Fekete JD, Fellner DW (2011) Visual analysis of large graphs: state-of-the-art and future research challenges. In: Computer graphics forum, vol 30. Wiley Online Library, pp 1719–1749
27. Wang YT et al (1985) Load sharing in distributed systems. IEEE Trans Comput 100(3):204–217
28. Weiser M (1984) Program slicing. IEEE Trans Softw Eng SE-10(4):352–357. https://doi.org/10.1109/TSE.1984.5010248
29. Wolfe M, Banerjee U (1987) Data dependence and its application to parallel processing. Int J Parallel Programm 16(2):137–178

30. Xu Y, Salapaka SM, Beck CL (2014) Aggregation of graph models and Markov chains by deterministic annealing. IEEE Trans Autom Control 59(10):2807–2812
31. Zegura EW, Calvert KL, Donahoo MJ (1997) A quantitative comparison of graph-based models for internet topology. IEEE/ACM Trans Network 5(6):770–783

Index

A
AAL, *see* Ambient assisted living (AAL)
ABC algorithm, *see* Artificial bee colony (ABC) algorithm
Access Service Network Gateway (ASN-GW), 289
Activity recognition (AR)
  accelerometer sensor, 163
  evaluation, 166–167
  localization method, 164–165
  probability distributions model, 164
Adam *vs.* Adagrad optimizer, 174, 175
Adaptive uplink scheduling (AUS), 292
ADC, *see* Adenocarcinoma (ADC)
Adenocarcinoma (ADC), 256, 258
Ambient assisted living (AAL), 160–161
Analytical hierarchy process (AHP), 414
Annual energy production, 26
Ant colony optimization algorithms, 221
Ant colony system (ACS), 280
AR, *see* Activity recognition (AR)
Area spectral efficiency (ASE), 145
Artificial bee colony (ABC) algorithm
  base, 343
  classification, 342–344
  flowchart, 343, 344
  hyperparameter optimisation, 340
  LHABC, 346
  meta-heuristic optimisation, 340
  modification rate, 341
  neural network, 339, 341
  NMABC, 345
  orthogonal learning strategy, 341
  simulation and performance analysis
    CIFAR-10 multinomial classification colouriser, 354–356
    CIFAR-10 regression colouriser, 353–354
    Iris dataset, 347–350
    MNIST digit classification, 350–352
Artificial neural networks, 339
ASE, *see* Area spectral efficiency (ASE)
ASN-GW, *see* Access Service Network Gateway (ASN-GW)
AUS, *see* Adaptive uplink scheduling (AUS)
AVA cognitive services platform, 271–272
Azure cognitive services
  APIs, 274–275
  neural text to speech, 274
  oral communication transcript, 274
  services, 274

B
Bayesian personalized ranking-based rank prediction scheme (BPR-RPS)
  cloud services, 203
  cost function, 210, 211, 213–215
  DDoS attack, 204–205
  issues, 204
  optimal personalized ranking, 206–207
  QoS parameters of cloud application, 206
  ranking prediction, 207–208
  response time, 208–211, 215
  throughput, 208, 209, 211, 212, 214

A. Haldorai et al. (eds.), *2nd EAI International Conference on Big Data Innovation for Sustainable Cognitive Computing*, EAI/Springer Innovations in Communication and Computing, https://doi.org/10.1007/978-3-030-47560-4

Big Data
  de-recognizable proof procedures, 181
  heterogeneous sources, 181
  information economy, 181
  privacy by design, 181, 184–186
Binary discrete memoryless symmetric (B-DMS), 362
Binary phase shift keying (BPSK), 365
BL, *see* Brand loyalty (BL)
Body sensor network (BSN), 158
Bootstrap analysis, 198
BPR-RPS, *see* Bayesian personalized ranking-based rank prediction scheme (BPR-RPS)
Brand loyalty (BL), 193, 195
Brand satisfaction (BS), 193, 195–196
Broadcast energy, 147, 149
Broadcast ratio, 147, 151, 152
Brute force analysis, 49
BS, *see* Brand satisfaction (BS)
BSN, *see* Body sensor network (BSN)

C
CAD, *see* Computer-aided design (CAD)
CAD mesh model (CMM), 116, 120, 121, 128, 134–138
Cassandra
  architecture, 228–230
  contribution, 228
  data integrity, 227
  experiment results
    accuracy, 238
    analysis of replication time, 232–234
    prediction of replication time, 234–237
    significant parameters, 234–237
  experiment setup
    parameters, 230–231
    replication time, 231–232
    workloads, 230
  query path, 229
  resource utilisation, 228
Cambridge Brain Sciences (CBS)
  administer custom assessments, 269
  brain health trials to dementia, 269–270
  clinics, 269
  cognitive deficits, lifestyle changes, 270
  large/small run trails, 270
  working, 269
CART, *see* Classification and regression trees (CART)
CAVS, *see* Context-aware video summarization (CAVS)
CBE, *see* Customer brand engagement (CBE)
CBS, *see* Cambridge Brain Sciences (CBS)
CDF, *see* Cumulative distribution function (CDF)
CEB, *see* Customer engagement behavior (CEB)
CFS, *see* Correlation-based feature selection (CFS)
CGPA, *see* Clustering-based gateway placement algorithm (CGPA)
Channel-aware scheduler, 290
Channel-unaware scheduler, 290
Chaotic optimization, 312–314
Classification and regression trees (CART), 264
Clinical prognostic factors (CPF), 259
Cloud computing, 203, 204, 279
  collaboration of, 51
  parameters, 48
  security aspects, 204
  SG environment, 49
Cloud data centers
Clustering-based gateway placement algorithm (CGPA), 222
Clustering methodology, 281
CMM, *see* CAD mesh model (CMM)
CNNs, *see* Convolutional neural networks (CNNs)
Cognitive computing, 272
Cognitive parallel access channel (C-PAC), 385
Cognitive radio (CR), 144, 145, 149
Cognitive radio network (CRN), 384–386
  architecture, 311
  components, 310
  cycle, 309
Cognitive semantic sensor network (CoSSN), 162
Comatose servers, 279–280
Communicate interact and operate (CIO) protocol, 47, 272
Component reputation fault segregation-based swift factor technique (CRFS-SFT), 208, 209, 211
Computationally robust smart grid
  availability reliability, 55–56
  consistency, 53
  decentralization, 51
  high-assurance computing, 55–56
  privacy and security, 55
  real-time analytics, 53–54
  scalability, 52–53
Computer-aided design (CAD)
  CAD-CAM integration, 115, 116
  CMM, 116, 134–138

Context-aware video summarization (CAVS), 85
Contrast features, 261
Convolutional neural networks (CNNs), 260
Convolutionary wireless network
  comparison table, 462
  and DL, 260
  garbage image classification problem, 451
  neural network, 17
  ten layer model, 457–460
  workspace, 67
Correlation, 261
Correlation-based feature selection (CFS), 262–263
Cosmos, 275–276
CoSSN, *see* Cognitive semantic sensor network (CoSSN)
Cost of energy (CoE)
  estimation, 30–31
  probability distributions, 26
  rotor radius, turbine, 27–28
  wind
    power model, 29
    speed models, 28–29
  See also Wind turbines
Coverage
  assessment, 122
  calculation, 122
  percentage, 124
  SINR, 380
  test cases, 133
  WMN, 218
CPF, *see* Clinical prognostic factors (CPF)
CR, *see* Cognitive radio (CR)
CRFS-SFT, *see* Component reputation fault segregation-based swift factor technique (CRFS-SFT)
CRM, *see* Customer relationship marketing (CRM)
Cumulative distribution function (CDF), 389
Customer brand engagement (CBE)
  conceptual model, 196
  discriminant validity, 197
  literature review, 192–193
  methodology, 196
  online retail businesses, 191
  path coefficients, 198
  reliability results, 197
  structural model, 198
  See also Online brand experience (OBE)
Customer engagement behavior (CEB), 192
Customer loyalty (CL), 192
Customer relationship marketing (CRM), 193
Cybercrime
  JNTUHJAC, 108, 111
  nation building, 3
  remedy solution
    analysis of RTI Act 2005 documents, 6, 8–9
    detailed analysis and interpretation, 6–9
    RTI request-response communication, 6, 7
  Telangana society, 5
  website, 1, 2
Cyberpolicing, 108

D

Dagum distribution, 28, 30, 31
Data broadcast, 148, 149
Data centers, 279–280
DCNN architecture configuration
  Adagrad optimizer, 173
  Adam optimizer, 172
Decentralization, 50, 51
Decision support system (DSS), 331
Decision tree classifier, 263–264
Deep convolutional neural network (DCNN)
  architecture (*see* DCNN architecture configuration)
  block diagram, 171
  confusion matrix, 172
  evaluation
    for Adagrad optimizer, 174
    for Adam optimizer, 173–174
  performance measurement, 174
Deep learning (DL)
  architecture, 170
  cardiac arrhythmia, 332
  detecting pattern, 16
  ECG signal, 333
  extraction of pattern, 16
  hyperparameter, 169
  MPDL (*see* Max pooling deep learning (MPDL))
  multi-layered neural network, 67
  multispectral data processing (*see* Multispectral data)
  and neural networks, 16
  performance based on dataset, 18
Deep neural networks (DNN), 332, 333
Deficit weighted round robin (DWRR), 290–293, 296–299
Demosaicing algorithm
  area analysis of output image, 325–327
  block diagram, 318, 319

Demosaicing algorithm (*cont.*)
  core *vs.* thermal power dissipation, 325, 326
  demosaicked image, 321, 322
  hexadecimal value, 322, 323
  output
    image, 328
    text file, 322, 324
  power analysis, 325, 327
  PSNR, 320
  reconstructed image, 322, 324
  signal-to-noise value, 328
  true input image, 320–321
  Xilinx output for edge detection, 322, 323
Dense Web, 171
Derivative superposition (DS), 242–244
DFS, *see* Disease-free survival (DFS)
Diabetic retinopathy, 93
Digital imaging, 259
Digital signal processing (DSP), 470
Dimenx_Strategy algorithm, 475–480, 484, 486
Discriminant validity, 197
Disease-free survival (DFS), 259
Distributed database storage, 157
DL, *see* Deep learning (DL)
DNA methylation, 256–257
DNN-based decision support system
  architecture, 333
  ECG statistics, 336
  generalized monitoring system framework, 332
  innovative architectures, 331
  modules, 334
  performance evaluation, 336–337
  pseudo-code, 335
  remote control-based smart technology, 332
DS, *see* Derivative superposition (DS)
DSP, *see* Digital signal processing (DSP)
DSS, *see* Decision support system (DSS)
Dual resonant mode, 33–45
DWRR, *see* Deficit weighted round robin (DWRR)
Dynamic spectrum access (DSA)
  frequency bandwidth, 310
  utilizations, 309

E
Earliest deadline first (EDF), 291
EDBA, *see* Entropy distance-based approach (EDBA)
Edgefit_Strategy algorithm, 478, 484, 487
EGFR, *see* Epidermal growth factor receptor (EGFR)
EHRs, *see* Electronic health records (EHRs)
Electronic health records (EHRs), 258
ELVIRA Biomedical Data Set Repository, 260
End-to-end delay, 146, 149, 151
Energy and thermal-aware scheduling algorithm (ETAS), 281
Energy distribution, 147, 148
Energy-efficient architectures, 144
Energy-efficient radio technologies, 144
Energy-efficient resource management, 144
Energy utilization, 147, 151, 152
Energy value, 261
Entropy, 261
Entropy distance-based approach (EDBA), 413
Epidermal growth factor receptor (EGFR), 259
ETAS, *see* Energy and thermal-aware scheduling algorithm (ETAS)
Extractive text summarization, 84
Exudate detection, 98–99
  classification of hard exudates, 99–101
  fuzzy logic, 94
  macular edema recognition, 94
  RGB fundus images, 93
  supervised classifier, 95
  trained classifier, 95

F
Face-based emotion detection, 15
Facet-based region growing (FBRG), 120, 132–137
Fair Information Practices (FIPs), 184
Fake Government of Telangana (FGoT), 8–10
Fast dynamic multiframe demosaicing (FDMD) process, 320, 321
FBRG, *see* Facet-based region growing (FBRG)
Feature recognition (FR), 115–119, 123, 125, 134, 135
Feature pyramid network (FPN), 303–304
FGoT, *see* Fake Government of Telangana (FGoT)
FIPs, *see* Fair Information Practices (FIPs)
Fifth-generation New Radio (5G NR), 361
5G wireless communication network
  frozen bits and significance, 364–365
  improved SCL decoder, 371–373
  low-density parity check, 361–362
  polar code, 361–362
  SC decoder, 365–368, 373–374

SCL
decoder, 368–371
vs. SC decoder, 375
with threshold 1.24, 377–378
with threshold 1.35, 376
system architecture, 363–364
transmission chain, 363
Fog computing
Brute force analysis, 49
century-old legacy power grid, 49
challenges
compatibility, 58
complexity, 58
performance, 59–60
reliability, 59–60
scalability, 58
security, 58–59
CIO, 47
computationally robust smart grid, 49–56
opportunities
compute and connectivity convergence, 56
flexibility in application, 56–57
network equipment vendors, 57
new revenue stream, 57
SG, 47, 48
storage/computation, 48
FR, *see* Feature recognition (FR)
Fundus camera, 93, 102

G

Garbage segregation, 450
Gateways, 217
Genetic algorithm (GA), 220–221, 340
GLCM feature extraction, *see* Grey-level co-occurrence matrix (GLCM) feature extraction
Graphcut_Strategy algorithm, 476, 481, 486
Graph summarization, 469
Greedy method, 219–220
Grey-level co-occurrence matrix (GLCM) feature extraction, 261–262
GoT, *see* Government of Telangana (GoT)
Government of Telangana (GoT), 2, 8, 107

H

Healthcare monitoring
connected devices with IoT, 156
HERA system, 158
high reliability, 156–157
HRO, 159–160
IoT-based applications (*see* Internet of Things inspired approach)
predictive and preventive device, 158
reliability metrics, 155
TuneWalk, 157
wireless body sensor network, 158
High reliability organization (HRO), 159–160
HMS, *see* Hybrid mesh segmentation (HMS)
Homogeneity, 262
HRO, *see* High reliability organization (HRO)
Hybrid mesh segmentation (HMS)
architectural framework, 119–121
coverage, 131, 133
extraction, primitive, 131, 133
FBRG, 133, 135
illustrative examples, 121–122
overall timing, 134
primitive quality for test cases, 131, 132
quantitative performance analysis
comparisons, 124–125
measurement, 122–123
segmentation algorithms evaluation, 123–124
Hyperparameter tuning, 170, 356
Hyperspectral method, 64–68, 80

I

IBM Watson, 272–273
iCBT, *see* Internet-based cognitive behavioral therapy (iCBT)
Identity theft, 1, 107
IDWRR, *see* Improved DWRR (IDWRR)
IEEE 802.16 networks, 290–291
IIHE, *see* Index based histogram equalization technique (IIHE)
Image enhancement, 96–97
CNN, 94
damaged capillaries, 93
evaluation parameters, 101
exudates detection, 98–99
fundus camera, 93, 94
hard exudate classification, 99–101
IIHE-RVE, 97
multi-channel intensity features, 94
optic disc elimination, 98
results and discussions, 102–103
SNR, 328
work flow depiction, 95
Image segmentation, 301
Improved DWRR (IDWRR), 293–299
Improved PSO (iPSO), 312, 314, 315

Index based histogram equalization technique (IIHE)
  performance matrix, 103
  RVE, 97
  work flow, 95
Indoor positioning systems (IPS), 395–396
Information and communication technologies (ICT), 2, 155
Information assurance, 177–180
Instance segmentation, 301, 302
Institute of Electrical and Electronics Engineers (IEEE), 241
Interacting feature, 116, 119, 125, 127, 131, 135
International Quality Improvement Program, 157
Internet-based cognitive behavioral therapy (iCBT)
  AVA cognitive services platform, 271–272
  Azure cognitive services, 273–275
  CBS, 269–270
  Cosmos, 275–276
  IBM Watson, 272–273
  mental disorders, 267
  mobile networks, 268–269
  proactive approach, healthcare, 268
  research, 276
  short- or longer-term neurologic disorders, 268
  therapeutic alliance, 267, 276
Internet of things (IoT)
  connected devices, 156
  deployments, 52
  effective activity recognition
    AR, 163
    Bayes' theorem, 164
    localization method, 164–165
  and healthcare, 155, 161–163
  network layer system, 332
  pay-per-use resources, 48
  reliability
    factor, 155
    lacking, 161
  SG application, 50, 58
Internet of Things inspired approach
  AAL system, 160–161
  activity recognition, 163–165
  data transfer, 161–163
  lacking of reliability, 161
Internet retailing, 191
Intrusion detection system, 16, 21
IoT healthcare network (IoThNet), 332
IPS, *see* Indoor positioning systems (IPS)

J

JNTUHJAC website, 2, 9, 10, 107, 108, 110

K

K-nearest neighbour (KNN), 258

L

LASSO, *see* Least absolute shrinkage and the selection operator (LASSO)
Layered-hybrid ABC (LHABC), 346
Learning motivation, 192
Least absolute shrinkage and the selection operator (LASSO), 259, 260
Lecture summary
  evaluation and results, 89–90
  methodology, 87–89
  online video sharing sites, 83
  proposed
    approach, 87
    method, 84
  related work, 85–87
  teaching and learning, 83
  video summarization, 84
Levenberg-Marquardt (LM) algorithm, 341
LLR, *see* Log likelihood ratio (LLR)
LNA, *see* Low-noise amplifier (LNA)
Localization algorithms
  angulation, 401–402
  circular lateration, 399–400
  differential detection, 402–403
  hyperbolic lateration, 400–401
Localization method, 164–165
Local search method, 220
Log likelihood ratio (LLR), 367–358
Long short-term memory (LSTM)
  embedding, 22
  proposed framework, 19–21
  sentiment detector framework, 18, 19
  train and test sequences, 21
Low-noise amplifier (LNA)
  802.11 wireless RF receiver, 241
  feedback, 242
  gain enhancement techniques, 242
  layout, 249
  linearity with differential MOS loads and auxiliary source stage
    1-dB compression point, 250, 251
    input matching, 248
    noise figure, 246, 247, 250–251
    residual nonlinearity components, 246

S parameter, 246, 247, 250–251
stability, 251
voltagc gain, 249–250
MDS method, 243, 246
MOS transistor drain current, 244
performance summary, 253
third- and second-order nonlinearities, 243–245
LSTM, *see* Long short-term memory (LSTM)
Lung cancer
diagnosis, 255
histology, 256
mortality rate, 255

M

Machine learning (ML), 271, 272, 257
Machine learning techniques (MLT), 258
Malaria, 169, 170, 174
Mask network, 304
Mask R-CNN
core idea of, 302
experimental results, 305–306
image segmentation, 301
loss after 16 epochs, 305, 306
proposed system
backbone, 303
ROI classifier and bounding box regressor, 304
RPN, 303–304
segmentation masks, 304
segmentation task, 302
Maxima_Strategy algorithm, 474–476, 483–484
Maximum relevance and minimum redundancy (MRMR), 262
Max pooling deep learning (MPDL)
algorithms, 69–70
instruction collection, 73
mathematical analysis, 69–70
satellite picture, 73
MDS, *see* Modified derivative superposition (MDS)
Mean squared error (MSE), 238
Message-oriented middleware (MOM), 162
Mesh router placement scheme
heuristic and meta-heuristic approaches
ant colony and particle swarm optimization algorithms, 221
genetic algorithm (GA), 220–221
Greedy method, 219–220
local search method, 220
multi-objective optimization, 221–222
objectives, 218, 219
taxonomies, 220
Mesh routers, 217, 218
MIC, *see* Mother India Consciousness (MIC)
Microsites, 144
Millimeter wave (mmWave) technology
association techniques, 383–384
association with cell and carrier, 382–383
cognitive radio networks, 384–386
hierarchical access model, 380
licensed and unlicensed users, 380
optimum selection of parameters, 386–392
signal to noise ratio, 384–386
simulation environment, 386
spectrum sharing, 379
system model, 381–382
Minima_Strategy algorithm, 474, 475, 483
ML, *see* Machine learning (ML)
MLT, *see* Machine learning techniques (MLT)
Mobile networks, 268–269
Modification rate (MR), 341
Modified best fit decreasing (MBFD) algorithm, 281, 282
Modified derivative superposition (MDS), 252
MOGA, *see* Multi-objective genetic algorithm (MOGA)
MOM, *see* Message-oriented middleware (MOM)
Mother India Consciousness (MIC), 108, 112
MPDL, *see* Max pooling deep learning (MPDL)
MSE, *see* Mean squared error (MSE)
MTW, *see* Multi-hop traffic-flow weight (MTW)
Multi-criteria decision-making
accreditation, 411
AHP-based weight derivation, 416–417
attributes and contributing parameters, 415
literature survey, 412–414
problem description, 415–416
PROMETHEE calculations, 417–420
ranking schemes, 411, 412
Multi-hop traffic-flow weight (MTW), 222–224
Multi-objective genetic algorithm (MOGA), 292
Multi-objective optimization, 221–222
Multispectral data
existing methods
DL, 67
hyperspectral method, 67–68
RFs, 66
SRCs, 66

Multispectral data (*cont.*)
    imaging devices, 63
    parameter analysis table, 76, 80
    proposed method, 68–70
    results
        colour model of crop, 71, 72
        contrasted image, 73, 75
        distorted and original image, 71, 72
        filtered image, 72, 74
        histogram comparison, 71
        image selection, 73, 76
        message windows, 71, 72, 73, 76
        satellite image, 71
        segmented data, 72, 75
        selected region, 71, 73
        slice selection, 72, 75
        work flow, 73, 77–79
    RGB, 63
    satellite image, 64
    uncertainties, 65–66
Municipal solid waste (MSW), 450

N
Naïve Bayes' classifier, 264–265
Narrow slot, 34, 37, 41, 43–45
National Lung Screening Trial (NLST), 260
NativeTransportRequest, 238
Network density, 271
Neural-modified ABC (NMABC), 345
NLST, *see* National Lung Screening Trial (NLST)
Non-small cell lung cancer (NSCLC)
    ADC, 256
    defined, 256
    DNA methylation, 256–257
    literature survey, 258–260
    methodology, 261–265
    oligometastatic, 259
    radiomics, 257–258
    smoking cessation, 256
Non-uniform memory access (NUMA), 423
NSCLC, *see* Non-small cell lung cancer (NSCLC)

O
OBE, *see* Online brand experience (OBE)
OFDMA, *see* Orthogonal frequency-division multiple access (OFDMA)
Online brand experience (OBE)
    activation, 194
    affection, 194
    cognitive processing, 193
    customer engagement behavior, 194–195
    learning performance factors, 192
    loyalty, 193, 195
    satisfaction, 193, 195–196
    See also Customer brand engagement (CBE)
Optima_Strategy algorithm, 478, 482–483, 486–487
Orthogonal frequency-division multiple access (OFDMA), 290
Otsu method, 170

P
Packet-by-packet generalized processor sharing (GPS), 291
Packet delivery ratio, 146, 148, 150, 295, 298
Parallel machines
    aggregate scores, 444
    classification, 423
    cost equations, 434–440
    direct graphs, 424
    interconnection networks, 467
    load balancing, 423, 424
    non-uniform memory access, 423
    performance results, 440–446
    software workload, 467
    TAP, 425–426
    task assignment, 424
    topology mapper tool, 426–434
    uniform memory access, 423
Parametric study
    current distribution, 43–44
    effect of position of slot, 41–43
    narrow slot width and length, 41
    patch length and width, 40, 41
    pin position, 41
    radiation pattern, 43–45
    return loss characteristics, 41
Partial least square (PLS), 196
Particle swarm optimization (PSO) algorithms, 221, 283–284, 340
Patch antenna
    conventional, 35
    MPA, 37
    narrow bandwidth, 33
    odd-order modes, 34
    optimized parametric value, 36
PbD, *see* Privacy by design (PbD)
PDR, *see* Pedestrian dead reckoning (PDR)
Pedestrian dead reckoning (PDR), 164
Peer-to-peer architecture, 227

People Governance Forum (PGF), 2, 3, 11
homepage screenshot, 107, 108
related work, 108
research methodology, 108–110
Telegram, 110, 111
Wayback Machine web crawler tool, 107
WhatsApp, 110, 111
Performance Assessment Tool, 157
PGF, *see* People Governance Forum (PGF)
PIFA, *see* Planar inverted-F antenna (PIFA)
PIOs, *see* Public Information Officers (PIOs)
Planar inverted-F antenna (PIFA)
analysis, 37–40
configuration of, 34, 35
EM energy, 33
final design, 37–40
low-profile antennas, 33
parametric study, 40–44
resonant frequencies, 34
working mechanism, 35–37
PLS, *see* Partial least square (PLS)
PNs, *see* Pulmonary nodules (PNs)
Power utilization, 143, 148, 150
Privacy by design (PbD)
Big Data, 181, 184–186
commandments
alter resistant audit logs, 186–187
analytics on anonymized data, 186
data tethering, 186
false-negative favouring methods, 187
full attribution, 186
information transfer accounting, 187–188
self-correcting false positives, 187
context accumulating systems, 178
data protection and services offered, 178
information insurance by plan, 180
information security, 179
'plan in' data assurance, 177
sense making, 182–184
Processor interconnection topologies, 470–473
Processor Interconnect Topology, 467
Processor topology graph (PTG), 425, 473
Processor topology table (PTT), 426, 429
Program dependence graph (PDG), 468
Public Information Officers (PIOs)
All India Council Technical Education, 4
Andhra Pradesh Public Service Commission, 5
and Chief Secretary, 5–6
Council of Scientific and Industrial Research, 5
Lok Sabha Secretariat, 3–4
Principal Secretary, 4, 6
Pulmonary nodules (PNs), 260
Punctuations, 87–91

Q

QoS, *see* Quality of service (QoS)
QOS-RPS, *see* Quality of service in cloud using reputation potential services (QOS-RPS)
Quality of service (QoS), 144, 290–291
Quality of service in cloud using reputation potential services (QOS-RPS), 208, 209, 211
Quantitative performance analysis
CAD-CAM integration, 115, 116
discussions
coverage, 131, 133
FBRG, 133, 135
overall timing, 134
primitive extraction, 131, 133
primitive quality for test cases, 131, 132
error analysis, 128–130
literature review, 117–119
mechanical engineering parts, 116
mesh density, 128, 131
SDM, 116
segmentation results evaluation, 125, 127–128

R

Radiomics, 257–258
Random forests (RFs), 66
Real-time analytics, 53–54
Real-time monitoring, 155
Red, green, blue (RGB), 63, 67, 93–96, 303
Region proposal network (RPN), 303
Reliability
and high-assurance computing, 55–56
performance, 59, 60
Remote sensing, 79, 307
Replication lag, 228
Resonant frequency, 34, 35, 37, 40, 43, 45
Retinal vessel enhancement (RVE), 95, 97, 103, 104
RGB, *see* Red, green, blue (RGB)
Risk stratification, 259
RTI-Act 2005
cybercrime remedying, 6–9
FGoT, 10
JNTUHJAC website, 1, 2
MHRD, 10
motivation, 2
PGF, 3, 11

RTI-Act 2005 (*cont.*)
PIOs, 3–6
road map, 108
RVE, *see* Retinal vessel enhancement (RVE)

S
Scalability, 50, 52–53, 58, 160–162, 228, 292
Scan derived mesh (SDM), 116
Scarce image classifiers (SRCs), 66
SCLC, *see* Small-cell lung cancer (SCLC)
SDM, *see* Scan derived mesh (SDM)
SDN, *see* Software defined networking (SDN)
Sedition law, 1
Segmentation problem strategy, 170
SEM, *see* Structural equation modelling (SEM)
Semantic interoperability, 162
Sense-making systems, 182–184
Sentiment recognition
DL, 16
emotion
detection, 15
intelligence, 16
intrusion detection system, 16
literature review, 17–18
LSTM, 18–21
personal web blog, 15
system model, 18–21
Service-oriented architecture (SOA), 157, 158
Shorting pin
couple of, 45
gain and bandwidth, 34, 40
position effect, 42
radiating edge, 34
Signal processing in-node environment (SPINE), 158
Signal-to-interference-plus-noise-ratio (SINR), 144, 380
Signal-to-noise ratio (SNR), 292, 328, 371, 373–375, 384–386
Simulation constraint, 147, 148
SINR, *see* Signal-to-interference-plus-noise-ratio (SINR)
Small-cell lung cancer (SCLC), 256
Smart grid (SG), 47, 48
consistency need, 53
data center-based, 51
electrical power, 47
global consensus, 47
infrastructure, 59
latency hierarchy, 54
mission-critical computing needs, 49
mobile and static nodes, 48
utility vendors, 55–56
Smart trash segregator (STS)
Fastai and Keras frameworks, 459–461
flowchart, 455
hardware model, 453–457
ResNet50 model, 461
sensor module and DNN module, 461–464
sixteen layer CNN model, six class classification, 458
software model, 451–453
solid waste collection, 449
waste segregation, 450
SMI, *see* Social media intelligence (SMI)
SNR, *see* Signal-to-noise ratio (SNR)
SOA, *see* Service-oriented architecture (SOA)
Social media intelligence (SMI), 109
Software defined networking (SDN), 205
Software reliability growth models (SRGMs), 413
Spectrum allocation, 309
Spectrum handoff, 312
Spectrum particle swarm optimization (SpecPSO), 312, 315
Spectrum sensing, 311
Spectrum sharing, 379, 381
SPINE, *see* Signal processing in-node environment (SPINE)
SRCs, *see* Scarce image classifiers (SRCs)
SRGMs, *see* Software reliability growth models (SRGMs)
Structural equation modelling (SEM), 196
Subtitles
lecture videos, 87
preprocessing module, 88
with punctuations, 87, 90
ripped file, 88
text summarization, 87
Support vector algorithm (SVM), 85, 258, 657

T
TASA, *see* Thermal-aware scheduling algorithm (TASA)
Task assignment graph (TAG), 425, 432, 474
Task assignment problem (TAP), 425–426, 468, 473–474
Task communication graph (TCG), 425, 473
Task communication table (TCT), 426–428, 430, 431
Task computation table (TMT), 426, 427, 431
Task mapping
algorithms, 474–482
complexity analysis, 482–487
dynamic profiling, 470

graph summarization, 469
inter-task communication demands, 468
preliminary results, 487–488
processor interconnection topologies, 470–473
query processing, 470
spectral filtering techniques, 469
topology representation, 468
TBDCOs, *see* Twin Big Data Cybercriminal Organizations (TBDCOs)
Telegram, 108–111
Term frequency-inverse document frequency (TFIDF), 87–89, 91
Text summarization, 84, 87–89
TFIDF, *see* Term frequency-inverse document frequency (TFIDF)
Thermal-aware scheduling algorithm (TASA), 281
Throughput ratio, 146–147, 151
Twin Big Data Cybercriminal Organizations (TBDCOs), 3–6

U
Ultrasonic-based systems, 396
Uniform memory access (UMA), 423

V
Video summarization
CAVS, 85
description, 84
students and teachers, 84
Virtualization, 279
Virtual machines (VMs)
active number of servers, 285, 286
consolidation, 280
double threshold strategy, 281
DVFS algorithm, 280
energy consumed by racks, 287
live migration, 280
power consumed after initial placement, 286
proposed system, 281–285
robust optimization technique, 281
Visible light communication (VLC)
algorithm error plot, 405, 406
coordinate plots for algorithms, 405, 406
demonstration error plot, 406, 407
hardware description, 404–405
IPS, 395–396
localization algorithms, 399–403
localization/positioning, 395
measurement, 407–408
received power distribution, 408
system configuration, 397–399
VMs, *see* Virtual machines (VMs)
Voxel classification, 94

W
Waste segregation awareness, 450
WBASNs, *see* Wireless body area sensor networks (WBASNs)
Wearable blood pressure sensors, 158
Web robot, 465
Weibull transfer function, 25, 28–30, 220, 221
Weighted fair queuing (WFQ), 291
Weighted round robin (WRR), 291
WhatsApp, 15, 108–111
Wi-Fi technologies, 241
WiMAX networks, *see* Worldwide Interoperability for Microwave Access (WiMAX) networks
Wind power model
analysis of cost minimization, 29
linear function, 29, 31
Wind speed model distribution
Dagum, 28
Gamma, 28
Weibull, 28–29
Wind turbines
configurations, 26
cost minimization, 26–29
data description, 30
estimation of CoE, 30–31
meager selection, 25
Weibull transfer function, 25
Wireless body area sensor networks (WBASNs), 158
Wireless mesh network (WMN)
architecture, 217, 218
design metrics
distance, 224
MTW, 224
node cluster, 224
throughput calculation, 225
performance comparison, 222–225
routers, 217, 218
See also Mesh router placement scheme
Wireless networks
CR, 144
energy utilization, 143
microcells and picocells, 144
microsites, 144
packet delivery, 145
performance metrics

Wireless networks (*cont.*)
broadcast ratio, 147
end-to-end delay, 146
energy utilization, 147
packet delivery ratio, 146
throughput ratio, 146–147
simulated results, 147–152
Wireless sensor network (WSN), 157
WLAN-based system, 396
WMN, *see* Wireless mesh network (WMN)
Worldwide Interoperability for Microwave Access (WiMAX) networks
DWRR, 292–293, 296–299
IDWRR, 293–299
IEEE 802.16 networks, 290–291
literature survey, 291–292
QoS, 289
simulation parameters, 295
WRR, *see* Weighted round robin (WRR)
WSN, *see* Wireless sensor network (WSN)